Novus XVII
Inferioris Germaniæ
Provinciarum Typus
GERMANICUS
De Noord Zee
TEXEL
VLIELANDT
DER SCHELLING
ZUYDER ZEE
PROVINCIA

I0878506

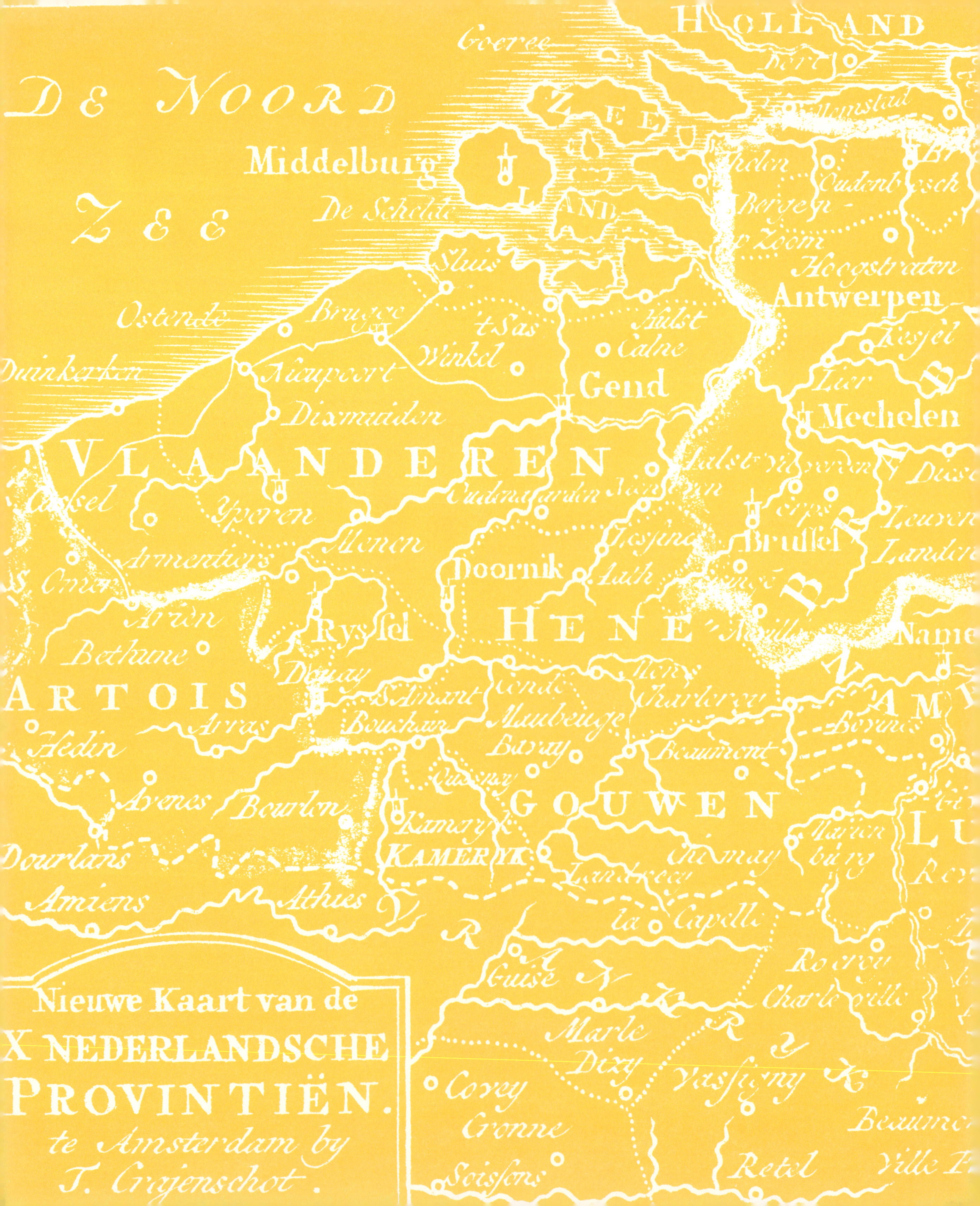
Nieuwe Kaart van de
X NEDERLANDSCHE
PROVINTIËN.
te Amsterdam by
T. Crajenschot.
DE NOORD ZEE
HOLLAND
Goeree
Dort
ZEELAND
Middelburg
De Schelde
Bergen op Zoom
Oudenbosch
Hoogstraten
Antwerpen
Sluis
Ostende
Brugge
t Sas
Hulst
Calne
Winkel
Duinkerken
Nieuport
Dixmuiden
Gend
Lier
Mechelen
VLAANDEREN
Yperen
Oudenaarden
Cassel
Menen
Armentiers
Doornik
Aath
Brussel
Leuven
St. Omer
Ryssel
HENE
Arien
Bethune
Douay
ARTOIS
Arras
Hedin
S. Amant
Bouchain
Condé
Maubeuge
Charleroy
Bavay
Beaumont
Quesnoy
Avenes
Bourlon
GOUWEN
Kameryk
KAMERYK
Chimay
Marienburg
Dourlans
Landrecy
Amiens
Athies
la Capelle
Rocroy
Guise
Charleville
Marle
Dizy
Vassigny
Corey
Gronne
Soissons
Retel

GROUNDBREAKERS

REMARKABLE MAPS
FROM THE LOW COUNTRIES,
1500-1900

THE PHOEBUS FOUNDATION

ANNE-RIEKE VAN SCHAIK

GROUND BREAKERS

REMARKABLE MAPS FROM THE LOW COUNTRIES, 1500-1900

HANNIBAL

TABLE OF CONTENTS

FOREWORD

Antipodeans

In his *Commentary on the Dream of Scipio*, the fifth-century philosopher Macrobius riffed on Cicero's telling of the *Somnium Scipionum*, the eponymous dreamer being the celebrated Roman general Scipio Aemilianus. This hugely influential contemplative treatise explained the celestial spheres and the inaudible yet cosmic music they make. He rounded off with a description of the world, one that was spherical, not flat, and consisted of northern and southern zones. The two halves were, however, tragically destined never to meet because of the torrid zone that separated them.

Macrobius's *Commentary* became an instant classic, assiduously copied and studied in many a medieval monastery. The invention of the printing press allowed the text to be disseminated more widely, with educational diagrams added to what had become a treatise for the more advanced reader. Best of all, though, was the map of that wondrous world, in which Antipodeans dwelt somewhere 'down under'.

It might seem quaint to us, but a thousand years from now, someone might well think the same of our Google Maps. Because, as Abraham Ortelius (1527-1598) already realized as he compiled the first atlas of the world a millennium after Macrobius, show me your map and I'll tell you who you are. It is no coincidence that he gave his book the title *Theatrum Orbis Terrarum* – 'The Theatre of the World'. Because just as there is nothing objective about theatre, so too land maps, sea charts, globes and city views might appear neutral and factual, but mapmakers view the world through their own lens just as much as a modern documentary maker does. What do you show? And, in some cases more importantly, what don't you show? Where do you place borders and orientation points? And which 'centres' do you put at the centre? Is Jerusalem the centre of the world? Europe? Cartographers don't reproduce; they *construct*. A map does not show *the* image but merely *an* image of the world. Viewed this way, a map – any map, in fact – is a gateway into the knowledge and technology, norms and values, matters of wealth and power, and religious beliefs (or their decline) of the era in which it was made. A gateway, in short, into the world itself. Which is precisely what this book is all about.

More than a thousand words

We tend to believe that everything was more straightforward in the past: God created the universe with the earth bang in the middle. He was not the type to think 'that'll do more or less for the centre'. Although Macrobius's fans were better informed, as far as most medieval people were concerned, the earth was flat and had edges, a *finis terrae* you would fall off if you didn't take care. And that, of course, is where it all began: if you don't want to fall off the edge of the world, you're going to need a map.

Medieval mariners hugged the coast because you really didn't want to meet the kind of things that lurked further out to sea. The more they voyaged, the better they recorded the crinkly coastlines of Europe in what became known as 'portolan' charts – hand-drawn sea charts that allowed you to travel the known world.

Those same outlines, of course, contained the land on the other side. The land on which the Romans (where would we be without them?) built a network of roads, all of which conveniently led to Rome. Things could hardly get better. And they didn't, in fact, because before you knew it, the Roman Empire suffered its decline and fall, taking with it that centralized vision of the world. Stories of Rome were all very impressive but by then long ago and also far away. A feudal lord in the year 1000 or so had no interest as yet in an *Autoroute du Soleil* to get him to his Mediterranean holiday faster. He wanted to know precisely which territory fell under his control; where the rivers, plateaus and mountains were; which land belonged to God via his agent, the Church (not too much); and which belonged to himself (as much as possible). A vast, unbounded landscape was gradually hemmed in by boundary posts specifying jurisdiction. Medieval people turned out to be extremely clever lawyers: they developed an entire legal system focused on land ownership and its precise definition. A host of surveyors and notaries swiftly emerged, ready and willing to move patiently from one post and one goose to the next, capturing the entire landscape in deeds of land ownership. A picture, as we all know, says more than a thousand words, so it is actually slightly surprising that it was not until the fourteenth century that someone first came up with the idea of mapping the land.

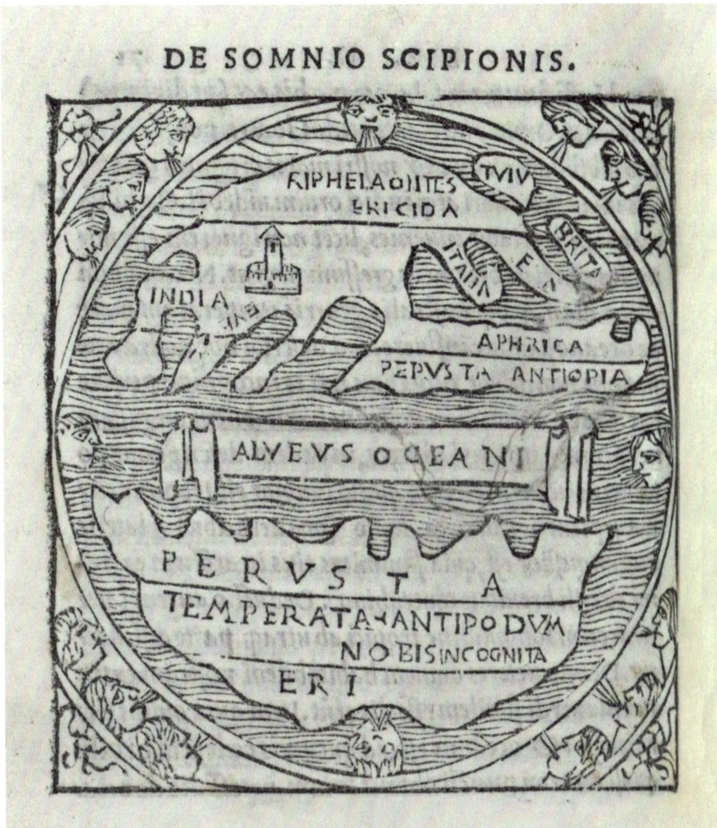

Zonal map
Detail from: Niccolo Angeli after Ambrosius Macrobius, *Hoc volumine continentur Macrobii interpretatio in somnium Scipionis a Cicerone confectum*, 1515
ANTWERP, THE PHOEBUS FOUNDATION

Which brings us to our starting point. While surveyors and notaries were measuring up and describing the terrain, painters began to produce *vues de lieu*, panoramic sketches of the terrain made from a nearby elevation. Back in the studio, these were drawn out more neatly and incorporated in maps. None of which might seem all that special to us now, but at the time it was revolutionary. The new cartography was born.

Manhattan on the Scheldt

Cartography naturally boomed in the late-medieval Low Countries, simply because everything was booming there. By around 1570, trade via the river Scheldt meant that Antwerp had become the wealthiest city in the known world, the sixteenth-century equivalent to Manhattan, courtesy of the cosmopolitan entrepreneurs who flocked there. Besides Dutch, you would hear more exotic languages in the street, such as Spanish, Portuguese, English, German, Italian and Yiddish. The place smelled of dung and filth, because that's just how it was in medieval cities. But in Antwerp the stench was offset by the wondrous scents of pepper and cinnamon, while mariners told tales of the vast horizon and what lay beyond. The celebrated printer Christophe Plantin had good reason for swapping Paris for Antwerp and becoming Christoffel Plantijn. Meanwhile, book printing was taking off, and the city by the Scheldt was becoming a centre for the dissemination of knowledge too: the place to go for any treatise on the New and the Old World alike. Painters worked furiously, producing for the open market for the first time in Western history rather than working strictly to commission. There was no resisting all these thrilling discoveries and novelties, which forced even the stuffiest of theologians into a U-turn: a wider world they had once fulminated against as the Devil's playground now became God's gift to humanity, in which everything under the sun was worthy of study. *Terra incognita* was no longer out of sight, out of mind and best left to the monsters; the world was steadily revealing all its facets at this bustling intersection of commerce, entrepreneurship, nascent science, technology and religion. And that included maps.

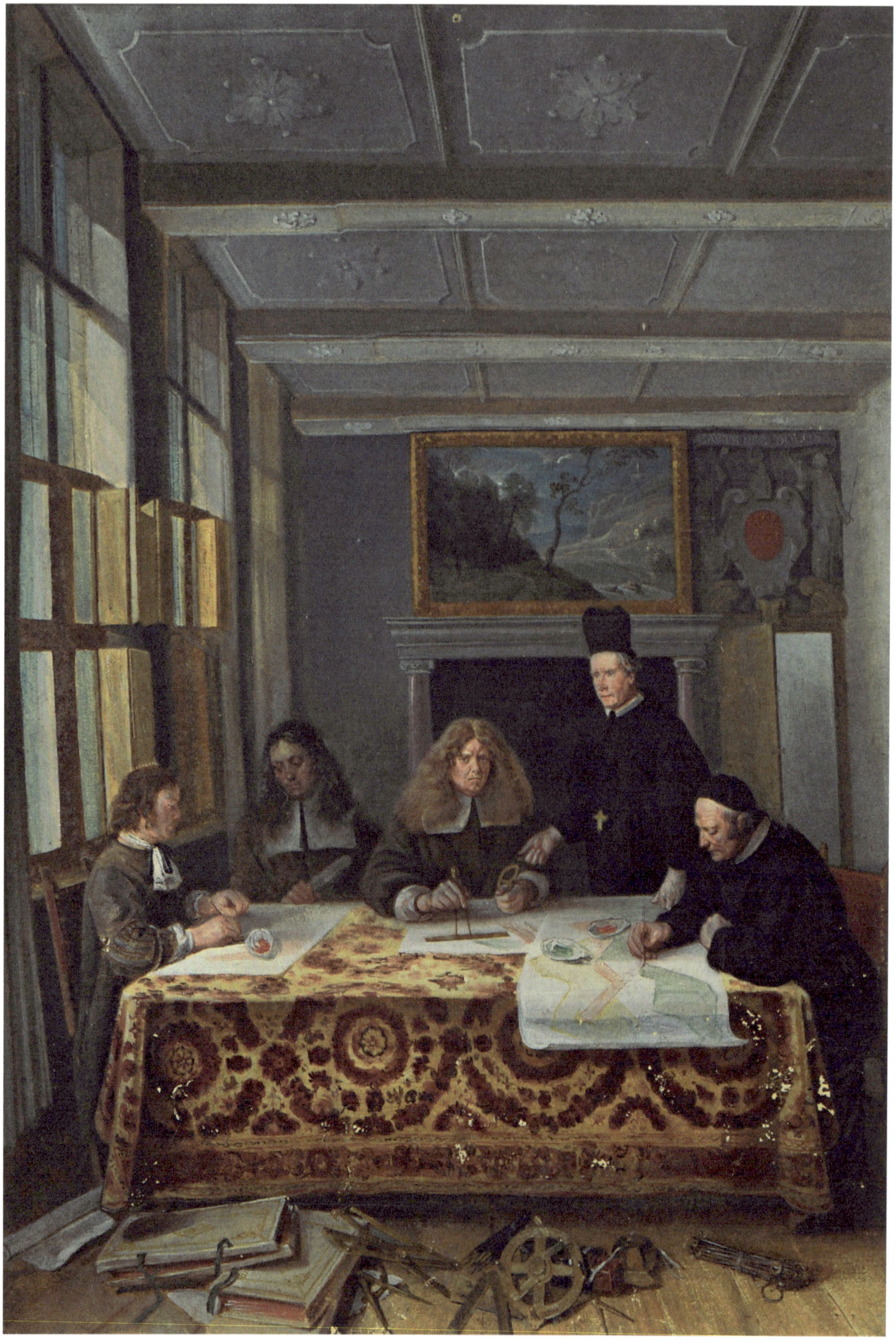

Anonymous master
Portrait of Four Surveyors with Abbot Antoon De Loose of the St Saviour Abbey in Ename, 1658
Oil on paper, 45.5 × 32.5 cm
ANTWERP, THE PHOEBUS FOUNDATION

To measure is to know and *Ordnung muss sein*, a need for order that was swiftly applied to the very universe itself, giving rise to the discipline of cosmography. This views the earth as, well, some kind of cosmic given that has circles of latitude (albeit not meridians as yet) to show you where you are in the world and hence in the universe. Geographers then set about mapping that world, ideally in the wake of expeditions and exploration. Not to mention our old friend the surveyor, who could enthusiastically apply Euclid's art of triangulation to the world around him. We refer to this as chorography, which might have been good news for notaries but was not something to make an art historian's heart beat any faster. Happily for us, the step towards topography followed quickly. The topographer did, at least, show us what there was to see: plains and valleys, inaccessible forests, hard-to-reach heights and featureless plains were all flattened out, coloured in and/or encoded as lines. Towns show up in that landscape as oases and places to rest after a long journey. Each was different, with rivers, towers, ramparts and windmills. The silhouette of a town was its portrait.

With Antwerp as their base, Europeans drew a picture of the world with the combination of creativity and hard-headedness typical of the entrepreneur. It took a series of religious wars, the violent persecution of Protestants and the exodus of pretty much everyone with a brain, money or both to Amsterdam for that city to pick up the baton in the seventeenth century and for a new Antwerp to be born.

From meadow to universe

The Phoebus Foundation has an impressive collection of world maps, globes, atlases, measuring instruments and city views. From medieval 'T-O maps' centred on Jerusalem and the first, tentative images of 'The New and Unknown World' (*De nieuwe en onbekende weereld*) to the bustling roadstead of Antwerp and New Amsterdam, the collection tells the history of the world's land and water in words and images.

To the historian Anne-Rieke van Schaik, a map is all this and more. She is fascinated by the different ways in which a map can be read and by the debates that the medium triggers. Anne-Rieke has further honed this expertise as a member of the Explokart research group affiliated with the Allard Pierson and the University of Amsterdam and is currently preparing a thesis on narrative cartography in the early modern Low Countries.

Between November 2021 and February 2022, Anne-Rieke van Schaik was The Phoebus Foundation's first Research Fellow, focusing on our Topography and Cartography sub-collection. Besides preparing a scholarly inventory of these holdings, she also pursued the stories behind the objects. The result is this book, in which Anne-Rieke places this facet of the Phoebus collection firmly on the map for the first time. With the greatest of ease, she zooms in and out from meadow to universe, drawing in the brightest stars in the cartographic firmament such as Mercator and Blaeu without overlooking unsung heroes like our worthy surveyor. Together, they literally and figuratively pushed the frontiers.

Groundbreakers takes you on a global journey through time and space. Navigation is not by GPS, although the maps created by the leading players in this book unquestionably laid the foundations for that technological *tour de force*. So while you might not be able to view our planet with quite the awestruck gaze of the first cartographers, you will never look at Google Maps in the same way again after reading *Groundbreakers*.

I wish you an immensely pleasurable read.

Dr Katharina Van Cauteren

Chief of Staff of The Phoebus Foundation Chancellery
Executive Director of The Phoebus Foundation Public Benefit Foundation | Stichting van Openbaar Nut

INTRODUCTION

A vast number of maps from the Low Countries are worthy of study, and there are endless different ways in which the history of cartography in the region might be written. We could, for instance, describe the techniques and methods used to produce maps; recount the lives and times of important surveyors, cartographers, engravers and publishers; or zoom in on the most widespread maps, atlases, globes, instruments, city views and panoramas. We could also take the opposite approach and actively seek out the rarest and most obscure objects. While studying the maps, we could examine the correctness of their content, how accurately they represent the area in question, but we could also consider how certain elements can impose a particular idea on viewers, how they criticise society and guide and manipulate viewers.

I do a little of all these things in this book, based on The Phoebus Foundation's collection of maps, and with several questions in mind: what narratives does the map convey about a location, a country or the world? How does a map shape our picture of reality? What stories does a map *not* tell? My choices are drawn from the Antwerp-based collection and consist of objects that reflect that specific ensemble, bearing in mind that this too is merely a selection and one, moreover, that is personally coloured. Having completed my fellowship at The Phoebus Foundation, I embarked on a doctoral research project into 'story maps' (maps combining narrative and spatial information) from the early modern Low Countries – an interest that resonates in the texts and maps contained in this book.

The selected material dates from roughly the sixteenth to the nineteenth centuries, although the time span of 1500-1900 ought to be taken broadly, since the oldest object is an edition of Ptolemy's 1486 *Cosmographia*, while the most recent is drawn from the archives of Wim Strecker, which contain items up to around 1960. The focus, however, is on the early modern era, when cartography gained a foothold in the Low Countries: a great many, relatively innovative maps were produced in the region – in Flanders and Holland particularly – in that period. The eleven chapters discuss famous highlights, such as Joan Blaeu's *Atlas Maior*, as well as rare, anonymous works. They include both classic geographical maps and topographical city views, while addressing themes such as water management, the Eighty Years War and the struggle for Belgian independence. From an early, unfamiliar view of Antwerp to a monumental wall map several metres high, each document offers a glimpse into how people sought to grasp the complex world around them. The nineteenth century is given somewhat less attention than it deserves, but this too reflects the fact that the collection features more material from the early modern period (1500-1800) than it does from the modern (1800-2000).

There are three pillars to this book, beginning with an object-oriented approach: The Phoebus Foundation's collection was the point of departure when it came to writing the chapters. A second pillar comprises the attempt to show how the Southern and the Northern Netherlands (roughly modern-day Belgium and the Netherlands, respectively) came to be intertwined in the map business. The book highlights the role that mapmakers and other professionals from the Southern Netherlands played in the history of cartography, from Gerard Mercator and Abraham Ortelius to the many anonymous cartographers and surveyors. However, it also situates this cartographic cradle in the broader historical context of the Low Countries, Europe and the world. What we present here, thirdly, is less a chronological than a kaleidoscopic view of the collection and of the history of early maps. Eleven themes are addressed, each exploring a particular view or facet of cartography in terms of its production, content, appearance, style or use. Topics such as surveying, science, navigation, news, and politics form different angles from which to highlight the diversity of The Phoebus Foundation's collection, while also painting a richly nuanced picture of the history of its objects.

Several artefacts drawn from outside the Topography and Cartography subcollection are also incorporated at various places in the text, including paintings and prints by old masters as well as old books, almost all of which likewise belong to The Phoebus Foundation's collection. Because far from being isolated, standalone objects, maps firmly reflect the cultural, intellectual and artistic context of their times. The heyday of Southern Netherlandish cartography coincided with that of Flemish art in the sixteenth and seventeenth centuries. Economic interests, political developments and artistic trends all helped shape the production and consumption of maps.

In the course of writing this book, I made grateful use of the substantial secondary literature available on the various topics. A concise list of key publications can be found at the end of each chapter, along with suggestions for readers wishing to delve into a topic in greater depth. Existing studies thus formed the basis for interpreting the objects from The Phoebus Foundation and situating them within the history of cartography. But there are fresh insights and findings here too, sparked by the unique characteristics of the objects or where my personal enthusiasm led me to carry out additional research. Examples include the unfinished manuscript military map for Louis XIV (⟶ pp. 164-165) and the links between the narrative maps of Claes Jansz. Visscher and Pieter Verbiest II in Chapter IV. It is to be hoped, therefore, that this publication might also inspire future researchers to carry out further in-depth research on the objects. Since the book itself is not arranged chronologically, a summary now follows of some of the key figures, works and aspects in the history of cartography. This short history will help readers to place the names and concepts raised in the themed chapters within the context of their time and the evolution of cartography.

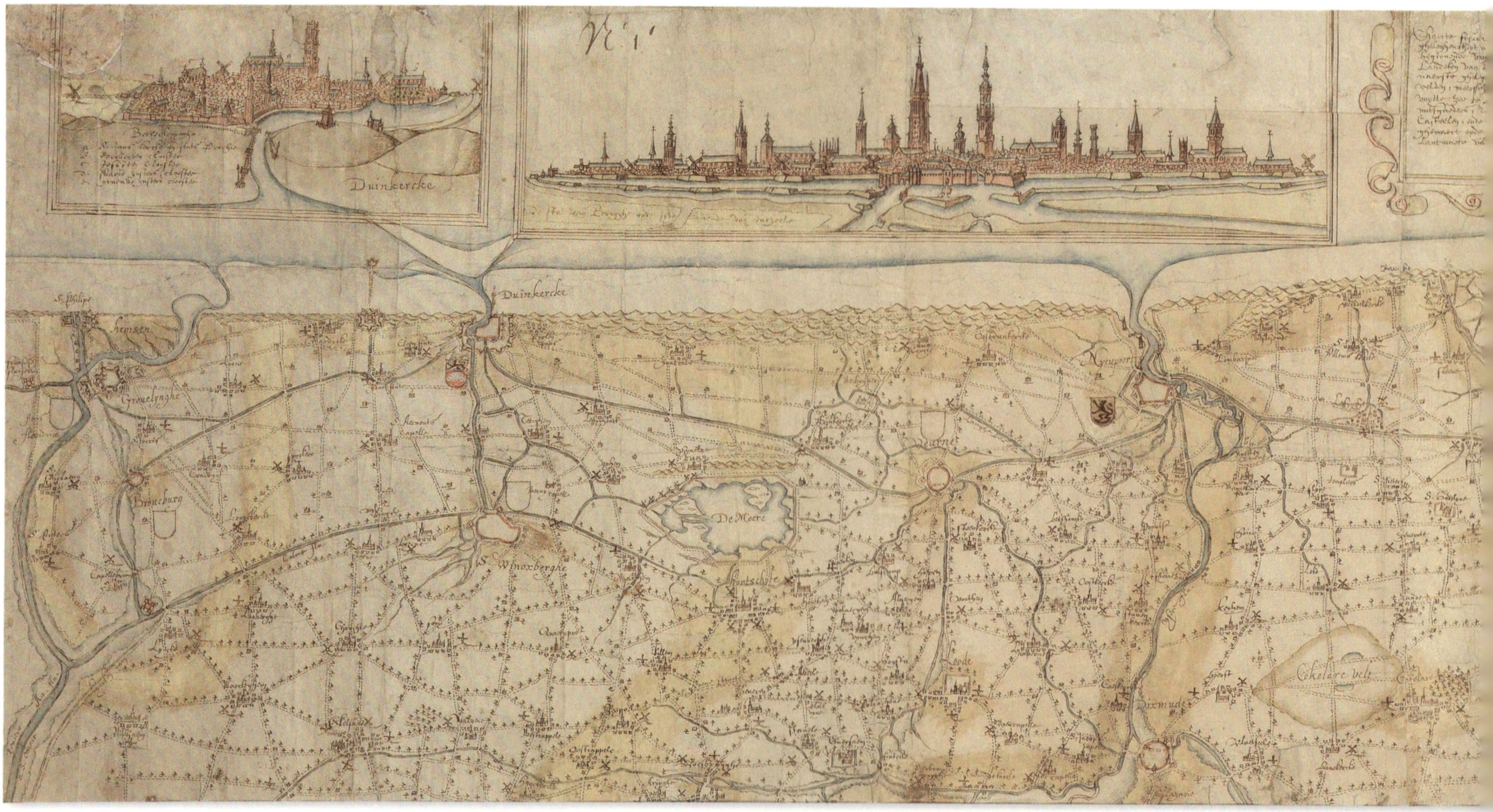

Attributed to Adriaan Villegas
Chaerte figuratyve vande rechten gheleghentheyt vande Vlaemse zeecosten, after 1640
Manuscript map of the North Sea coast
Pen in brown ink on paper, 485 × 1775 mm
ANTWERP, THE PHOEBUS FOUNDATION

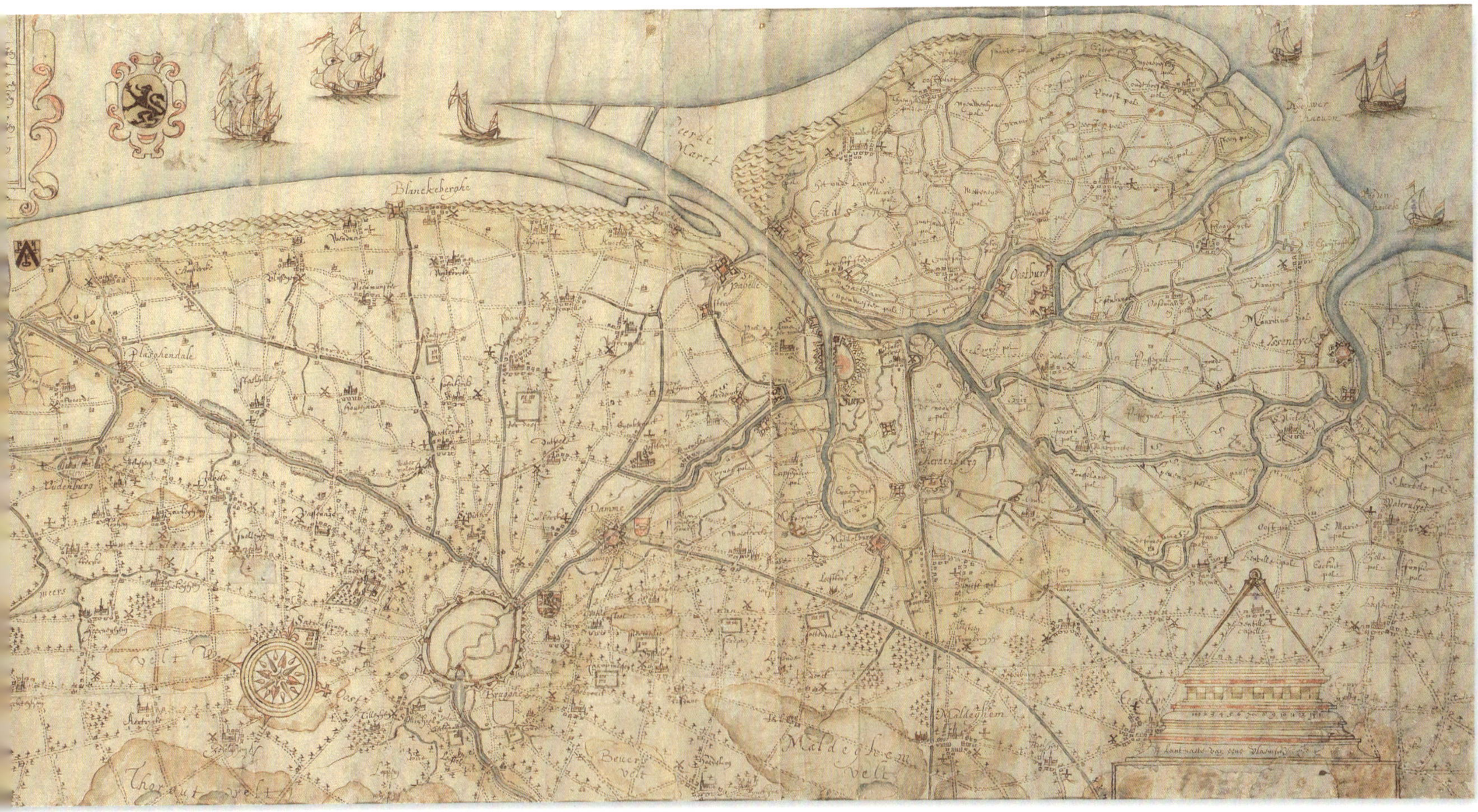
Blanckeberghe
Damme

A BRIEF HISTORY OF CARTOGRAPHY IN THE LOW COUNTRIES

The span of this book begins as far back as antiquity when, in Alexandria around the year 100 CE, Claudius Ptolemy penned a treatise on how to draw maps, to which he appended a lengthy list of the longitudes and latitudes of some eight thousand locations worldwide. His text was one of the principal sources on which maps of the world would be based for centuries to come, right through to medieval and early modern times. Countless scholars in the Renaissance revised and commented on the classical geographer's standard work, each eager to have their say on what was known about the world, combining the ancient texts with the new knowledge and measurement methods being developed at the time. The Ptolemaic picture of the world persisted as late as the eighteenth century, although in the meantime locations had been surveyed in person, more parts of the world were known, and distances had been measured more accurately.

So when did mapping begin in the Southern Netherlands? It is hard to say, not least because we immediately encounter the problem of how precisely to define a map. In the broadest sense, a map is a visual representation of geographic objects and the spatial relationships between them. But not one by any means that has to be based on precise measurements or calculations. The criteria to be met by a map depend on its function, the mode of production, the knowledge and skills of its intended users, and the political or social interests at play.

The earliest maps often served an administrative purpose: from time immemorial, there has been a need to record the ownership of land in order to avoid disputes. Local surveyors measured the terrain and drew or painted a large-scale map (one in which the depicted objects are represented relatively big). This allowed landowners to tax properties, settle disputes or draw up building plans. The rapid growth of towns and fortifications lent fresh impetus to these practices, as did the need to manage water or drain land for farming. The landscape of the Low Countries was constantly changing in the Middle Ages and the early modern period, and maps likewise needed to be updated regularly. Maps in The Phoebus Foundation's collection showing the Waasland area and other polders on the left bank of the river Scheldt recall this history of water management, while offering a special insight into regional cartography. One exceptionally detailed large-format manuscript map shows the North Sea coast from Grevelingen to Biervliet (⟼ pp. 12-13). Attributed to Adriaan Villegas (1594-1669), it has been dated to shortly after 1640. This 'figurative map' (one that depicts an area in a visual, picturesque way) can be read as an expression of patriotic awareness and might have been linked to the Peace of Münster, which brought an end to the Eighty Years War (1568-1648). That conflict, in which the Low Countries sought to extricate themselves from Spanish-Habsburg rule, is one of the main threads running through this book. In the course of the war – from which only the Northern Netherlands would emerge independent – the various conflicts fought out locally and further afield fuelled demand for maps, which could in turn trigger further political tensions.

A revived interest in and translation of classical texts, the development of triangulation, a newly empirical approach to science, voyages of exploration, the invention of the printing press, commerce, and nascent capitalism all contributed to the growth and flourishing of cartography in the Southern Netherlands in the sixteenth century. Comprehensive world maps, useful sea charts, lavishly illustrated city maps, informative news maps, giant globes, and precise instruments were widely produced and distributed.

The result was a fruitful interaction between commercial cartography on the one hand, and scientific, mathematical and astronomical advances on the other. Measuring the terrain was not enough in itself to map the world: to fully comprehend the earth, knowledge of the universe was needed too. The epicentre of geographic, cosmographic and astronomic learning in the sixteenth century was Leuven (Louvain), where scholars from the region and further afield came together to exchange ideas at the university. An important factor was the printing press, which was (re)invented in Mainz around 1450 and swiftly spread from German cities to centres all over Western Europe. Scientific knowledge thus found its way readily to the Low Countries in the form of printed texts (and images).

Several illustrious figures were active at Leuven university, including the mathematician and cosmographer Gemma Frisius, who made his name setting out and developing the (by then centuries-old) principles of triangulation in a printed treatise. He was tutor to Jacob van Deventer, among others, who toured the Low Countries in the service of the government to map the regions and then the towns, doing so by means of triangulation. Frisius also made globes and other instruments in Leuven and trained Gerard Mercator, who focused on methods for calculating longitude and latitude and projecting the spherical shape of the earth onto a flat surface. This resulted in his monumental map of the world, in which he applied cylindrical projection – better known as the Mercator projection – which is still the best-known Western method for depicting the earth.

The work carried out by Van Deventer and Mercator was not necessarily intended for lay people, but the invention of the printing press nevertheless allowed cartography to be disseminated to a wider audience. Maps were included in books, with one of the earliest maps of Flanders in The Phoebus Foundation's collection (⟼ p. 16) appearing in Sebastian Münster's description of the world, the *Cosmographia Universalis* (first edition Basel, 1544). While the map is not especially detailed nor accurate, it does emphasise the strong communication links in the area, with the major towns, identifiable by a castle, connected by rivers and canals. Other early maps in the collection include woodcuts, such as the *mappae mundi* – medieval maps of the world – made for hefty volumes on world history.

When woodcuts gave way to copper engravings in the sixteenth and seventeenth centuries, maps began to sell well too. They could now reach a wider audience. While the scholars of Leuven racked their brains over the mathematics underpinning the science of geography, publishers in Antwerp were making serious money through the sale of commercial maps and atlases. Bruges had been an important trading hub earlier, but by the sixteenth century it found itself eclipsed by Antwerp, where a substantial trade had grown up in all sorts of goods, including books, maps and prints.

The multinational publisher Christophe Plantin produced books, prints and also maps for a large international readership, while the collector, map illuminator and future mapmaker Abraham Ortelius had settled in Antwerp too. Plantin and Ortelius – each an excellent networker with a fine nose for business – did not take long to find one another. Ortelius would have the distinction of publishing the first modern world atlas in 1570, of which Plantin printed several editions. The maps were engraved by Frans Hogenberg from Mechelen, who ran a workshop and business of his own in Cologne, from where he published news prints on the Eighty Years War. Hogenberg also co-produced a world city atlas – the *Civitates Orbis Terrarum* (Cologne, volume one: 1572) – with Georg Braun, which drew on examples like the Italian Lodovico Guicciardini's city atlas focusing on the Low Countries, which was enjoying immense success.

The capture of Antwerp by Spanish Habsburg forces in 1585 was a tipping point in the development of cartography in the Southern Netherlands. Many of the mapmakers and publishers previously active in the city now fled to the more tolerant North. The centre of the map market shifted from Antwerp to Amsterdam, and numerous scholars relocated from Leuven to Leiden. Among others, Petrus Plancius, Pieter van den Keere, Cornelis Claesz. and Jodocus Hondius I made the transition. The Dutch East India Company (VOC) and West India Company (WIC), the new Republic's large-scale trading ventures, generated increasing wealth, mostly acquired from the colonies, for their investors.

Petrus Plancius became a regular cartographer for the VOC and studied the shipping routes to the East Indies. He built on Mercator's work but also incorporated information from Portuguese and Spanish sources. The Antwerp-born WIC governor Johannes De Laet contributed to knowledge of the Americas, even though he never set foot there. The seafarer Jacob le Maire, also from Antwerp, on the other hand, completed a major voyage around the world, in the course of which he identified a previously unknown passage to the Pacific Ocean. For centuries afterwards, his route was given pride of place on globes and maps. Various Walloons also sought their fortune in the New Netherlands colony on the other side of the Atlantic.

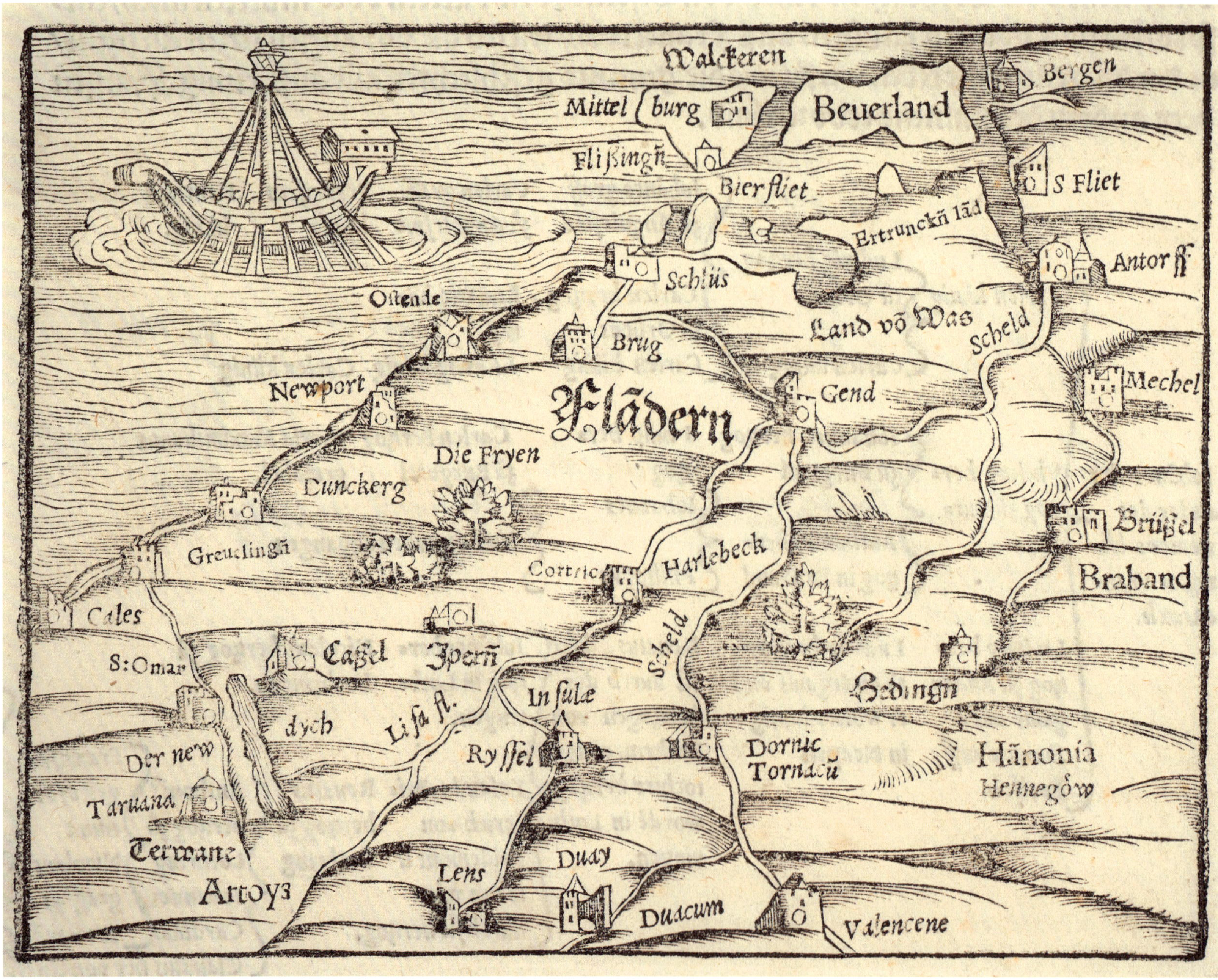

Sebastian Münster
Fla[n]dern, 1550 or after
Woodcut map of Flanders
Woodcut, 125 × 160 mm
ANTWERP, THE PHOEBUS FOUNDATION

These overseas adventures were, however, anything but altruistic. Both the mariners themselves and their promoters back home had just one thing in mind: to make money and become lords and masters of the new territories. Voyages of 'discovery' and colonization were inextricably linked with violence, oppression and enslavement. Cartography too was thoroughly implicated in all this, since maps either arose in the colonial context or were a means of pursuing such practices at sea.

The melting pot of the previously mentioned southern mapmakers (such as Petrus Plancius and Jodocus Hondius), northern ones (such as Johannes Janssonius, Lucas Jansz. Waghenaer and the Blaeu family), together with other visitors from abroad, brought about the apogee of cartography in the Low Countries, in terms of both quantity and quality. Despite all this activity by southerners in the Dutch Republic or on the other side of the world, things had not fallen entirely silent in the Southern Netherlands themselves. There were some, for instance, who made the reverse migration from the North to the Catholic South, including the Van Langren family, while some Flemings also returned after a few years, Pieter Verbiest I and his son Pieter Verbiest II among them.

In the course of the eighteenth century, the leading centres of cartography shifted to Britain, France and the Holy Roman Empire. Important names in this regard include Johann Baptist Homann, the Sanson family, Guillaume Delisle and the Delamarche firm in Paris, which produced globes. Just as the Low Countries became less leading in European map production, a flood of 'chorographic' works in the Guicciardini tradition began to appear, designed to promote and boost patriotic pride in Flanders, Brabant or the Low Countries as a whole. The phenomenon, which lasted far into the eighteenth century, might have been a response to the declining prosperity in the region in the second half of the seventeenth. These lavishly illustrated books, running to several volumes, were amply furnished with maps and detailed topographical prints of the south's villages, towns, castles, country houses and monuments. The tone was set by Antonius Sanderus with his *Flandria Illustrata* (Amsterdam, 1641-1644) and *Chorographia sacra Brabantiae* (Brussels, 1659-1663), but important imitators followed, such as Jacques Le Roy with his work devoted to Brabant (⟼ pp. 18-19), Nicolaas Despars with his chronicle of Flanders, and Jean-Baptiste Christyn with his description of the whole of the Low Countries.

The first truly systematic mapping of the territory corresponding with modern-day Belgium was the gigantic, large-scale map that Joseph Jean François, Count de Ferraris, produced in 1777-1778 for Maria Theresa, empress of Austria. The Ferraris Map meticulously charted what were then the Austrian Netherlands. The fashion for large-scale, nationwide topography of this kind continued into the nineteenth century – the era of imperialism and nation states. Orders came down from government for entire territories to be mapped precisely in a uniform and systematic manner. This became a particular imperative when Belgium broke away from the Kingdom of the United Netherlands in 1830. The newly forged nation felt a powerful need – in terms of identity as well as cartography – to set itself down on paper and present itself as a unified whole. Philippe Vandermaelen had the opportunity to produce the lion's share of this official cartography.

In addition to formal documents of this kind, there were of course maps for ordinary users too. The middle classes saw their mobility expand in the nineteenth century, with families setting out on their travels – primarily in their own country but also with the occasional trip abroad. Travelogues and guidebooks became more common, and convenient pocket-sized atlases grew in popularity. There was also an explosion in this period in the publication of city views, portraits, and pictures of favourite spots, attractions and buildings in Belgium and abroad. The invention of lithography meant that these could be printed in very large runs. Topographical and cartographical material – in the form of maps, text and images – became more accessible and affordable, no longer the preserve of a wealthy, intellectual elite. Maps now became part of everyday life and leisure in several layers of society.

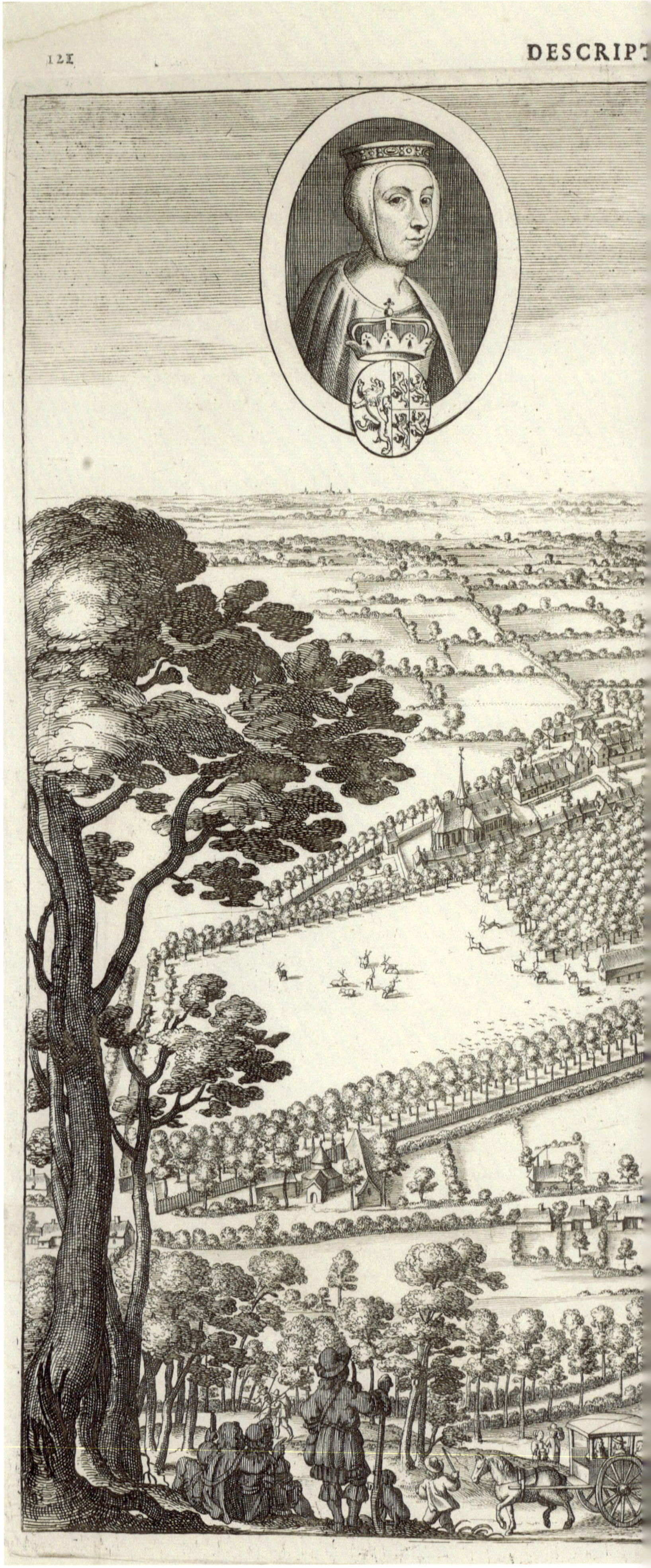

Turnhout celebre taxandriae municipium
Bird's-eye view of Turnhout
From: Jacques Le Roy, *Castella & Praetoria Nobilium Brabantiae*, 1699
ANTWERP, THE PHOEBUS FOUNDATION

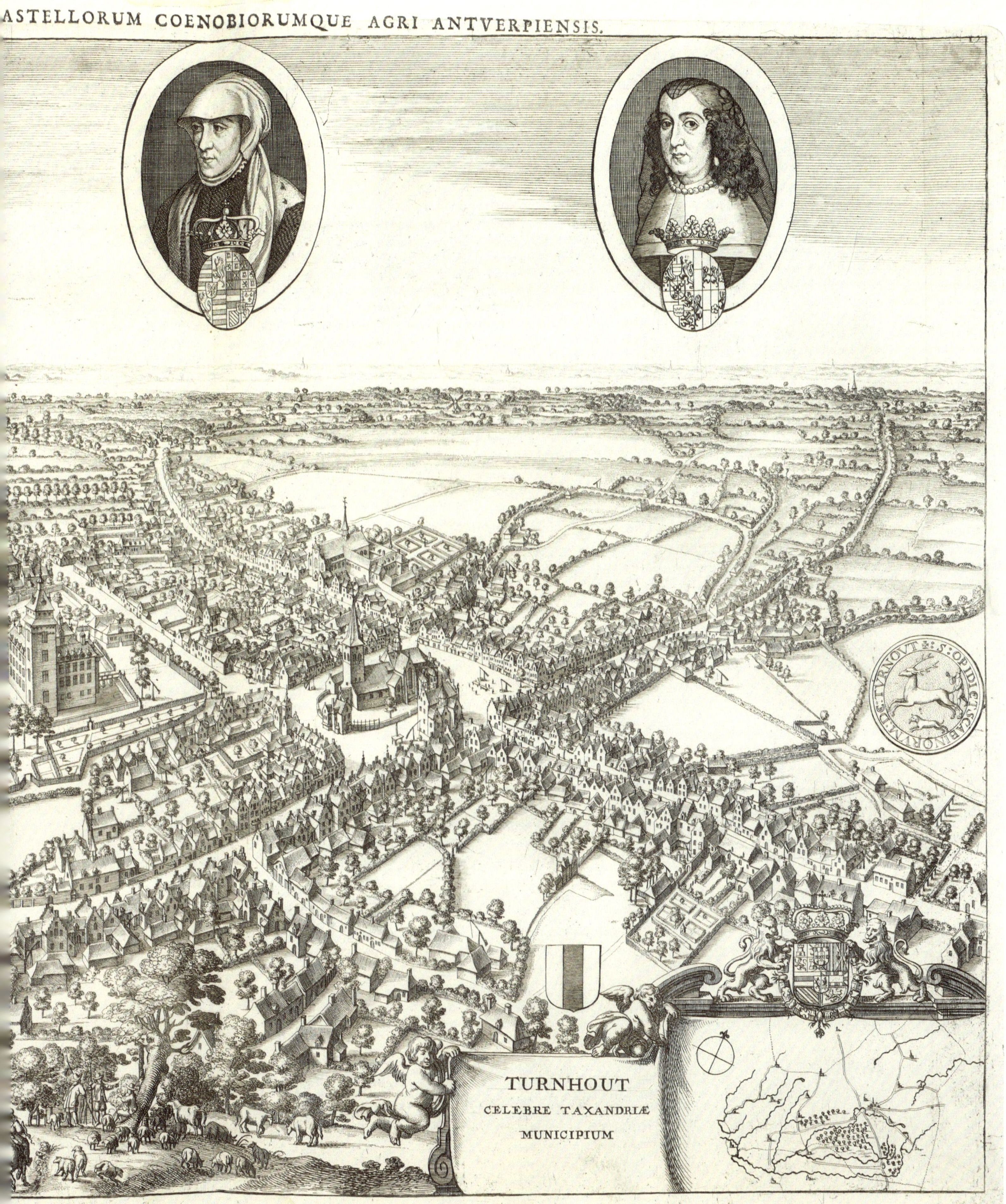
ASTELLORUM COENOBIORUMQUE AGRI ANTVERPIENSIS.
TURNHOUT
CELEBRE TAXANDRIÆ
MUNICIPIUM

A FLURRY OF PRINTING

Copies, states, editions, impressions, reprints, and variants are all technical terms that give an idea of the complexity of map printing. In some cases they are used more or less interchangeably, while in others they are highly specific. Although surveyors' maps often continued to be drawn or painted by hand until the nineteenth century, most maps were printed. After the printing press had made its appearance in Western Europe (from 1450 onwards), it became possible to cut a map from a block of wood or to engrave or etch it onto a copper plate, and to then print and distribute it in large editions. The earliest maps in this book are woodcuts, where the part of the drawing not to be printed was cut from a block of wood and the part left behind was then printed. The technique does not allow for much detail but lent itself to large editions and to illustrations in books. In the course of the sixteenth and seventeenth centuries, copper engraving and etching became more common for maps. Engraving is done with a sharp tool called a burin, which is used to cut lines into a copper plate, while etching uses a finer needle to draw the design into a layer of wax covering a similar plate. Maps from the early modern period often used a combination of etching and engraving. Engraving was suitable for lettering, for instance, while etching allowed very precise map elements to be drawn as well as decorations such as cartouches (ornamental frames). These techniques gradually gave way in the nineteenth century to planography, the best-known form of which is lithography, in which the design is drawn on a stone and could also be readily printed in multiple colours.

It is possible, therefore, to have several copies of a printed map in circulation (identical copies of some manuscript maps were also made, although examples of this kind are rarer). The printer or publisher decided on the size of the edition and pulled a specific number of prints from the wood block, the copper plate or the lithographic stone. These were then sold, distributed, or otherwise disseminated. One copy need not be identical to another: each ended up in a different place and was used, annotated or stamped by different people. A particular owner might also ask for their copy to be coloured in. If further examples were needed, a fresh print run could be set up, resulting in a reprint, in which the map might be amended or left unchanged. Where a mapmaker made adjustments to the wood block, copper plate or lithographic stone – deliberately or otherwise and no matter how small – we refer to a new 'state'.

The study of books distinguishes between impressions and editions. An impression refers to one and the same print run, with little difference between a first and a second impression in terms of typesetting or composition: at the very most, an occasional spelling mistake might be corrected. In the case of a new edition, by contrast, a large number of adjustments have been made to the set type, so that the composition of the pages, sections, text or illustrations is different. A new edition will often have been expanded or improved too.

The situation with atlases is a little more complicated. An atlas consists largely of graphic material – printed maps – and its composition is considerably more variable than that of a textbook. This is reflected in the way seventeenth-century map publishers like Hondius, Blaeu and Janssonius issued numerous different editions of their atlases in a short space of time, often adjusting them slightly as they did so: adding extra maps, replacing one map with another, similar one, or adjusting the order in which they appeared.

This makes it hard for us to speak of an 'impression' or an 'edition' of an atlas, and so these terms are rarely applied to early modern atlases. To further complicate things, the maps in atlases could also be distributed separately or else removed from an atlas – whether at the time or at some later date – making it even harder to reconstruct the precise context in which an individual map first appeared. The same goes for many maps that had originally appeared in books or were intended for that purpose. 'Cartobibliography' sets out to classify all these different editions and the individual maps published within them.

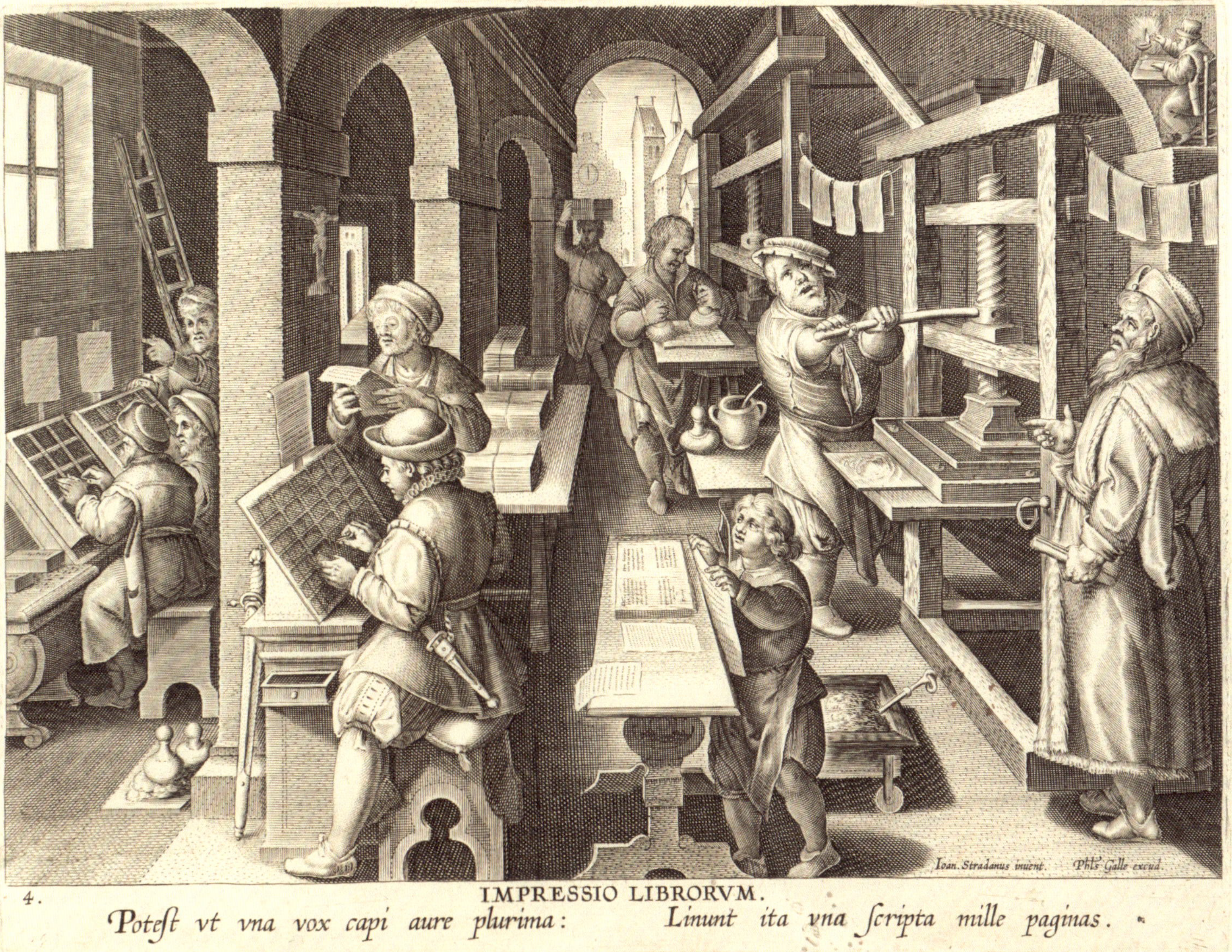

Hans Collaert and Theodoor Galle after Johannes Stradanus
Nova Reperta: Impressio Librorum, c.1590
The invention of book printing
Copper engraving, 263 × 340 mm
ANTWERP, THE PHOEBUS FOUNDATION

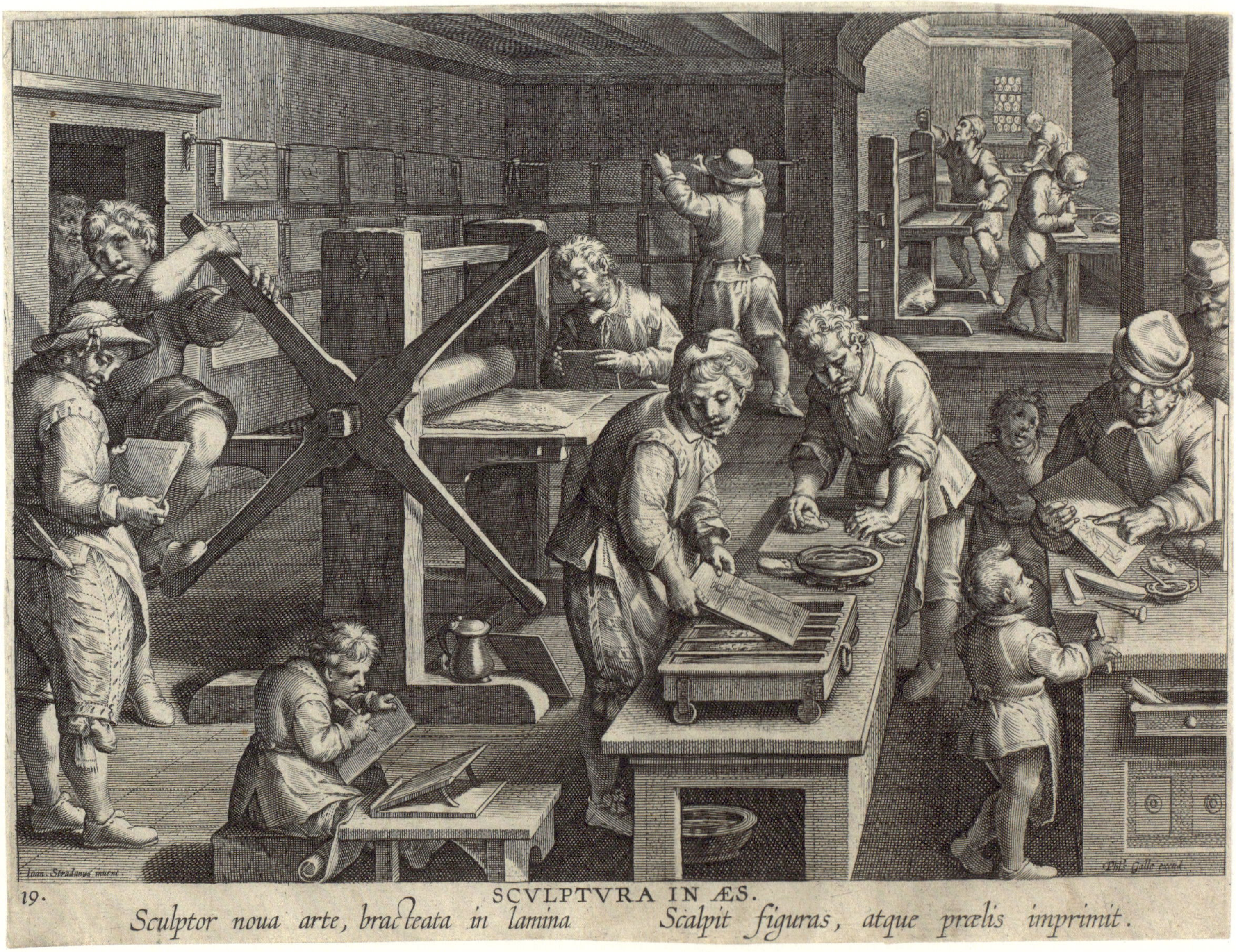

Hans Collaert and Theodoor Galle after Johannes Stradanus
Nova Reperta: Sculptura in aes, c.1590
The invention of copper engraving
Copper engraving, 263 × 340 mm
ANTWERP, THE PHOEBUS FOUNDATION

CONCLUSION

This book invites you to discover for yourself the history of cartography in the (Southern) Netherlands, as briefly summarised above. The first chapter begins with local cartography and topography, based on surveying and fieldwork. We then travel from the countryside to the urban landscape, where a variety of city 'portraits' and maps are featured. This is followed by several chapters on the role of cartography in news culture and historiography, with particular reference to the Eighty Years War. Chapter III discusses the different news media, after which a specific case is examined in Chapter IV. Chapter V looks at how the Low Countries reflected on their own history and on the relationship between North and South, while Chapter VI deals with the role played in historiography and mapping by foreigners – Italians in particular. In Chapter VII we take to the water: how are maps used to plot a course and navigate at sea? The next chapter focuses on where vessels from the Low Countries sailed to, which 'new worlds' were visited and mapped, and for what reasons. We then zoom out again to consider the different world views that held sway through the centuries, before examining world atlases, including Blaeu's *Atlas Maior*. The book rounds off by bidding farewell to the earth, using astronomical instruments and cosmography to gaze out into the universe and back.

The chapters are structured by scale, steadily zooming out as we go, yet also simultaneously zooming in on the objects from The Phoebus Foundation's collection and their various creators, each of whom contributed to the story of Netherlandish cartography. The highlighted thematic texts provide extra information about certain subjects, specific makers or important events. The interspersed texts marked in yellow point out special objects from the collection. The people and works that are examined along the way shed light on glorious moments in cartographic history while not shrinking from the dark side that often lies hidden. There is more that maps *do not* show than they *do*, while at the same time they seek to present us with particular ideas and world views. An early map not only represents a geographical area and a historical situation; it is also not easy to pin down and can prompt an infinite number of new questions, discussions and interpretations. *Groundbreakers* invites you to explore and expand the horizons of your own world through cartography.

LITERATURE

J. Bossu, *Vlaanderen in oude kaarten: drie eeuwen cartografie*, Tielt, 1982.

W. Bracke & E. Leenders, *Vlaanderen in 100 kaarten*, Leuven, 2015.

P. De Maeyer, 'Cartografie, de zee en de rol van de Vlamingen', *VLIZ, De Grote Rede*, 31, 2011: 11-16.

P. De Maeyer, M. Galand, B. Vannieuwenhuyze et al., *De geschiedenis van België in 100 oude kaarten*, Tielt, 2021.

D. Duncker & H. Weiss, *Het hertogdom Brabant in kaart en prent*, Tielt, 1983.

C. Koeman, *Atlantes Neerlandici. Bibliography of terrestrial, maritime and celestial atlases and pilot books, published in the Netherlands up to 1880*, Amsterdam, 1967-1971, 5 vols.

C. Koeman, *Geschiedenis van de kartografie van Nederland: zes eeuwen land- en zeekaarten en stadsplattegronden*, Alphen aan den Rijn, 1983.

C. Koeman, G. Schilder, M. van Egmond et al., 'Commercial Cartography and Map Production in the Low Countries, 1500-ca. 1672', in: D. Woodward (ed.), *The History of Cartography, vol. 3. Cartography in the European Renaissance*, Chicago / London, 2007, vol. 2: 1296-1383.

C. Koeman & M. van Egmond, 'Surveying and Official Mapping in the Low Countries, 1500-ca. 1670', in: D. Woodward (ed.), *The History of Cartography, vol. 3. Cartography in the European Renaissance*, Chicago / London, 2007, vol. 2: 1246-1295.

P. van der Krogt (ed.), *Koeman's Atlantes Neerlandici. New Edition*, Houten, 1997 et seq., 6 vols.

G. Schilder & M. van Egmond, 'Maritime Cartography in the Low Countries during the Renaissance', in: D. Woodward (ed.), *The History of Cartography, vol. 3. Cartography in the European Renaissance*, Chicago / London, 2007, vol. 2: 1384-1432.

H. van der Heijden, *Het Hertogdom Brabant in oude kaarten, 1536-1785*, Eersel, 1991.

B. Vannieuwenhuyze, M. Griffioen & A. van Schaik, *Oude kaarten lezen. Handboek voor historische cartografie*, Zwolle, 2023.

K. Zandvliet, *De groote waereld in 't kleen geschildert: Nederlandse kartografie tussen de middeleeuwen en de industriële revolutie*, Alphen aan den Rijn, 1985.

K. Zandvliet, 'Mapping the Dutch World Overseas in the Seventeenth Century', in: D. Woodward (ed.), *The History of Cartography, vol. 3. Cartography in the European Renaissance*, Chicago / London, 2007, vol. 2: 1433-1462.

WITHOUT LAND, there would be no maps. While you will also find maps of the sea, the heavens and imaginary realms in this book, representing land has always been the most obvious task of the cartographer. Boundaries, plots and roads, from cultivated land to pristine nature: these are what a map shows us. Maps are the medium of choice for registering land and have been so since time immemorial. The need to correctly administer land taxes and the rights and duties of landowners dates back centuries. The history of the Dutch and Belgian landscape, characterised by a constant struggle with water, required its inhabitants to be flexible and inventive. They reclaimed land from the sea and, if the dykes were breached, drained it all over again. All these changes to a landscape viewed as 'malleable' demanded the creation of maps: for administrative purposes, tax collection, the settlement of disputes and claiming ownership of a piece of land.

By no means were the local surroundings always mapped using the same yardstick – whether literal or figurative. What methods and techniques, then, *did* mapmakers use in the Southern Netherlands? The earliest measurements were made using a chain and a surveyor's cross. Later, triangulation allowed larger distances to be measured, and the method continued to be used until the nineteenth century, after Belgium had become an independent nation. An official land register was introduced in 1834 (two years later than in the Netherlands), influenced by recent Napoleonic innovations and the French revolutionary occupation of the Low Countries (1794-1814). Henceforward, the land registry would administer real estate and the associated rights and duties in the form of taxes, leading to the uniform, systematic mapping of the entire landscape of the Low Countries. Steadily zooming out from local cartography, accurate surveying was raised to a national level.

CHAPTER I

THE ART OF SURVEYING

FROM LOCAL TO NATIONAL

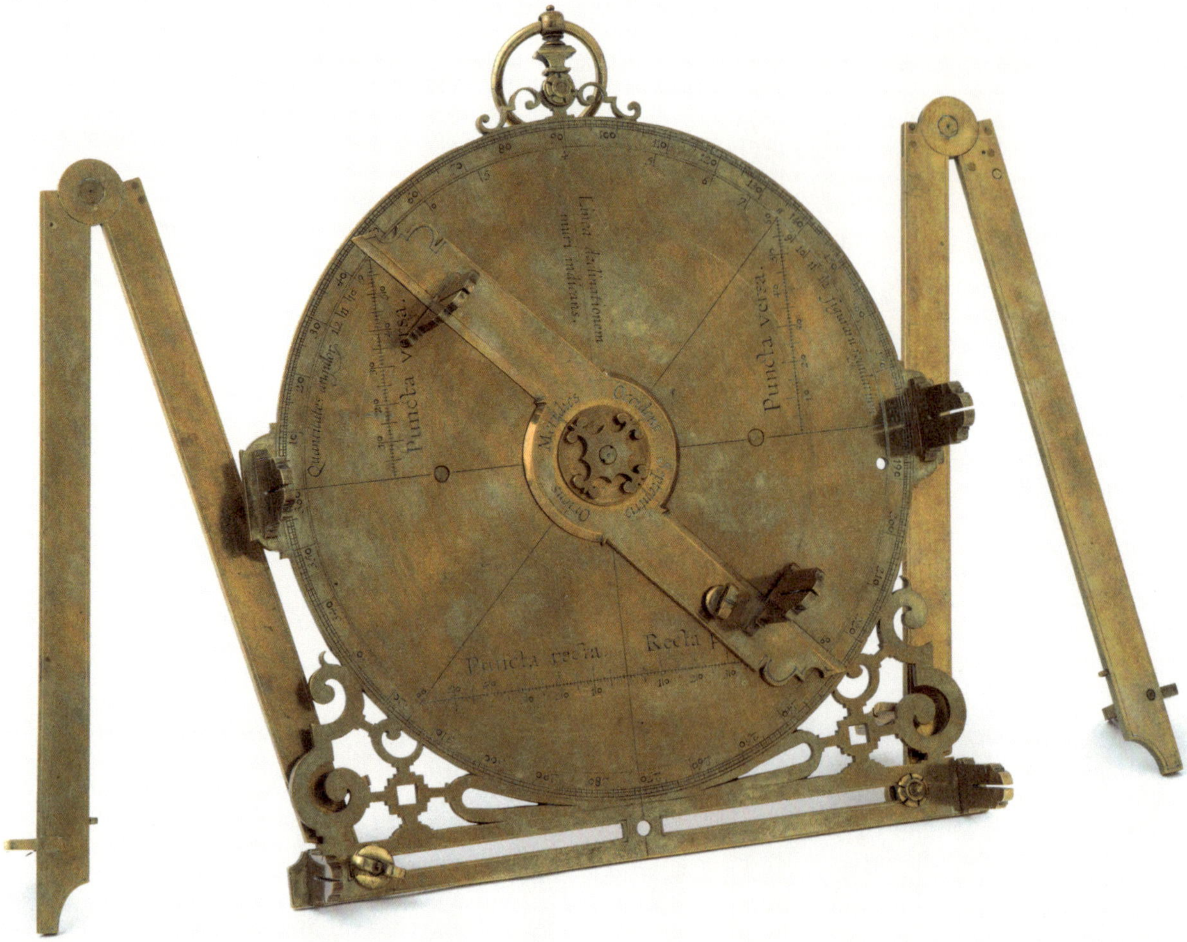

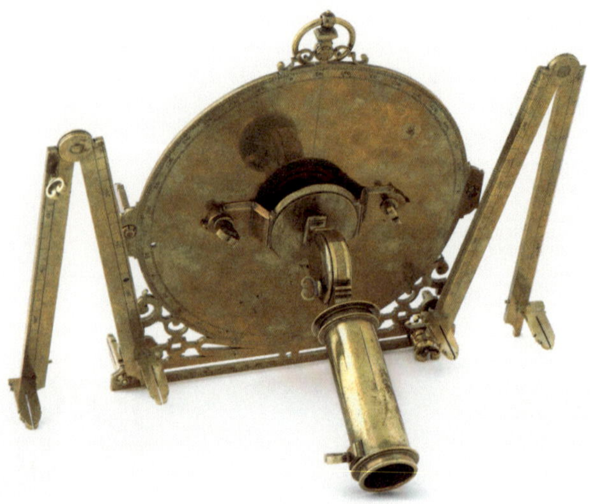

Attributed to Michel Coignet
Double-sided surveyor's circle, c.1610
Polished brass, 24.5 × 19.2 × 0.2 cm
ANTWERP, THE PHOEBUS FOUNDATION

THE COIGNET AND DOU CIRCLES

For many years, triangulation was the most commonly used and accurate method for calculating the distance between places. It relied on measuring the angles between specific locations. What kind of tools did surveyors take with them to measure those angles and lengths around the year 1600 or so? Their instruments were traditionally fairly simple and would not change drastically until the nineteenth century. A rod and chain, for instance, were used to measure relatively short distances. The physical rod could be half or a whole rod in length, while a surveyor's chain was equal to five or ten rods. A rod was an ancient unit of measurement, the precise size of which varied between around three and six metres depending on the location. The Rhenish rod was the most common in the Low Countries, measuring 3.767 metres. Surveyors sometimes worked with 'chain pullers' – assistants who laid out the chain for them. The surveyor's cross was a right-angled device with a fixed sight at the end of each arm through which the surveyor could look in order to plot right angles.

A more sophisticated instrument arrived with the surveyor's circle, a polished-brass version of which in The Phoebus Foundation's collection is attributed to Michel Coignet (1549-1623). This Antwerp mathematician, cartographer and instrument maker spent part of his career as a scientist at the Spanish-Habsburg court of Archduke Albert of Austria (1559-1621) and the Infanta Isabella Clara Eugenia (1566-1633), governors of the Southern Netherlands. Working with the engraver Ferdinand Arsenius (active 1589-1628), Coignet caused the centre of scientific instrument-making to shift from Leuven to Antwerp.

The surveyor's circle developed by Coignet is similar to the well-known 'Holland circle' or 'Dou circle', the latter named for Jan Pietersz. Dou (1573-1635), one of the most influential and well-known surveyors of the Low Countries. Dou built his reputation through the circle he invented around 1610, for which he ingeniously combined several existing principles. Taking the basic surveyor's cross, he added a single rotating sight, which allowed angles greater than ninety degrees to be measured as well. A neat feature of this device was that it was now possible to triangulate using just one instrument. Because of its widespread use in the Low Countries, his invention later became known as the 'Holland circle'.

Coignet's instrument is similar to Dou's, consisting of a disc with a fixed and a rotating sight. The circumference of the circle is engraved with 360-degree marks, allowing angles to be read off directly and easily. Two rotating, fold-out rulers are fixed to the back, with further sights at the ends, allowing the highly accurate measurement of angles over large distances. The rules have a scale of their own too. The device can also be suspended by the ring at the top, so that vertical angles as well as horizontal ones can be measured.

All the same, the surveyor's circle that Coignet designed lacks some of the clever touches that Dou added to his version, such as the recesses in the disc to reduce the weight of the instrument and allow a larger size. Nor does it have the socket for placing on a long stick, so that the surveyor could take sightings at eye level, or the compass at the centre for use in orientation.

Gemma Frisius and triangulation

Surveying made its appearance in medieval Western Europe by way of the Arab world and treatises by Roman surveyors (*agrimensores*). While simply counting footsteps is one way to measure distances, the triangulation method would change this forever. 'Forward intersection' is based on Pythagoras's theorem. A triangle has three angles and three sides, and if at least three of these elements are known, including the length of one or more sides, the remaining dimensions can be calculated. If, say, you know the length of one side of the triangle and two of its angles, you can swiftly work out the rest.

But how do you go about measuring the angles between places located some distance from each other? To do this, you need to stand on a high spot, such as a church tower, hill or some specially installed base point. The angle between two other elevations further away can then be measured, or alternatively the angle relative to the north. These angles are then plotted on paper, following which the length of the sides of the triangle can be calculated. In the course of the eighteenth and nineteenth centuries, imaginary triangles between places were created in the Low Countries to facilitate this process – 'triangle networks' that allowed large-scale topographical maps to be drawn up for bigger areas.

There is evidence that the principles of triangulation were already being applied in ancient times, and one of the earliest Dutch cartographers, Jacob van Deventer (1505-1575), is known to have used the technique in his early regional maps of the Low Countries, which are among the oldest of their kind. It is now widely accepted, however, that the first establishment and dissemination of the age-old method can be attributed to the Dutchman Gemma Frisius (1508-1555). Fries, as he was called in his native Dokkum, was active from around 1525 in Leuven, where he studied before going on to teach mathematics and medicine. He also designed globes and instruments there, drawing on the work of all manner of important scientists. It was in Leuven that Frisius published an amended reissue of *Cosmographicus liber* (first edition: Landshut, 1524) by Peter Apian (⟶ Chapter XI). He added an appendix on surveying entitled *De locorum describendorum ratione* ('Concerning the method of describing places'), in which he introduced the triangulation method in such a practical way that it could be applied to mapmaking. Frisius's reissue first appeared in February 1533. The Phoebus Foundation owns the improved second edition, translated into French, of 1581. It is thanks to this widely disseminated publication that Frisius is now viewed as the inventor of triangulation. Apian and Frisius's treatise lived on in some thirty translations and reprints until 1609.

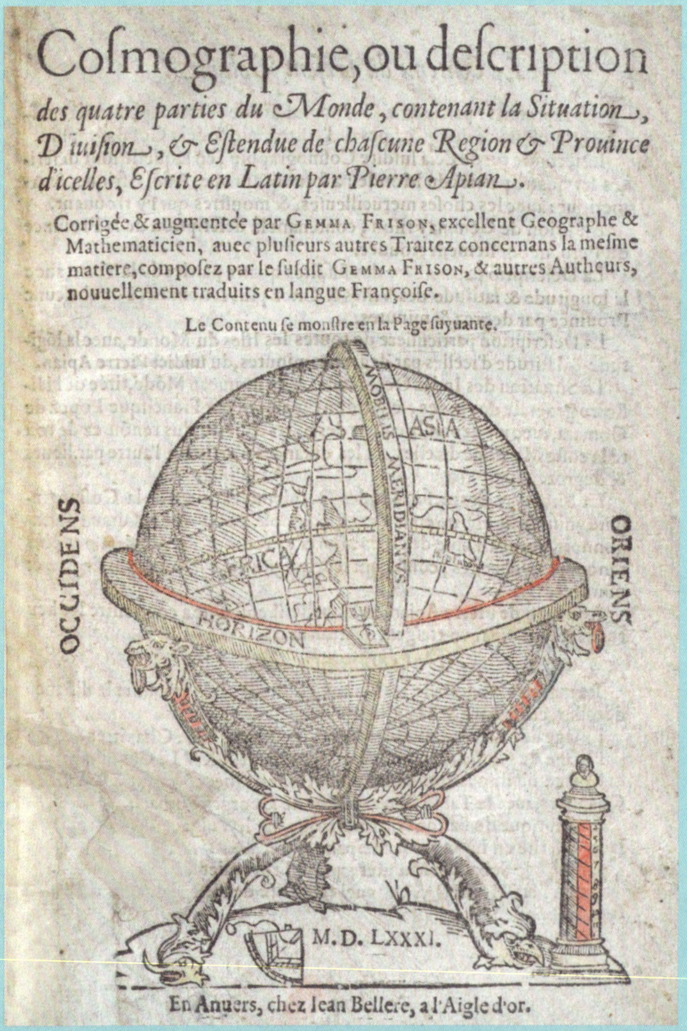
Cosmographie, ou description des quatre parties du Monde, contenant la Situation, Diuision, & Estendue de chascune Region & Prouince d'icelles, Escrite en Latin par Pierre Apian.

Corrigée & augmentée par Gemma Frison, excellent Geographe & Mathematicien, auec plusieurs autres Traitez concernans la mesme matiere, composez par le susdit Gemma Frison, & autres Autheurs, nouuellement traduits en langue Françoise.

Le Contenu se monstre en la Page suyuante.

En Anuers, chez Iean Bellere, a l'Aigle d'or.

Title page
From: Peter Apian and Gemma Frisius, *Cosmographie, ou description des quatre parties du Monde*, 1581
ANTWERP, THE PHOEBUS FOUNDATION

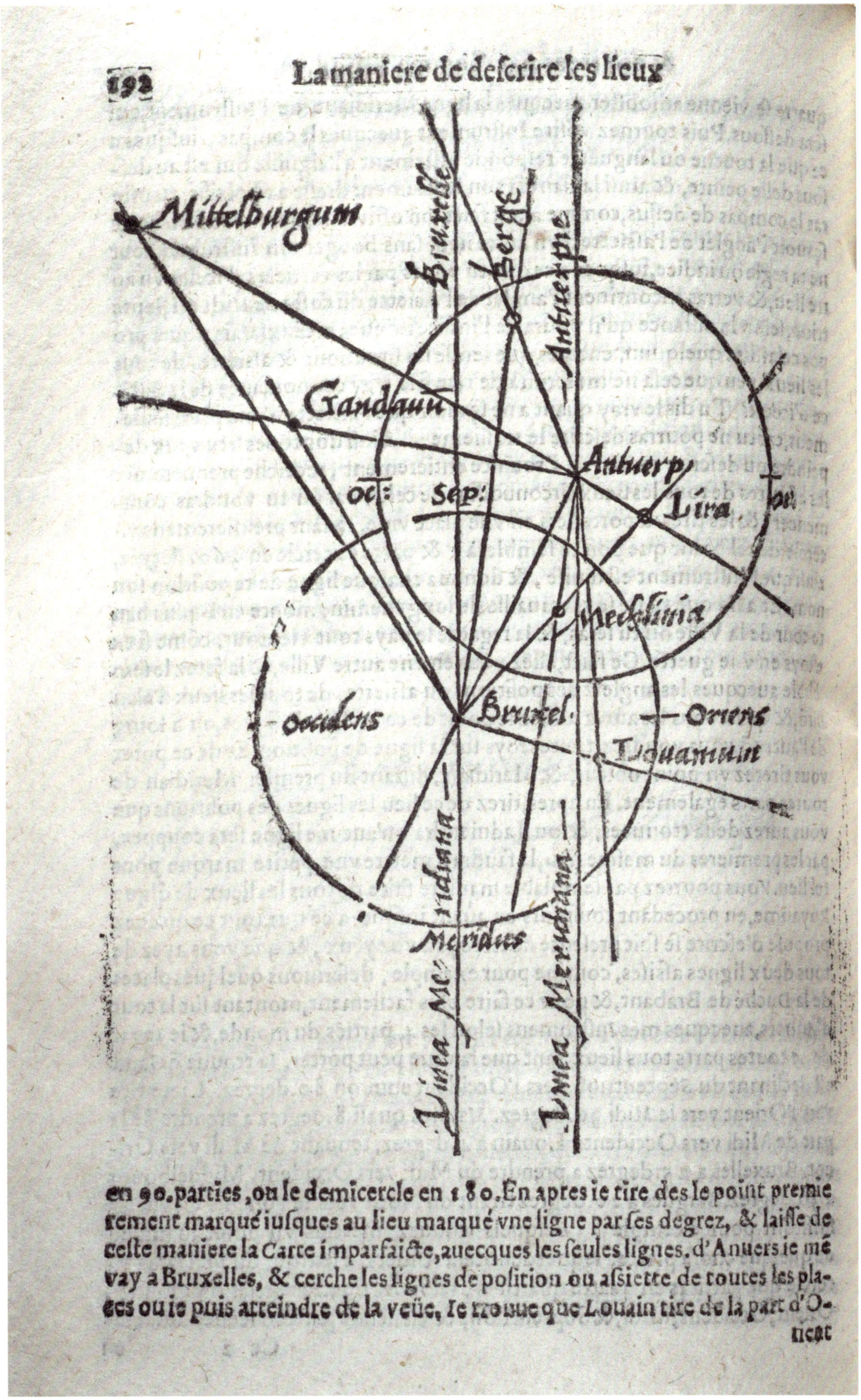

192 La maniere de descrire les lieux

en 90.parties, ou le demicercle en 180. En apres ie tire des le poinct premie
rement marqué iusques au lieu marqué vne ligne par ses degrez, & laisse de
ceste maniere la Carte imparfaicte, auecques les seules lignes. d'Anuers ie mē
vay a Bruxelles, & cerche les lignes de position ou assiette de toutes les pla-
ces ou ie puis atteindre de la veüe, Ie trouue que Louain tire de la part d'O-
rient

Illustration of the triangulation method
From: Peter Apian and Gemma Frisius, *Cosmographie, ou description des quatre parties du Monde*, 1581
ANTWERP, THE PHOEBUS FOUNDATION

MAPPING AND PICTURING THE WAASLAND

The flat polders of the Low Countries readily lent themselves to triangulation: no matter how far apart they were, it was relatively easy to pick out church towers and other tall buildings in the distance across the sprawling landscape. Newly reclaimed land was divided into parcels, which meant it had to be registered. There are references to surveyors in Flemish coastal regions from as early as 1190, and the Waasland in what is now East Flanders also boasts a long tradition of surveying in connection with land reclamation, particularly from the fifteenth century onwards. Accurate maps needed to be drawn up before, during, and after a new polder was drained. The Phoebus Foundation's collection includes a series of figurative polder plans of the Waasland which vividly map and depict this changing, malleable landscape.

The manuscript series was drawn up by an unidentified surveyor and is undated, although the dates of several polder projects are given, the earliest being 1688. Each map shows the subdivision of the polders and includes a scale indicator of 100 to 1,000 rods. The unit in question might have been the Waas or the Ghent rod (both of which were equal to 3.854 metres). The maps contain numerous details about roads (in brown), dykes (green), windmills, marshes, salt marshes and farmland. Houses and larger buildings are drawn three-dimensionally, viewed vertically from the front, and some are labelled (the Singelberg, for instance). The surveyor might have used some of these tall buildings as triangulation points, including the *capel op gaverland* shown bottom right or the church next to Fort Verrebroek, on the opposite side of the mapped area.

Local maps like this are a rich source of information, not least for regional toponyms. Next to the Singelberg, for example, we see the words *de cleyne verre* ('little Verre') to the right of the *groote verre* ('big Verre'), to identify the polders. Another manuscript map in The Phoebus Foundation's collection zooms in further on this small piece of polder, located at the Singelberg and Beveren Castle. With a scale of 100 Ghent rods, it shows the precise parcels making up the *cleyne verre*, as drawn and painted by the Beveren surveyor Henricus Joppen on 3 January 1771. The handwritten text at the bottom explains that Petrus De Kever, prior of the nearby Wilhelmite monastery, commissioned the map, which is a copy of one showing the Beveren polder in the year 1692, made by another local surveyor called Baudewijn Speelman.

The monastery's records show that it was indeed headed by a prior Petrus Jozef De Kever from 1762 to 1770. Speelman's original 1692 map has yet to be located, but we do know that he prepared maps for several estates belonging to the aristocratic Arenberg family, including one of the plot division of Oud-Arenberg-polder, reclamation of which was completed in 1688. Once a piece of land had been fully drained, the polder was divided into parcels, which were precisely mapped out and their areas stated. The plots were then assigned by lot and sold to subscribers. The new owners were often specified by name on parcel maps, which formed part of the register or 'land book'.

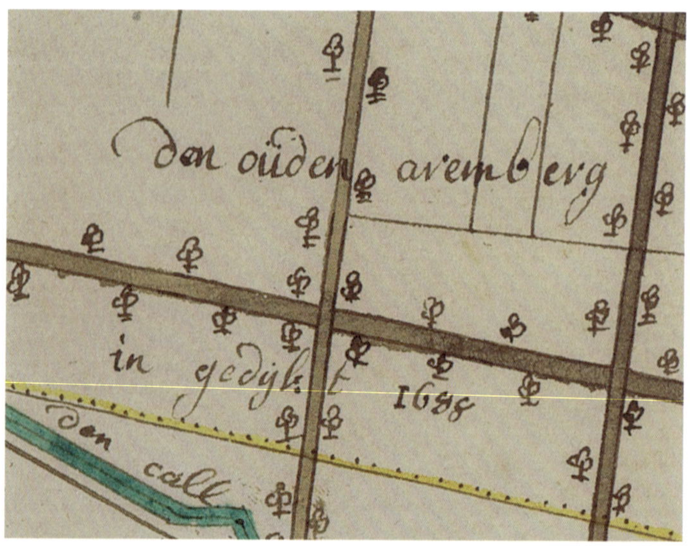

Den ouden Aremberg
Detail from: Manuscript map of the Waasland, with the Kallo polder, Doel, the Melschen polder, Verrebroek and the river Scheldt, after 1688 (p. 31)

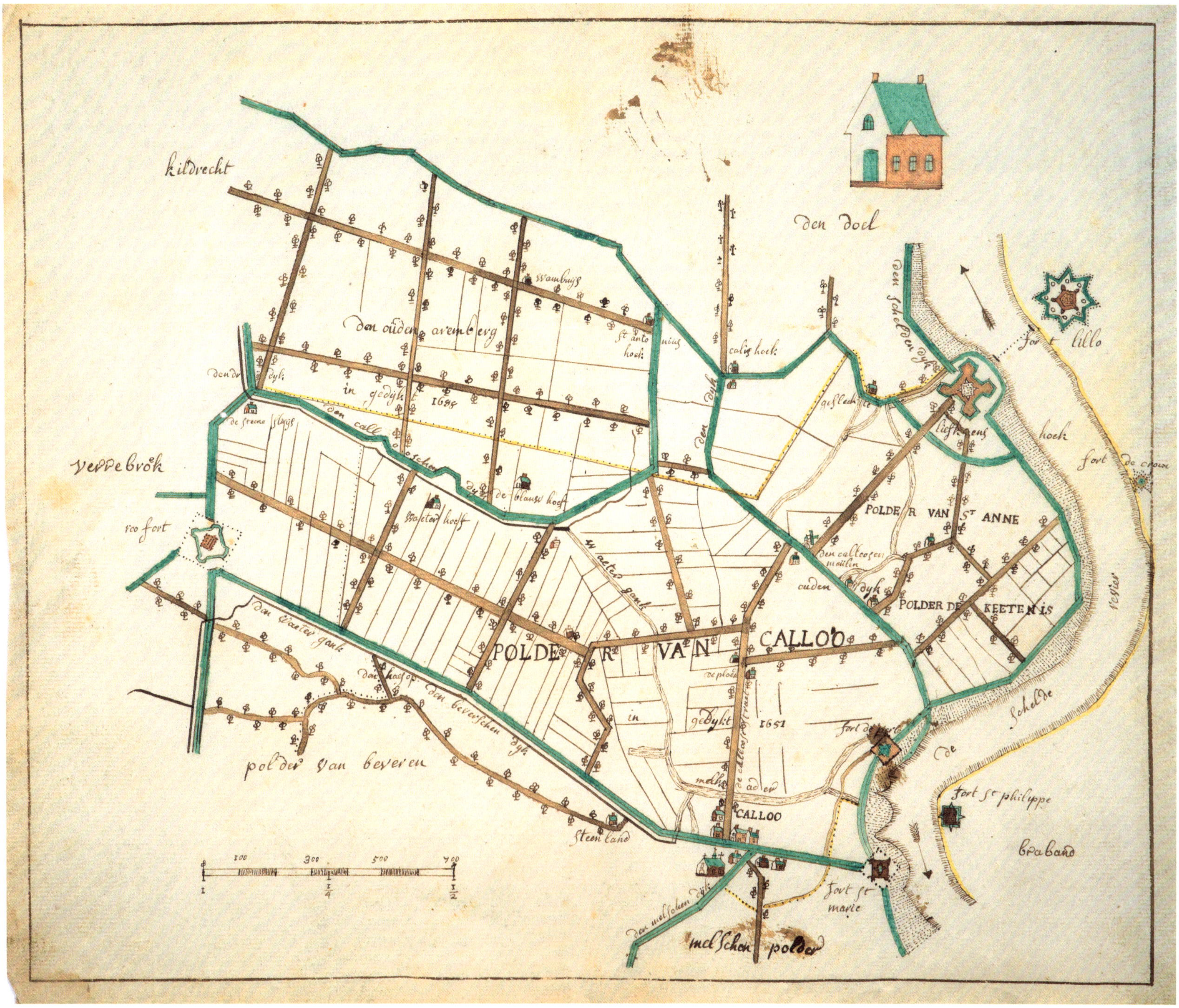

Manuscript map of the Waasland, with the Kallo polder, Doel, the Melschen polder, Verrebroek and the river Scheldt, after 1688
Ink and watercolour on paper, 314 × 376 mm
ANTWERP, THE PHOEBUS FOUNDATION

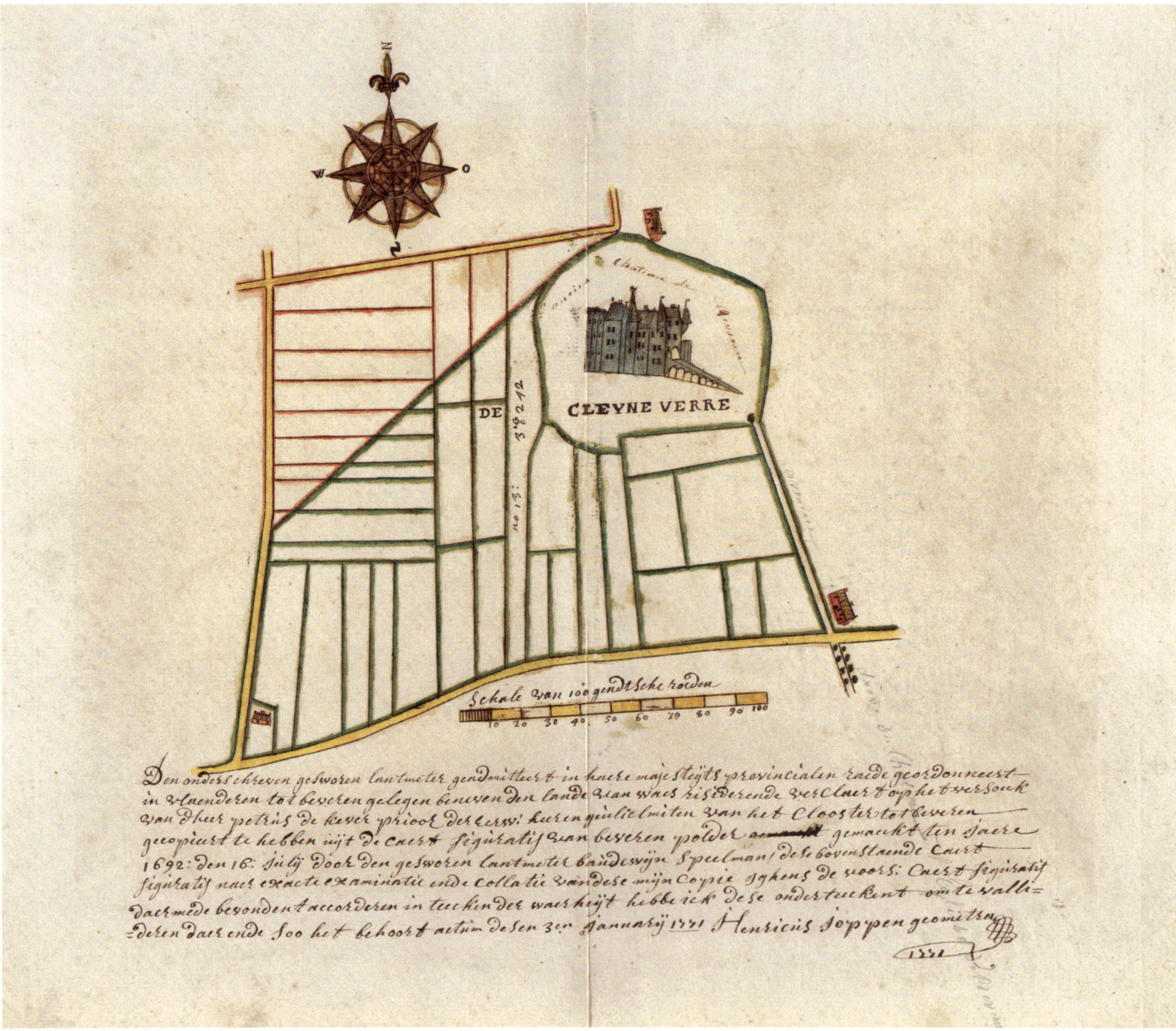

Henricus Joppen after Baudewijn Speelman
Manuscript map of part of the Beveren polder known as 'cleyne verre', 1771
Ink and watercolour on paper, 330 × 427 mm
ANTWERP, THE PHOEBUS FOUNDATION

Manuscript map of the Waasland, with the Beveren polder,
Fort Verrebroek and Kallo, after 1688
Ink and watercolour on paper, 315 × 372 mm
ANTWERP, THE PHOEBUS FOUNDATION

Capel op Gaverland and *Verrebroek fort*
Details from: Manuscript map of the Waasland, with the Beveren polder,
Fort Verrebroek and Kallo, after 1688 (p. 33)

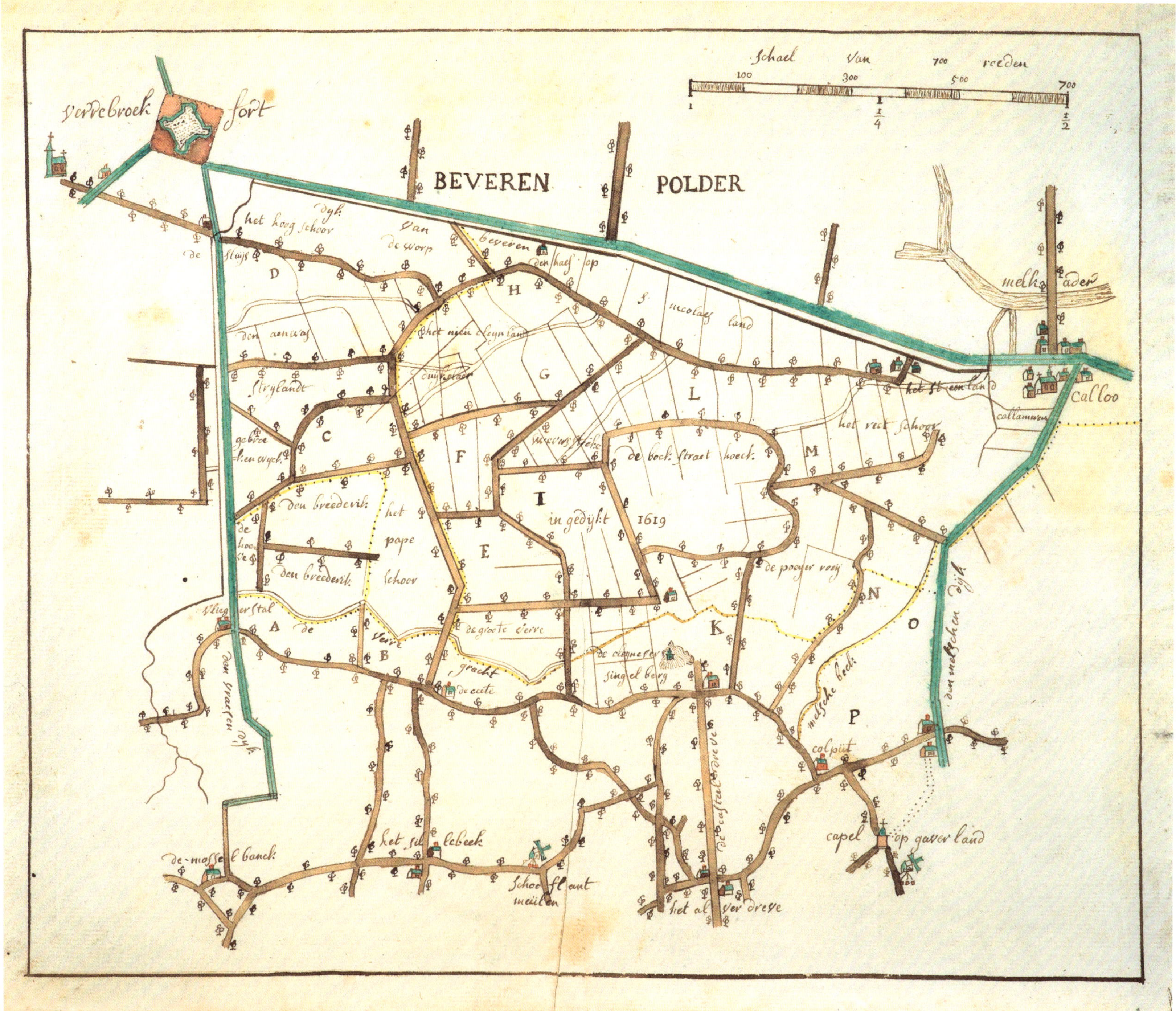
BEVEREN POLDER
Schael van 700 reeden
100 300 500 700
verdebroek fort
Calloo
melk ader
in gedykt 1619
capel op gaver land

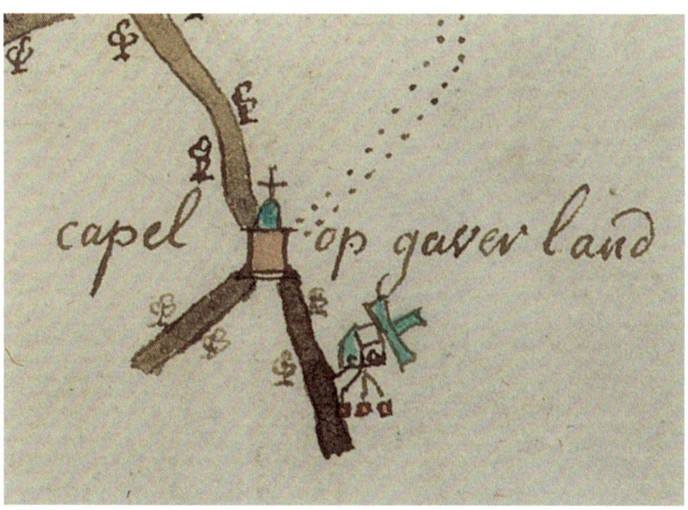
capel op gaver land

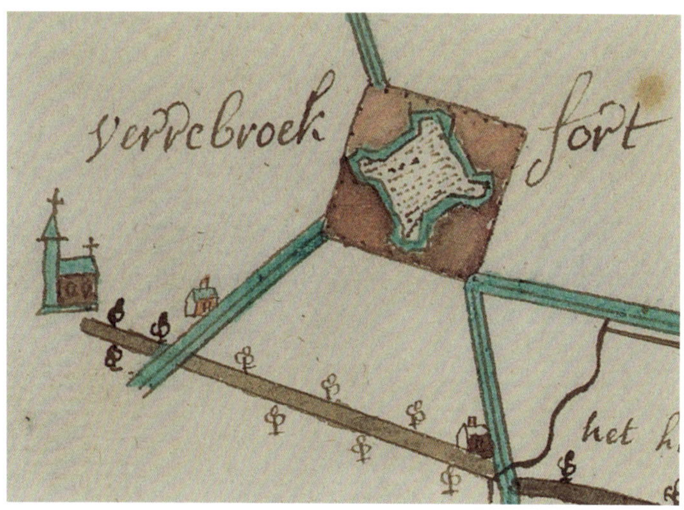
verdebroek fort

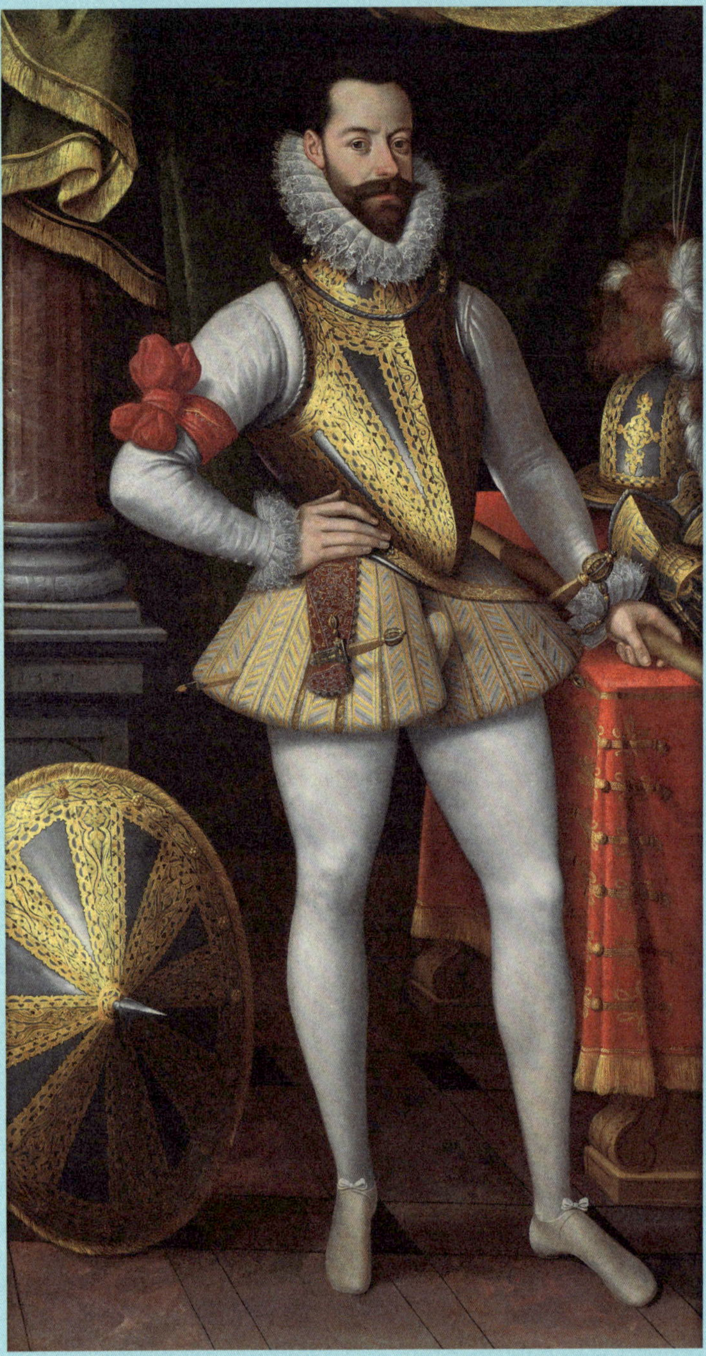

Beveren Castle

Beveren Castle stands out in Henricus Joppen's polder map, which contains a large depiction of it, even though the building on the Singelberg near Beveren had been demolished in the middle of the seventeenth century. It had featured prominently during the revolt against Spanish rule as the headquarters of Alexander Farnese (1545-1592), Duke of Parma and governor of the Low Countries (⟶ p. 68). It was here that representatives of Antwerp officially agreed the city's capitulation to the Spanish in 1585. The castle and surrounding estates had come into the hands of the Arenberg family in 1622 but had since fallen into disrepair. It was demolished in 1652, and the materials were reused during the construction of churches in Beveren, Verrebroek and Kallo.

Joppen's map keeps the memory of the castle alive by depicting it large and centrally. The picture recalls the bird's-eye view of Beveren in the chorographic book *Verheerlijkt Vlaandre* ('Exalted Flanders'; The Hague, 1735), a reissue of Antonius Sanderus's chorography of Flanders (⟶ p. 60), which also shows Beveren Castle, even though nothing but ruins remained of it at the time. The print was made by the Dutch engraver Reynier Blokhuysen (1663/1683-1724/1744), who did not visit his locations to record them in person but will have looked instead at seventeenth-century images, maps and text descriptions. The coats of arms of the duke of Arenberg (left) and of Beveren (right) are featured in the map, which shows all the villages and churches in the area, as well as its windmills, gallows and fortresses. The most important buildings are explained in the legend, including the Wilhelmite monastery (B), which was suppressed in 1784, and Beveren Castle (C), which is also depicted in the insert at the bottom. Could it be that the eighteenth-century mapmakers failed to do their homework and were simply unaware that Beveren Castle no longer existed here? Despite being demolished, the memory of the castle was preserved for many years after, living on in works of art and maps.

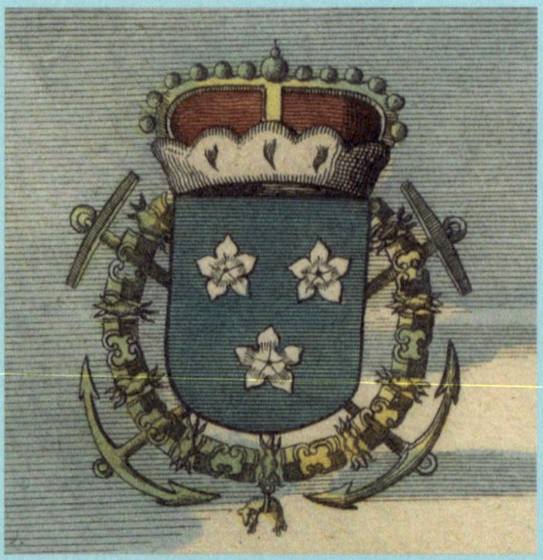

Jan Baptist Saive
Portrait of Duke Alexander Farnese, Governor of the Spanish Netherlands, 1592
Oil on canvas, 203 × 108 cm
ANTWERP, THE PHOEBUS FOUNDATION

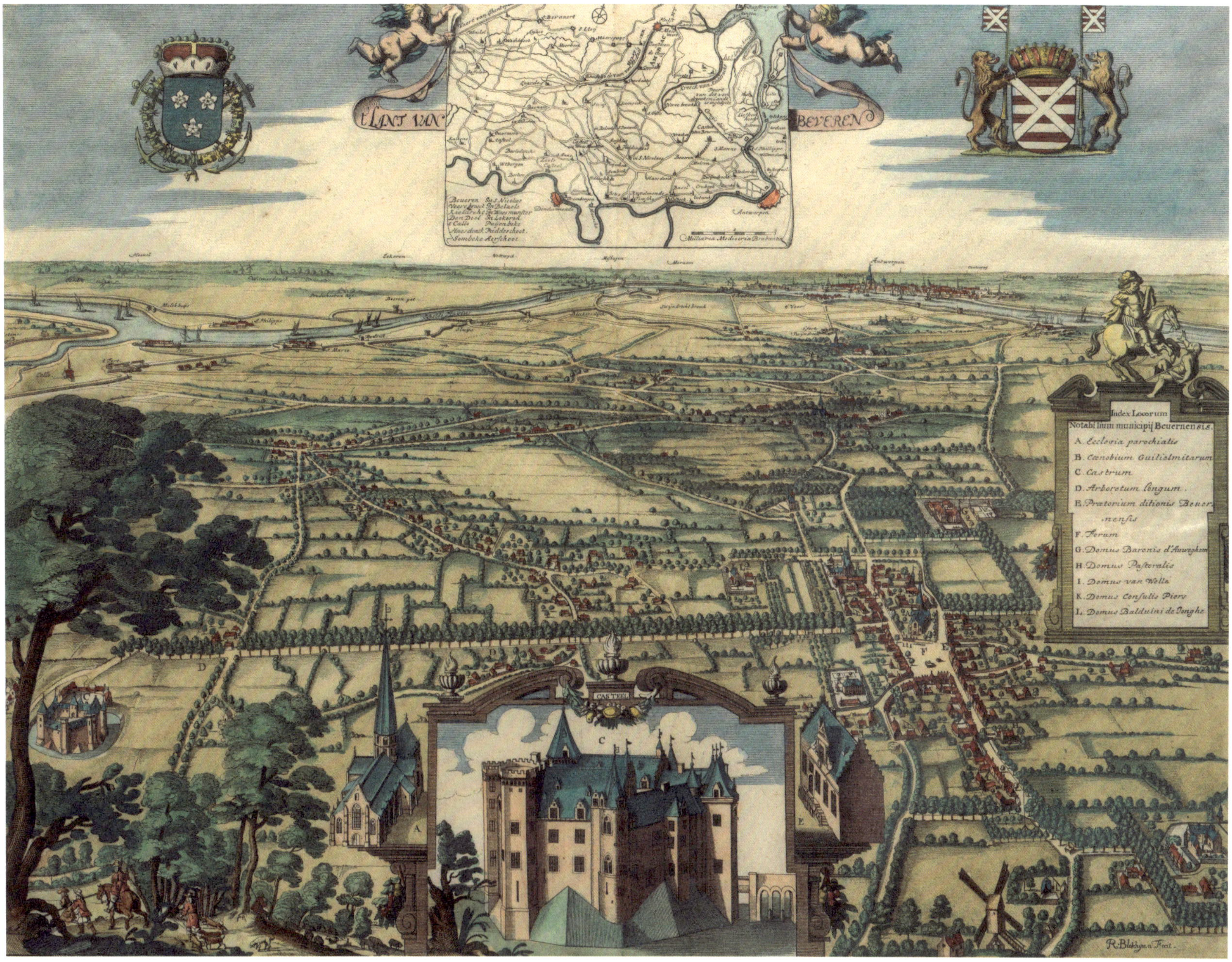

Reynier Blokhuysen
't Lant van Beveren, 1735
Bird's-eye view of the Beveren area
Hand-coloured copper engraving and etching, 470 × 525 mm
ANTWERP, THE PHOEBUS FOUNDATION

The Arenberg family arms
Detail from: Reynier Blokhuysen, *'t Lant van Beveren*, 1735

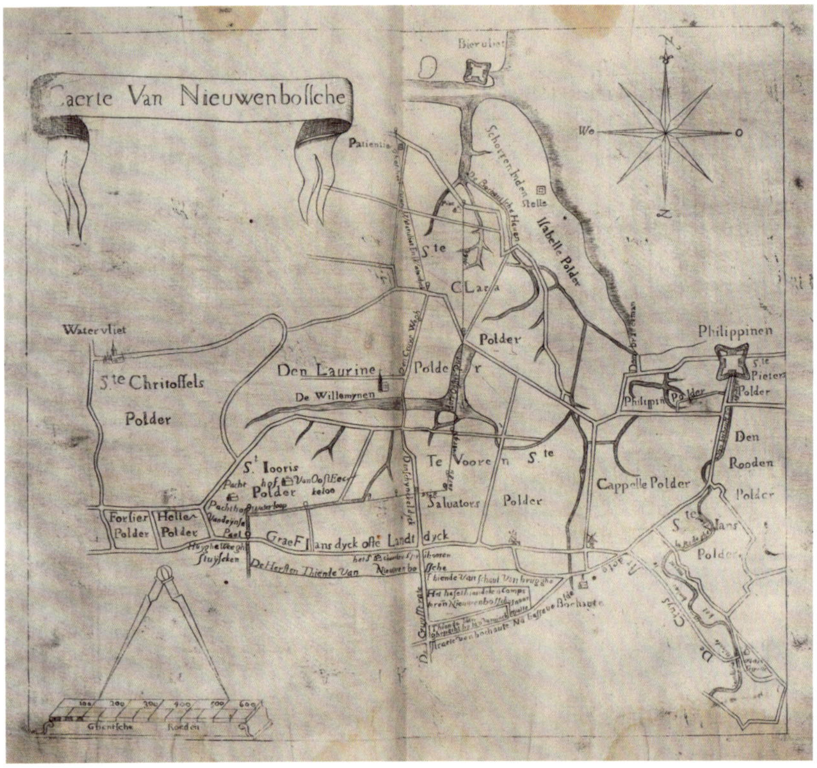

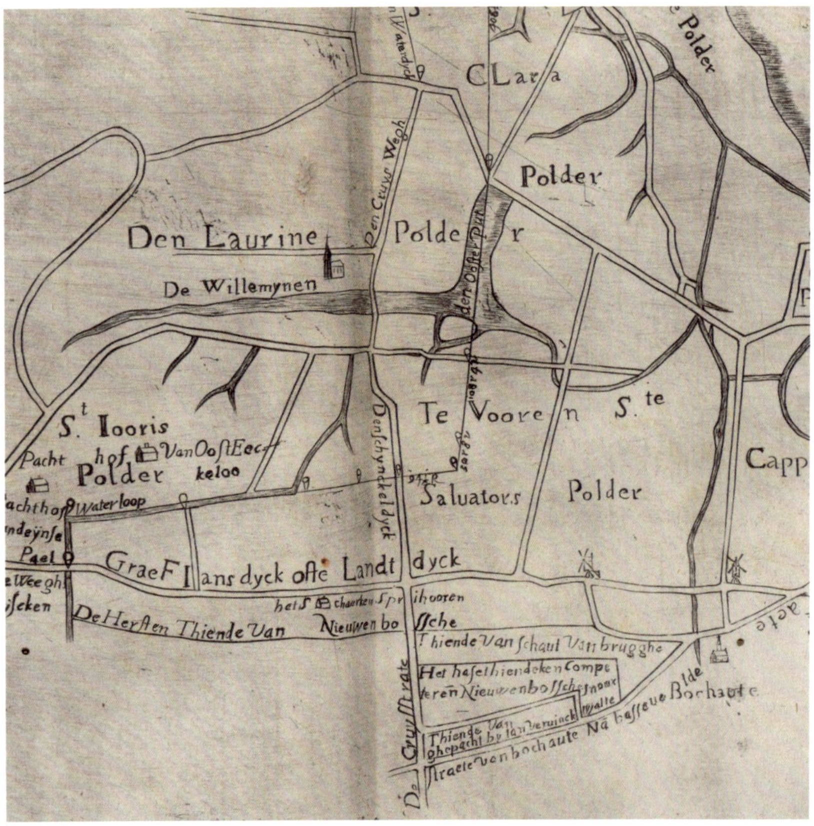

Jan Bale
Caerte van Nieuwenbossche
Dispute map
From: J. Vander Beke, *Applicat, ende employ van preuve, omme d'eerw. vrauwe abdesse van Nieuwenbosghe, H.ghe. Jeghens den eerweer. heere prelaet van S. Pieters, neffens Ghendt*, 1653
ANTWERP, THE PHOEBUS FOUNDATION

Detail from: Jan Bale, *Caerte van Nieuwenbossche*, 1653

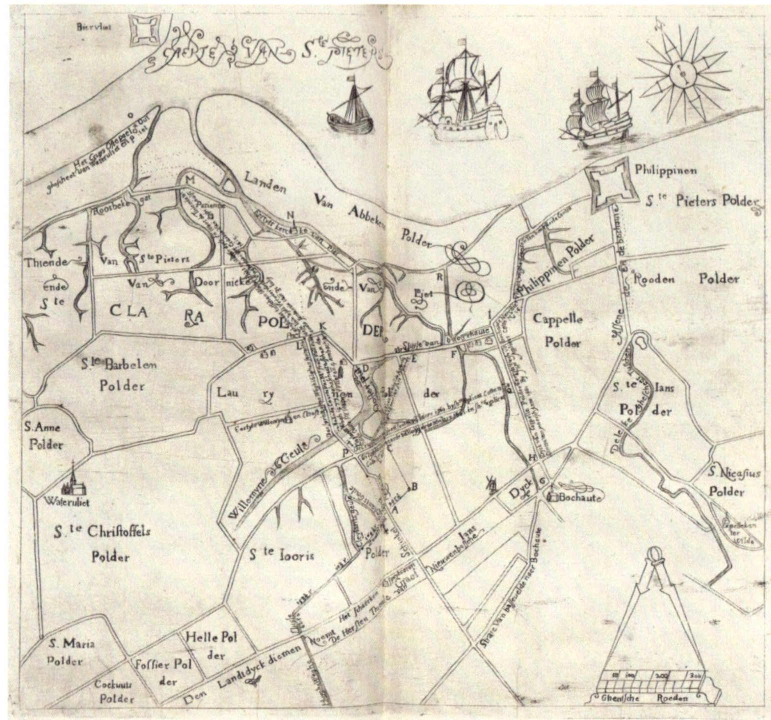

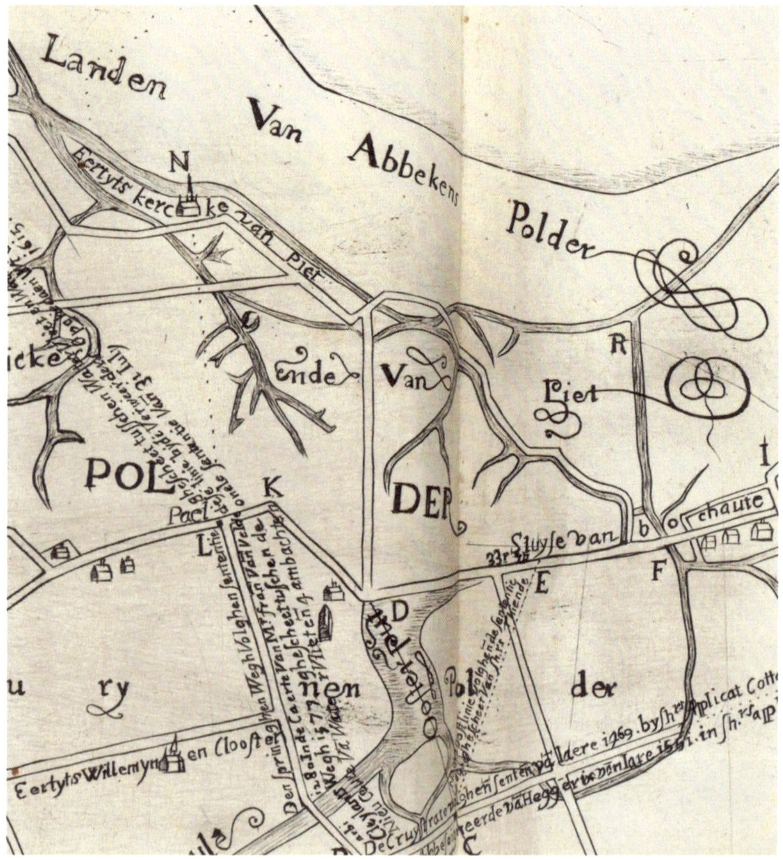

Jan Bale
Caerte van Ste. Pieters
Dispute map
From: J. Vander Beke, *Applicat, ende employ van preuve, omme d'eerw. vrauwe abdesse van Nieuwenbosghe, H.ghe. Jeghens den eerweer. heere prelaet van S. Pieters, neffens Ghendt*, 1653
ANTWERP, THE PHOEBUS FOUNDATION

Detail from: Jan Bale, *Caerte van Ste. Pieters*, 1653

DISPUTE MAPS

It was not unusual for the prior of a monastery to commission a map from a surveyor, as Petrus De Kever did from Henricus Joppen, since Church ownership of land required the production of a great many maps. Long before the creation of the land registry, it was primarily religious institutions like this that owned and leased out parcels of land. To that end, monasteries, abbeys and parishes had surveyors measure and map their real-estate holdings to enable them to collect taxes and keep records. It was not unusual for disputes to arise between major landowners about which land belonged to whom or the precise size of a particular plot. Existing surveyors' maps were useful when seeking to settle such cases, but judges sometimes also commissioned an impartial surveyor to map the land for a specific judicial procedure. One of the parties to the dispute might dispatch a surveyor to the contested area too, as seems to have occurred with the maps opposite.

The record of a court case dated 5 April 1653 between St Peter's Abbey near Ghent and Nieuwenbos Abbey, penned by J. Vander Beke, includes a dispute map of this kind. The abbess of Nieuwenbos claimed the right to tithes raised on the Clara and Isabelle Polders. Tithes were a type of Church tax which required the tenant farmers to hand over a tenth of their produce (crops, livestock, grain, and so on) to the landowner. The entire area around Ghent had been deliberately flooded in 1621 and 1622 for strategic military reasons during the Eighty Years War. With the return of peace in 1648 and the granting of permission to drain the land once more, a renewed discussion arose as to who was entitled to levy tithes on it.

What makes the case interesting is that the arguments presented by the abbess of Nieuwenbos were supported by two maps (most likely partial copies of existing ones) included in the printed report of the hearing. The first is the *Caerte van Nieuwenbossche*, showing several polders in the vicinity of the abbey. The *kercke van Piet* (St Peter's Church) was a particularly important element in the dispute, as was the point where the land ended (*ende van Piet*). The second map, a *Caerte van Ste Pieters*, shows the same polders, and a few more besides. The *thienden* (tithes) are explicitly inscribed in several places, and notes were added to particular features, such as the year in which the polder was drained. The map will have been used to determine the precise boundaries and who was to receive how much of the proceeds. Surveyors might also have indicated the boundaries with stones erected in situ.

The maps are not signed, but have been linked with the Ghent surveyor and mapmaker Jan Bale (or Jan Van Bale, active in the second half of the seventeenth century), who was the regular surveyor for St Peter's Abbey and had made similar maps for the dispute between the same parties a year earlier. That case too had pitted the abbot of St Peter's and the pastor in Boekhoute, Arnulf Borcquelmans, against the abbess of Nieuwenbos. If it was indeed Jan Bale who prepared the maps, it is not impossible that he skewed his survey of the parcel in favour of St Peter's Abbey, his employer.

PHILIPPE VANDERMAELEN

'To measure is to know', but as the aforementioned dispute maps testify, to measure land can also be to appropriate it. Determining areas and fixing boundaries would occur two centuries later for a much larger territory, now with different interests at play. In 1830, Belgium gained its independence from the United Kingdom of the Netherlands following a series of uprisings. The brand-new nation needed to be documented and mapped as swiftly as possible. The task fell to Philippe Vandermaelen (1795-1869) who, around that same year of independence, founded the Établissement Géographique de Bruxelles – a scientific institute and intellectual centre that also boasted its own lithographic printing works, a large library and a museum. Vandermaelen set about improving the topographical map drawn up by Count de Ferraris (1762-1814) and overseeing the production, among other things, of a series of 250 topographical maps of the new kingdom. He was given access to the cadastral plans and triangulations that had previously been made of what was now Belgian territory.

Philippe Vandermaelen is viewed as the greatest Belgian cartographer since Abraham Ortelius (⟶ p. 126) and Gerard Mercator (⟶ p. 202), partly because his company enjoyed a monopoly on mapmaking for a substantial part of the nineteenth century. In the course of that period, he became highly influential: topographical cartography was rendered practicable, scale was applied uniformly and consistently, and lithographic printing was introduced. The Établissement continued to operate for several years after Vandermaelen's death in 1869.

The *Plan de la Wateringue de Calloo* (1871) was published by the institute in this latter period – a hydrographic map of the polders near Kallo, for which Vandermaelen's *Cartes topographiques de la Belgique* (1846-1854) served as the basis. The scale was the same, but the map was adapted to its new subject: hydrology and watercourses. Some sections were greatly simplified, with houses and other buildings, numerous roads, and details of subdivision and land use all omitted. The focus was firmly on the polders, the course of the water and other elements relevant to water management, such as dykes (the thick black lines), salt marshes, windmills and sluices. The water itself is coloured blue and stands out immediately.

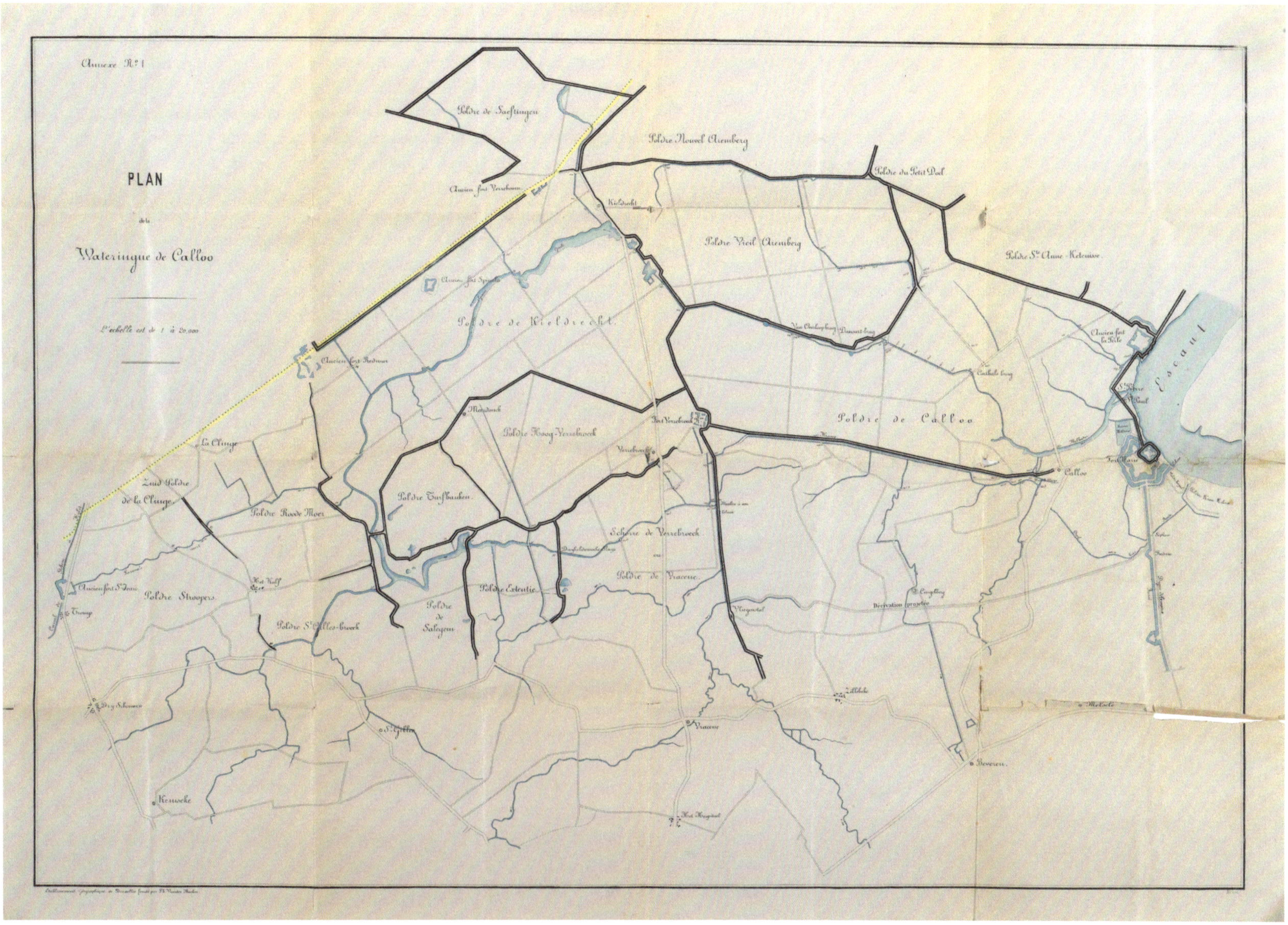

Philippe Vandermaelen
***Plan de la Wateringue de Calloo*, 1871**
Map of the polders belonging to the Kallo water authority
Colour lithograph, 698 × 990 mm
ANTWERP, THE PHOEBUS FOUNDATION

Ferraris's topographical map

In 1777 and 1778, Holy Roman Empress Maria Theresa commissioned Joseph Jean François, Count de Ferraris (1726-1814), to prepare a huge, large-scale topographical map, running to 275 sheets, of her territory in the Low Countries. The Austrian Netherlands (*Belgium Austriacum*) is the collective name given to the eleven provinces of the Southern Netherlands during the period of Austrian Habsburg rule between 1715 and 1795 (with a brief hiatus in 1790), when the region became an autonomous part of the Habsburg monarchy. It is sometimes referred to as the 'Imperial Netherlands' to distinguish it from the Republic of the Seven United Netherlands (*Belgium Foederatum*), roughly the modern-day Netherlands.

The Ferraris Map is viewed as the first detailed topographical survey of what is now Belgium, drawn at a scale of 1:11,520. De Ferraris created the maps by hand and in a military context: he was a general of the Austrian artillery and a field marshal of the Austrian Netherlands. Only three original versions still exist, but a commercial printed version was also sold as a limited edition, a copy of which is in The Phoebus Foundation's collection. The scale of the printed Ferraris map has been reduced and the 275 map sections brought back to 25 sheets, bound together into a book. The trade edition too was made under the supervision of Count de Ferraris and was engraved by Louis-André Dupuis (active 1769-1787) and Jean-Baptiste Poirson (1761-1831). Besides these two mapmakers, many other engravers, etchers and draughtsmen contributed to its production.

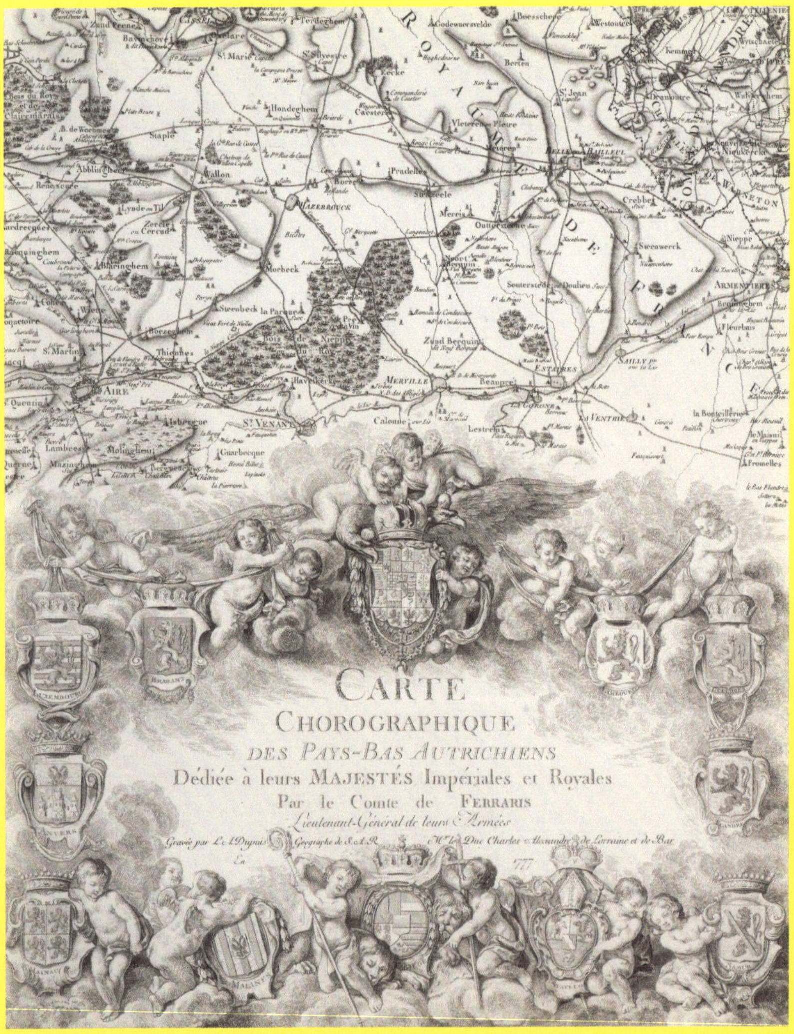

Map sheet 11: title cartouche
From: Joseph Jean François de Ferraris, *Carte Chorographique des Pays-Bas Autrichiens Dédiée à leurs Majestés Impériales et Royales*, 1777 and 1797
ANTWERP, THE PHOEBUS FOUNDATION

Map sheet 9: Brabant and Limburg
From: Joseph Jean François de Ferraris, *Carte Chorographique des Pays-Bas Autrichiens Dédiée à leurs Majestés Impériales et Royales*, 1777 and 1797
ANTWERP, THE PHOEBUS FOUNDATION

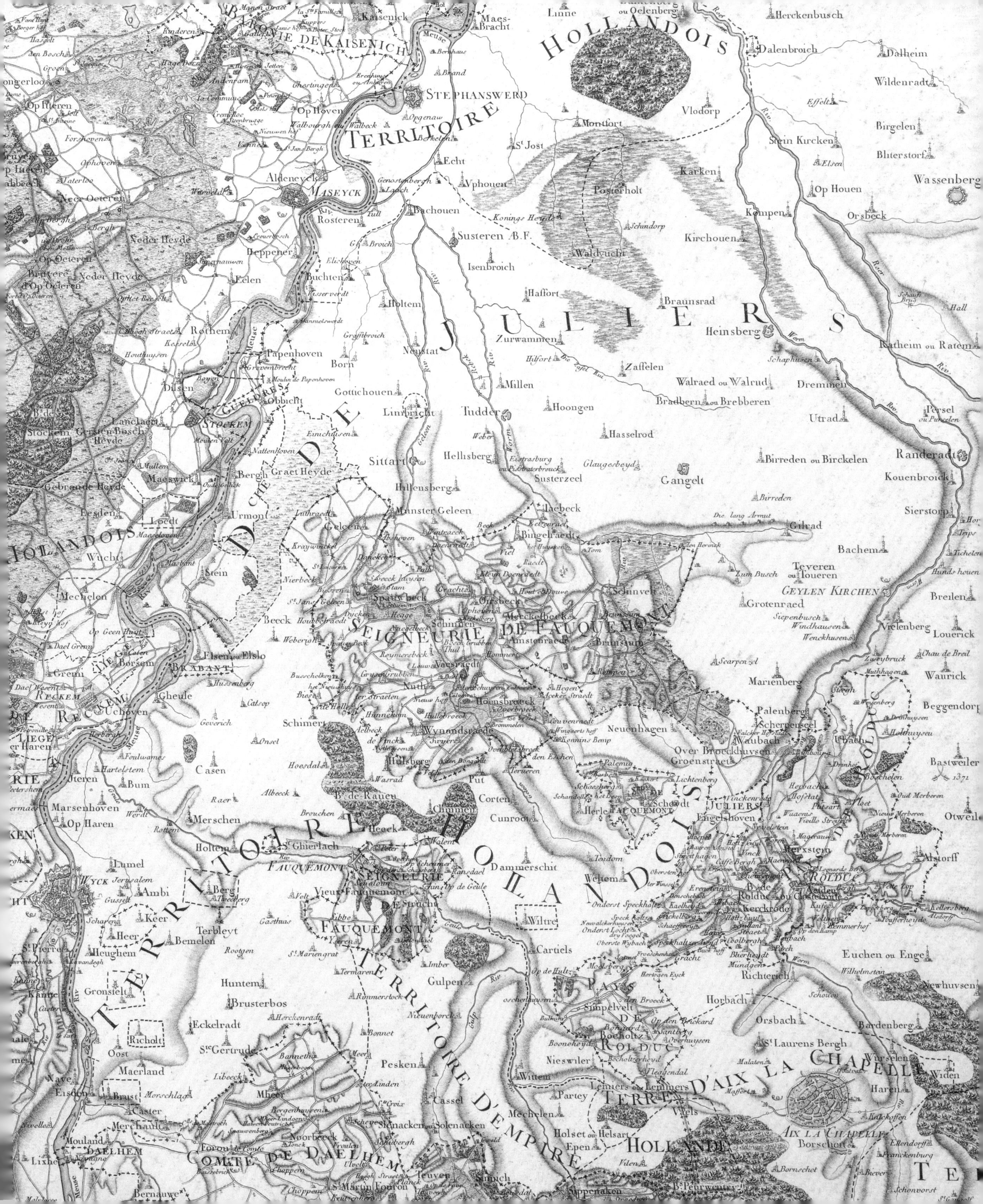

HOLLANDOIS
BARONIE DE KAISENICH
TERRITOIRE
STEPHANSWERD
MASEYCK
JULIERS
D. DE GUELDRES
STOCKEM
SEIGNEURIE DE FAUQUEMONT
TERRITOIRE HOLLANDOIS
TERRITOIRE D'EMPIRE
FAUQUEMONT
PAYS DE ROLDUC
TERRE D'AIX LA CHAPELLE
AIX LA CHAPELLE
HOLLANDE
COMTÉ DE DAELHEM
DAELHEM
BRABANT
LIEGE
RECKEM
GEYLEN KIRCHEN
WYCK
Sittart
Heinsberg
Maes-Bracht
Linne
Dalenbroich
Herckenbusch
Dalheim
Wildenradt
Birgelen
Wassenberg
Vlodorp
Montfort
Echt
Susteren
Posterholt
Kirchouen
Karken
Randeradt
Gangelt
Tudder
Millen
Limbricht
Munster Geleen
Geleen
Stein
Urmont
Elsen ou Elslo
Maeswick
Mechelen
Teveren ou Toueren
Hoensbroeck
Wynandsraede
Amstenraede
Brunsum
Schinvelt
Merckelbeck
Scherpenseel
Palenberg
Marienberg
Ubach
Herzstein
Kerckrode
Rolduc ou Closterode
Herle
Schaesberg
Heerlen
Valckenberg
Gulpen
Wittem
Simpelvelt
Bocholtz
Vaels
Richterich
Horbach
Orsbach
Bardenberg
Wurselen
Haren
Borscheit
Eupen
Epen
Mechelen
Cassel
Slenacken ou Solenacken
Noorbeeck
Mheer
Eckelradt
Oost
Maerland
Eisden
Lixhe
Mouland
Heer
Keer
Ambi
Terbleyt
Bemelen
Berg
Gronsfelt
Rieholt
Brust
Caster
Merschen
Houtem
Heugem
Rothem
Papenhoven
Dusen
Born
Neustat
Hasselrod
Hoongen
Birreden ou Birckelen
Kouenbroick
Sierstorp
Gilrad
Bachem
Grotenraed
Roermond
Waurick
Beggendorp
Bastweiler
Otweiler
Alstorff
Euchen ou Enge
Newhuysen
Gevenich
Wylre
Imber
Cartiels
Nieswiler
Partey
Lemiers
Holset ou Helsart
Teuven
Sinnich
Fouron
Bernauwe

MILITARY TOPOGRAPHICAL MAPS

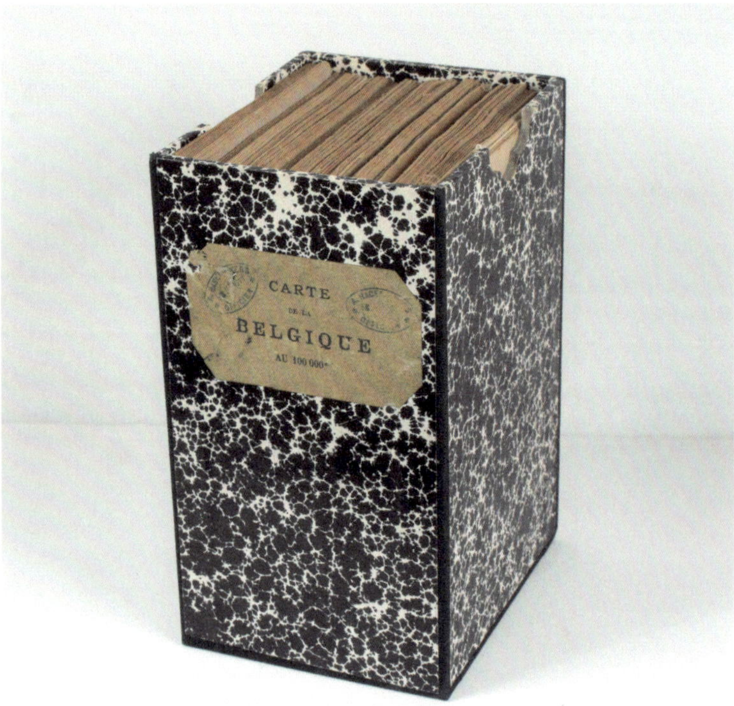

Vandermaelen did not go out mapping the country in person: he dispatched a team of topographers to take on-site measurements and make sketches. By the mid nineteenth century, this kind of fieldwork had become an official, military affair, carried out in Belgium by the Dépôt de la Guerre (established in 1831). Given that it was in the national interest to have a standard set of maps of Belgian territory, the army was called on to draw up accurate ones. The first topographical maps to encompass the entire country were produced by the Dépôt de la Guerre et de la Topographie (later renamed the Institut cartographique militaire) on a 1:20,000 scale.

The Phoebus Foundation has a 1908 map published by the same institution on the smaller scale of 1:100,000 and with the *ex libris* of an army officer called 'A. Rademakers', of whom nothing further is known. It includes an overview map in which the individual sheets are laid out in a grid shape, allowing the user to place the individual map sections within the larger whole. The arrival of this official nationwide form of topography meant that the days of local surveyors like the previously mentioned Henricus Joppen of Beveren or Jan Bale of Ghent were over. Anyone wishing to become a surveyor in the nineteenth century required centralised academic training before going to work for the land registry or some other government agency. This is not to say, however, that small hand-drawn surveyors' maps of a municipality or polder disappeared entirely: documents of this kind continue to turn up regularly in local archives and (private) collections.

Institut cartographique militaire in Brussels
Carte de la Belgique au 100000e, 1908
Storage box for the 1:100,000 scale map of Belgium
The box contains 26 map sheets in colour lithography, paper glued to fabric, measuring 490 × 525 mm (open sheet) and 112 × 175 × 7 mm (folded)
ANTWERP, THE PHOEBUS FOUNDATION

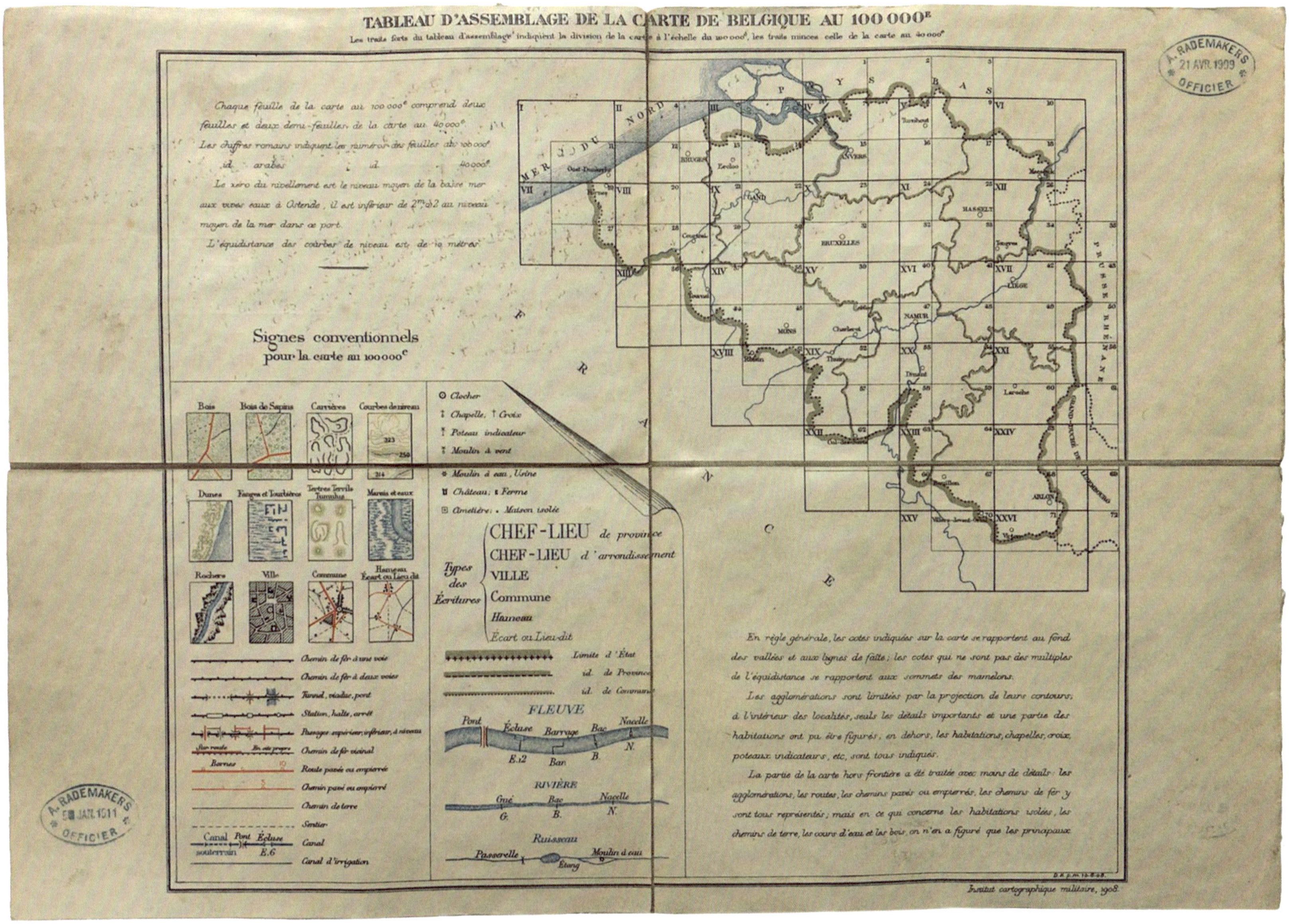

Index map to the 1:100,000 scale map of Belgium
From: Institut cartographique militaire, Brussels, *Carte de la Belgique au 100000e*, 1908
ANTWERP, THE PHOEBUS FOUNDATION

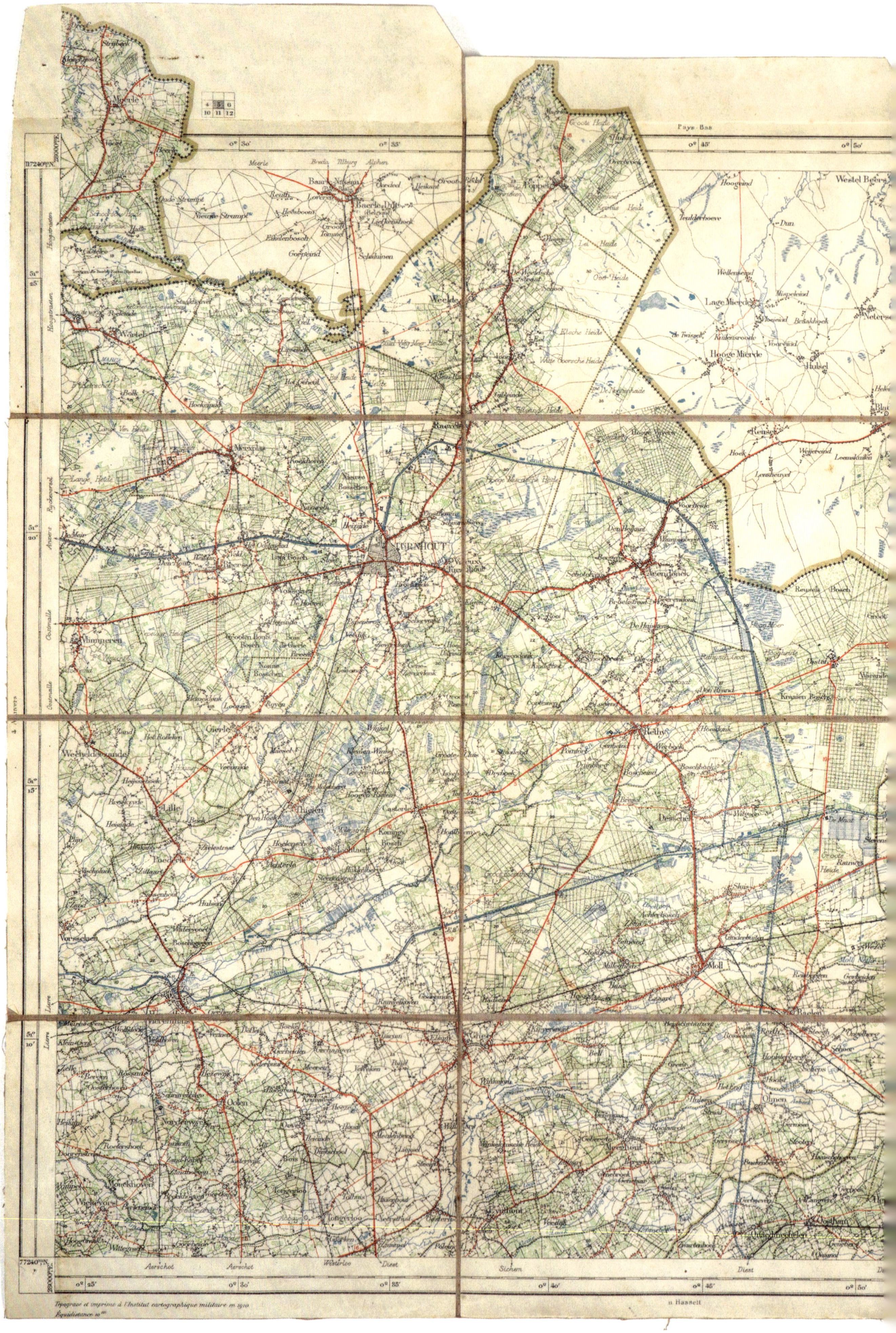

Map sheet 5: Turnhout
From: Institut cartographique militaire, Brussels, *Carte de la Belgique au 100000e*, 1908
ANTWERP, THE PHOEBUS FOUNDATION

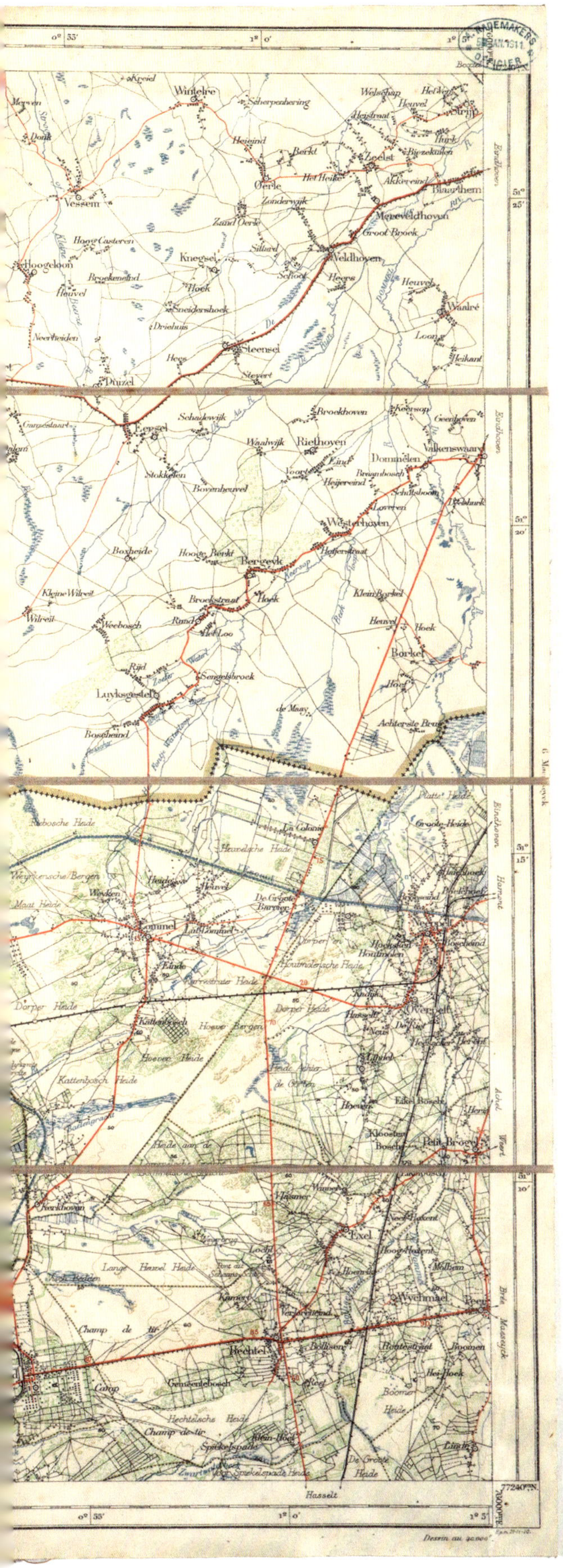

LITERATURE

I. Jongepier & T. Soens, 'Poldercartografie in het digitale tijdperk: een nieuwe blik op het historische polderlandschap', in: M. Carnier, P. Alkhoven et al. (ed.), *Cartografie. Visie op de kaart*, The Hague, 2016: 188-196.

C. Koeman & M. van Egmond, 'Surveying and Official Mapping in the Low Countries, 1500-ca. 1670', in: D. Woodward (ed.), *The History of Cartography, vol. 3. Cartography in the European Renaissance*, Chicago / London, 2007, vol. 2: 1246-1295.

A. Meskens, *Wiskunde tussen Renaissance en Barok: aspecten van wiskundebeoefening te Antwerpen tussen 1550 en 1620*, Antwerp, 2009.

H.C. Pouls, *De landmeter Jan Pietersz. Dou en de Hollandse Cirkel*, Delft, 2004.

M. Silvestre, 'Philippe Vandermaelen, de Mercator van Het Jonge België', *Brussels Studies* [Online], General collection, nr. 106, 14 November 2016.

M. Silvestre & M.-B. Fincoeur, *Inventaire raisonné des collections cartographiques Vandermaelen*, Brussels, 1994.

K. Van Cleempoel, 'De Leuvense school van instrumentenmakers in de 16de eeuw', in: R. Halleux, C. Opsomer & J. Vandersmissen (ed.), *Geschiedenis van de wetenschappen in België van de Oudheid tot 1815*, Brussels, 1998: 217-228.

V. Van de Kerckhof, H. Bussers, D. Buisseret et al., *Met passer en penseel. Brussel en het oude hertogdom Brabant in beeld*, exhib.cat. Brussels, 2000.

L. Wellens-De Donder & A. De Smet, *Philippe Vandermaelen, 1795-1869*, exhib.cat. Brussels, Koninklijke Bibliotheek Albert I, 1969.

M. Wilssens (ed.), *Singelberg: het kasteel en het land van Beveren*, Tielt, 2007.

THICK EXPANSES OF CLOUD above sprawling polders, cows grazing in endless pastures, and the amiably rotating sails of a windmill on a dyke: this is the image of the Low Countries evoked by the surveyors' maps in the previous chapter. Yet the densely populated region had actually been characterised for centuries by its crowded streets, bustling market squares, formidable ramparts and the many roads converging on its city gates. The rise of the city from around 1300 sparked admiration, love, identity and pride that were manifested in all sorts of ways. Richly illustrated urban histories were written, poets sang the city's praises in amateur literary societies known as 'chambers of rhetoric', and artists painted and drew the faces, plans, elevations and panoramas of the city. City 'portraits' appeared in the early modern era in all shapes and sizes – in visual art, maps and writing. Just as the social elite had their portraits done by a famous painter, staged in an idealised setting and with the appropriate attributes, so too mapmakers and engravers produced endless portraits of cities. A changing urban landscape, the desire to record and document one's individual surroundings, and the flourishing of cartography meant that townscapes and city maps flourished.

Several cities of the Southern Netherlands are represented in The Phoebus Foundation collection, including Bruges, Brussels, Ghent, Leuven and Mechelen, each with their own characteristic features and highlights incorporated in their respective portraits. It is the city of Antwerp, however, that accounts for the largest share of this extensive corpus of city images and maps. Numerous mapmakers settled in the city on the river Scheldt over the centuries, making it one of Europe's most important cartographic centres for many years. From a small view of the harbour painted in the sixteenth century to postcards of the Boerentoren landmark from more recent times, the city has displayed many different faces through the ages. But which sides of Antwerp were shown and which were not? How was the important Southern Netherlandish port and trading metropolis depicted and mapped?

CHAPTER II

THE MANY FACES OF A CITY

ANTWERP AND THE URBAN CULTURE OF THE SOUTHERN NETHERLANDS

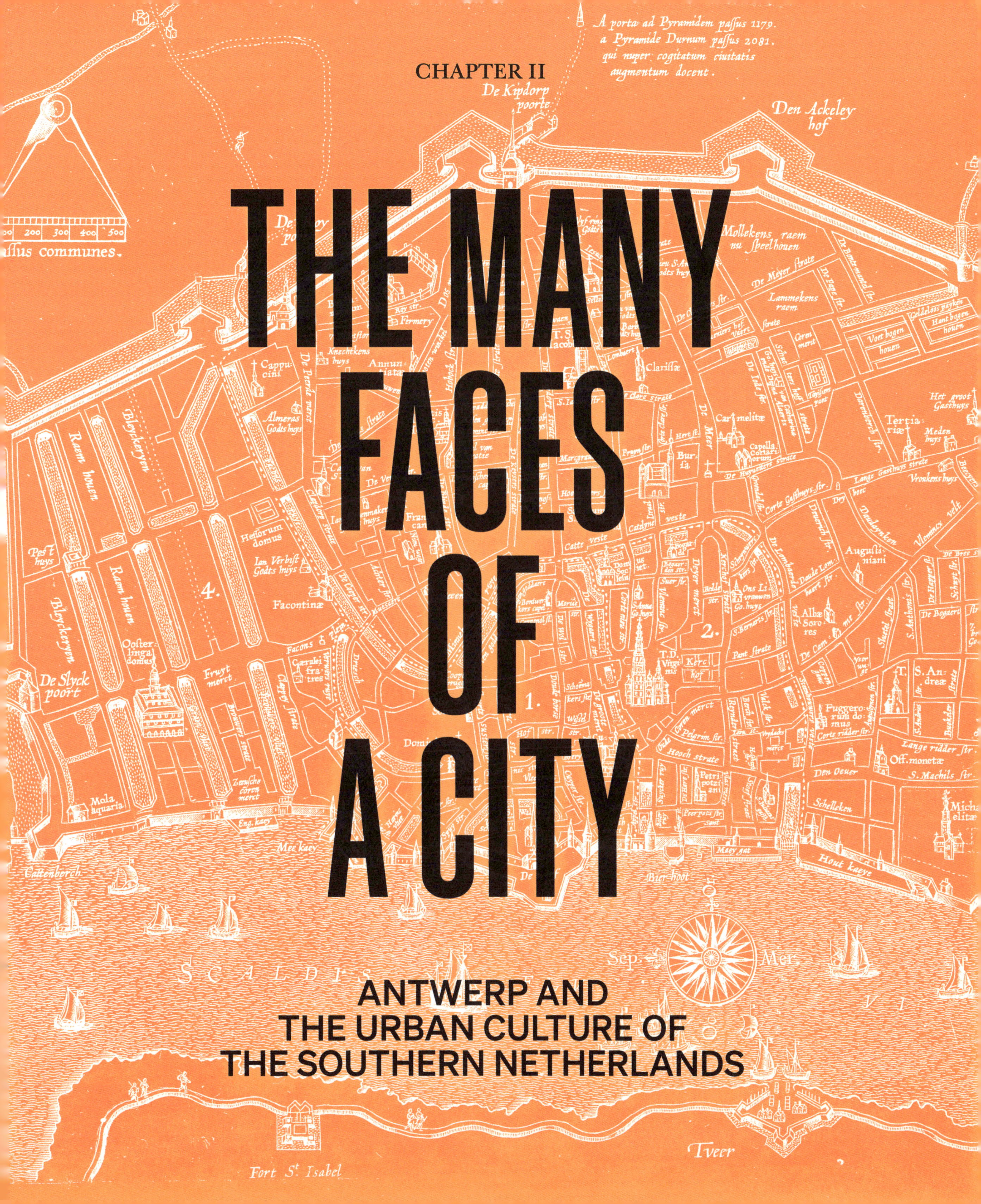

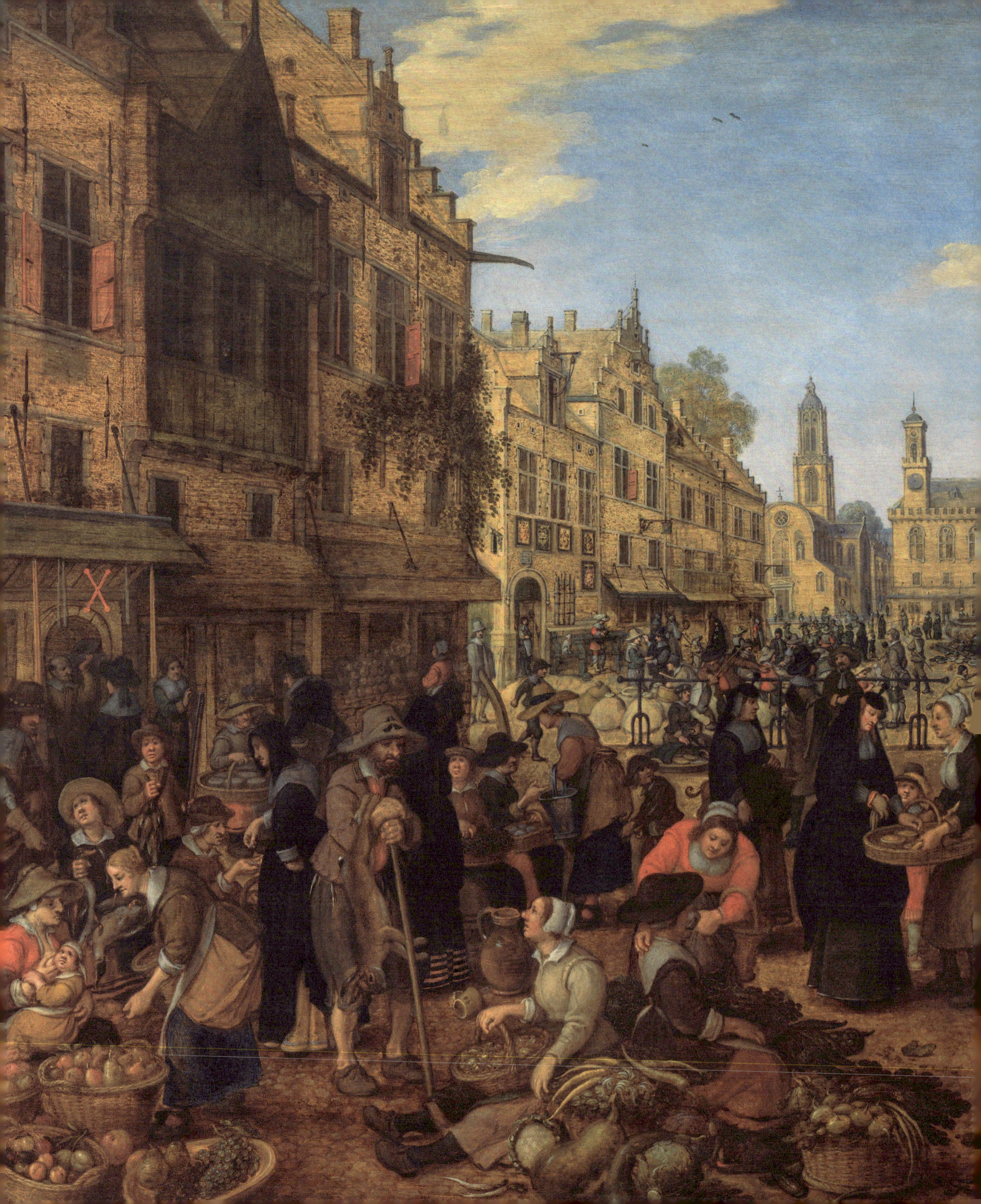

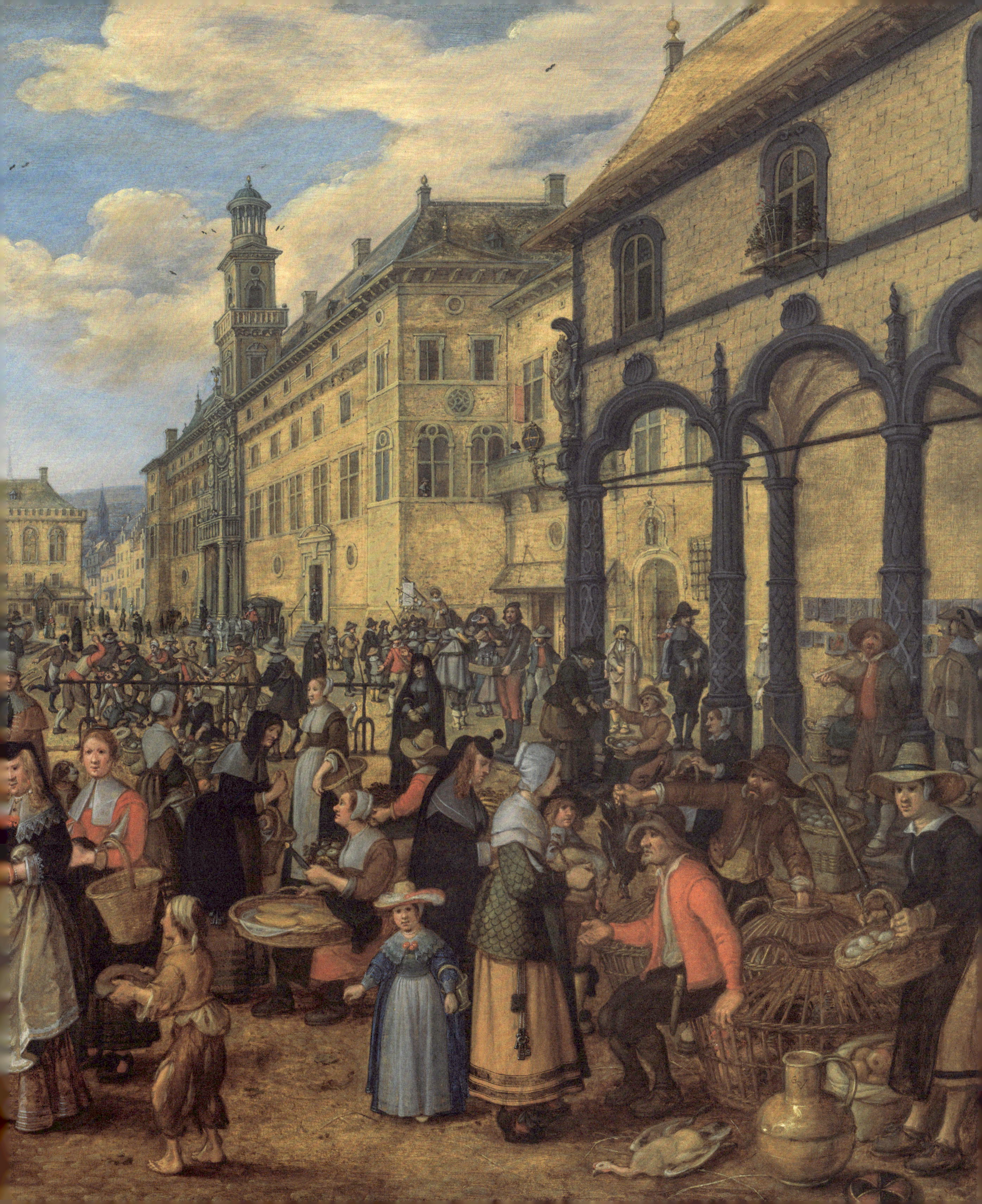

A NEW CONNECTION BETWEEN BRUSSELS AND ANTWERP

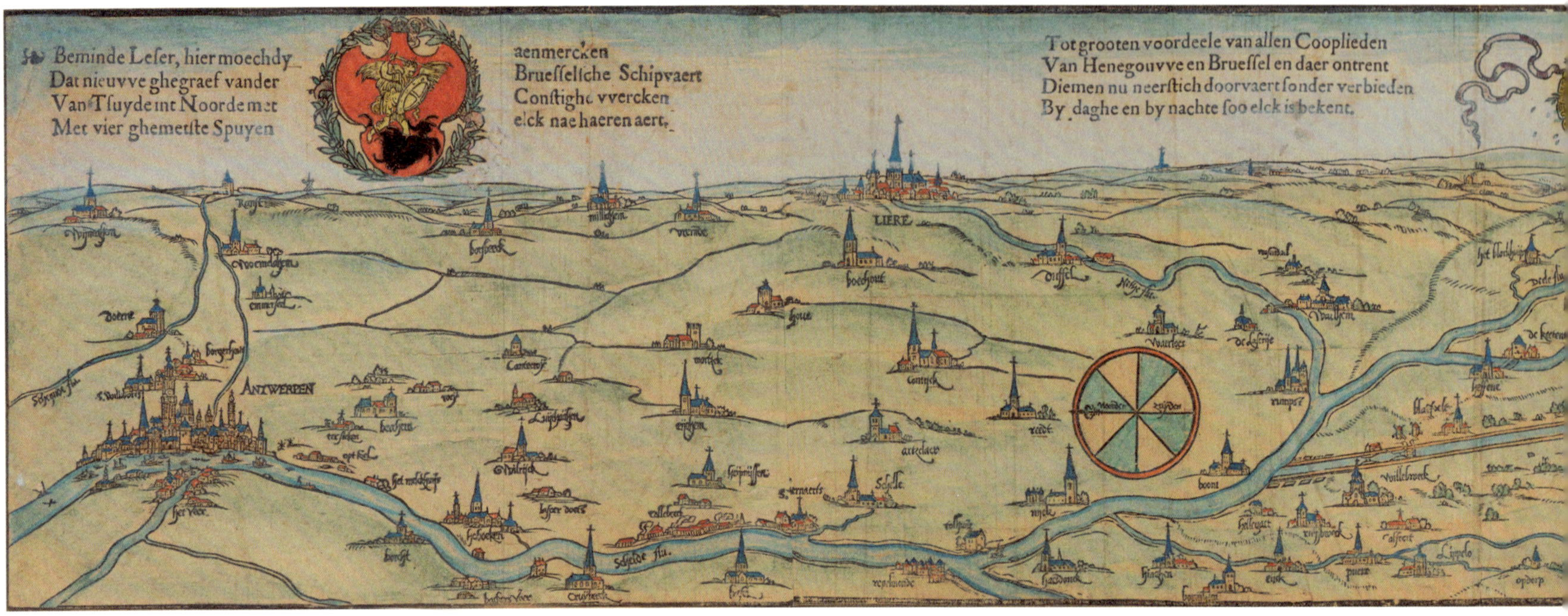

Successive waves of urbanization swept the Low Countries, both north and south, from the eleventh century onwards, and the process would continue for hundreds of years more. By the middle of the fifteenth century, roughly one in three people in the region lived in a town, making Flanders and Brabant some of the most densely populated areas in Europe. A city acted as a link between its immediate surroundings and its hinterland, connecting patricians and farmers, shipowners and merchants. A good network of roads, waterways and a navigable connection to the sea were the key to success. The landscape of branching rivers meant that most towns had the potential to form an effective hub, although cities located at the mouth of an estuary obviously had an additional advantage as international maritime trade expanded.

Hence the celebrations when a new canal was completed between Brussels and Willebroek in 1561, connecting Brussels with the rivers Rupel and Scheldt and ultimately the North Sea. This brought an end once and for all to the challenging and unreliable route to Antwerp via the river Zenne, punctuated by mandatory passage through Mechelen. The civic authorities in Brussels had long desired a direct canal of roughly thirty kilometres in length, and this was finally excavated between 1550 and 1561. A celebratory map drawn up around 1565 praises the completion of the new *Bruesselsche schipvaert* (Brussels Ship Canal), which rendered the North Sea readily accessible from inland. The woodcut map is remarkably large (19.4 centimetres high by 114.9 centimetres wide) and also coloured in. The course of the new canal is clearly visible but, due to the bird's-eye view (obliquely from above), the map also offers a panorama of the Duchy of Brabant, incorporating views of the principal cities. We make out the cities of Antwerp, Mechelen, Lier, Vilvoorde, and Brussels, but also the smaller towns of Schaerbeek, Grimbergen and Rupelmonde – birthplace of Mercator (⟶ p. 202).

pp. 48-49
Sebastiaen Vrancx
Market Scene, c.1600
Oil on panel, 85.7 × 146.7 cm
ANTWERP, THE PHOEBUS FOUNDATION

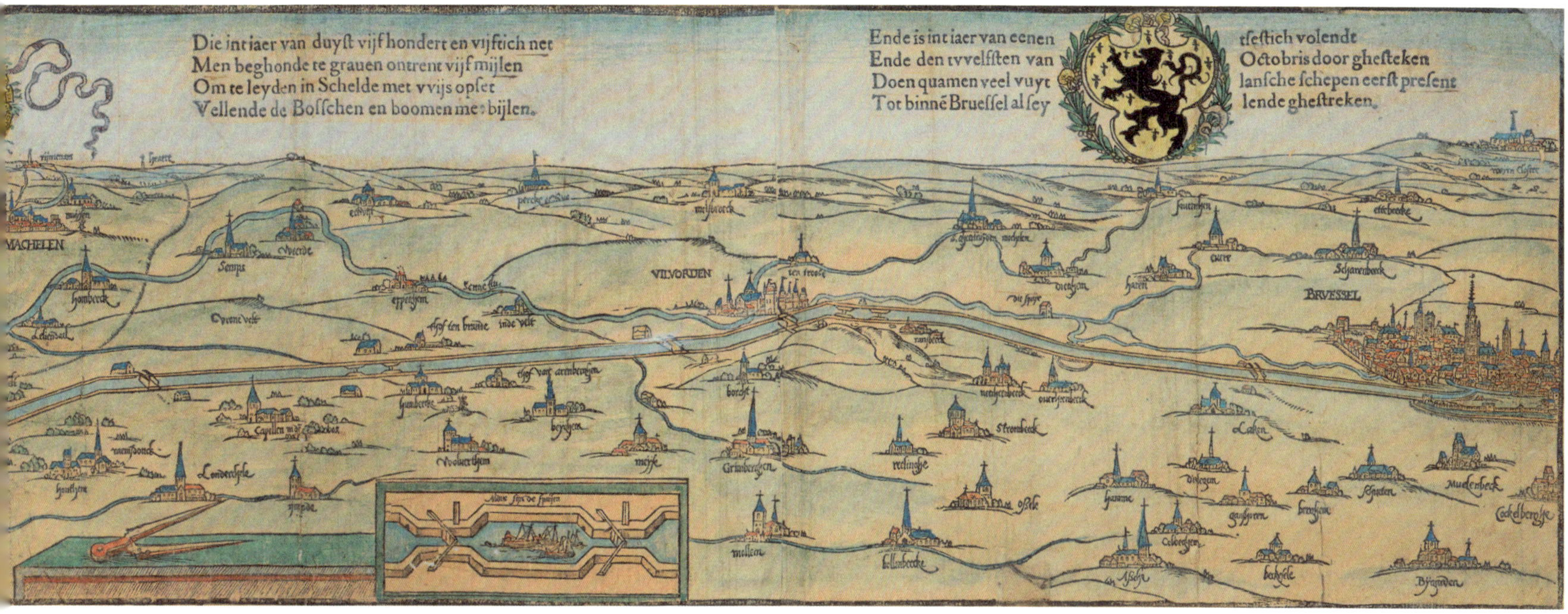

The document served as a pullout map in *De nieuwe chronijcke van Brabandt* ('New Chronicle of Brabant'; Antwerp, 1565), a history book that also includes woodcuts by the mapmaker Bernard Van De Putte (1528-1580), who might have made the canal map too. Copies still containing the map are rare, as it is likely to have been removed to allow it to be admired separately, just as occurred with this loose coloured version.

The lines of poetry are another noteworthy element: they come from a lengthy paean of praise running to a total of over four hundred verses, penned by Gysbrecht Mercx (1492-after 1565), who belonged to a chamber of rhetoric in Brussels. He wrote his triumphal ode to mark the spectacular and festive opening of the canal. The whole city spent a week in celebratory mood: shipowners could now stow far more cargo in their vessels – a fact marked in playful fashion by contests to see which ship could carry the biggest load. Together, the map and the poetic tribute hailed the longed-for new connection and the flourishing trade that awaited the area, 'to the great benefit of all the merchants of Hainaut and Brussels and beyond'.

Bird's-eye panorama of the canal from Brussels to Willebroek, c.1565
Hand-coloured woodcut, 215 × 1089 mm
ANTWERP, THE PHOEBUS FOUNDATION

Duchy of Brabant

This map shows the southern part of the Duchy of Brabant and the area around Brussels. Antwerp had been part of Brabant for centuries, one of its four major cities along with Leuven, Brussels and 's-Hertogenbosch (Den Bosch, now in the Netherlands), which served as the respective legal and administrative centres of the 'Four Quarters'. Each of the main Brabant cities had its own special features and interests. Leuven was a true university town and the duchy's intellectual centre (⟶ Chapter XI), Brussels was the capital and seat of the sovereign's court, Antwerp was a renowned trading metropolis, and Den Bosch was home to the archdiocese. Leuven, Mechelen and Brussels are all visible in this part of the map.

It was engraved after a design by Michael Floris van Langren (1598-1675), regular cartographer to the king of Spain. The copy in The Phoebus Foundation's collection comes from a French edition of Joan and Willem Jansz. Blaeu's *Atlas Novus*, published in Amsterdam in the 1630s (⟶ Chapter X). The coats of arms in the title cartouche have been left blank in this instance. From 1642, they were filled in with the escutcheons of Nijvel and Vilvoorde – two other towns that also feature in this map.

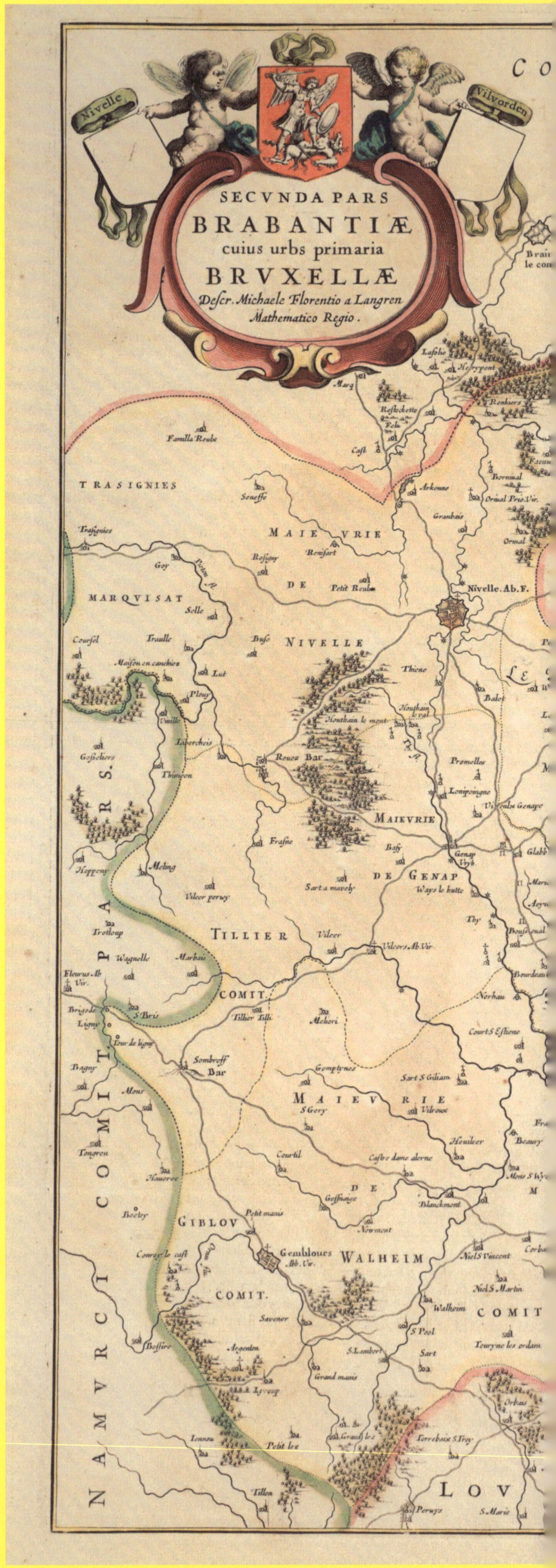

Michael Floris van Langren
Secunda pars Brabantiae cuius urbs primaria Bruxellae, 1635/1638
Map of the southern part of the Duchy of Brabant and the area around Brussels
Hand-coloured copper engraving, 500 × 595 mm
ANTWERP, THE PHOEBUS FOUNDATION

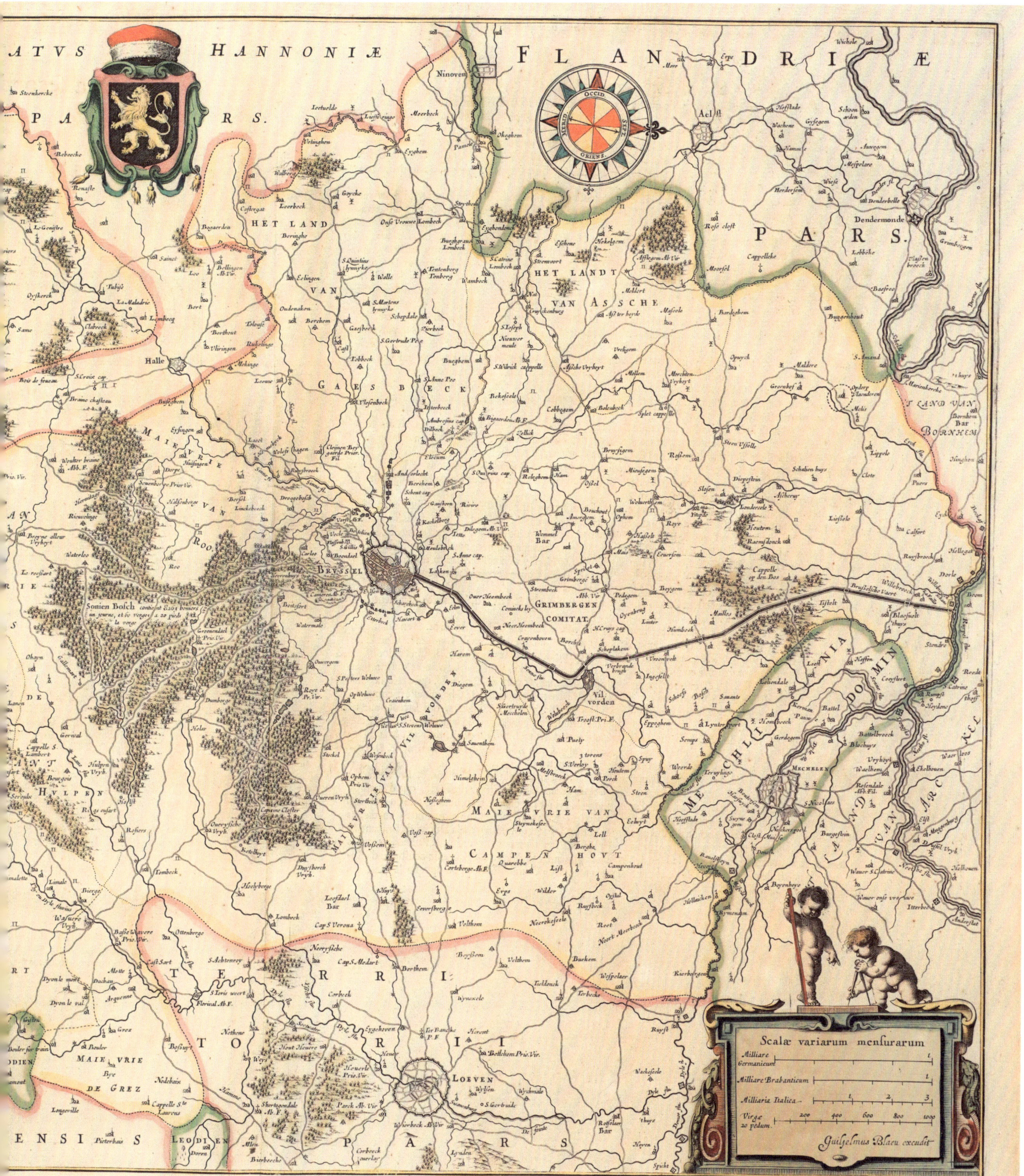

ATVS HANNONIÆ
PARS.
FLANDRIÆ
PARS.
HET LAND VAN GAESBECK
HET LANDT VAN ASSCHE
Ninoven
Aelst
Dendermonde
Halle
BRVSSEL
Sonien Bosch
GRIMBERGEN COMITAT.
Vil vorden
MAIE VRIE VAN CAMPEN HOVT
MECHELEN
TERRITORII LOEVEN PARS
LOEVEN
MAIE VRIE DE GREZ
Occid.
Septe.
Meridi.
Oriens
Scalæ variarum mensurarum
Milliare Germanicum
Milliare Brabanticum
Milliaria Italica
Virgæ 20 pedum.
200
400
600
800
1000
Guiljelmus Blaeu excudit

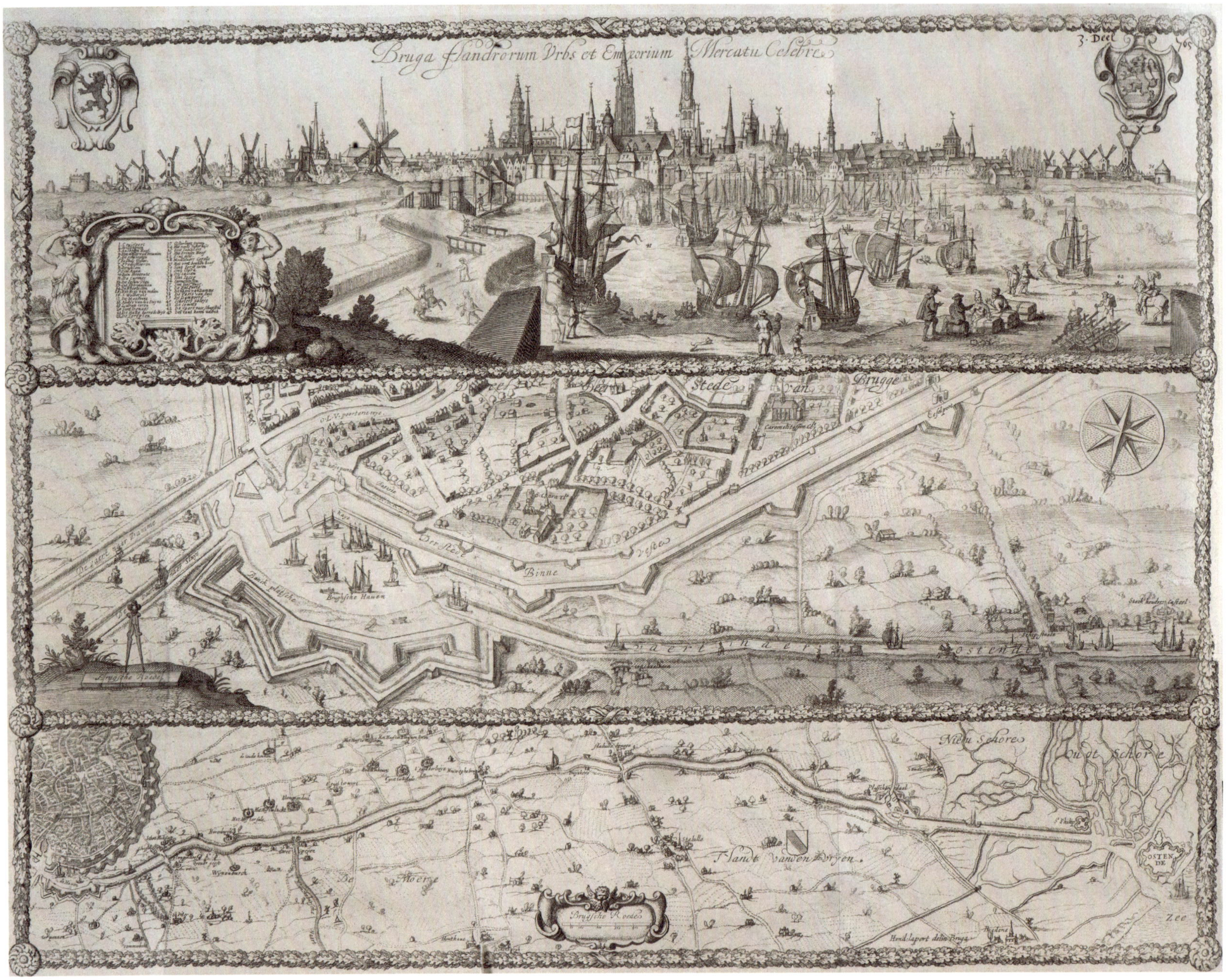

Bruga Flandrorum Urbs et Emporium Mercatu Celebre
View and map of the port of Bruges and the Ghent-Bruges canal
From: Nicolaas Despars, *Chronyke van Vlaenderen*, 1736
ANTWERP, THE PHOEBUS FOUNDATION

CITIES IN THE SOUTHERN NETHERLANDS

The region's biggest towns in the Middle Ages were in the county of Flanders. By around 1300, Ghent – the largest city with some 65,000 inhabitants – had grown into a major textile city and an international hub for the wool and cloth industries. Bruges came second but was celebrated from the thirteenth century in particular as an important entrepôt for goods including textiles. The city on the river Zwin benefited, for instance, from the emergence of seaborne trade. Maritime routes were far preferable to transport over land in terms of logistics, safety and costs, and Bruges's position on the Zwin offered a connection not only with the sea but also further inland. The city enjoyed a favourable status as an important member of the Hanseatic League, a medieval trading alliance of merchants from cities around the Baltic and North Sea, founded on jointly agreed rights and duties. The League facilitated the trade of goods, which were then distributed from the coast to smaller towns and villages in the hinterland.

This relatively modern mode of commercial organisation brought major benefits to Flanders as a region, and the city of Bruges liked to identify itself with shipping. In 1665, the civic authorities commissioned Cornelis van Caukercken (1626-1680) to make a three-part print of Bruges to mark the opening of the Handelskom docks (visible in the upper part), which had been excavated alongside the Ostend canal (shown in the lower part). The city's pride in its upgraded port infrastructure is also apparent from the fact that the print is included in *Chronyke van Vlaenderen* (Brugge, 1736), the posthumously printed and published manuscripts of Bruges mayor Nicolaas Despars (1522-1597).

A city like Bruges might have been successful economically, but this did not necessarily make it a centre of administrative, political and legal authority too. In fact, the Southern Netherlands did not really have a single, centralised capital: individual cities were relatively autonomous, held privileges of their own and more or less governed themselves. Mechelen briefly became the capital of the Habsburg Empire in 1507. The governors of the Low Countries resided in palaces there, and the city was also the base of the High Council, the region's supreme court. In 1559, Mechelen also became the seat of the archdiocese. In 1531, however, the Habsburg government had relocated to Brussels, which became the seat of the ruler's court, though without going on to dominate the entire realm in the manner of a Paris or London. The ducal palace, built as early as the eleventh and twelfth centuries, was now adopted as the residence of the region's governors. Lucas Vorsterman II depicted it in bird's-eye view as part of Antonius Sanderus's chorography of holy places in Brabant, *Chorographia sacra Brabantiae* (Brussels, 1659).

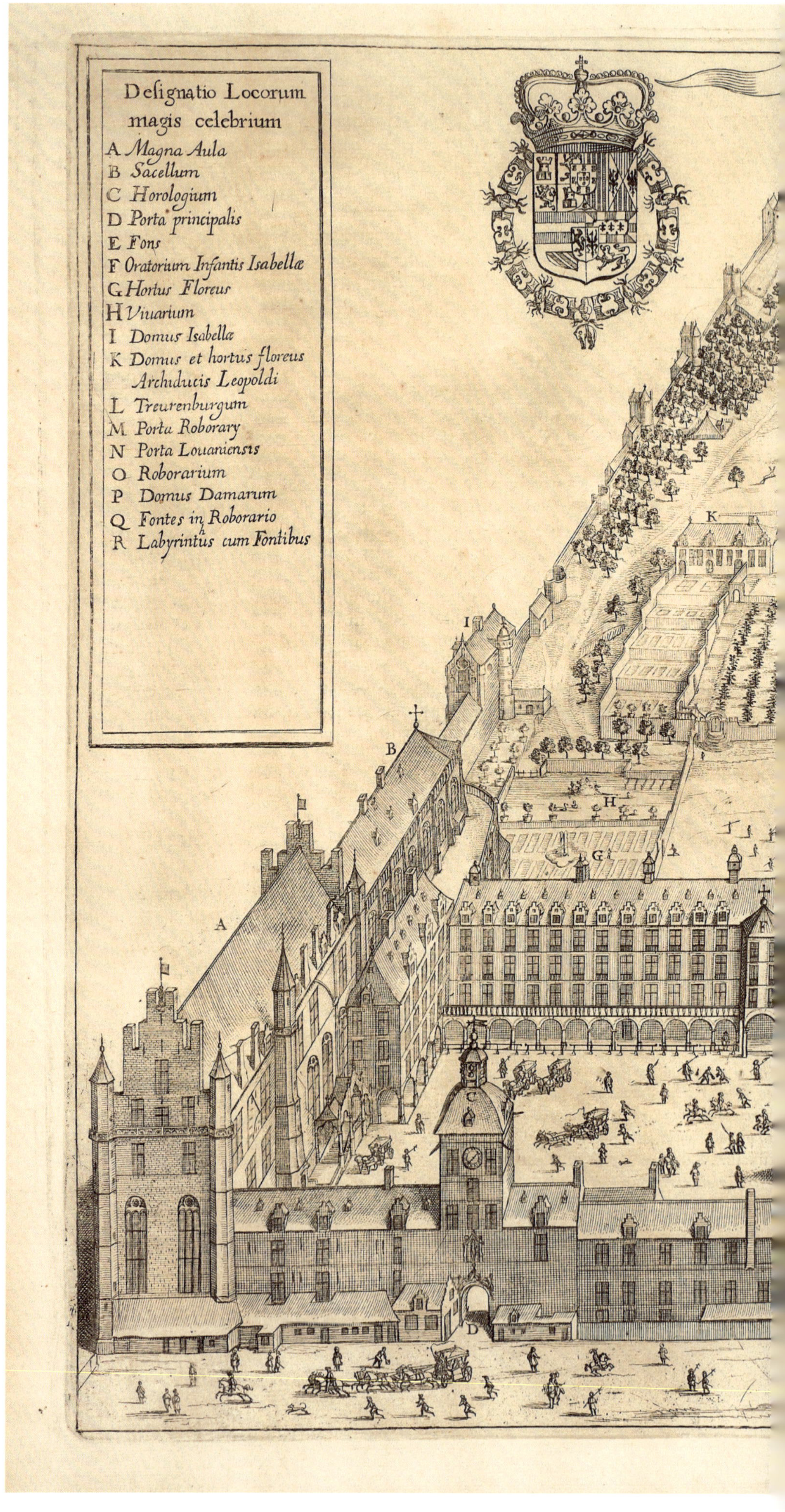

Lucas Vorsterman II
***Palatium Bruxellense Ducis Brabantiae*, 1659**
Bird's-eye view of the palace in Brussels
Etching, 447 × 540 mm
ANTWERP, THE PHOEBUS FOUNDATION

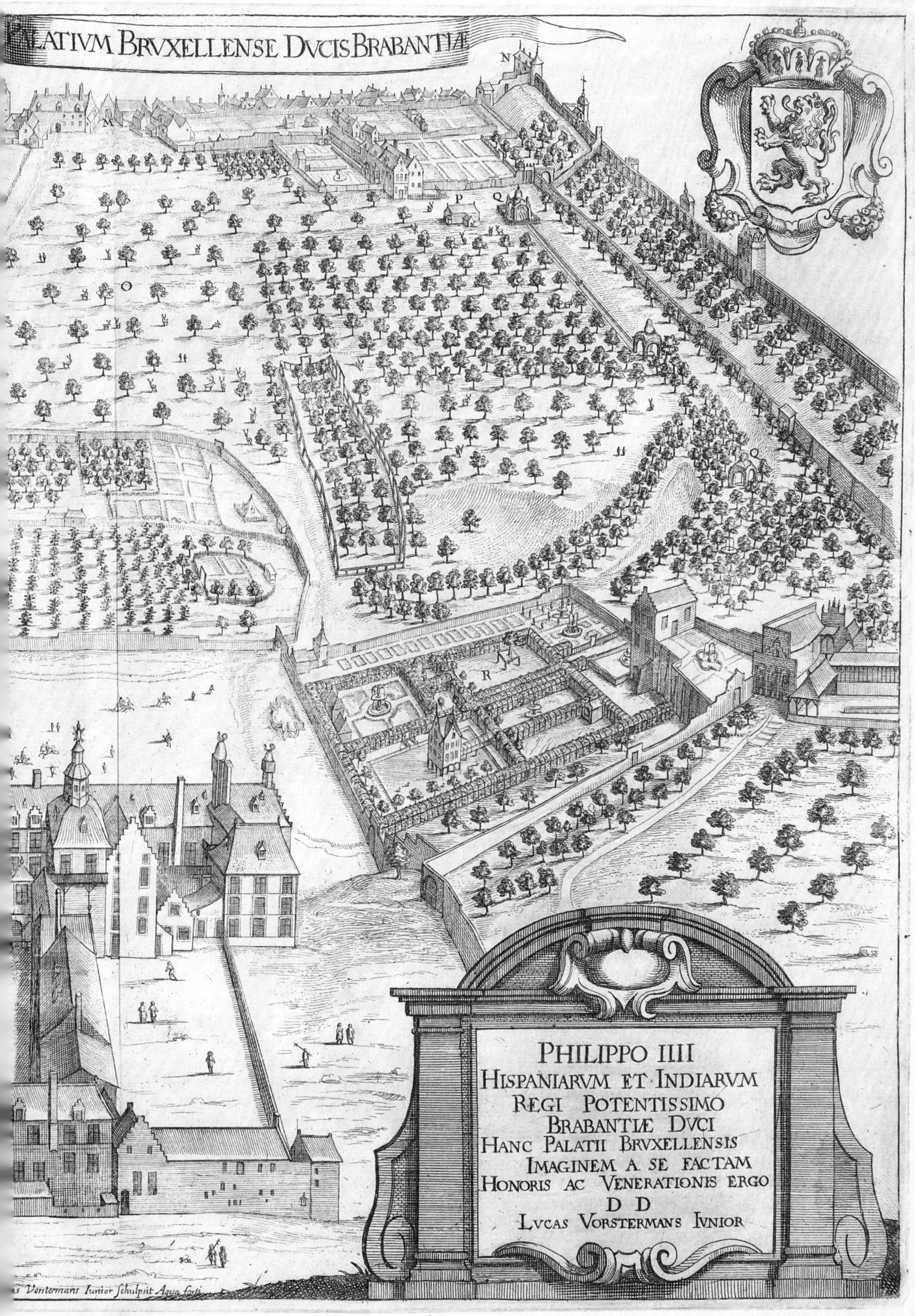
ALATIVM BRVXELLENSE DVCIS BRABANTIÆ
PHILIPPO IIII
HISPANIARVM ET INDIARVM
REGI POTENTISSIMO
BRABANTIÆ DVCI
HANC PALATII BRVXELLENSIS
IMAGINEM A SE FACTAM
HONORIS AC VENERATIONIS ERGO
D D
LVCAS VORSTERMANS IVNIOR

Den Bosch, Leuven, and Mechelen all in a row

This triptych of city views of Den Bosch, Leuven and Mechelen in the duchy of Brabant appeared in the first volume of the city atlas compiled by Georg Braun (1541-1622) and Frans Hogenberg (1535-1590), *Civitates Orbis Terrarum* (Cologne, 1572-1617). Hogenberg was the engraver, Braun wrote the texts. This hand-coloured copy from The Phoebus Foundation's collection belongs to the 1599 Latin edition printed by Bertram Buchholtz in Cologne.

's-Hertogenbosch – literally 'the Duke's Forest' – owes its name to the former residence of the duke of Brabant and its leafy, wooded surroundings. The immense church in the middle is St John's Cathedral, seat of the archdiocese of Den Bosch. The panorama of Leuven, meanwhile, clearly shows the old city walls and the castle (*Borcht*, left) and watchtower (*Verloren cost*). The text praises the university and the size of its student body.

While Hogenberg himself came from Mechelen, he did not personally draw the view at the bottom as he was living and working in Cologne by then. He probably based the engravings instead on the work of the well-travelled artist Joris Hoefnagel (1542-1601). The administrative, legal and ecclesiastical roles that the city played in the sixteenth century are clearly visible. The Keizershof and Hof van Hoogstraten (two royal residences) stand out alongside several churches and monasteries. Braun mentions Mechelen's churches in the accompanying text: one is the Church of Our Lady and the other St Rumbold's Church, which has a 'very tall and fine tower'.

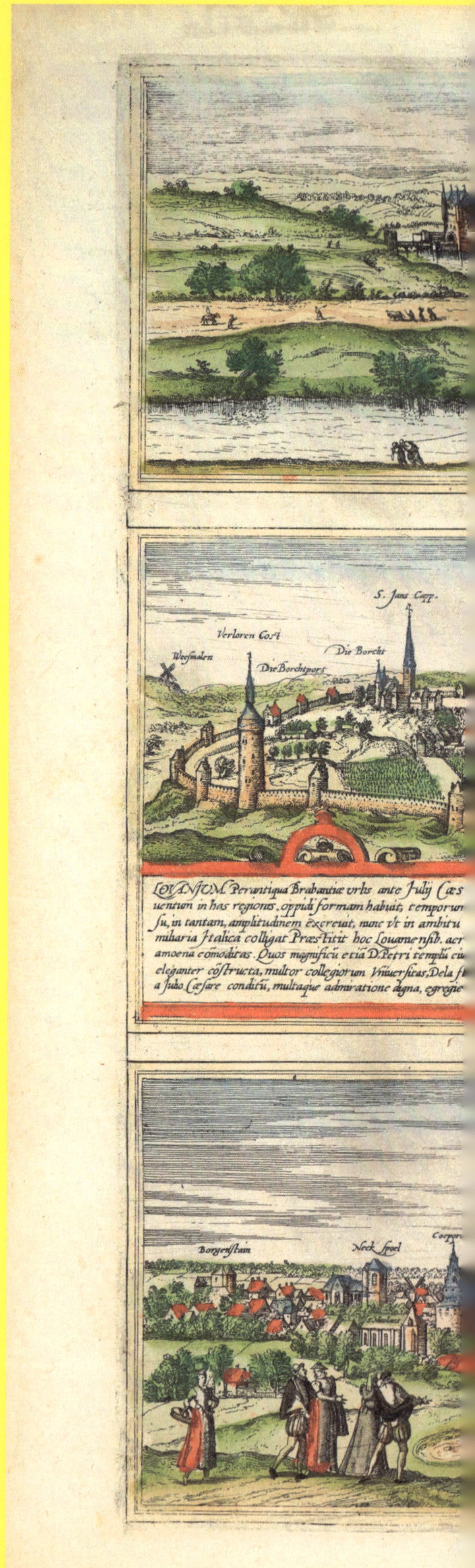

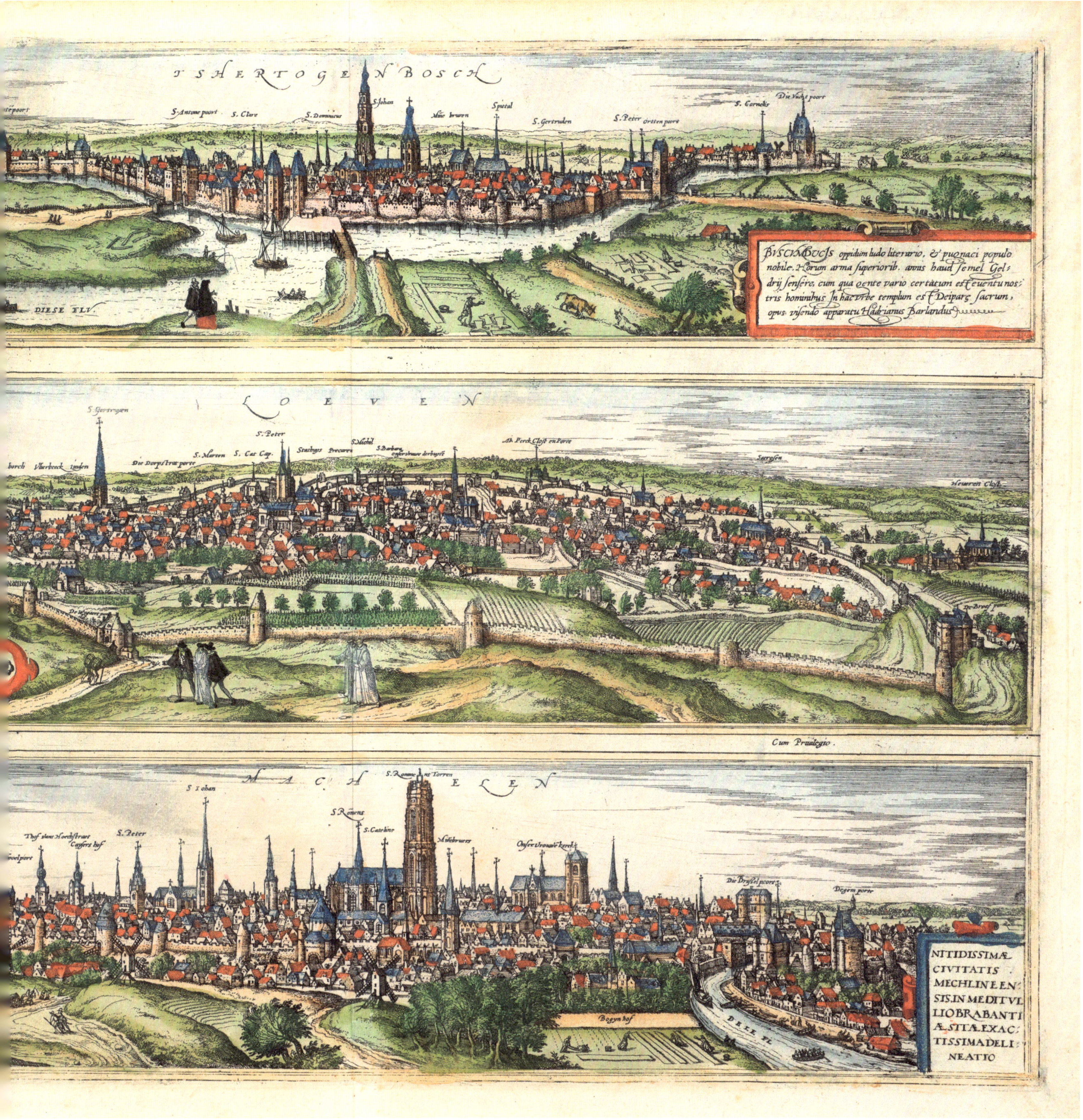

Frans Hogenberg
Tshertogenbosch, Loeven, Machelen, 1593
Three panoramic bird's-eye views of 's-Hertogenbosch, Leuven and Mechelen
Hand-coloured etching, 411 × 524 mm
ANTWERP, THE PHOEBUS FOUNDATION

Chorography: from city to region

'Chorography' is a term that arose in antiquity, when scholars like Claudius Ptolemy (⟶ Chapter IX) set out to describe the world and the cosmos. If topography is the description of places by the naked eye, chorography studies a region of the inhabited world. Geography describes the earth as a whole, while cosmography focuses on the universe. Chorography considers a place, area or region as a self-contained entity. In contrast to cosmography and geography, that location is not described in purely scientific terms, but rather in a subjective and emotive manner, in words, maps and/or pictures. It entails a strong dash of pride, identity and imagination.

Chorographic books emerged during the Renaissance, with the trend set in the Low Countries by the Antwerp-based Italian Lodovico Guicciardini's *Descrittione di tutti i Paesi Bassi* ('Description of all the Low Countries'; Antwerp, 1567; ⟶ Chapter VI). A famous example in the Northern Netherlands is the *Theatrum sive Hollandiae Comitatus et urbium nova descriptio* by Marcus Zuerius Boxhorn (Amsterdam, 1632). The Phoebus Foundation has a copy of the first edition of the Dutch translation of that work, *Toneel ofte Beschryvinge Der Steden van Hollandt* ('Theatre or description of the towns of Holland'; Amsterdam, 1634). The volume contains maps, profiles and plans of Dutch cities, such as Dordrecht, Leiden and Haarlem, with a text influenced by Guicciardini. The illustrations in this copy have been richly coloured by a contemporary hand.

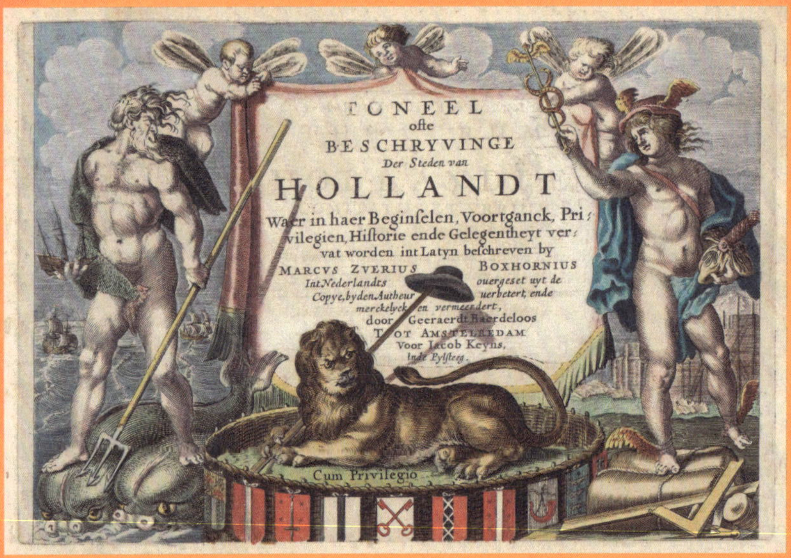

Title page
From: Marcus Zuerius Boxhorn, *Toneel ofte Beschryvinge Der Steden van Hollandt Waer in haer Beginselen, Voortganck, Privilegien, Historie ende Gelegentheyt vervat worden*, 1634
ANTWERP, THE PHOEBUS FOUNDATION

A century later, the Antwerp-born theologian and historian Antonius Sanderus or Antoon Sanders (1586-1664) followed in Boxhorn's footsteps with his two-volume *Flandria Illustrata* (Amsterdam, 1641-1644), which offers descriptions and pictures of the towns, villages, abbeys, monasteries and castles of the county of Flanders. Sanderus's work as a canon at St Martin's Church in Ypres required him to consult the archives of numerous monasteries and castles in the area, on which he would later draw for this substantial publication. The first edition of *Flandria Illustrata* was published by Joan and Cornelius Blaeu in Amsterdam, the first volume in 1641, the second in 1644. The title page features a false imprint, however, that of a publisher in Cologne – a ploy by Sanderus to prevent his book, which was actually published in Protestant Amsterdam, from being banned in Catholic Flanders.

Sanderus wrote the texts himself, drawing on a number of historical sources. He had the maps and prints made by all manner of local surveyors, draughtsmen and engravers. The result was a substantial tome with a multitude of prints – vistas, plans, city panoramas, maps, bird's-eye views and portraits. New editions were published in 1732 and 1735, along with a Dutch translation, *Verheerlijkt Vlaandre* ('Exalted Flanders'; The Hague, 1735).

Sanderus later published a chorography of Brabant too, which focused in particular on religious sites: *Chorographia sacra Brabantiae* (Brussels, 1659-1663). The work comprises historical descriptions of abbeys, monasteries and churches in the duchy. Lucas Vorsterman II (1595-1675) and Jacob Neeffs (1610-after 1660) engraved a substantial proportion of views of church buildings in bird's-eye perspective. Sanderus worked on the book at the Benedictine abbey in Affligem, where he later died.

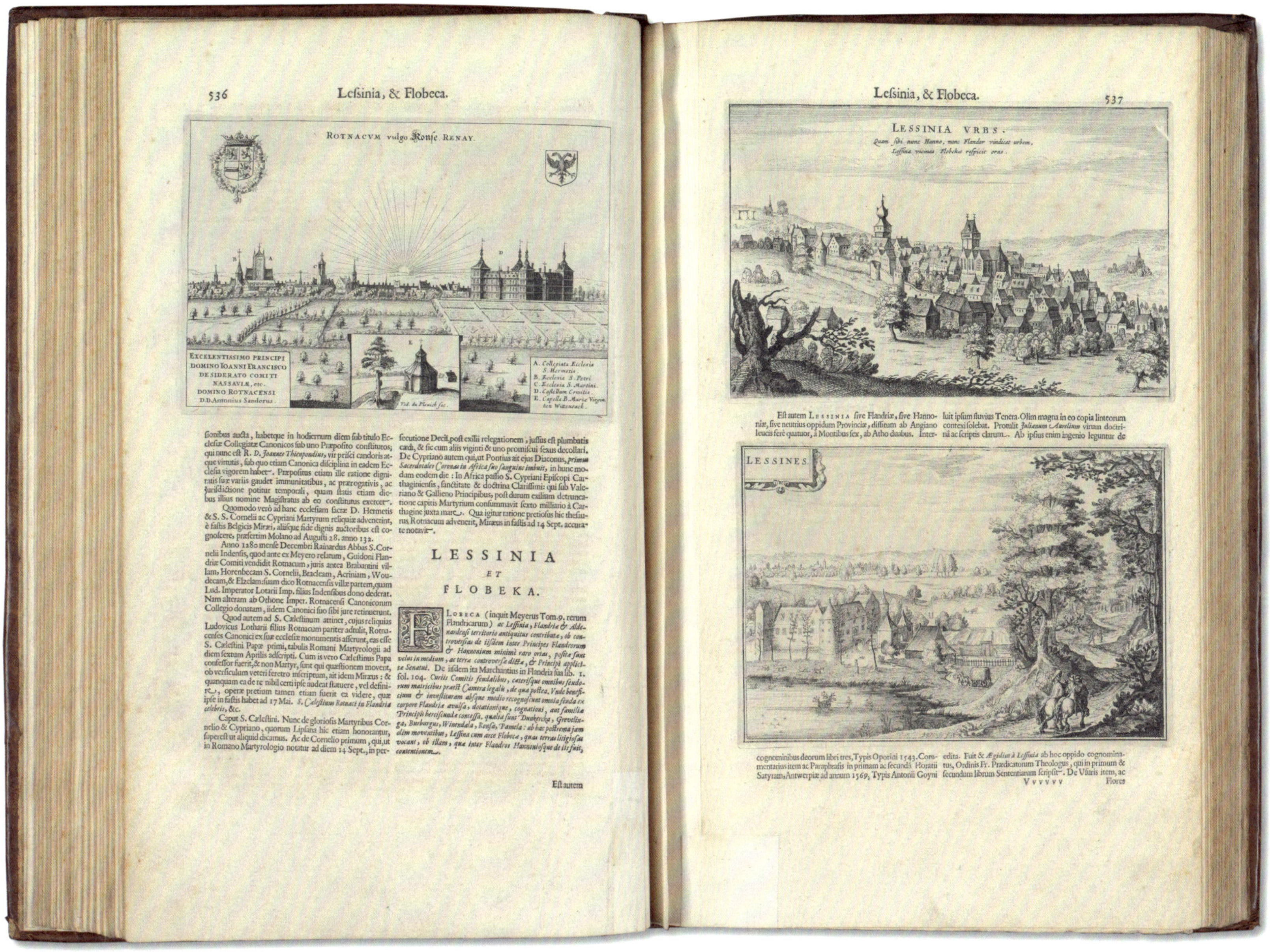

Lessinia & Flobeca
Pages with views of Ronse and Lessen
From: Antonius Sanderus, *Flandria Illustrata sive Descriptio comitatus istius per totem terraru[m] orbem celeberrimi*, 1641-1644
ANTWERP, THE PHOEBUS FOUNDATION

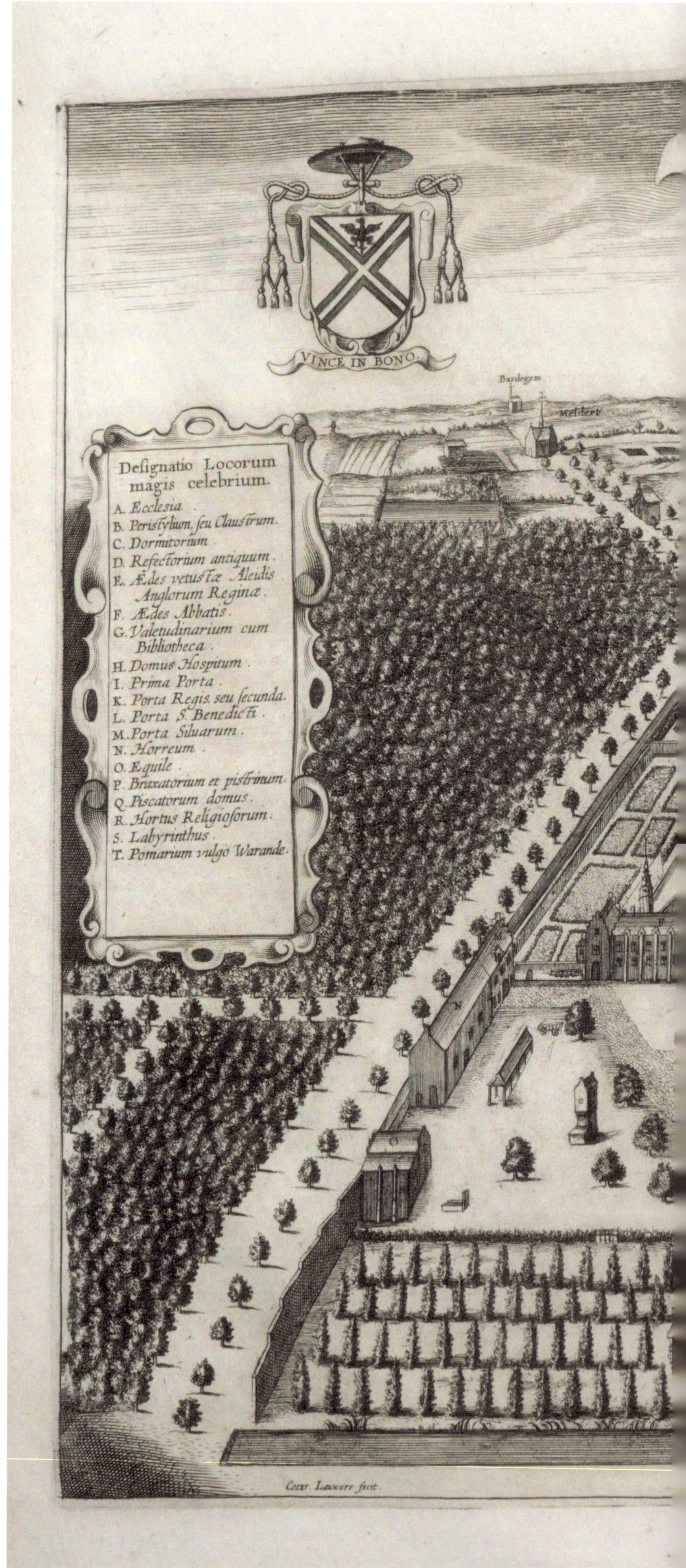

Coenraad Lauwers
Affligenium vulgo Affligem Primaria ac nobilissima in Ducatu Brabantia
Bird's-eye view of Affligem Abbey
From: Antonius Sanderus, *Chorographia sacra Brabantiæ*, 1659
ANTWERP, THE PHOEBUS FOUNDATION

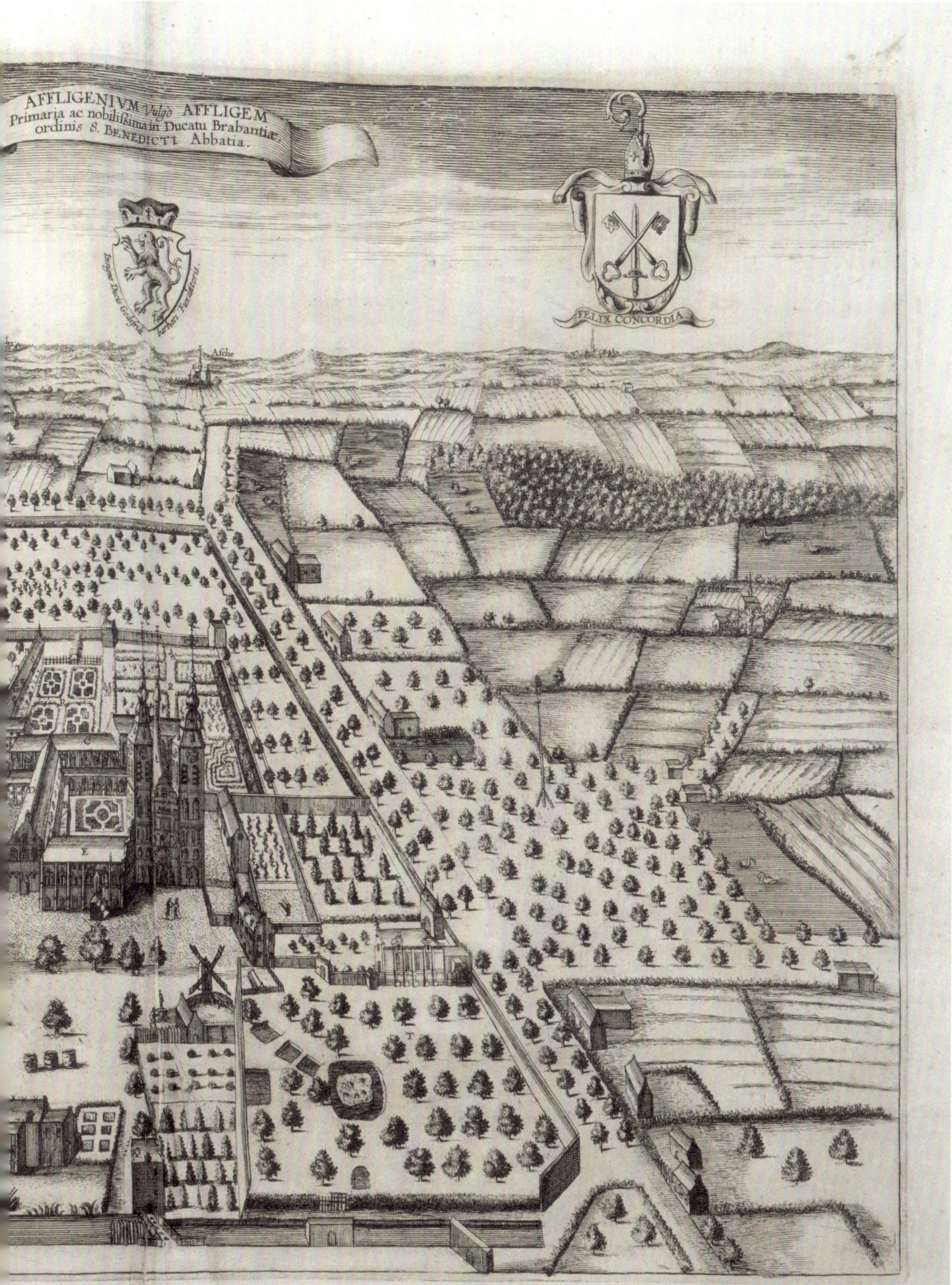
AFFLIGENIVM Vulgò AFFLIGEM
Primaria ac nobilissima in Ducatu Brabantiæ,
ordinis S. BENEDICTI Abbatia.
FELIX CONCORDIA
Asche

ANTWERP AS A TRADING HUB

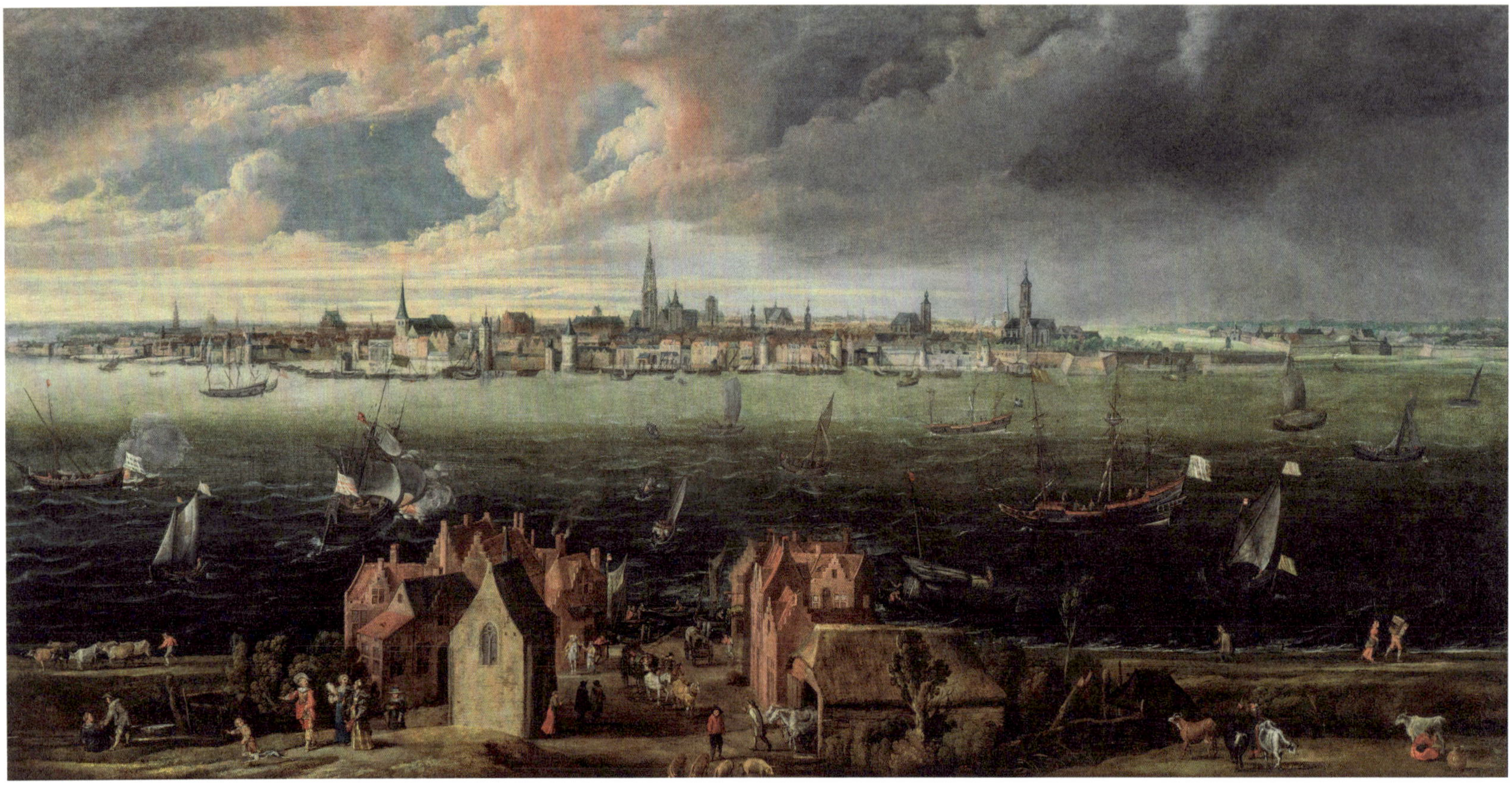

In the late fifteenth century, the economic centre shifted from Bruges to Antwerp in response to a complex interplay of political, economic and geographical factors. The sixteenth century came to be known as Antwerp's 'Golden Age'. Its commercial exchange brought together spices from Portugal's colonies, cloth from England and grain from the Baltic. Yet this financial success was not destined to last either. Antwerp, in turn, had to pass the baton to Amsterdam after it was captured in 1585 by Spanish troops commanded by Alexander Farnese and the river Scheldt was blockaded (⟶ p. 68). During the city's heyday, but also in the nostalgic period that followed, any number of artists, engravers and draughtsmen were eager to create pictures glorifying its status.

The Antwerp artist Jan Wildens (1585/1586-1653) painted a panorama of the city in which we make out the walls and the many towers and prominent buildings, including the town hall, the Church of Our Lady and St James's Church in the centre. Wildens painted some of the buildings with such accuracy that he probably viewed them from close by. He blended his sketches into a comprehensive composition, combining a frontal profile of the city in the background and a bird's-eye view of the foreground. He enlivened the left bank of the river with little village scenes, groups of peasants, milkmaids and fishermen, in the style of the light-hearted village views of his Antwerp colleague Jan Brueghel I (1568-1625).

Jan Wildens
Panorama of the Antwerp roadstead from the left bank, c.1635
Oil on canvas, 120 × 236.5 cm
ANTWERP, THE PHOEBUS FOUNDATION

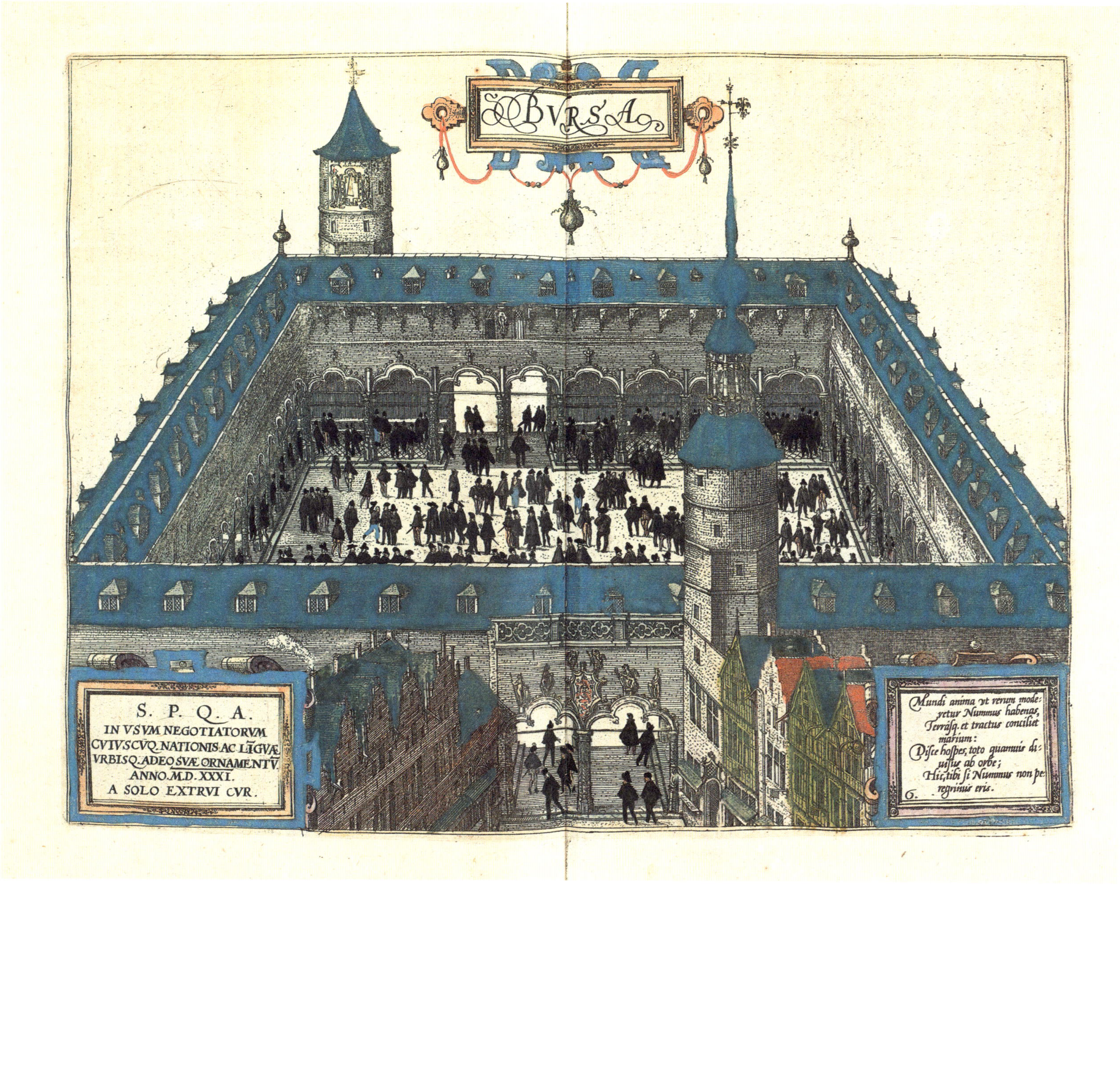

Bursa
Bird's-eye view of the Antwerp Stock Exchange
From: Lodovico Guicciardini, *Descrittione di tutti i Paesi Bassi*, 1588
ANTWERP, THE PHOEBUS FOUNDATION

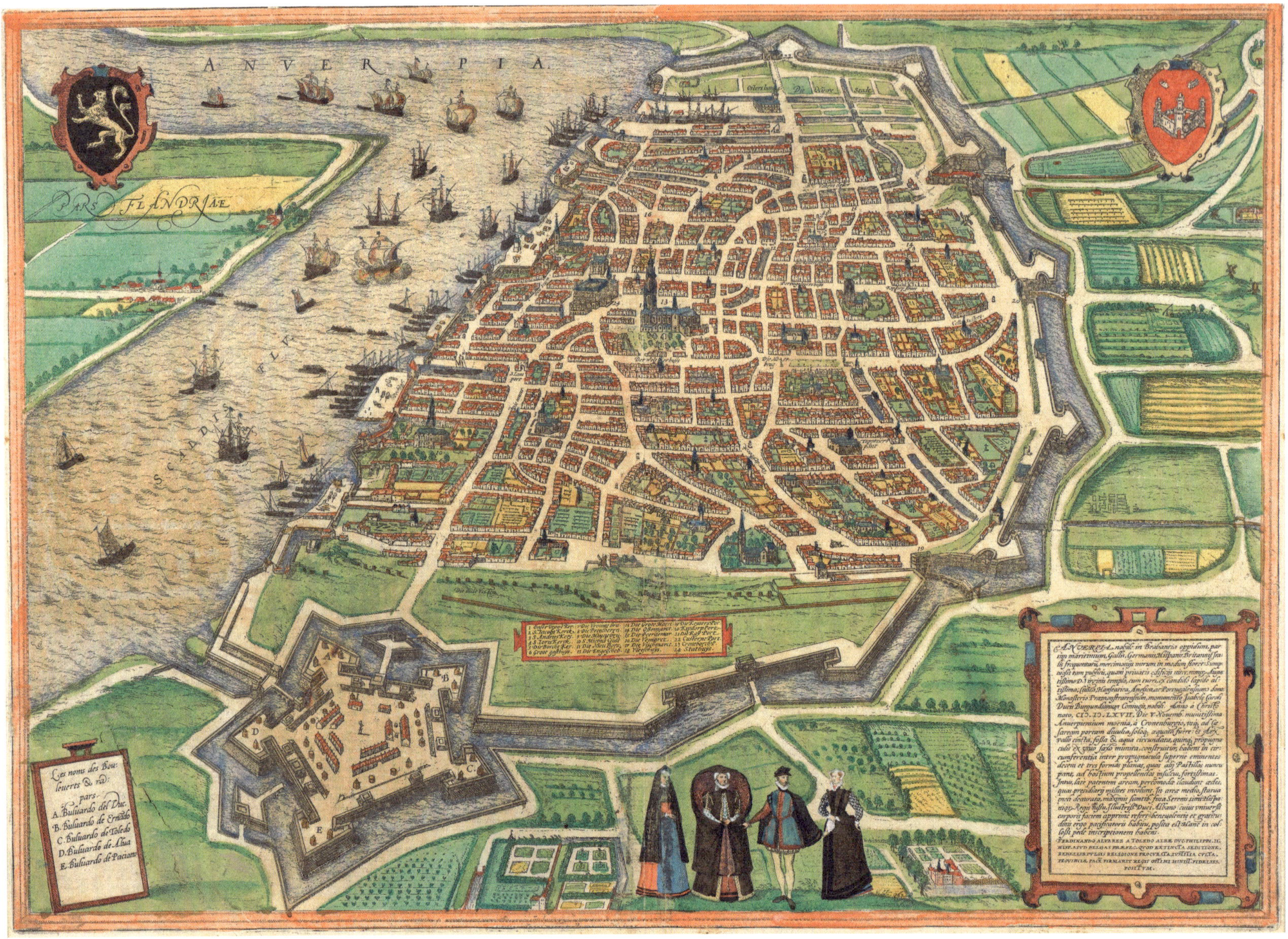

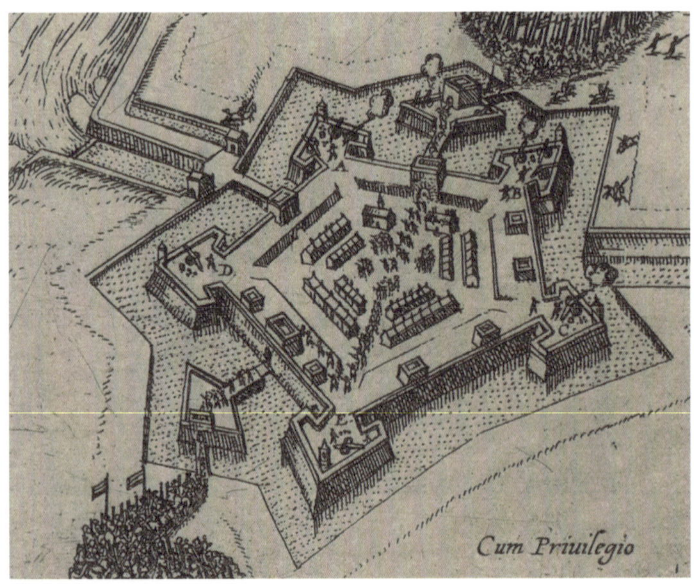

Frans Hogenberg
Anverpia, nobile in Brabantia oppidum, 1572
Bird's-eye view of Antwerp
Hand-coloured copper engraving and etching, 401 × 543 mm
ANTWERP, THE PHOEBUS FOUNDATION

The citadel in Antwerp
Detail from: *The 'Spanish Fury' in Antwerp*, c.1576 (pp. 84-85)

BIRD'S-EYE VIEW OF THE CITY

Antwerp's prosperity also made it vulnerable. Emperor Charles V, ruler of the Spanish Empire, to which the Low Countries belonged, ordered the building of a long heavily fortified wall around the city between 1542 and 1553. These modern defences proved useful during the revolt against Spanish rule, when the city came under repeated siege. Following a wave of Calvinist iconoclasm, the ramparts were extended to include a fortress on the south side under the new governor of the Spanish Netherlands, the Duke of Alva (1507-1582). The citadel became a new icon for Antwerp civic pride and identity. This is apparent in the bird's-eye map that Frans Hogenberg (1535-1590) engraved for the first volume of his six-part city atlas *Civitates Orbis Terrarum* (Cologne, 1572-1617), which he compiled with Georg Braun (1541-1622).

This was the first atlas to focus entirely on city views and maps, and contained information on cities from all over the world. Where Antwerp had mostly been depicted in the past from the east and hence from the land, Hogenberg now opted for a view from the south, placing the citadel or *casteel* as a prominent feature in the foreground. The cartouche on the right, Alva's statue at the centre of the fortress and the legend alluding to the various fortifications all express an appreciation of the military road that Antwerp had recently taken.

But this image of the city as a powerful bastion ready for war did not last for long: the citadel swiftly became a symbol of Spanish oppression and of the misery of the Revolt. In 1576, not long after the *casteel* was completed, Antwerp suffered what came to be known as the 'Spanish Fury', when troops garrisoned in the citadel mutinied and went on the rampage, plundering the city and setting it on fire.

In addition to the city atlas, Hogenberg's Cologne workshop also produced a sizeable body of news prints of events during the Revolt, including a series of pictures depicting the horrors of the Spanish Fury. To do so, he merely had to adapt the bird's-eye map of Antwerp in his *Civitates* by adding Spanish soldiers entering the city from their fortress in the lower left (⟼ pp. 84-85). Hogenberg listed the streets and houses that the Spanish torched and the councillors they murdered in two cartouches. The streets are now filled with the running figures of soldiers killing and setting fires as they go.

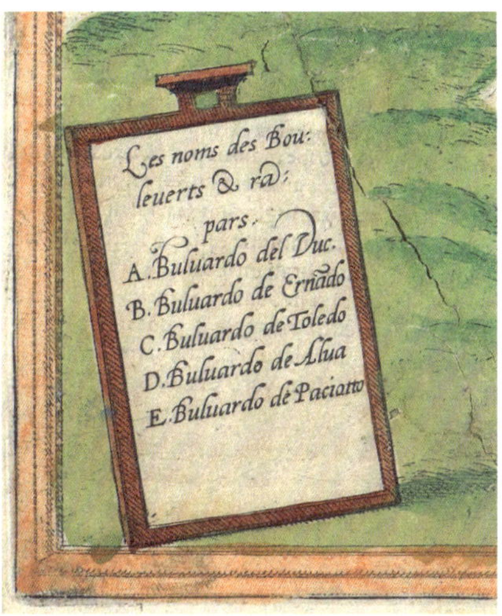

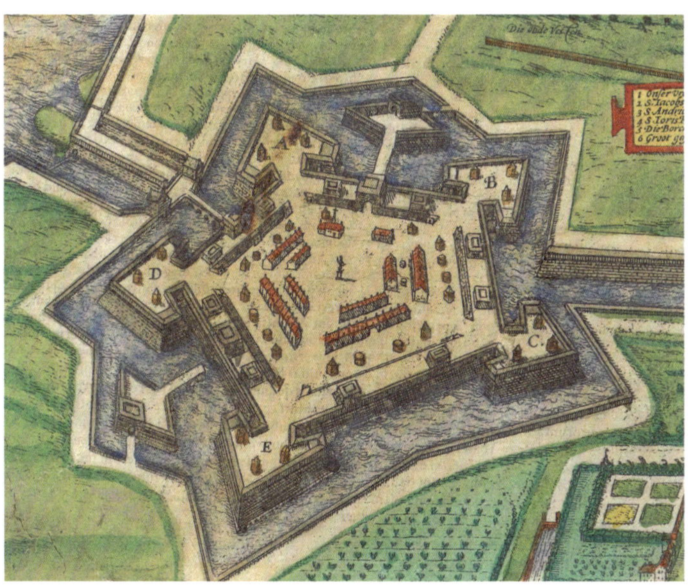

Cartouche and citadel
Details from: *Anverpia, nobile in Brabantia oppidum*, 1572 (p. 66)

The Dutch Revolt

When Charles V (1500-1558) annexed the duchy of Guelders in 1543, the Low Countries became a more or less coherent set of territories. The Habsburg royal line ruled over the area known variously as the Habsburg Netherlands, the Seventeen Provinces and the Spanish Netherlands, and which functioned in practical terms as an entity of its own within the Holy Roman Empire. The situation became fraught, however, when Charles V abdicated in favour of his son Philip II (1527-1598), just as the influence of the Reformation was making itself felt from the Empire's German territories. Philip was a harsh, absolutist Catholic who ruled the Low Countries with an iron hand from faraway Madrid, sparking tension and discontent among his Netherlandish subjects. This was exacerbated by the onerous taxation imposed by the Spanish sovereign and his representatives, who also sought to curb the traditional privileges of the hitherto autonomous cities in favour of more centralised control.

It all culminated in a revolt against the Habsburg Spaniards – a conflict that would become known as the Dutch Revolt or the Eighty Years War. Violence erupted in the late 1560s. A key early event was the so-called 'Iconoclastic Fury' (1566), in which Protestant insurgents in the Low Countries destroyed images of saints in churches and other religious buildings. Protestantism was a major driving force for the rebellious Netherlanders, who were keen to break free from their tyrannical Catholic rulers. The Revolt also coincided with the period in which book printing and cartography were flourishing, making it one of the first major conflicts in which news of the war and maps illustrating its progress were widely printed and distributed, both in the Low Countries and beyond (⟶ Chapter III). Frans Hogenberg, with his print workshop in Cologne, was also a news publisher who depicted (and mapped) the uprising. A famous example of his output is a news print showing the Iconoclastic Fury.

The Union of Arras, signed in 1579, gave rise to the new Republic of the Seven United Netherlands, which officially came into being in 1581 and consisted of a federation of the provinces of Holland, Zeeland, Utrecht, Guelders, Friesland and Groningen. Antwerp and the Southern Netherlands, which had also rebelled, were reconquered in 1585 by Alexander Farnese and his Spanish-Habsburg army, splitting the Low Countries into a southern and a northern part (⟶ Chapter V).

The Twelve Years Truce that held sway between 1609 and 1621 more or less restored peace in the region, allowing it (and Spain) to recover. But hostilities resumed in 1621, now folding into the continent-wide Thirty Years War, which was fought out primarily in Central Europe. The Thirty Years and the Eighty Years War were brought to an end in 1648 by the Peace and the Treaty of Münster respectively, both of which formed part of the Peace of Westphalia. The cessation of hostilities was proclaimed and celebrated in Antwerp on the Grote Markt, the main square in front of the town hall. The spectacle is recalled in a print made after the example of the Prague artist Wenceslaus Hollar (1607-1677).

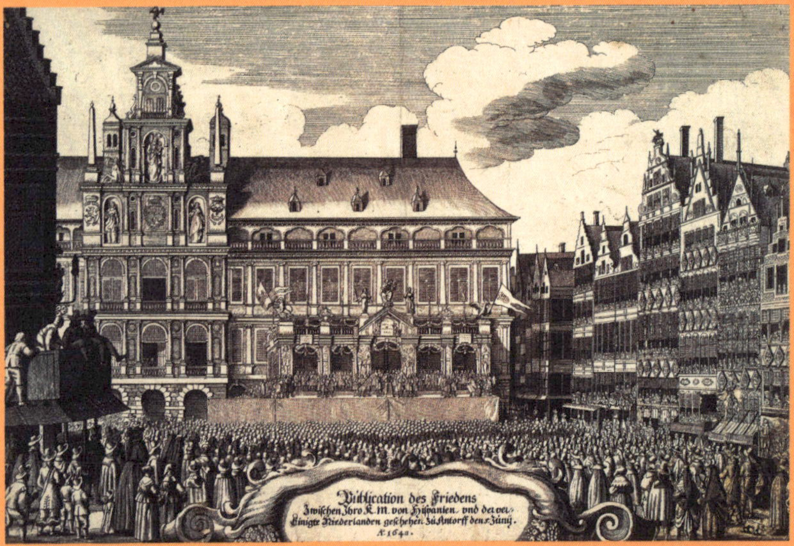

After Wenceslaus Hollar
Publication des Friedens Zwischen Ihre K.M. von Hispanien, und de[r] vereinigte Niederlanden. geschehen zu Antorff, den 5 Junij A.o 1648, c.1648-1652
Print of the town hall in Antwerp during the proclamation of the Treaty of Münster in 1648
Etching, 210 × 317 mm
ANTWERP, THE PHOEBUS FOUNDATION

Next fold-out map:
Jan Baptist Vrients
Antverpia, 1610
Panorama of Antwerp
Copper engraving and etching, four sheets, total 475 × 2280 mm
ANTWERP, THE PHOEBUS FOUNDATION

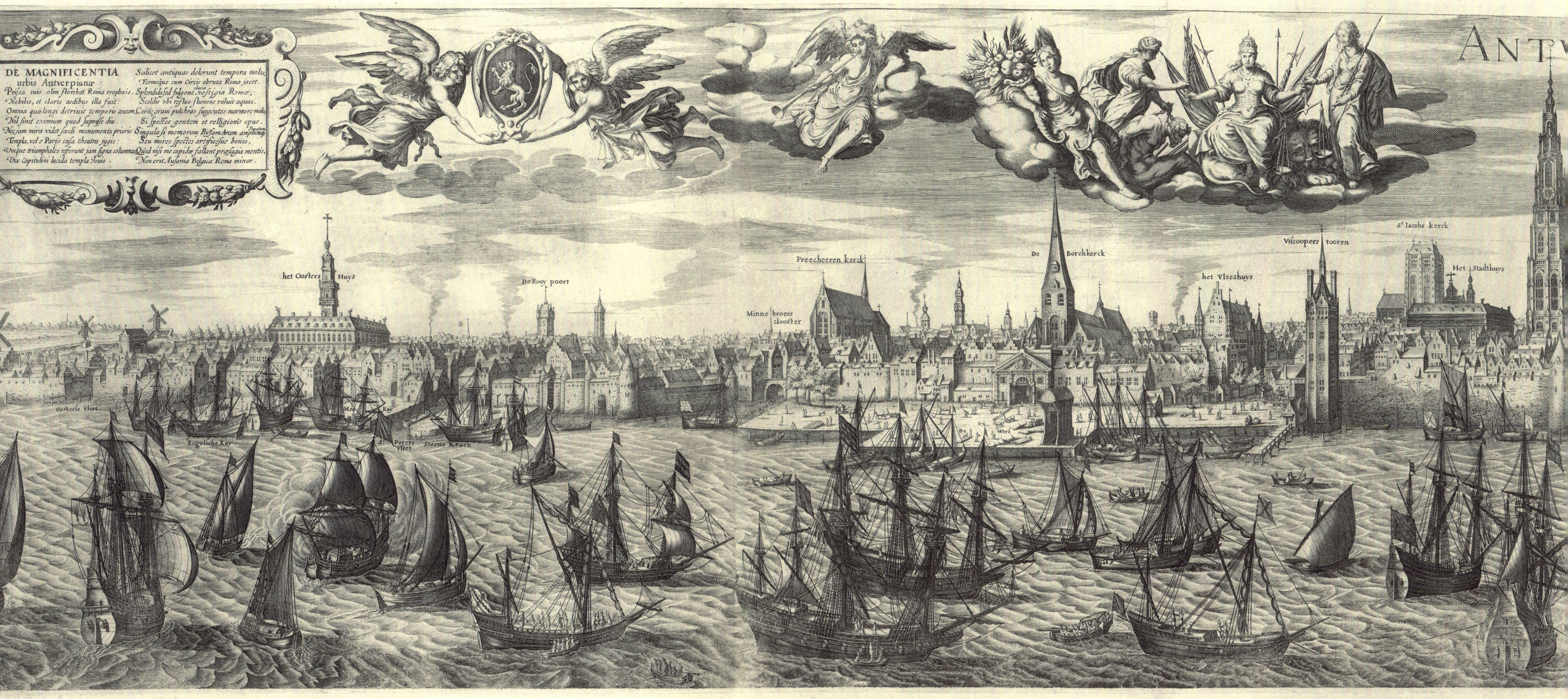
DE MAGNIFICENTIA
urbis Antverpianæ
Prisca suis olim florebat Roma trophæis,
Nobilis, et claris aedibus illa fuit:
Omnia quæ longi detriuit temporis æuum,
Nil sinit eximium quod superesse diu.
Nec jam mira videt sæcli monumenta prioris
Templa, vel e Pariis cæsa theatra jugis:
Vixque triumphales referunt jam signa columnas,
Vix Capitolini lucida templa Jouis.
Scilicet antiquas delerunt tempora moles,
Fermeque cum Circis obruta Roma jacet.
Splendida sed fulgent, claræ vestigia Romæ,
Scaldis vbi refluo flumine voluit aquas.
Cerne graui pulchras surgentes marmore moles,
Si spectes gentem et relligionis opus.
Singula si memorem Bursam Arcem amplumque Senatum,
Seu mores spectes artificesque bonos.
Quod nisi me cupidæ fallant præsagia mentis,
Non erit Ausonia Belgica Roma minor.
ANTV
het Oosters Huys
De Rooy poort
Preecheeren kerck
Minne broers clooster
De Borchkerck
het Vleeshuys
Viscoopers tooren
St. Iacobs kerck
Het Stadthuys
Oosterse vliet
Engelsche Kay
Mee Kay
St. Peters vliet
Steene Kraen
Werfpoort
De Werf

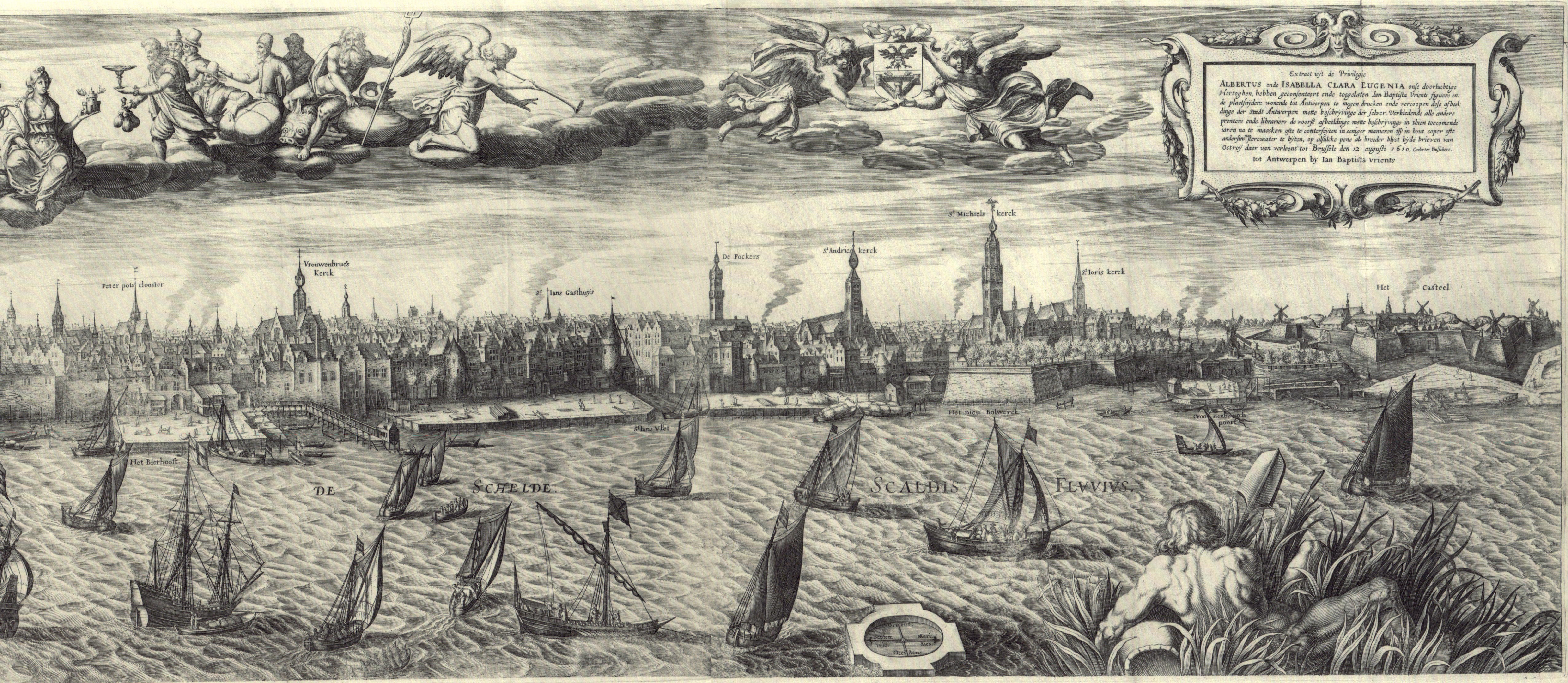
Extract uyt de Privilegie
ALBERTUS ende ISABELLA CLARA EUGENIA onse doorluchtige Hertoghen, hebben geconsenteert ende toegelaten Jan Baptista Vrients figuere en: de plaetsnydere wonende tot Antwerpen te mogen drucken ende vercoopen dese afbeel: dinge der Stadt Antwerpen mette beschryvinge der selver. Verbiedende alle andere prenters ende librariers de voorsz afbeeldinge mette beschryvinge in thien toecomende iaren na te maecken ofte te conterfeyten in eeniger manieren. tsy in hout coper ofte andersins sterewater te byten, op alsulcke pene als breeder blyct byde brieven van Octroy daer van verleent tot Brussele den 12 augusti 1610, Ondertec. Busschere,
tot Antwerpen by Ian Baptista vrients
Peter pots clooster
Vrouwenbrues Kerck
St. Ians Gasthuys
De Fockers
St. Andries kerck
St. Michiels kerck
St. Ioris kerck
Het Casteel
Het Bierhooft
St. Ians Vliet
Het nieu Bolwerck
DE SCHELDE.
SCALDIS FLVVIVS.

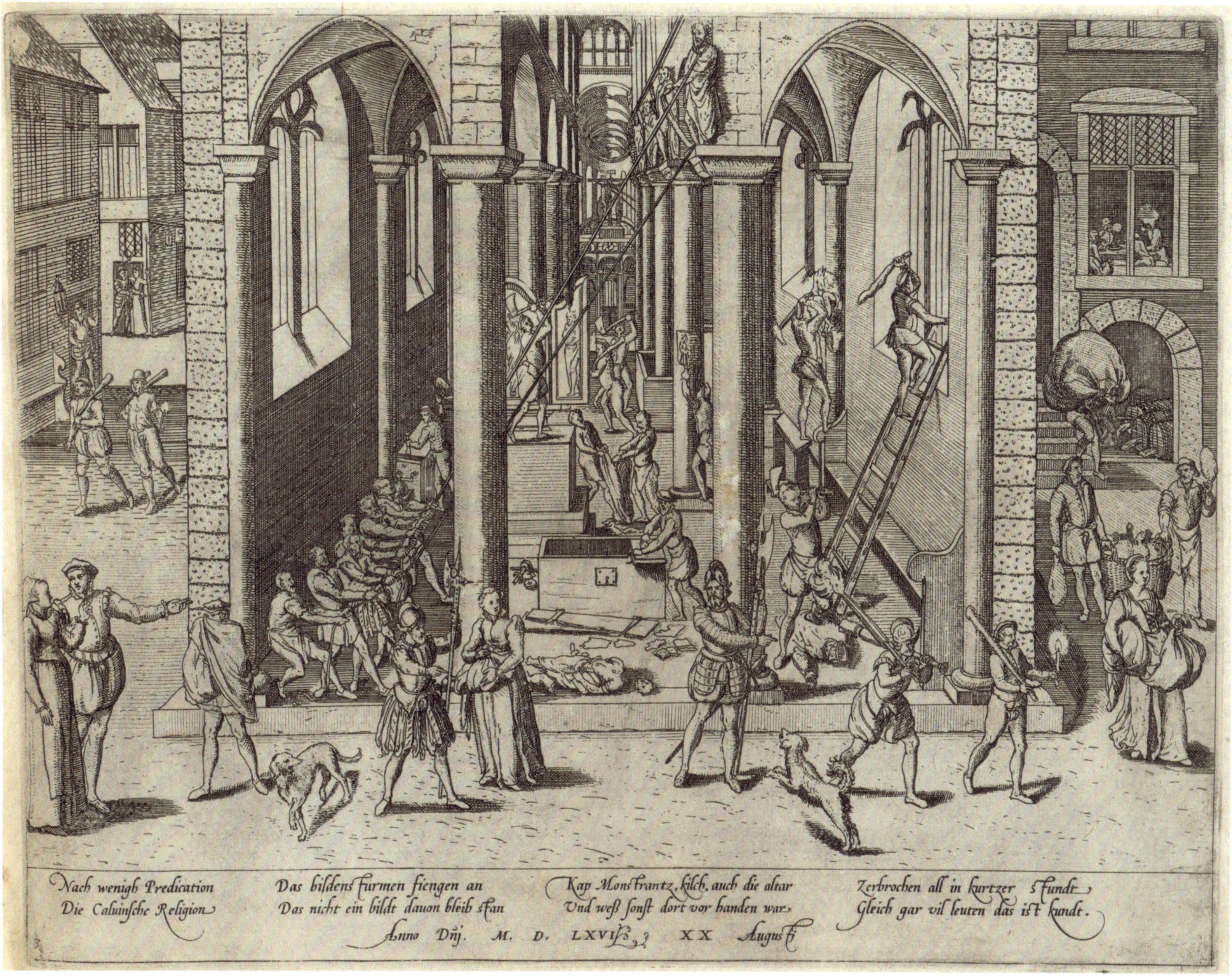

Frans Hogenberg
Iconoclastic rioting at the Church of Our Lady in Antwerp in 1566, after 1566
Etching, 210 × 280 mm
ANTWERP, THE PHOEBUS FOUNDATION

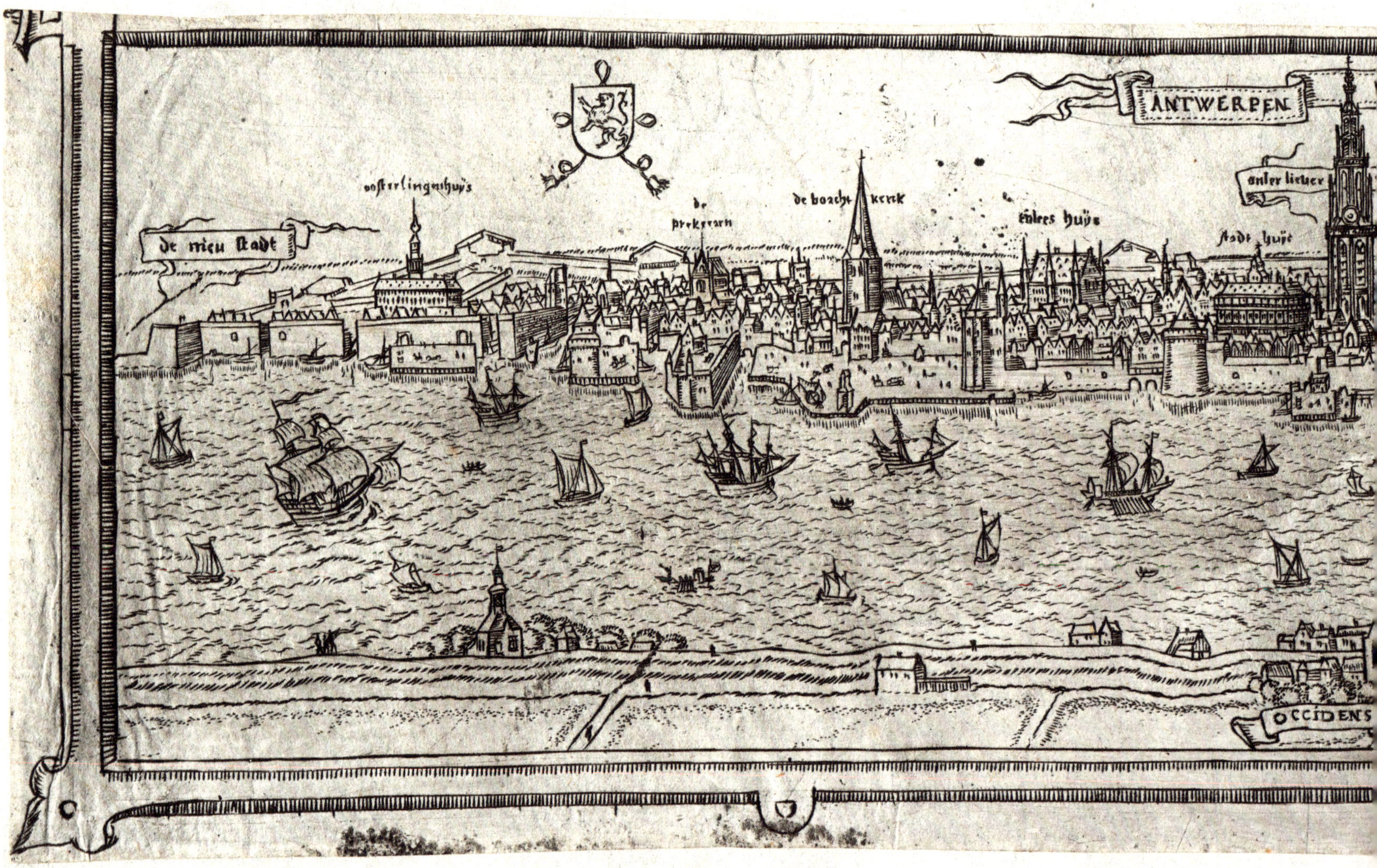

Melchisedech van Hooren
Antwerpen een coopstadt, after 1570
View of the Antwerp roadstead
Copper engraving, 94 × 330 mm
ANTWERP, THE PHOEBUS FOUNDATION

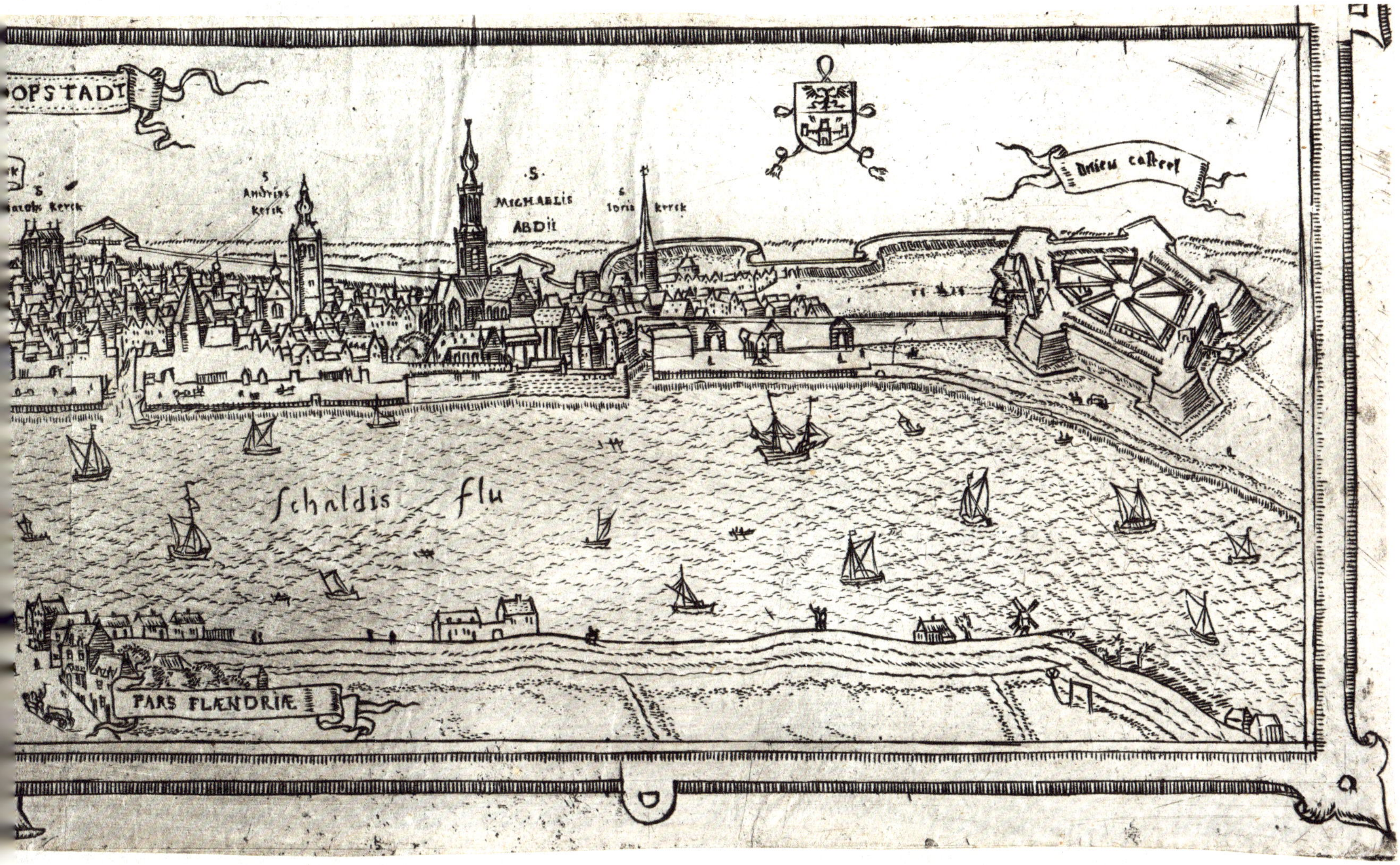
S.
MICHAELIS
ABDII
Schaldis flu
PARS FLÆNDRIÆ

LEFT BANK OF THE SCHELDT

Antwerp owed a great deal to the Scheldt. Of all the pictures made of the city over the years, the most popular was the view from the opposite left bank, with the river in the foreground. We find it, for instance, in the painting by Wildens, in which the viewer is offered an eye-level view of the skyline stretching out along the horizon – an iconic perspective that defined the image of the city on the Scheldt well into the eighteenth century.

The earliest roadstead view in The Phoebus Foundation's collection is probably the anonymous print attributed on stylistic grounds to Melchisedech van Hooren (active in Antwerp in 1552-1571). Judging by the buildings, it was made around 1570. It is not only a cityscape but also a profile depicting the whole town lengthways. One sees clearly how the new Spanish ramparts enclose the city, in the background too. The various tall buildings rise above it and are identified by text labels in the print, among them the Vleeshuis (the butchers' guild hall), town hall (built 1551-1564) and the Church of Our Lady. The novelty of the city's expansion is also highlighted: *de nieu stadt* (New Town) on the left and the *nieu casteel* (citadel) on the right. The print is titled *Antwerpen een coopstadt* ('Antwerp, a Mercantile City'), framing the city as a trading metropolis. Antwerp was known as one of the leading commercial ports in the sixteenth century, as also indicated by the busy waterborne traffic on the Scheldt.

Besides the countless small prints of the Antwerp roadstead that circulated in books, atlases, loose sheets or decorations on other prints and maps, there were impressively monumental pictures too. A panoramic view published in the city in 1610 by Jan Baptist Vrients (1552-1612) measures an unprecedentedly gigantic 47.5 × 228 cm. The engraving has been attributed to Johannes Rem (*c.*1566-*c.*1620/1623), who is thought to have made the equally monumental profile of Amsterdam which Willem Jansz. Blaeu (1571-1638) published a few years earlier in 1606 (⟶ Chapter X). Although the Antwerp panorama seems like an exceptionally accurate and lifelike snapshot, the port probably looked slightly different to what we see in the print. Following the Siege of Antwerp (1584-1585), the city – now under Spanish occupation – was cut off from overseas trade, causing its economy to stagnate.

The region enjoyed something of a revival after the appointment of Archduke Albert and the Infanta Isabella of Austria as rulers over the Spanish-Habsburg Netherlands. Their ceremonial reception ('Joyous Entry') in 1599 marked the beginning of a fresh period of prosperity for Antwerp, and Spanish forces went on to record several further victories. The beginning of the Twelve Years Truce (1609-1621) was likewise positive: the cessation of hostilities allowed the economy and trade to recover. The celebratory mood triggered by these developments might have prompted a desire to mark the promising new start with a monumental panorama of the city. The symbolic elements at the top of this view – Mercury, god of trade, the horn of plenty and foreign merchants bringing their wares – are perfectly in keeping with this optimistic feeling towards the future. Vrients did not organise the production of the print entirely on his own initiative: Albert and Isabella, the ruling couple, granted him an official privilege to publish it, encouraging this positive image of the city or perhaps even directly commissioning it.

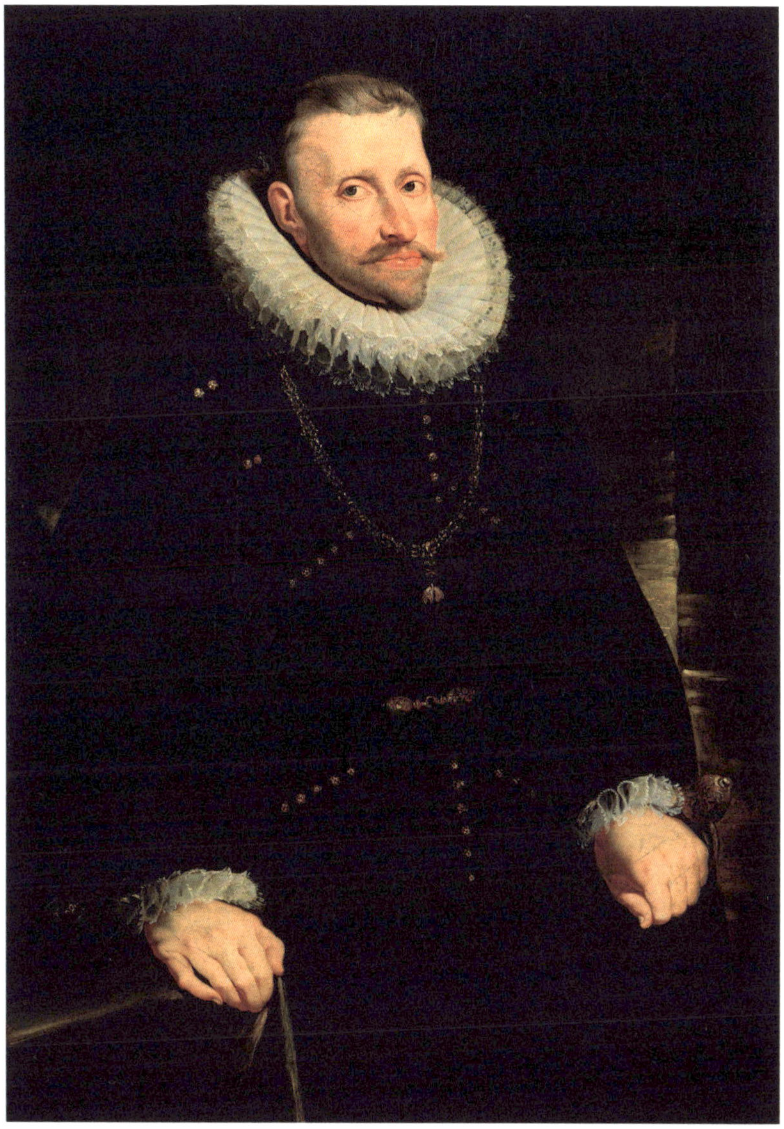

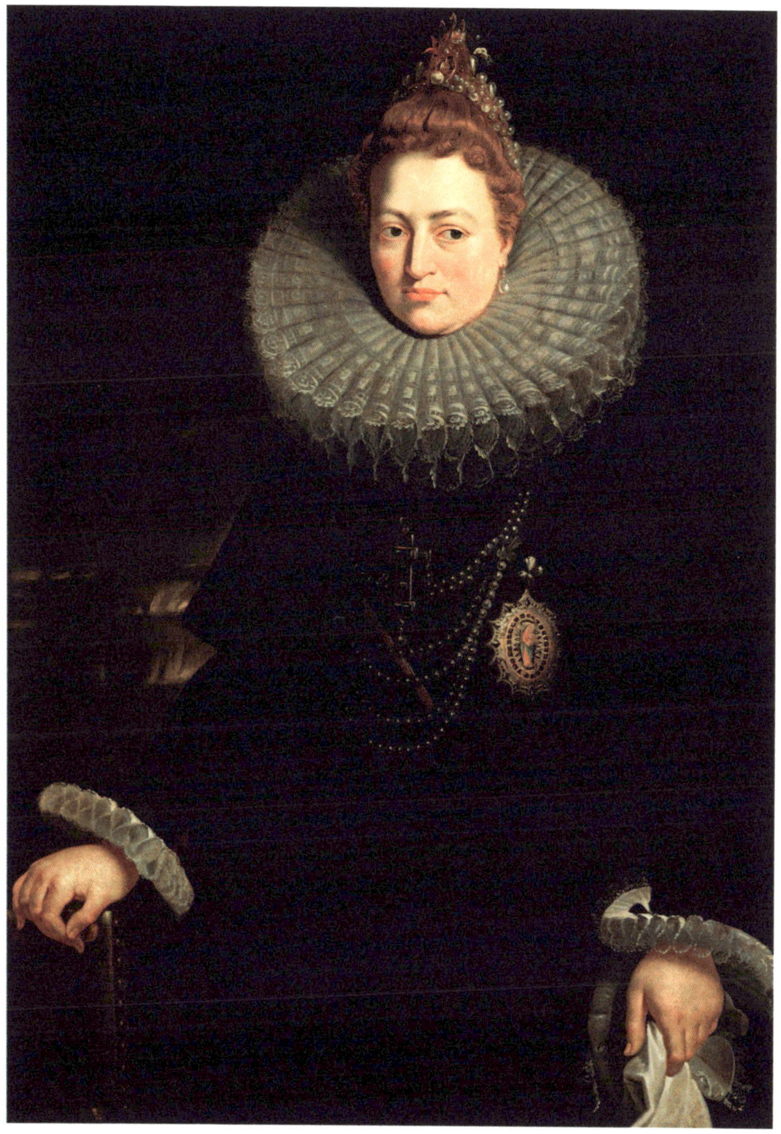

Studio of Peter Paul Rubens
Portrait of Archduke Albert of Austria, c.1617
Oil on panel, 126 × 100 cm
ANTWERP, THE PHOEBUS FOUNDATION

Studio of Peter Paul Rubens
Portrait of the Infanta Isabella Clara Eugenia, c.1617
Oil on panel, 126 × 100 cm
ANTWERP, THE PHOEBUS FOUNDATION

Panorama of Brussels

By the end of the seventeenth century, Jeremias Wolff (1663-1724) had become the biggest publisher in Augsburg, a centre of printing in Bavaria. His company's engravers were among the best in their country. Following Wolff's death, the business was taken over by his son-in-law, Johann Balthasar Probst (1673-1748), who, together with his sons, began to operate under the name 'Haeres Ieremiae Wolfii' ('the heirs of Jeremias Wolff'). Among other things, the family firm produced panoramas of European cities from Rotterdam, The Hague, Ghent and Leuven to London, Madrid, Paris and Constantinople (Istanbul). The Augsburg publishers discovered the topographical drawing talent of Friedrich Bernard Werner (1690-1776), whom they dispatched all over Europe to sketch. Wolff's heirs published a total of 118 different city views, ninety of which were based on Werner's designs. Werner too visited Southern Netherlandish cities like Antwerp, Ghent, Leuven and Brussels.

Probst's elder son, Johann Friedrich Probst (1721-1781), published Werner's monumental panorama of Brussels around 1729. It was drawn in bird's-eye view, so that we not only see the city walls from the front but also the buildings behind them. The legend at the bottom identifies the buildings. Compare the panorama with the map of Brussels by Jacob Harrewijn (⟼ p. 145), which was made around the same period. Do the buildings on the map tally with those in the panorama? Whoever engraved the latter seems to have manipulated the proportions a little so that all the buildings would fit in. The panorama of Antwerp includes the signature of Probst's other son, Georg Balthasar Probst (1732-1801). The portrait of the city was made once again from the iconic left bank, with its view across the Scheldt. The Phoebus Foundation's copy has been coloured in.

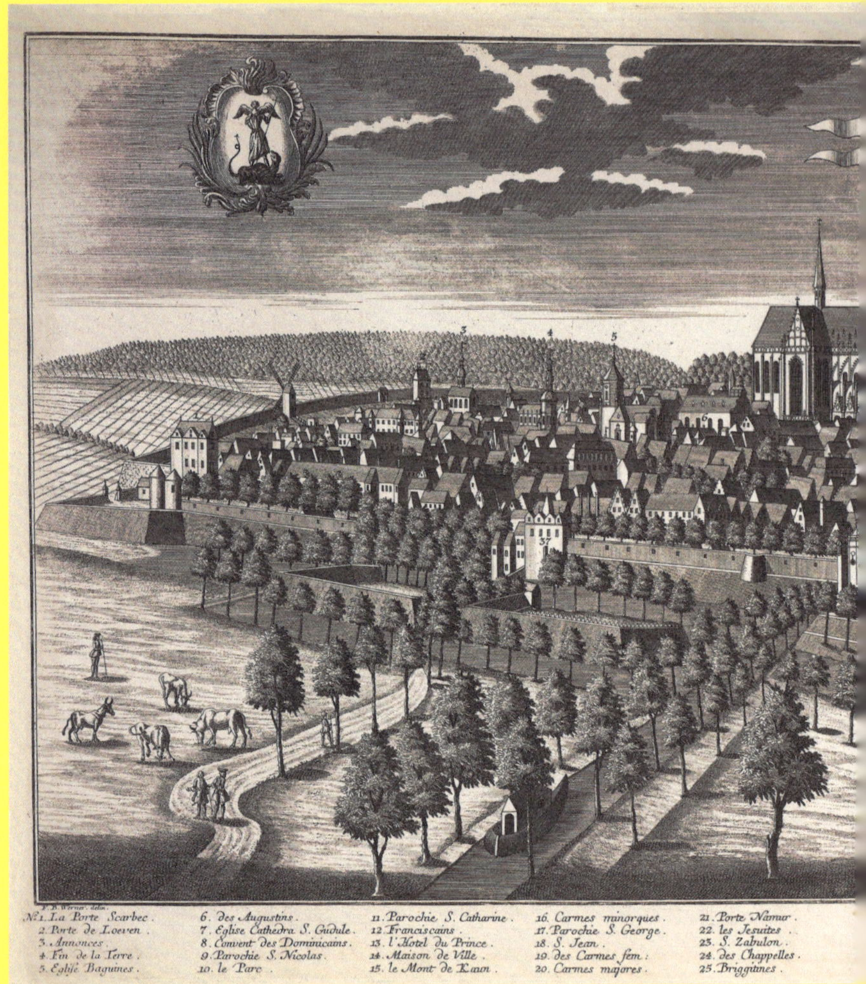

The prints published by the Wolff company were in the tradition of the seventeenth-century cityscapes of Dutch publishers like Willem Jansz. Blaeu, Claes Jansz. Visscher, Johannes Janssonius and Frederik de Wit. The panorama of Brussels is also comparable, however, to Jan Baptist Vrients's view of Antwerp. But where the Augsburg editions are printed on two large sheets rather than four and measured just one metre across, Vrients's panorama is over two metres wide. The panoramas continued to be produced in Augsburg until the end of the eighteenth century and were still sought after, as indicated by the many reissues from successive printers' addresses. With the dawning of the Romantic era (the late eighteenth and nineteenth centuries), interest shifted to the countryside, landscape and (untamed) nature, possibly explaining the waning interest in urban panoramas.

Friedrich Bernard Werner and Johann Friedrich Probst
Brüssel, c.1729
Bird's-eye panorama of Brussels
Hand-coloured copper engraving and etching, 434 × 1134 mm
ANTWERP, THE PHOEBUS FOUNDATION

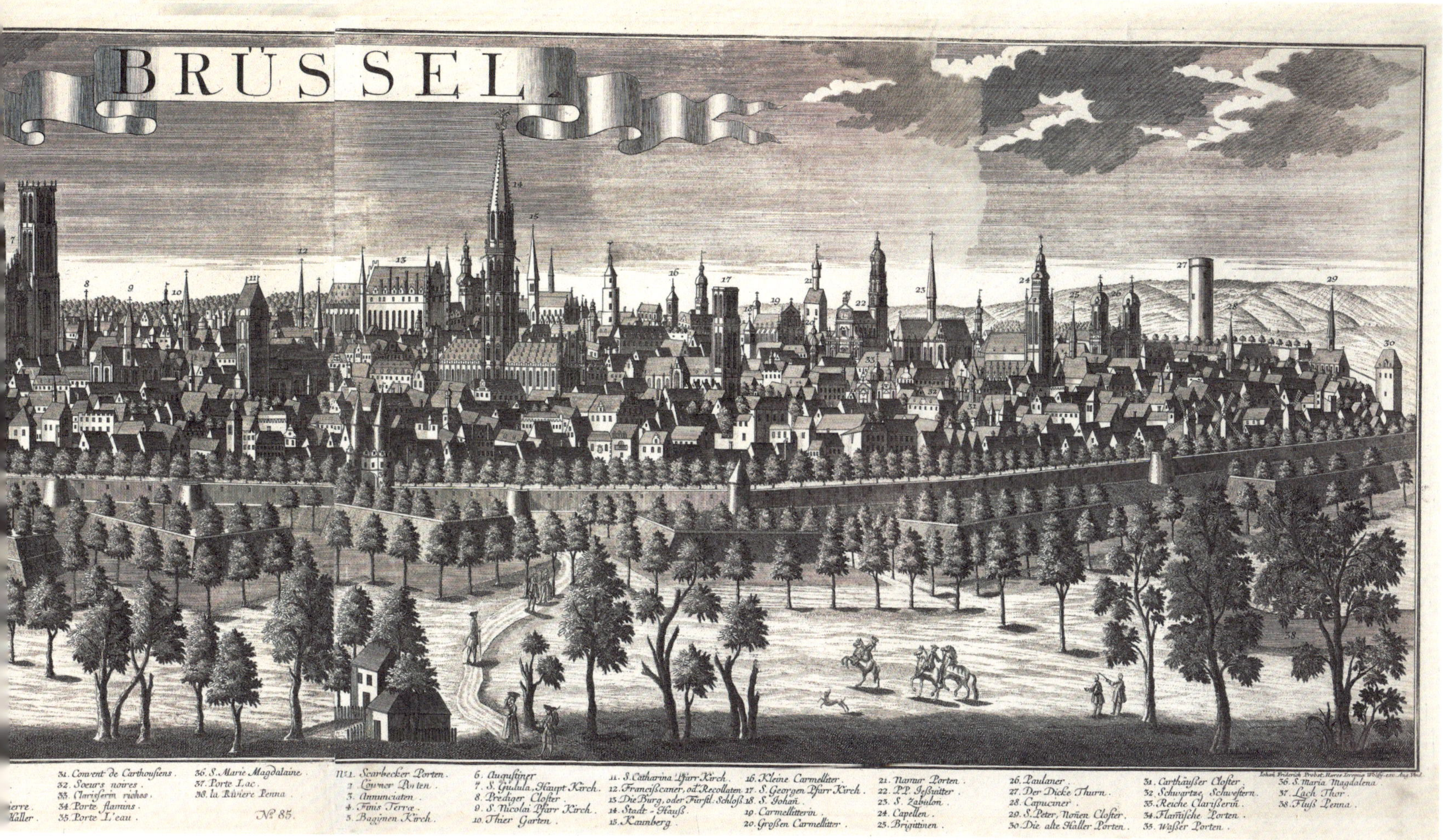

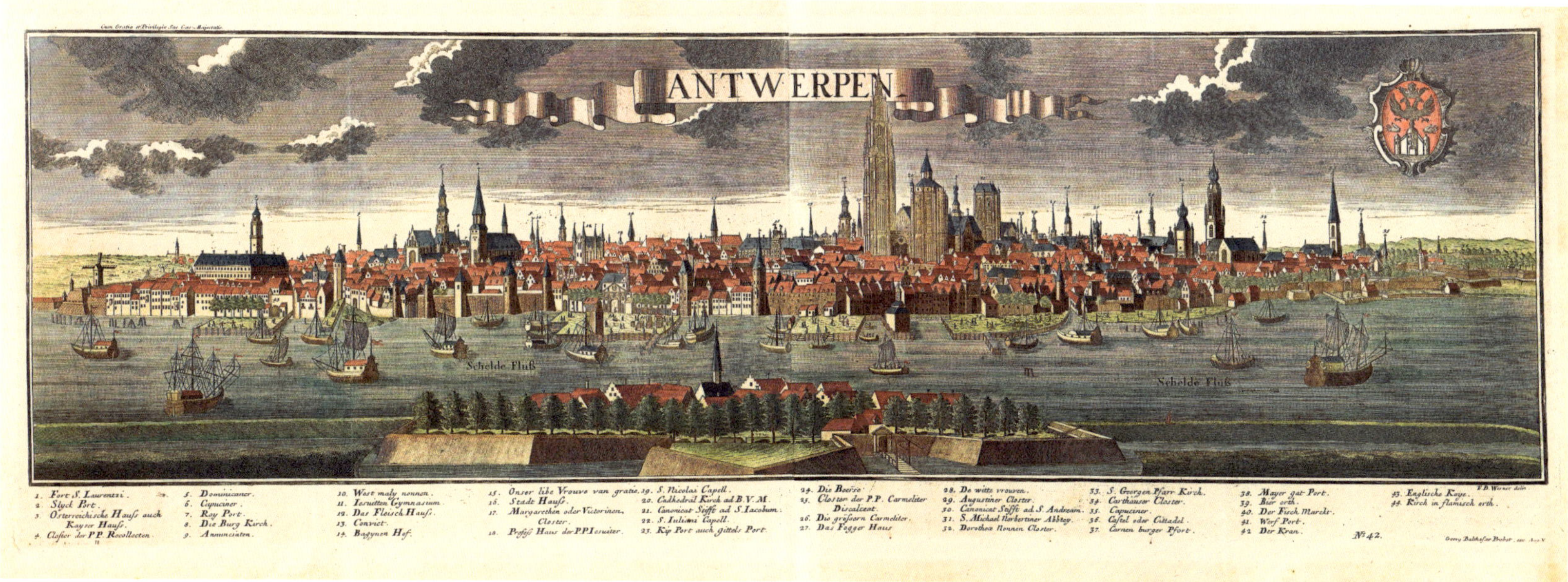

Friedrich Bernard Werner and Georg Balthasar Probst
Antwerpen, c.1729
Panorama of the Antwerp roadstead
Copper engraving and etching, 405 × 1100 mm
ANTWERP, THE PHOEBUS FOUNDATION

THE CITY IN A NUTSHELL

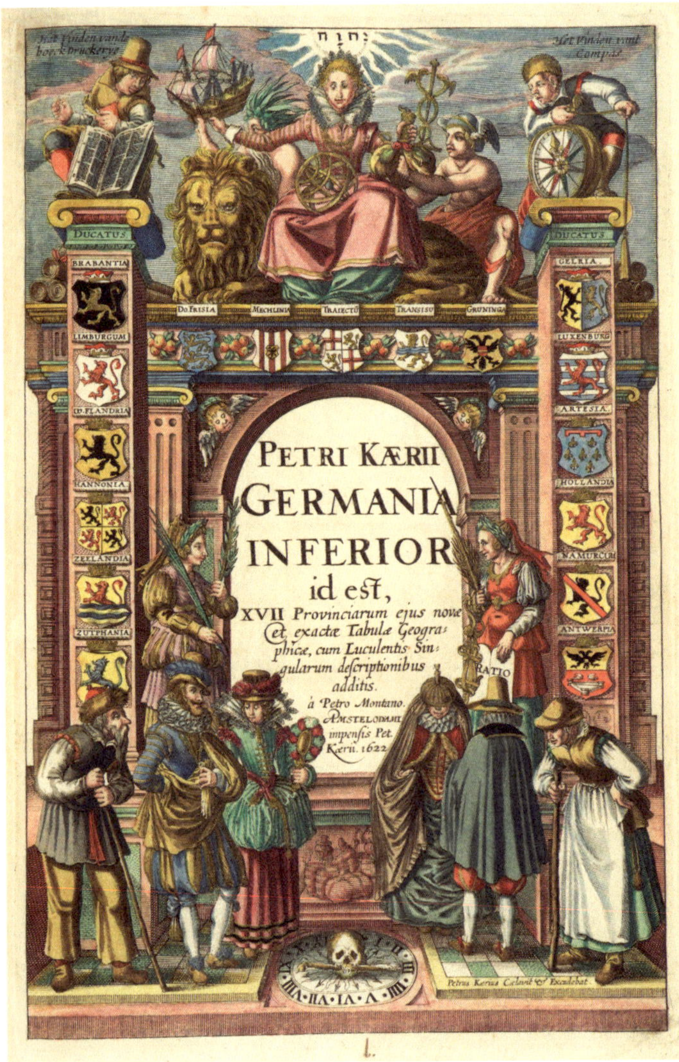

Title page
From: Pieter Van den Keere and Petrus Montanus, *Germania Inferior*, 1622
ANTWERP, THE PHOEBUS FOUNDATION

It was not only people from Antwerp who created portraits and maps of their city: Pieter Van den Keere (1571-*c.*1646) from Ghent designed the *Marchionatus Sacri Romani Imperii* map in 1617, showing Antwerp and the surrounding margravate. He included it in his atlas *Germania Inferior* (Amsterdam, 1617) – the first atlas of the Low Countries. Inspired by Van den Keere's map, Claes Jansz. Visscher (*c.*1587-1652) published a copy of it in 1624, producing a *carte-à-figure* of a kind typical of Amsterdam cartography. Maps like this were decorated around the edges with pictures of figures in costume, for instance, or views of cities (⟶ p. 267, on the map of the African continent the different figures represent the different peoples in stereotypical dress). Visscher added an inset map showing the Western Scheldt and several pictures of major buildings such as the Cathedral of Our Lady, St Michael's Abbey, St Mary's Gate, St Augustine's Church, the commercial exchange, the town hall, the Church of St Charles Borromeo, St George's Gate, the Church of the Brothers of Our Lady and the Hanseatic or 'Easterlings' House. Together, these represented the religious, secular and economic character of the city.

The tendency to combine views of prominent buildings with cartography was also reflected in sixteenth-century urban histories. The Brussels-born Carolus Scribani – philosopher, scholar and rector of Antwerp's Jesuit college – wrote a two-volume history of Antwerp entitled *Antverpia. Origines Antverpiensum* (Antwerp, 1610), a veritable ode to the city and its people, who had grown dear to him after twenty years. The first volume, *Antverpia*, is a sociocultural description of the inhabitants of Antwerp in what is an almost anthropological account. The second volume is a more thorough history extending further back in time to cover the city's origins, as well as discussing its contemporary institutions, administrative culture and buildings. The illustrations of the citadel, cathedral, town hall, exchange and Hanseatic House are all drawn from Lodovico Guicciardini's *Descrittione di tutti i Paesi Bassi* (Antwerp, 1567) (⟶ Chapter VI), but were retouched by Theodoor Galle (1570/1571-1633). An anonymous plan of Antwerp is included in the book, in which the buildings are depicted in elevation (three-dimensionally). It is not just any map: this was the first to contain the names of the roads, making it an early street plan of the city. Note also the dotted lines indicating successive waves of urban expansion.

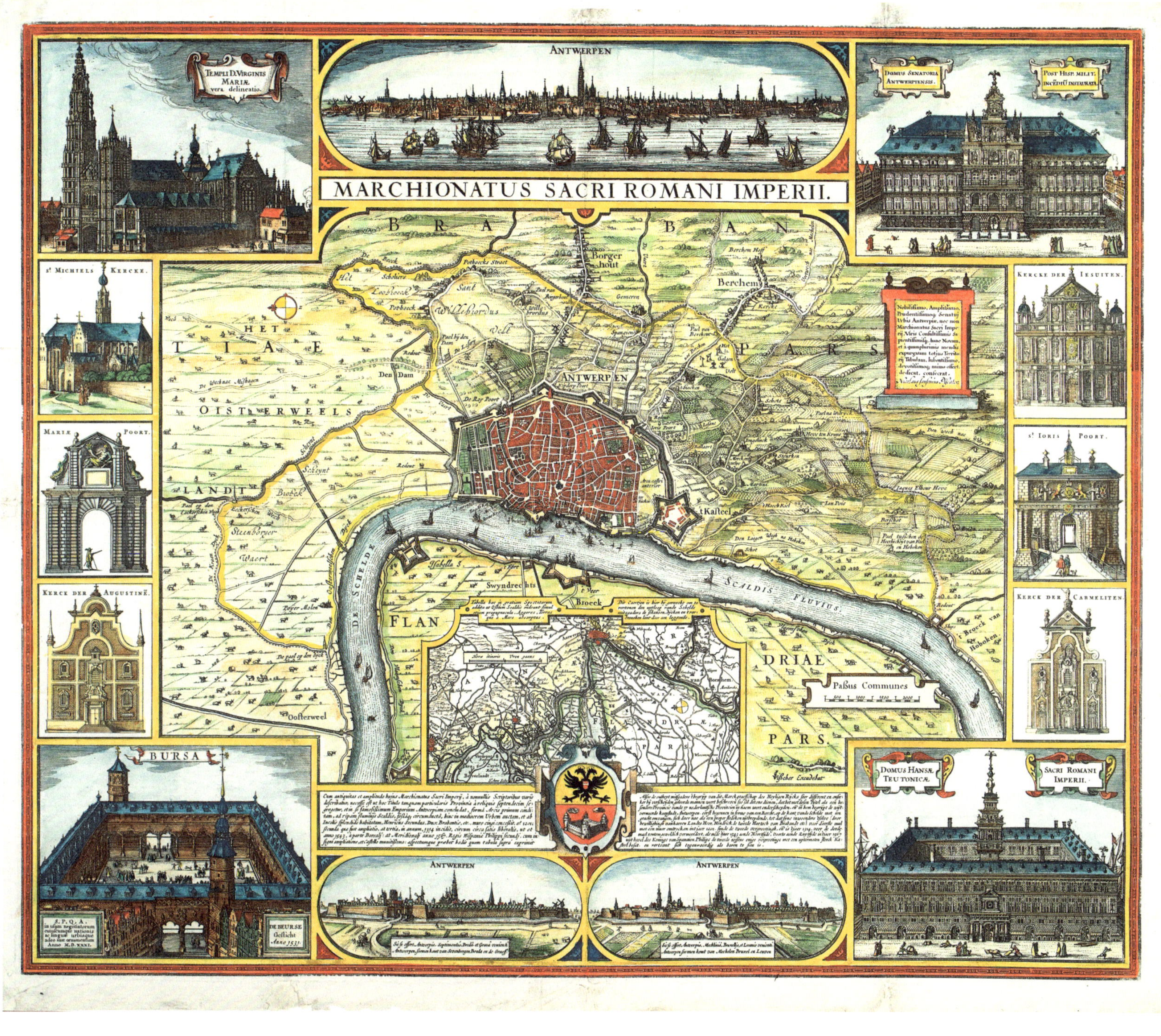

Claes Jansz. Visscher
Marchionatus Sacri Romani Imperii, 1624
Map of the Antwerp area with decorative border
Hand-coloured copper engraving and etching, 492 × 596 mm
ANTWERP, THE PHOEBUS FOUNDATION

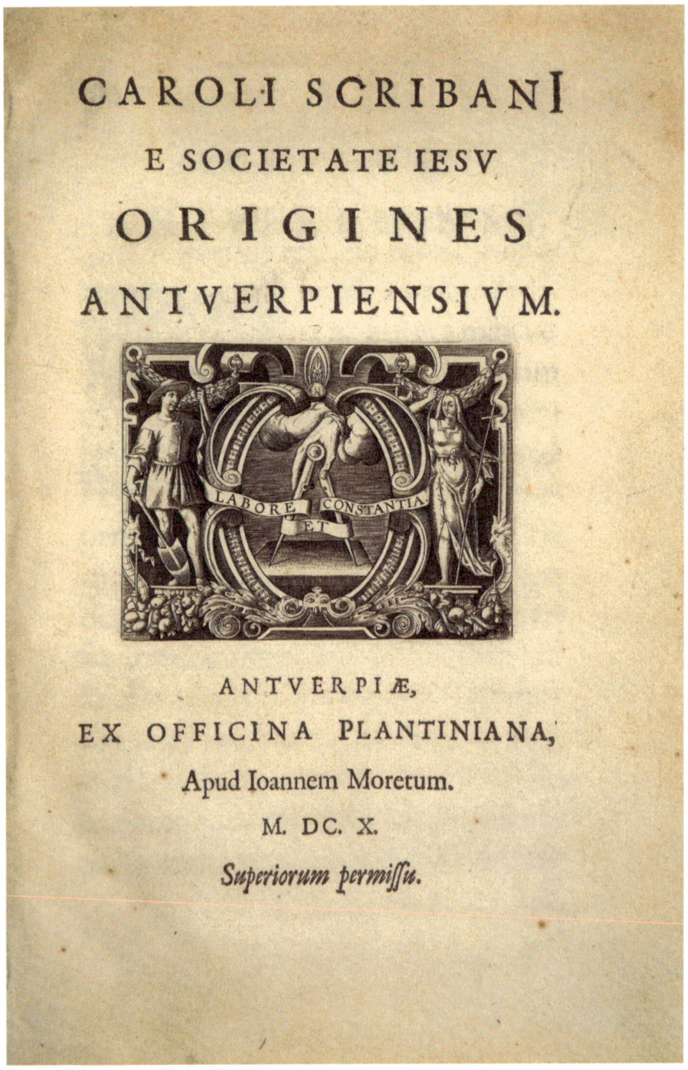

CAROLI SCRIBANI
E SOCIETATE IESV
ORIGINES
ANTVERPIENSIVM.

ANTVERPIÆ,
EX OFFICINA PLANTINIANA,
Apud Ioannem Moretum.
M. DC. X.
Superiorum permissu.

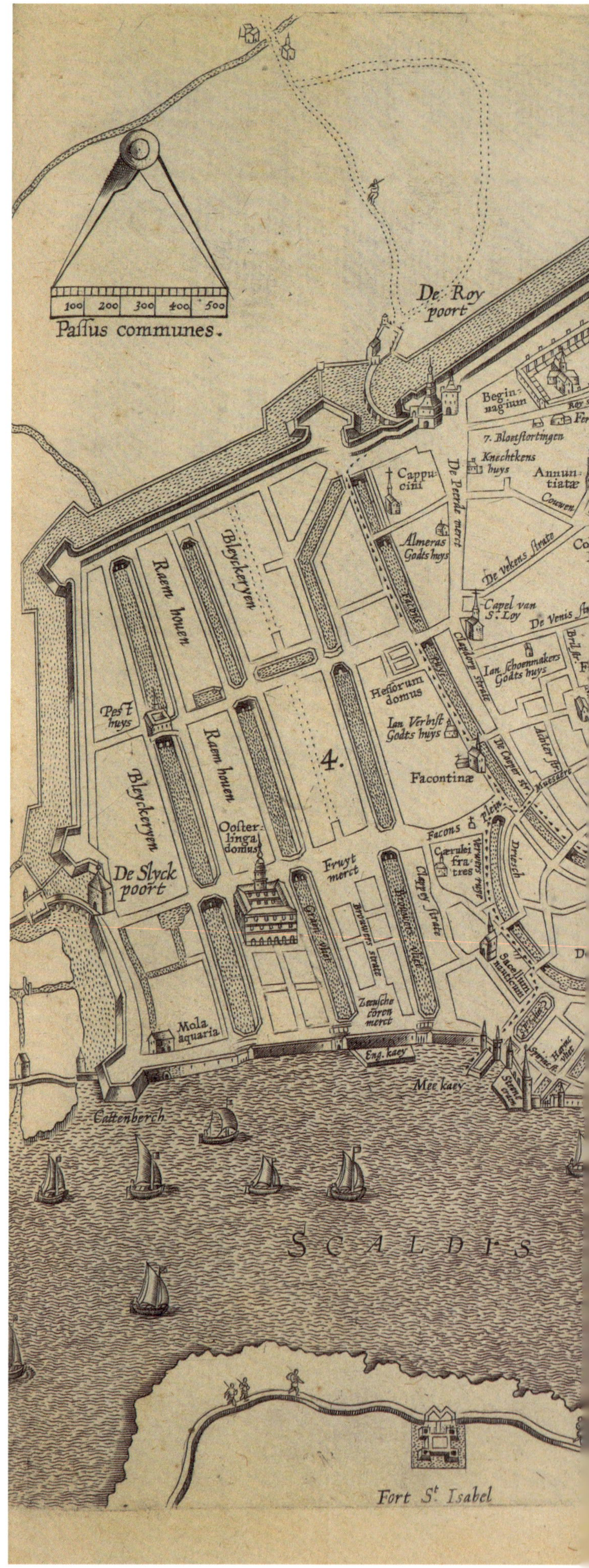

Title page
From: Carolus Scribani, *Antverpia. Origines Antverpiensum*, 1610
ANTWERP, THE PHOEBUS FOUNDATION

Antverpia
Map of Antwerp
From: Carolus Scribani, *Antverpia. Origines Antverpiensum*, 1610
ANTWERP, THE PHOEBUS FOUNDATION

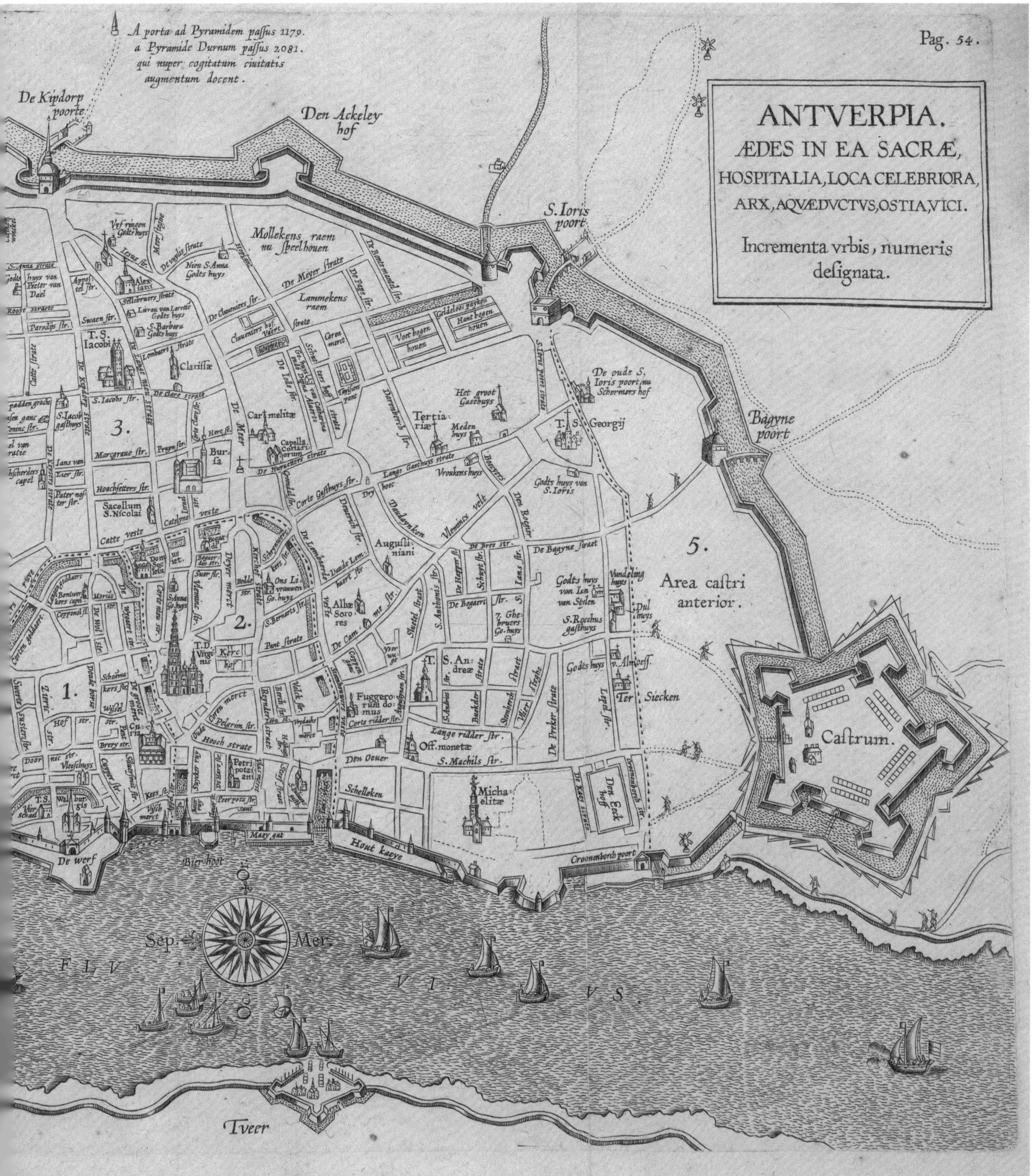

A porta ad Pyramidem passus 1179. a Pyramide Durnum passus 2081. qui nuper cogitatum ciuitatis augmentum docent.
Pag. 54.
ANTVERPIA.
ÆDES IN EA SACRÆ, HOSPITALIA, LOCA CELEBRIORA, ARX, AQVÆDVCTVS, OSTIA, VICI.
Incrementa vrbis, numeris designata.
De Kipdorp poorte
Den Ackeley hof
S. Ioris poort
Bagyne poort
Mollekens raem nu sheelhouen
Het groot Gasthuys
Tertiariæ
T. S. Georgij
T.S. Iacobi
Clarissæ
Carmelitæ
Augustiniani
Albæ Sorores
T. S. Andreæ
Fuggerorum domus
Off. monetæ
Michaelitæ
5.
Area castri anterior.
Castrum.
3.
2.
1.
De werf
Hout kaeye
Croonenborch poort
Sep.
Mer.
F L V
V I
V S.
Tveer

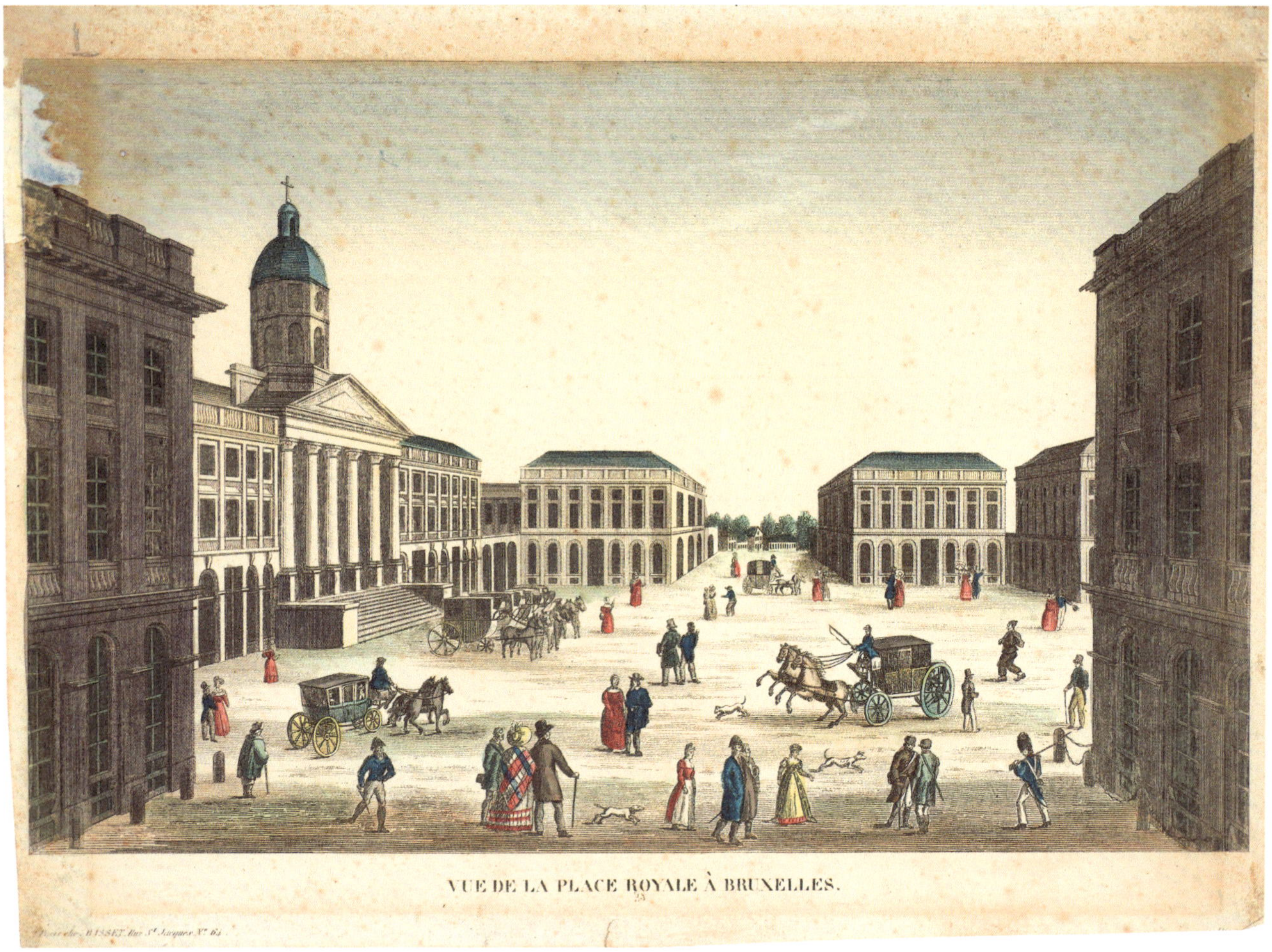
VUE DE LA PLACE ROYALE À BRUXELLES.
23

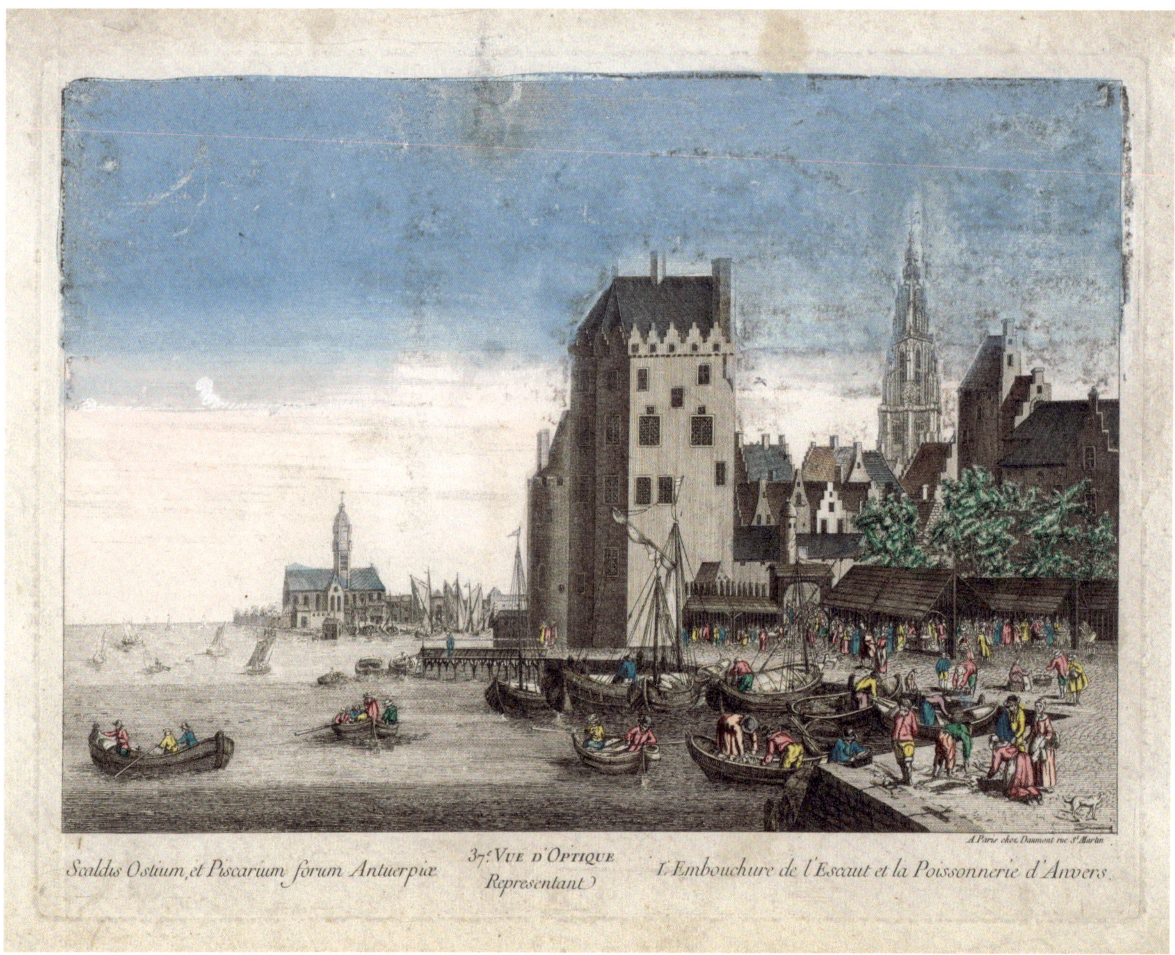
A Paris chez Daumont rue St. Martin
Scaldis Ostium, et Piscarium forum Antuerpiæ
37e. VUE D'OPTIQUE
Representant
L'Embouchure de l'Escaut et la Poissonnerie d'Anvers.

CITY TRIP ON PAPER

In the eighteenth century, people developed a growing taste for travel, including cities abroad. Topography (literally 'writing about a place') offered an alternative way to bring the world to the reader so they could imagine being there in person. The optical or perspective print was a technical innovation which enhanced that experience. The compositions of prints like this were structured according to a strict linear perspective, with all the lines converging at a single vanishing point. A view into the distance formed the central part of the print, which was then viewed using an optical mirror or an illumination cabinet, evoking a powerful depth effect. Viewing devices like this consisted of a single large convex lens with a diameter larger than the distance between the two eyes. The brain interprets the parallel images entering the eyes as a single distant one, creating the illusion of depth. Another angled mirror was often positioned behind it, so that the print could be laid on a flat surface. In this way, well-to-do eighteenth-century citizens at their kitchen table could suddenly imagine themselves at the Place Royale in Brussels, the fish market in Antwerp, St Peter's Square in Rome or the harbour of Manhattan. Prints like this were not very realistic, however, nor were they exactly artistic masterpieces. They were popular, idealised images which allowed the viewer to make an imaginary journey through the landmarks of the world.

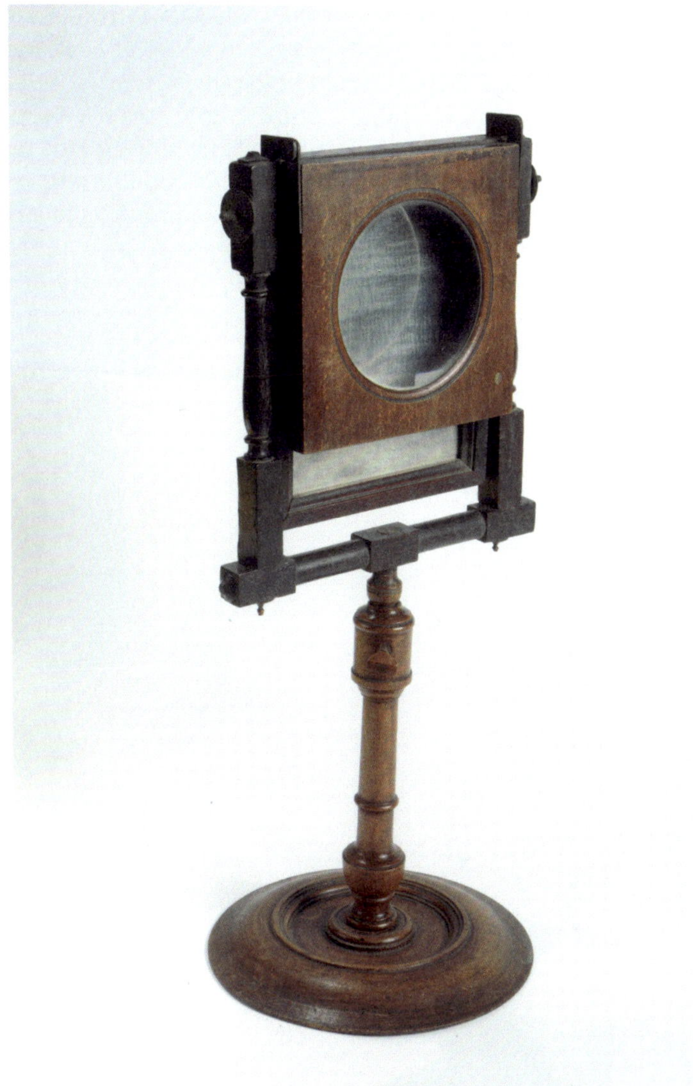

Optica mirror on wooden stand, eighteenth century
Glass, wood, metal, blown, turned, ground, 56.8 × 26.7 × 23.4 cm
LEIDEN, MUSEUM DE LAKENHAL

Basset
Vue de la Place Royale à Bruxelles, c.1730
View of the Place Royale in Brussels
Hand-coloured copper engraving, 300 × 412 mm
ANTWERP, THE PHOEBUS FOUNDATION

Jean François Daumont
Scaldis Ostium et Piscarium forum Antwerpiae, c.1770
Optica print of the Vismarkt in Antwerp
Hand-coloured etching, 350 × 504 mm
ANTWERP, THE PHOEBUS FOUNDATION

MAD ABOUT THE CITY

City trips all over the world – whether in real life or in print – might also have nurtured a love of one's own city. The Phoebus Foundation holds the impressive collection of postcards, city maps, photographs, prints, clippings and other archival material built up throughout his life by the Antwerp urban historian and photographer Wim Strecker (1952-2018). Strecker's passion for the city led him to collect and manage this material with immense care. Whether a picture of a church, a postcard of the roadstead or a tourist city map, they all date roughly from the period 1860-1960. In addition to building his collection, Strecker contributed to city history in several ways: he made a thorough study of statues of the Virgin Mary in the façades of houses and buildings, published an online *Chronicle of Antwerp* and wrote historical guides and descriptions. A veritable walking encyclopaedia, he was keen to share his knowledge with others. The collection is not only the valuable legacy of Strecker himself; it also expresses a love of one's home town that spans centuries.

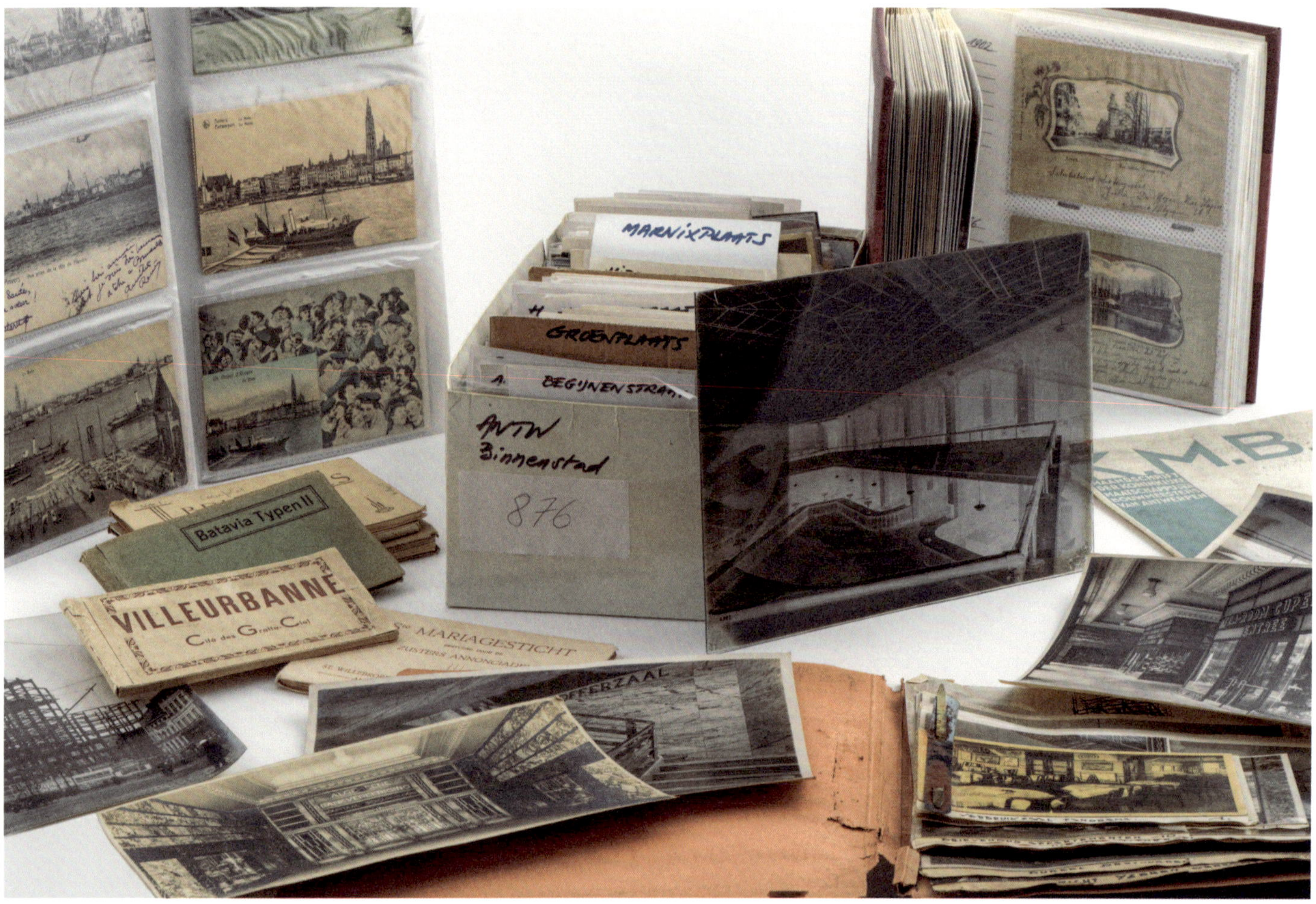

Wim Strecker
Collection of over 10,000 postcards, photographs and archival material from and concerning Antwerp, c.1860-1960
ANTWERP, THE PHOEBUS FOUNDATION

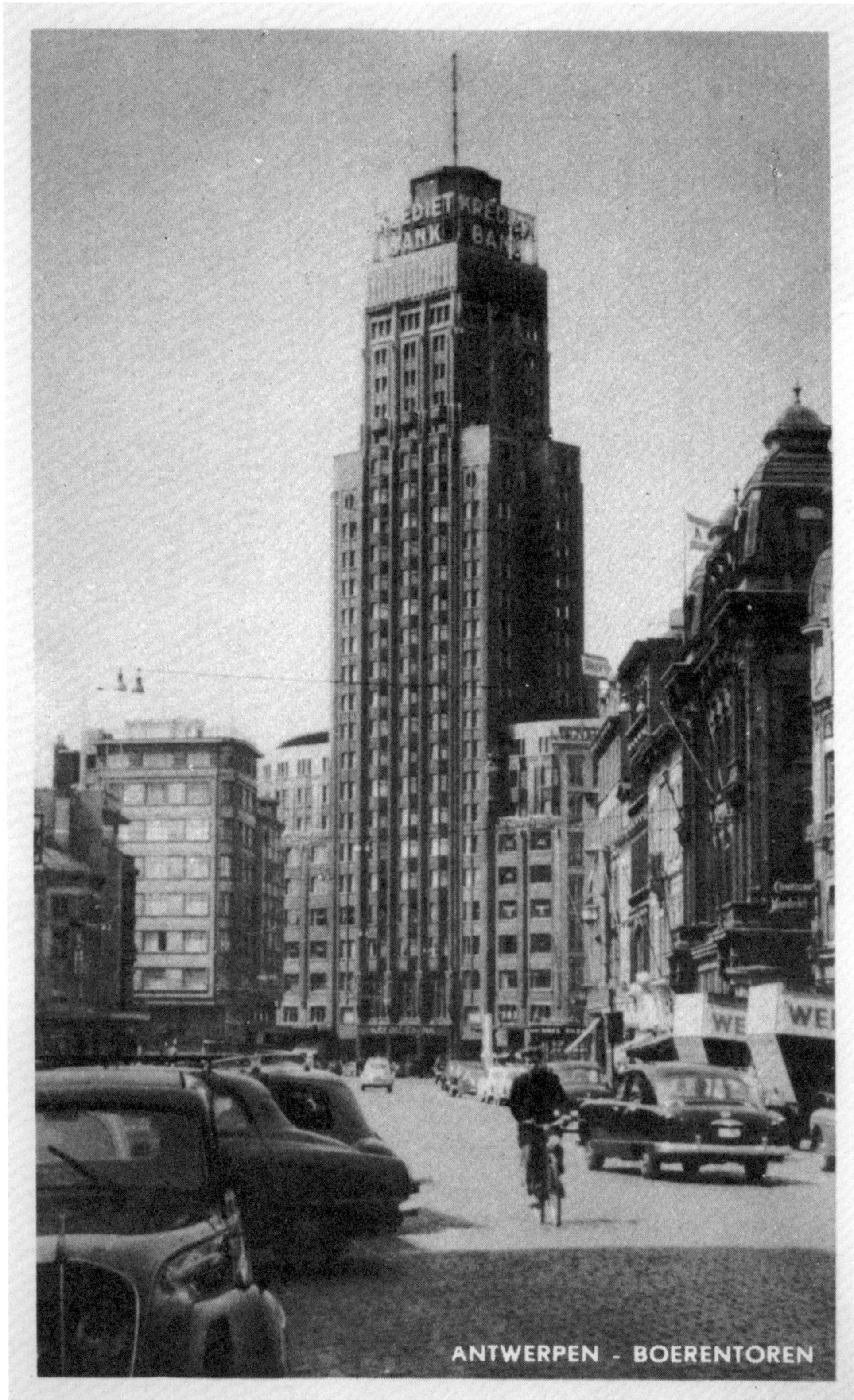

Postcard of the Boerentoren in Antwerp
From: Wim Strecker, collection of over 10,000 postcards, photographs and archival material from and concerning Antwerp, c.1860-1960
ANTWERP, THE PHOEBUS FOUNDATION

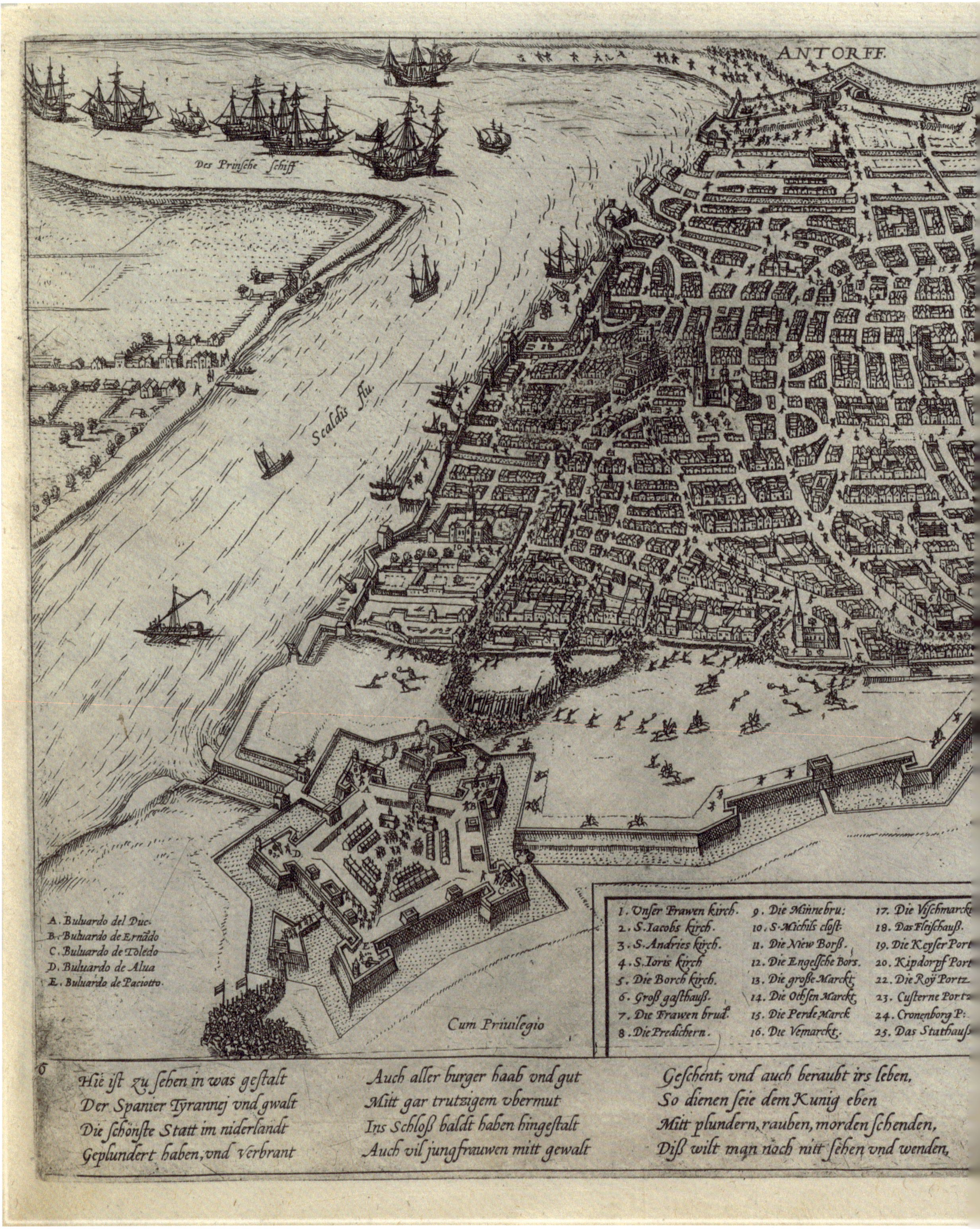
ANTORFF
Des Prinsche schiff
Scaldis flu.
A. Buluardo del Duc.
B. Buluardo de Ernādo
C. Buluardo de Toledo
D. Buluardo de Alua
E. Buluardo de Paciotto.
Cum Priuilegio
1. Unser Frawen kirch.
2. S. Iacobs kirch.
3. S. Andries kirch.
4. S. Ioris kirch
5. Die Borch kirch.
6. Groß gasthauß.
7. Die Frawen brud.
8. Die Predichern.
9. Die Minnebru:
10. S. Michils clost.
11. Die Niew Borß.
12. Die Engelsche Bors.
13. Die große Marckt.
14. Die Ochsen Marckt.
15. Die Perde Marck
16. Die Vemarckt.
17. Die Vischmarck
18. Das Fleischauß.
19. Die Keyser Port
20. Kipdorpf Port
22. Die Roÿ Portz
23. Custerne Portz
24. Cronenborg P:
25. Das Stathaus
6
Hie ist zu sehen in was gestalt
Der Spanier Tyrannej vnd gwalt
Die schönste Statt im niderlandt
Geplundert haben, vnd verbrant
Auch aller burger haab vnd gut
Mitt gar trutzigem vbermut
Ins Schloß baldt haben hingestalt
Auch vil jungfrauwen mitt gewalt
Geschent, vnd auch beraubt irs leben,
So dienen seie dem Kunig eben
Mitt plundern, rauben, morden schenden,
Diß wilt man noch nitt sehen vnd wenden,

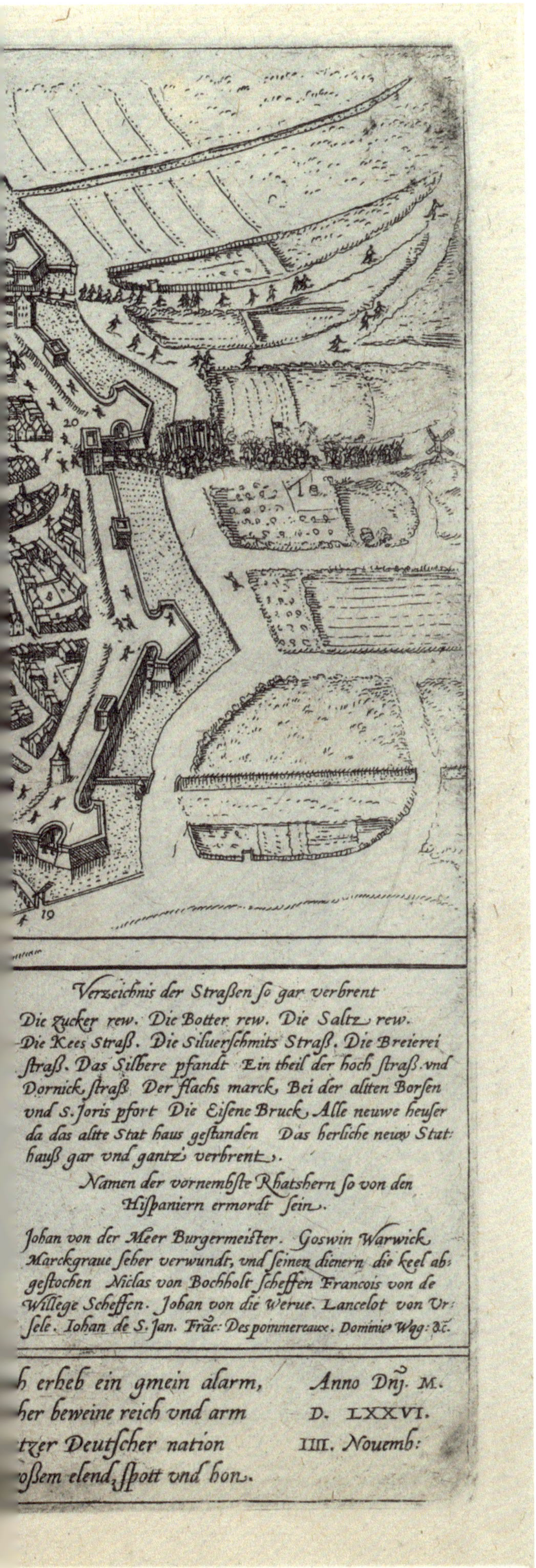

LITERATURE

L. Brouwers, *Carolus Scribani S.J., 1561-1629: een groot man van de Contra-Reformatie in de Nederlanden*, Antwerp, 1961.

G. De Brabander, *Na-kaarten over Antwerpen*, Bruges, 1988.

P. De Laet, *Schelde, Stroom van beelden*, Schilde, 2018.

C.J. Kaldenbach, 'De optica, een leerzaam kijkvermaak', *De Boekenwereld*, 1, 1984-1985, 2-12.

K. Kinds, *Kroniek van de opstand in de Lage Landen, 1555-1609: actuele oorlogsverslaggeving uit de zestiende eeuw met 228 gravures van Frans Hogenberg*, Wenum Wiesel, 1999.

K. Lichtert, J. Dumolyn & M.P.J. Martens (ed.), *Portraits of the City. Representing Urban Space in Later Medieval and Early Modern Europe*, Turnhout, 2014.

P. Lombaerde, *Antwerpen versterkt*, Antwerp, 2009.

P. Lombaerde, K. Geerts & M. Thys, *Antwerpen verbeeld: de Gouden Eeuw in kaart en prent / Antwerp portrayed: the Golden Age in maps and prints*, Wommelgem, 2015.

A. Van Bruaene, B. Blondé & M. Boone, *Gouden eeuwen: stad en samenleving in de Lage Landen (1100-1600)*, Ghent, 2016.

C. Van Lakerveld, *Opkomst en bloei van het Noordnederlandse stadsgezicht in de 17e eeuw*, Bentveld-Aerdenhout, 1977.

Frans Hogenberg
The 'Spanish Fury' in Antwerp, c.1576
Etching, 210 × 280 mm
ANTWERP, THE PHOEBUS FOUNDATION

PRINTED NEWS took off in the Southern and Northern Netherlands in the early seventeenth century. Publishers like Broer Jansz. and Caspar van Hilten in Amsterdam and Abraham Verhoeven in Antwerp were among the first *courantiers* in Europe, publishing their *nieuwstijdinghe* or newspapers several times a week. There was considerable interest in the progress of the Revolt of the Netherlands, in both the region itself and abroad. Besides these concise topical publications, which mostly ran to no more than one double-sided printed sheet, pamphlets were published that discussed specific domestic and international events in greater depth. News prints, periodicals and maps were produced too. The flow of news about all manner of events, in different forms and from various perspectives, cannot have made life easier for the seventeenth-century reader. There must already have been too much for anyone to absorb in its entirety. Perhaps they simply allowed the abundance of information to wash over them, picking out useful or interesting snippets along the way.

Maps too were directly deployed to communicate the news: a news map helped visualise a particular event (or sequence of events), with an accompanying text to expand on the content. They also appeared as illustrations in more comprehensive news publications, such as periodicals, which offered a review of events from a longer period and reflected on recent developments. Publications like this helped readers to digest the news and to think about current affairs, much like the longer in-depth background articles we find in newspaper supplements today.

CHAPTER III

NEWS IN ABUNDANCE

NAVIGATING THE MEDIA LANDSCAPE IN THE SEVENTEENTH CENTURY

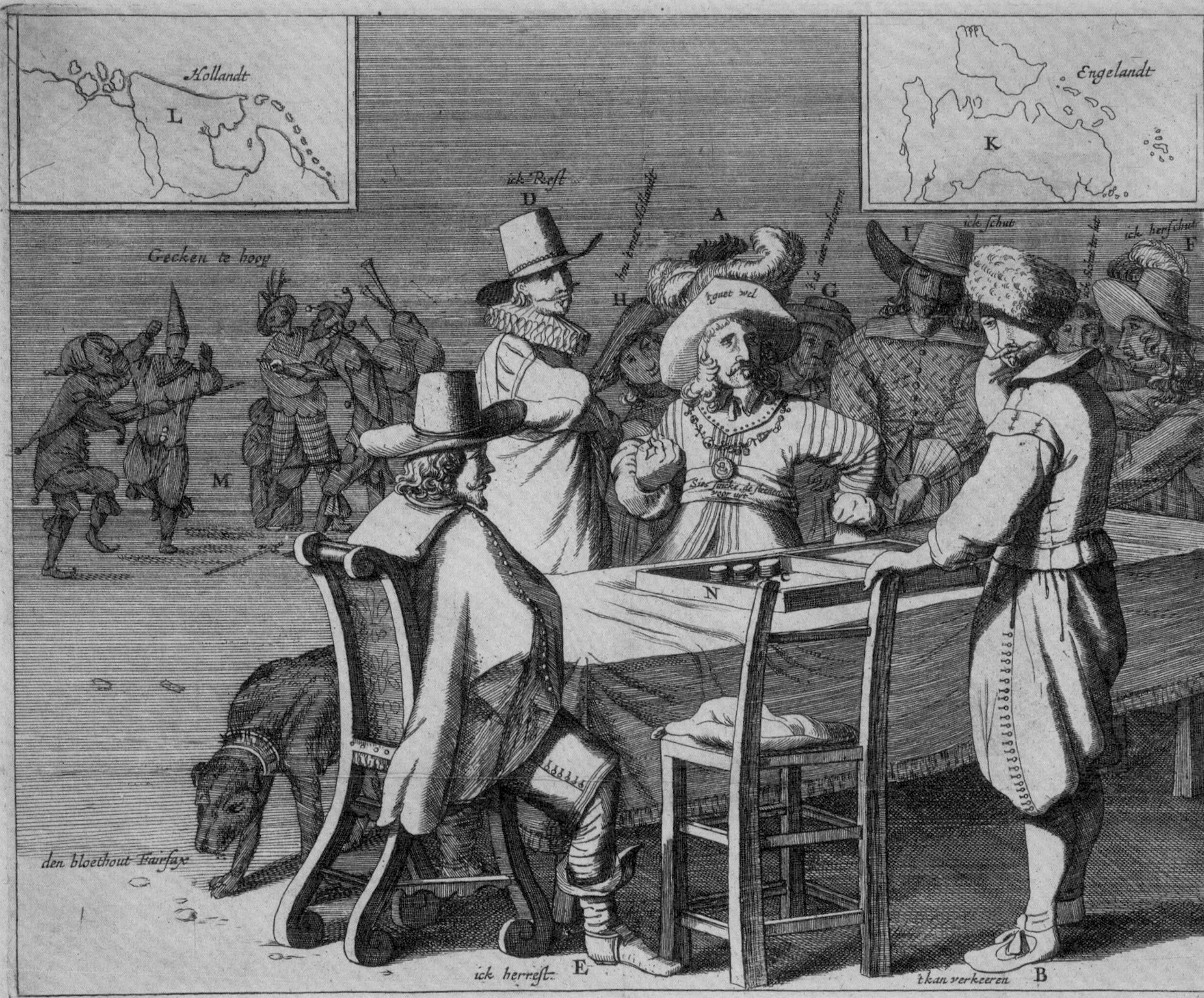

't WONDERLIK VERKEER-SPEL,

By gelegentheydt gepaſt op d'Engelſche en Nederlandſche Oorlogh;

Begonnen op den 22. van Auguſti, in het Jaar 1652.

AAnſchouwers van dit *Zinnebeeld*, ſta ſtil,
En hoor eens wat de Geeſt hier door verbeeld:
Dit is vry meer dan een verſierde gril,
Wijl elk zen deel in 't *Groot Verkeer-bort* ſpeeld;
Want ieder póógt naar winning waar hy kan;
Maar zelden brengt dien arbeidt rijkdom an.

Vraag *Ooſtenrijk*, en *Beemen*, en de *Pals*,
Hoe zuur 'er dit vervloekte Spel bequam.
Dit tijdverdrijf hilp KAREL om den hals;
En ſpoog een vloek op zijn geknotte Stam:
Dus kreeg de (a) *Knecht* de Schijven in zen handt,
En tartte voort het Vreêrijk Nederlandt.

(b) *Hollandtze Koen*, daar *Flip* de Vlag voor ſtreek,
En niet een hoek in 't Bort bezetten kon,
Stondt lang een poos op LEEUWEMONDT en keek,
Als ang voor 't Spel dat STAATSUGHT hier begon:
Maar toen men hem de STOKBEURS luyzen wouw,
Toen ſtak hy meê twee handen uit de mouw.

Den *Ientelman* ſmeet flukx *Sies-Sincq* voor uit;
En maakte zich zoo meeſter van den (c) *Hoek*.
(d) DIEGO, heel verlekkert op den buit,
Riep: *Koen, ik Schut*, den Baas valt u te klock:
Hier vliegt uw *Koe* voorzeeker in de kaars;
Maar KOENRAAD ſchonk den *Ezel* noch geen Vaars.

Hy kreeg terſtondt drie banden op een ry,
En ſloeg, met een, den *Snorkaart* op zen pens.
Door *Quator-Troy*, quam d'arme *Deux-aas* vry:
Voort wirp hy ſtaag Dubletten naar zen wens;
Des wierdt 'et Spel van (e) FRANSZE en (f) DEENE magt,
Met KOEN, HER-SCHUT; daar (g) *Schot*, en (h) *Yr* om lacht.

Den (i) PORTUGEES weet nauwlikx wat hy zal,
Wijl LEEUWEMONDT het *Bordt* ſchier ruimen moet:
Hy hoopt zen luk in KOENRAADTS ongeval;
Doch vrees weêrhoudt zen handen roodt van bloedt:
Want trof de ſchaê hem maar in 't minſte deel,
Hy hield gewis noch *Hoofd-* noch *Halsjuweel*.

Dus gaat dan hier 't begonne Spel zen gank:
Want KOENRAAD vreeſt noch *Loosheidt*, noch *Gewelt*.
De kans *Verkeert*: het *Luk* ruimt met een ſtank,
Daar LEEUWEMONDT zen hoop hadt op geſtelt.
Des wordt, terwijl elkx harte klopt, en jaagt,
Gants (k) ENGELANDT, en (l) HOLLANDT hier gewaagt.

Maar och! wie wenſcht alreê niet naar het endt;
Of dat me 't Spel, eer 't halzen koſt, liet ſteeken?
Het (m) *Zottewerk* van een pampiere prent,
Toont ons de Schets van 't al te ſchaadlik wreeken.
Men geve dan 't (n) VERKEER-BORDT an de vlam,
En ieder zy, in 't KRISTENDOM, een LAM.

Liefde verwint.

NEWS PRINTS

It was not only through the printed word that people responded to the latest events: in addition to newspapers, pamphlets, journals and news reports, social and political debate could also be sparked by news prints, which satirised particular figures or groups in a pictorial and often symbolic way, accompanied by an explanatory text or verse. *'t Wonderlik Verkeer-spel*, for instance, criticised the First Anglo-Dutch War (1652-1654). The man in the middle is England's Oliver Cromwell (A), who challenges Holland (B) to a game of backgammon, while Spain (D), France (E) and other European countries look on in dismay. The maker of the print was clearly not in favour of the war: 'for each seeks after victory, yet rarely finds prosperity'.

The two small maps inset at the top show that maps were already being used to spread news. The rise of commercial cartography in the Low Countries, with Amsterdam and Antwerp as the leading production centres, placed all manner of maps at the fingertips of printmakers and publishers, who unabashedly used and imitated each other's work. Maps – whether or not lifted from others – could serve as illustrations to depict relevant places but also to support and help explain the text.

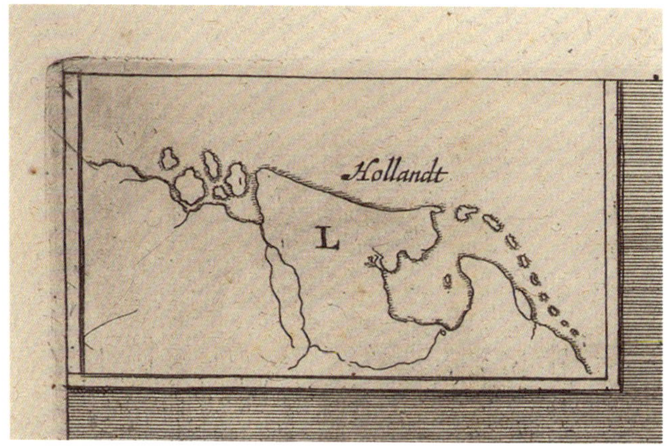

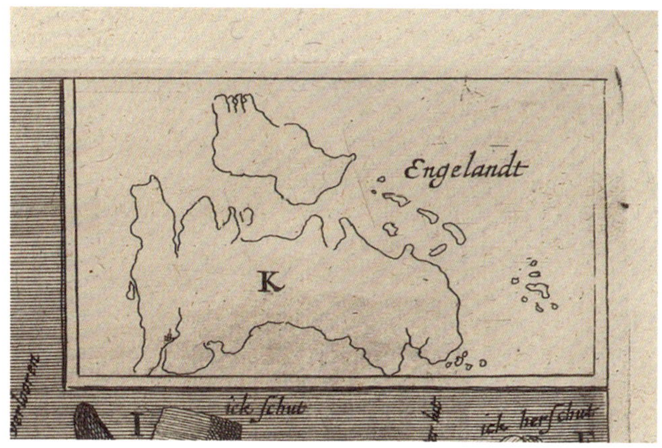

***'t Wonderlik Verkeer-Spel*, 1652**
Political news print on the Anglo-Dutch War
Copper engraving and etching, 460 × 535 mm
ANTWERP, THE PHOEBUS FOUNDATION

Details from: *'t Wonderlik Verkeer-Spel*, 1652 (p. 88)

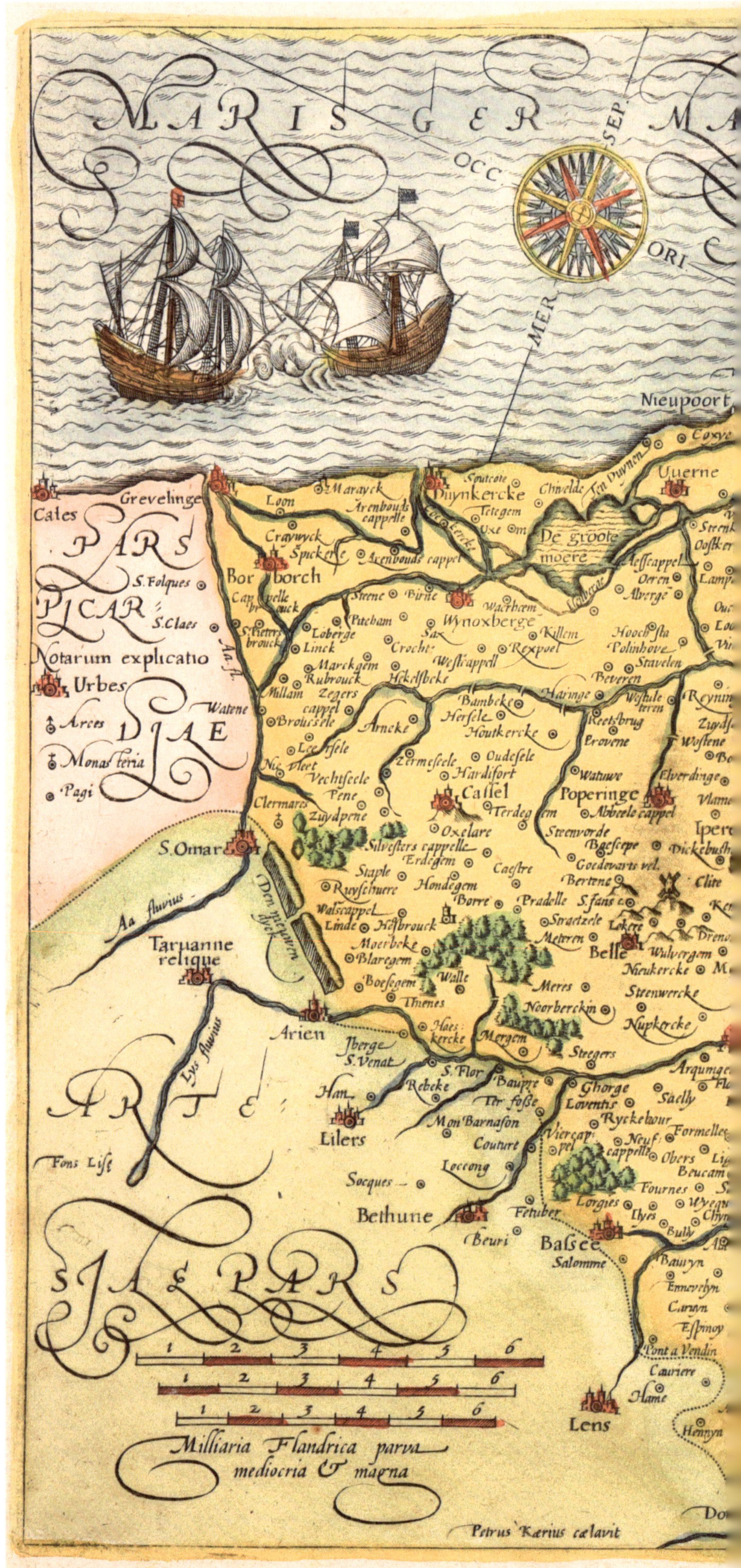

Pieter Van den Keere
Flandria, after 1612
Map of Flanders
Hand-coloured copper engraving, 315 × 374 mm
ANTWERP, THE PHOEBUS FOUNDATION

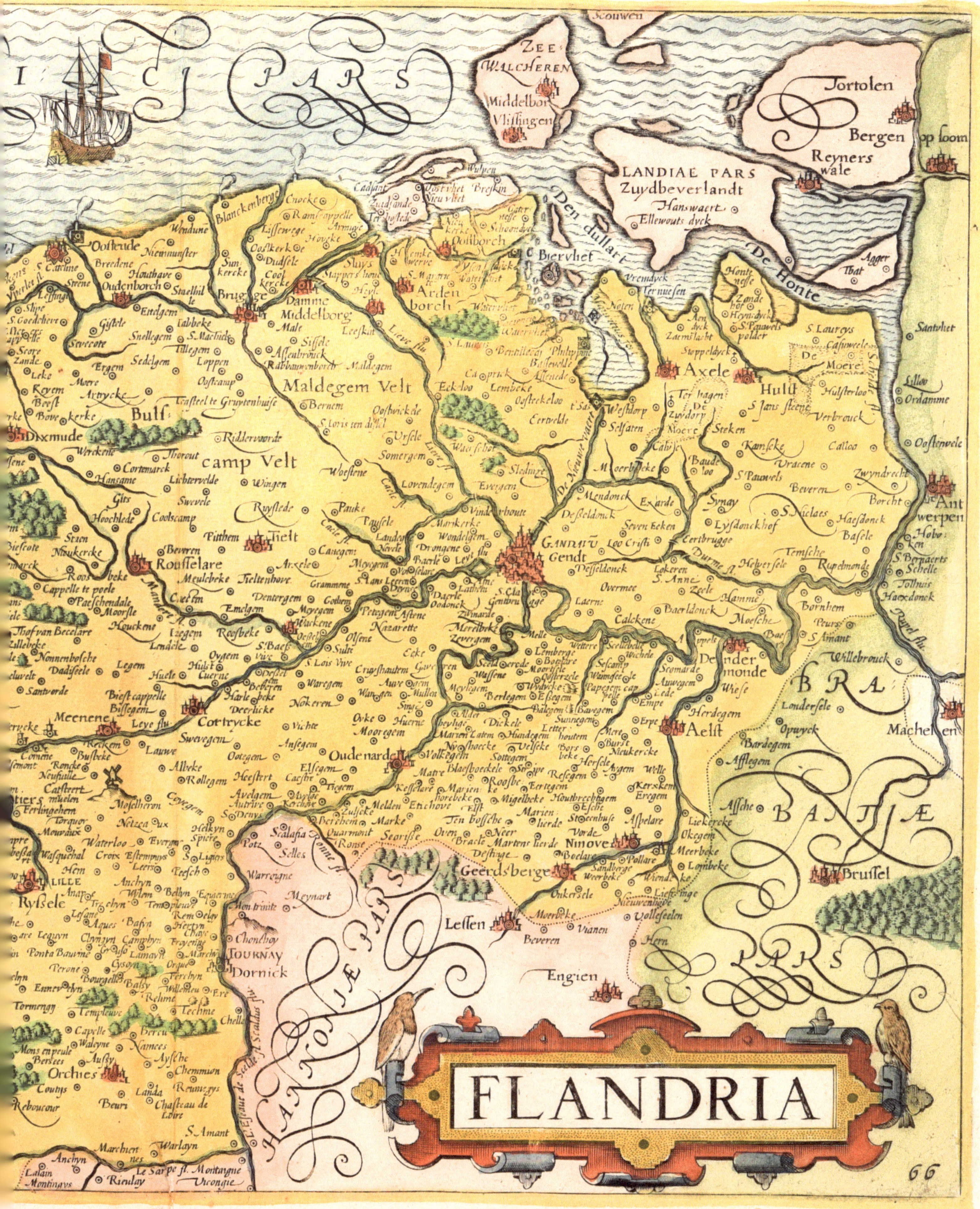
FLANDRIA
ZEE
WALCHEREN
Middelborch
Vlissingen
Scouwen
Tortolen
Bergen op Zoom
Reyners wale
LANDIAE PARS
Zuydbeverlandt
Hanswaert
Ellewouts dyck
Den dullart
De Honte
Biervliet
Axele
Hulst
Brugge
Damme
Middelborg
Maldegem Velt
Oostende
Oudenborch
Blanckenberge
Dixmude
Bulscamp Velt
Thielt
Roosselare
Meenene
Cortrycke
Oudenaerd
Gandavum
Gendt
Dendermonde
Aelst
Ninove
Geerdsberge
Lessen
Engien
Brussel
Antwerpen
Machelen
TOURNAY
Dornick
LILLE
Rysele
Orchies
HANNONIÆ PARS
BRABANTIÆ PARS
66

MAPS IN PAMPHLETS

Pamphlets were typically opinionated and polemical, and frequently served as an ideal medium through which to communicate a controversial standpoint and influence public opinion. A good example is the *Antwoorde op drie vraghen, dienstich en noodich in dese tijden van vervolginghe en Religie-twist* ('A helpful and necessary answer to three questions in these times of persecution and religious dispute'; 1621), which called for moderation in the face of religious conflict in the Dutch Republic between 'Remonstrants' and 'Counter-Remonstrants'. The author of the pamphlet seeks to reconcile the two factions who found themselves at odds in the period in question. He or she took the wise precaution of publishing this bold point of view anonymously.

Besides the polemics, however, pamphlets could also offer background and more in-depth analysis of news events. The maps of Flanders by Pieter Van den Keere, with a separately printed German text accompanying on the left, are taken directly from one such pamphlet. They were inserted in an anonymous German edition dealing with the Siege of Ostend between 1601 and 1604 (⟼ p. 94). Foreigners – including German-speaking merchants and other interested parties – were very much part of the readership, as testified by the large number of German-language news publications that were distributed. *Belägerung der Statt Ostende: Iournal: Tagregister und eigentliche Beschreibung* ('Siege of Ostend: Journal: Day-by-day report and factual description'; 1604) is amply illustrated with pictures added separately in two appendices, the first covering the beginning of the siege and the second its final year in 1604. The map of Flanders and that of the campaign of Maurice of Orange (1567-1625), stadholder and commander of the Republic's army, were both depicted by Van den Keere. They come from the second part of the appendix, where they were followed by several maps of the sieges of Sluis (1604) and of Ostend, a letter, and an account of the events. The main text of the pamphlet refers to the maps several times.

The *Flandria* map provides a large overview of the region in which it was all happening, since a German-speaker in the Holy Roman Empire who did not have direct experience of the events and was far removed from them would first have needed a picture of Flanders, where the Siege of Ostend took place. The explanatory accompanying text provides mostly general topographical and historical information about the county and a few individual Flemish cities, some of which dates from as far back as the twelfth century. The historical and topographical overview is also linked to current events: the description of the fortified town of Nieuwpoort refers to the battle of 1600. The map allowed readers to 'hover' above the place names and events in order to place recent happenings in their historical and geographical context. Those wishing to discover more about Flanders and its cities were referred in the pamphlet to the work of Lodovico Guicciardini (⟼ Chapter VI).

Van den Keere's next map is narrative in character, visualizing the story of the events on the map itself. The reader's attention is drawn to the Dutch Republic's fleet off Vlissingen (Flushing), which had set course for the island of Cadzand at the end of April with a huge army. The fleet heaves into the map with a 3D effect, edging its way through the inlet called the Zwarte Gat ('Black Hole'). We also see groups of soldiers landing, digging in and pitching camp south of Sluis – the town that the Dutch commander, Stadholder Maurits of Nassau, would besiege and capture later that summer.

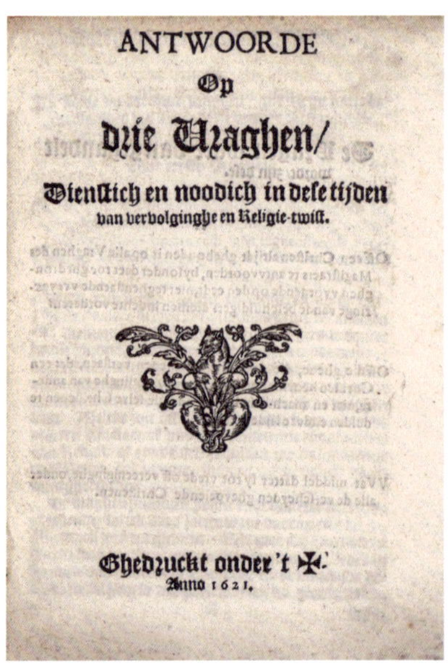
ANTWOORDE
Op
drie Vraghen/
Dienstich en noodich in dese tijden
van vervolginghe en Religie-twist.

Ghedruckt onder 't ✠.
Anno 1621.

Title page
From: *Antwoorde op drie Vraghen, Dienstich en noodich in dese tijden van vervolginghe en Religie-twist*, 1621
ANTWERP, THE PHOEBUS FOUNDATION

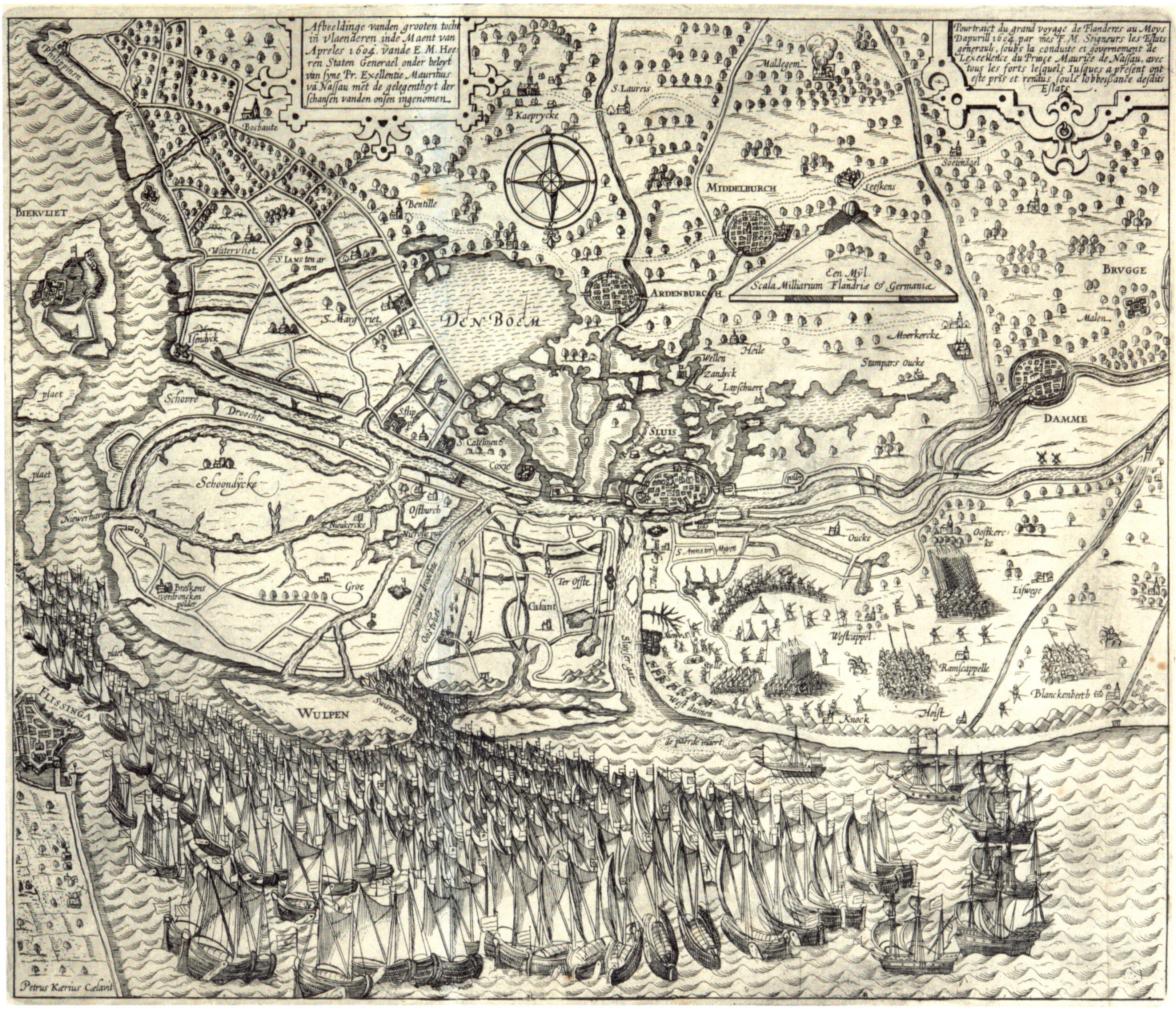

Pieter Van den Keere

Afbeeldinge vanden grooten tocht in Vlaenderen inde Maent van Apreles 1604, 1604

News map of the campaign of Maurice and the States army in Flanders in 1604

Etching, 345 × 500 mm

ANTWERP, THE PHOEBUS FOUNDATION

THE BLOODY SIEGE OF OSTEND

The Eighty Years War was, to a large degree, a campaign of successive sieges. Cities throughout the Low Countries were alternately besieged and relieved. In some cases, the attackers successfully encircled and captured the town, while in others, the defenders, possibly with allied troops coming to their aid, were able to ward off the attack. Numerous maps were created to tell the story of these events, offering a narrative depiction of the course of the siege.

One of the most notorious sieges was that of Ostend (1601-1604), because it lasted so long and cost so many lives. Campaigns of this kind were normally over within a few weeks or months, but this siege lasted as long as several years. The town on the North Sea coast was the only place in Flanders still held by the Republic of the Seven United Netherlands around 1600. When Archduke Albrecht and the Infanta Isabella arrived in the Southern Netherlands, one of the first major tasks facing them was to conquer Ostend too. They were eager to bring the Flemish town back into the Catholic camp, as it provided the Republic with a base from which to attack the area directly. Albert's 'Army of Flanders', consisting mainly of Spanish and Italian troops, surrounded Ostend in July 1601. The Republic had England as an ally, and given the many different interests that were at stake, the eyes of Europe were fixed on the siege.

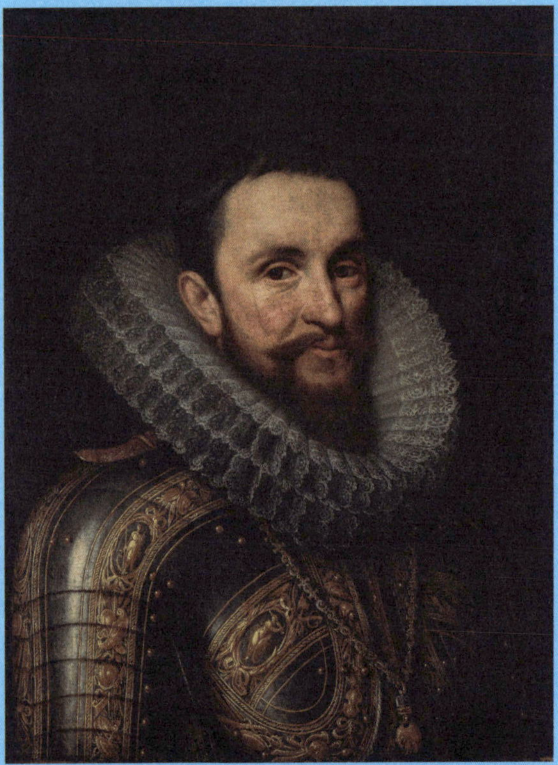

Portrait of Ambrosio Spinola as a Knight of the Golden Fleece, seventeenth century
Oil on canvas, 63.1 × 49.6 cm
ANTWERP, THE PHOEBUS FOUNDATION

King Philip III of Spain (1578-1621) began to lose faith in Albert's military capabilities and dispatched the Italian general and strategist Ambrosio Spinola (1569-1630) to the conflict zone, following which progress began to be made. Spinola successfully captured Ostend, partly by cutting off its supplies and also by assaulting and bombarding the town with greater ferocity. While the siege ended in victory for the Spanish Army of Flanders, it is hard to speak of a winner, given the tens of thousands of casualties on both sides and the damage running to millions of guilders – an astronomical sum for the time. The destruction inflicted on Ostend could not, anyway, be expressed simply in words or numbers: the battle came to be known not only as the bloodiest siege of the Eighty Years War but also as one of the costliest campaigns of attrition. In the aftermath of the event, the immensity of the losses assumed mythical proportions, capturing the imagination of many people in the years that followed. Contemporaries compared the campaign to the siege and fall of Troy in classical antiquity because of its prolonged and gruelling nature.

For all their consistency with the text, Van den Keere's maps were not made specifically for the pamphlet: they were multi-purpose creations that also featured in atlases, for example. The *Flandria* map is a copy of one that the Antwerp publisher Christophe Plantin (1520-1589) had already published in 1577 as part of his *Spieghel der wereld* ('Mirror of the World') atlas, and the same map of Flanders appears several times in French- and Dutch-language editions of Lodovico Guicciardini's famous description of the Low Countries (⟶ Chapter VI). The German engraver Georg Keller (1568-1634) was also involved in this evolving chain of maps: in 1605, he engraved one showing the iconic fleet at the bottom making its way to Sluis, but he also expanded the 'story' to include the area around Damme and continued the map on the right to Bruges and Ostend.

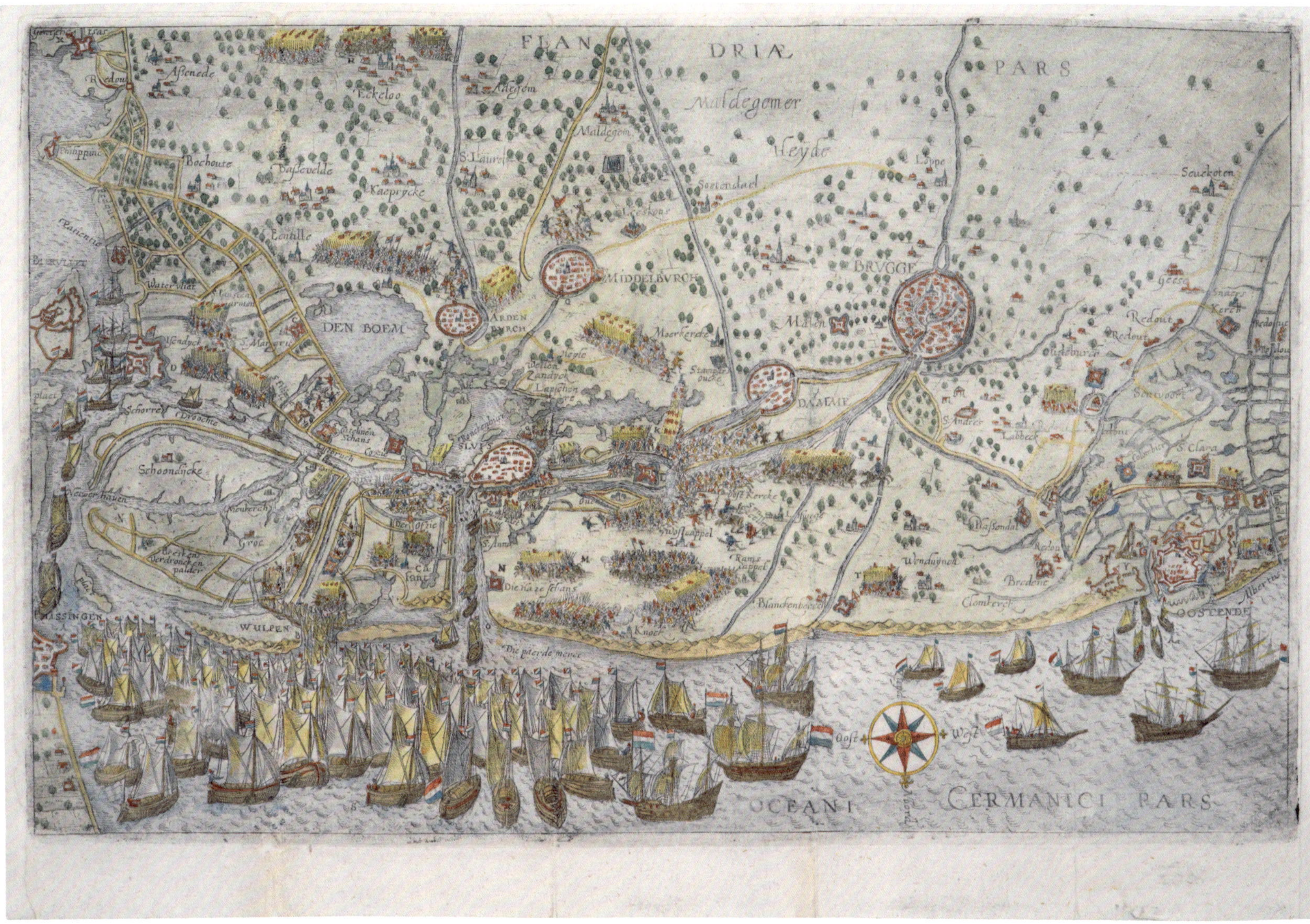

Georg Keller
Map of the Franc of Bruges from Biervliet to Ostend and the Siege of Sluis, 1605
Hand-coloured etching, 255 × 380 mm
ANTWERP, THE PHOEBUS FOUNDATION

PERIODICALS

Besides weekly reports in newspapers, periodicals were printed at monthly, six-monthly or annual intervals, offering a summary of news from a particular period. The best-known Dutch example in the seventeenth century was the *Hollandse Mercurius*, published once a year in Haarlem from 1651 onwards. Compiling periodicals was less subject to time pressures than publishing a newspaper, and so they were often illustrated: allegorical title prints, portraits of prominent figures and views of naval battles decorated their pages. Publishers drew on news prints and maps that had already appeared, sometimes adapting the images to create an overview and impose order on chaos.

Matthäus Merian (1593-1650), for instance – a German engraver based in Frankfurt and publisher of the periodical *Theatrum Europaeum* from 1633 onwards – engraved himself a summary illustration for his book. The Phoebus Foundation has a loose copy of the print of the Battle of Kallo in 1638 from the issue published in 1639 and covering the years 1633-1638. Merian based himself on two news prints by the Antwerp artist Bonaventura Peeters I (1614-1652), published by Joannes Meyssens (1612-1670). The States army had attempted to recapture Antwerp, but the assault had failed. A month after this resounding Spanish victory, the city commissioned Peeters to make maps of the campaign, the result of which was hung between the windows and the fireplace in the Hall of State in Antwerp town hall. The civic authorities immediately placed a follow-up order too: a large painting of the battle, which Peeters made in collaboration with his brother Gillis (1612-1653) and which was also given a prominent place in the Hall of State.

Peeters's prints were widely distributed and seemingly also came to Merian's attention. The publisher and engraver employed the neat trick of reducing the size of Peeters's print and map, and placing one above the other to form a single whole. The upper part of Merian's print – a view of the battle – is little changed compared with the original, but he did something remarkable with the map underneath. Peeters's original news map comprised a single large narrative document showing the fortifications and the river Scheldt where the battle took place, with a small overview map at the bottom featuring the Kruisdijk between Fort Verrebroek and Fort Kallo. Merian copied the latter map and inserted a drawing of the assault from the topmost print. In this way, and without too much difficulty, he turned two maps into one with all the essential information and an abbreviated legend by way of explanation.

Dramatic elements like Count William of Nassau-Siegen being literally dragged through the mud at the top by Spanish soldiers were omitted in the copy, possibly because sensational and provocative details of this kind were a little too partisan? It is possible that Merian, as an outsider, had a more neutral stance towards the events, or perhaps he considered the humiliating elements too specific for his more reflective news medium. Whatever the case, this example shows how the maps in news periodicals reflected both textually and visually on the proliferation of topical material in the seventeenth century, deliberately heightening some things while leaving others out.

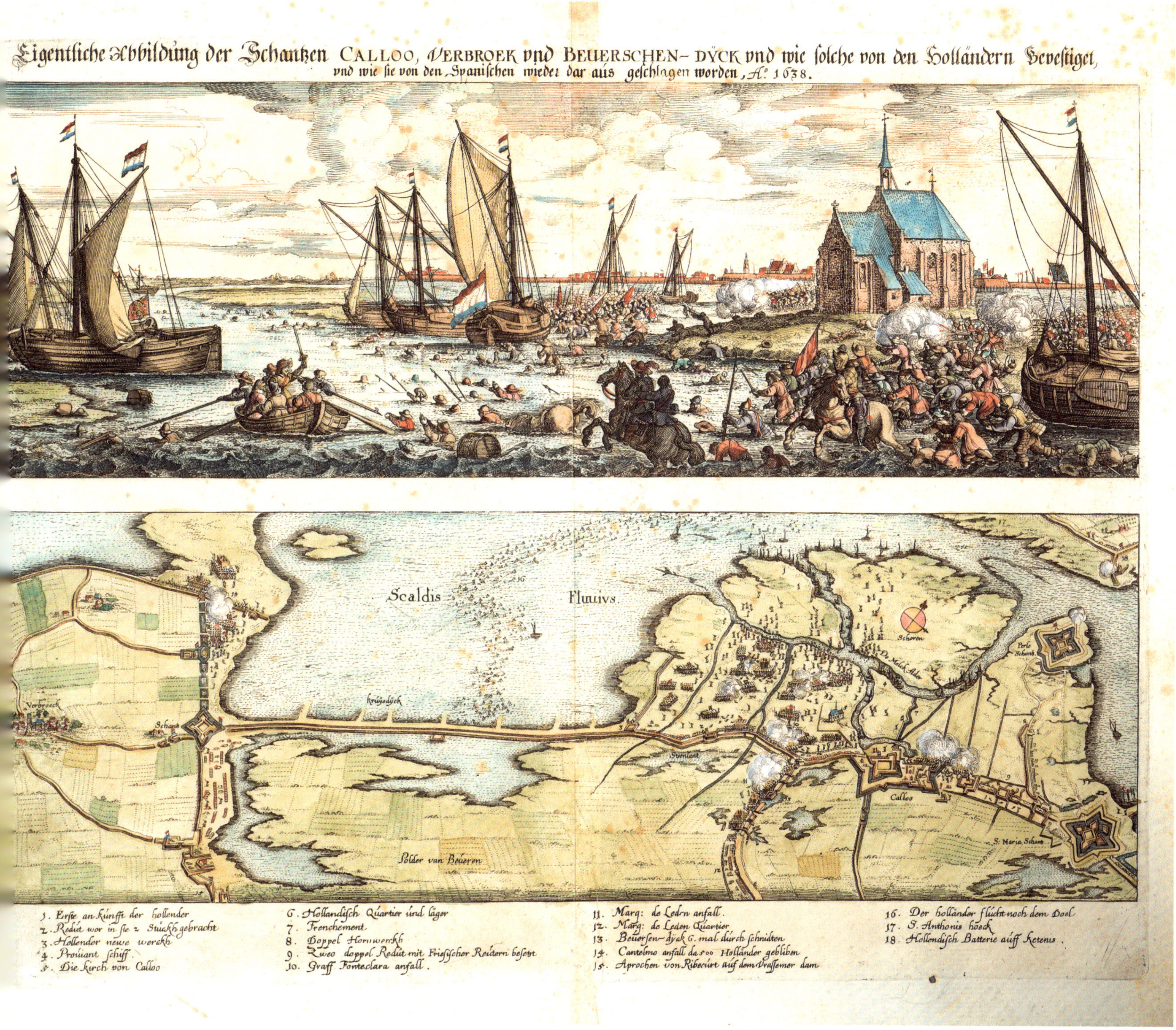

Matthäus Merian after Bonaventura Peeters I
Eigentliche Abbildung der Schantzen Calloo, Verbroek und Beverschen-Dyck, 1639
View and map of the Battle of Kallo in 1638
Hand-coloured etching, 330 × 387 mm
ANTWERP, THE PHOEBUS FOUNDATION

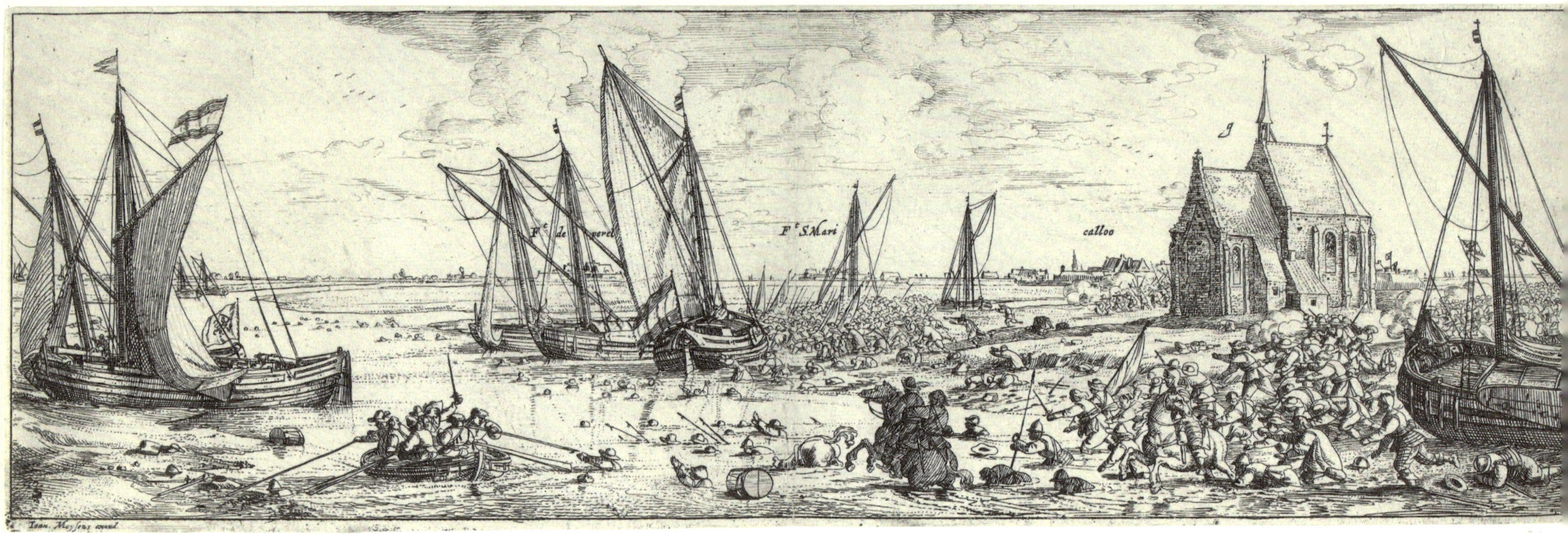

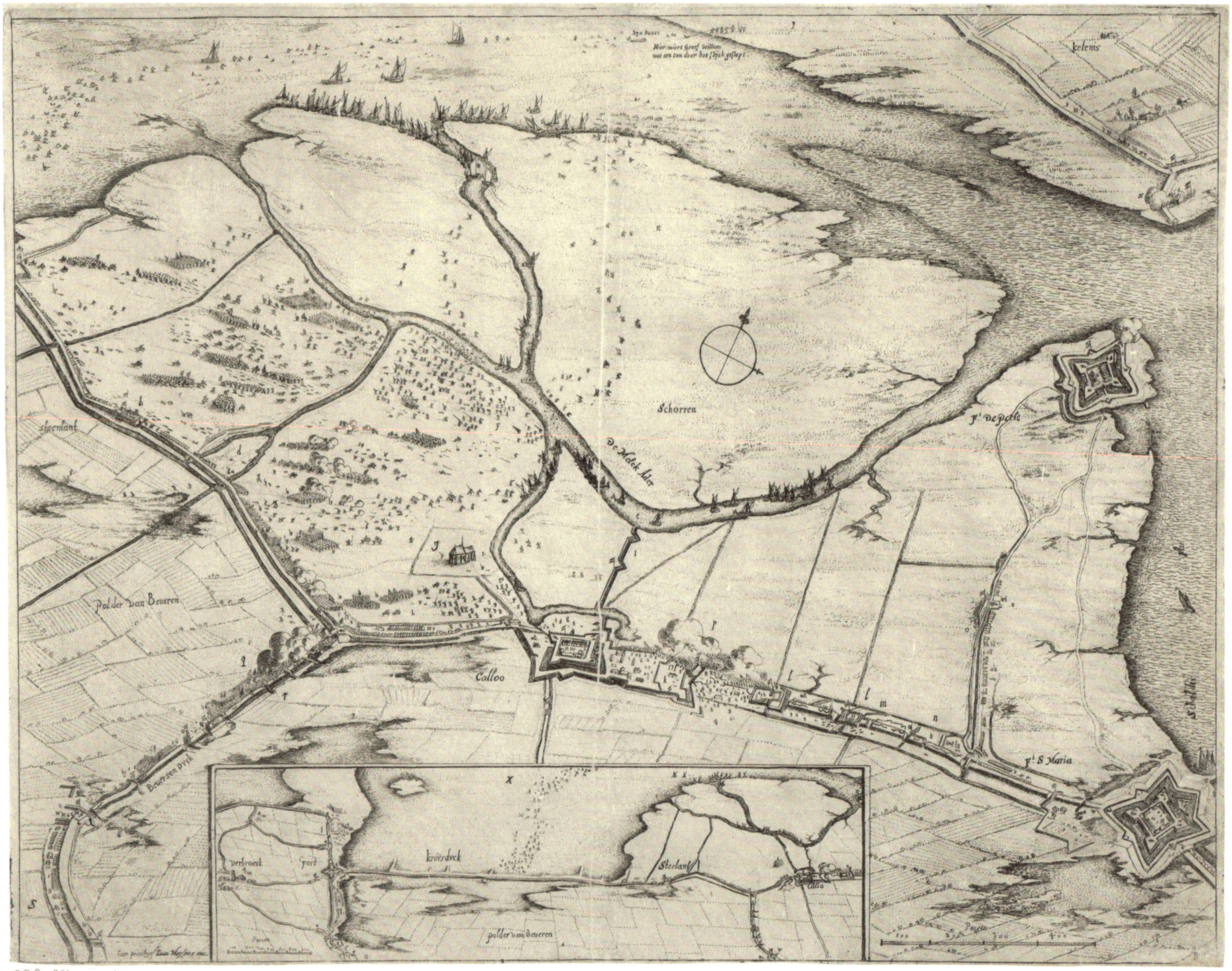

Bonaventura Peeters I
Perfecte Af-beeldinghe ende gheleghentheyt vande Forten Calloo, Verbroek en Beverschen-dijck, 1638
Map of the area around Fort Kallo, showing the failed assault by the States army on 21 June 1638, 1638
Etching, 367 × 480 mm
AMSTERDAM, RIJKSMUSEUM

Soldiers of the States army fleeing after the unsuccessful assault on the hill at Kallo, 21 June 1638
Part of: Bonaventura Peeters I, *Perfecte Af-beeldinghe ende gheleghentheyt vande Forten Calloo, Verbroek en Beverschen-dijck*, 1638
AMSTERDAM, RIJKSMUSEUM

LITERATURE

J. de Kruif & M. Meijer Drees, *Het lange leven van het pamflet: boekhistorische, iconografische, literaire en politieke aspecten van pamfletten 1600-1900*, Hilversum, 2006.

D. de Vries & C. van den Heuvel, *Oostende verloren, Sluis gewonnen, 1604: een kroniek in kaarten*, exhib.cat., Leiden, University Library Leiden, 2004.

H. Deys, M. Franssen, Vincent Van Hezik et al., *Guicciardini Illustratus. De kaarten en plannen in Lodovico Guicciardini's Beschrijving van de Nederlanden*, 't Goy-Houten, 2001.

K. Dillen & B. Vannieuwenhuyze, 'Bedrieglijke eenvoud: Flandria Borealis tussen kaart en historie, tussen afbeelding en uitbeelding', *Tijdschrift voor geschiedenis*, 130, 4, 2017: 521-543.

J. Koopmans, *Het nieuws verbeeld: Oorlog en vrede in de titelprenten van de Europische Mercurius (1690-1750)*, Hilversum, 2021.

M. Merian, *Theatrum Europaeum [...] Theatri Europaei Continuatio III: Das ist: Historischer Chronicken Dritter Theil, In sich begreiffend Ein kurze und warhaffte Beschreibung aller vornehmen, Denck- und Chronickwürdigen Geschichten, so sich hin und wieder in der gantzen Welt [...] von Anno 1633 biß 1638 [...]*, Frankfurt, 1639: 918.

A. Simoni, *The Ostend Story: Early Tales of the Great Siege and the Mediating Role of Henrick van Haestens*, 't Goy-Houten, 2003.

G. Verhoeven & S. van der Veen, *De Hollandse Mercurius; een Haarlems jaarboek uit de zeventiende eeuw*, Haarlem, 2011.

T. Werner, *De val van het nieuwe Troje: Het beleg van Oostende 1601-1604*, Leuven, 2004.

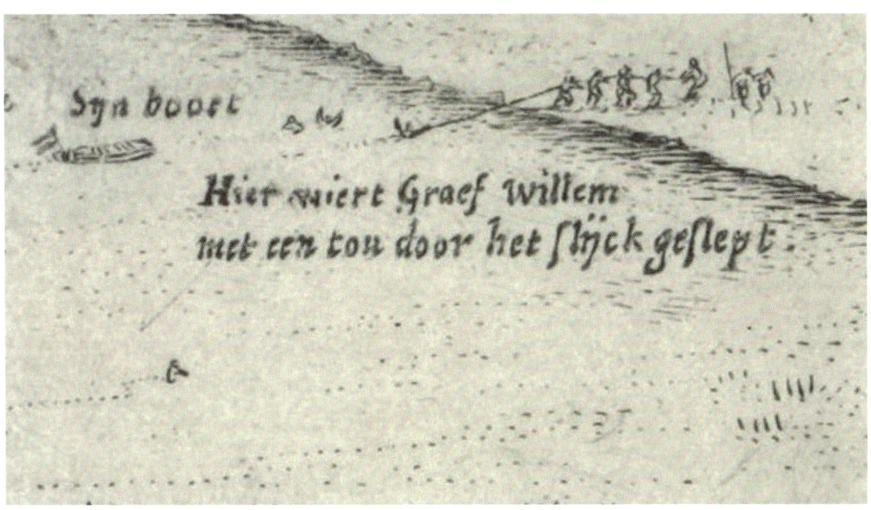

Count William dragged through the mire
Part of: Bonaventura Peeters I, *Perfecte Af-beeldinghe ende gheleghentheyt vande Forten Calloo, Verbroek en Beverschen-dijck*, 1638 (p. 98)

Wall map of Flanders after Fricx

This six-sheet, monumental wall map of the county of Flanders shows the area from the North Sea coast and the Western Scheldt to Thérouanne and Mons. According to the title, the map was produced by Etienne-Louis Crépy (died 1759), who published a series of copies by the Brussels map seller and printer Eugène Henri Fricx (1644-1730). Fricx was *imprimeur de sa Majesté* (printer to His Majesty), who used his royal privilege to make his name with city plans and topographical maps of the Low Countries. He was not a cartographer himself but one of the most important map publishers in the Southern Netherlands in the eighteenth century. Crépy expanded his map of Flanders to include the most recent observations made in 1744. It is decorated with a large cartouche including the coat of arms of the French king. Nine detailed town plans – Lille, Ypres, Tournai, Douai, Gravelines, Ostend, Oudenaarde, Nieuwpoort and Dunkirk – have been added at the bottom.

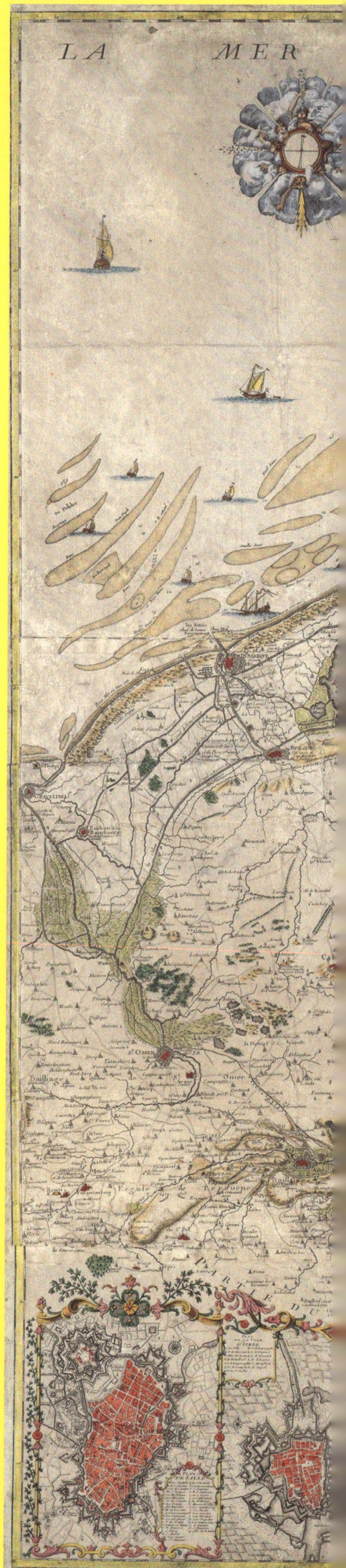

Etienne-Louis Crépy after Eugène-Henri Fricx
Comté de Flandre Dressé Sur les Memoires de Eugène Henry Fricx et Augmenté sur les Observations les plus Nouvelles en 1744, 1744
Wall map of Flanders
Hand-coloured copper engraving, 1125 × 1293 mm
ANTWERP, THE PHOEBUS FOUNDATION

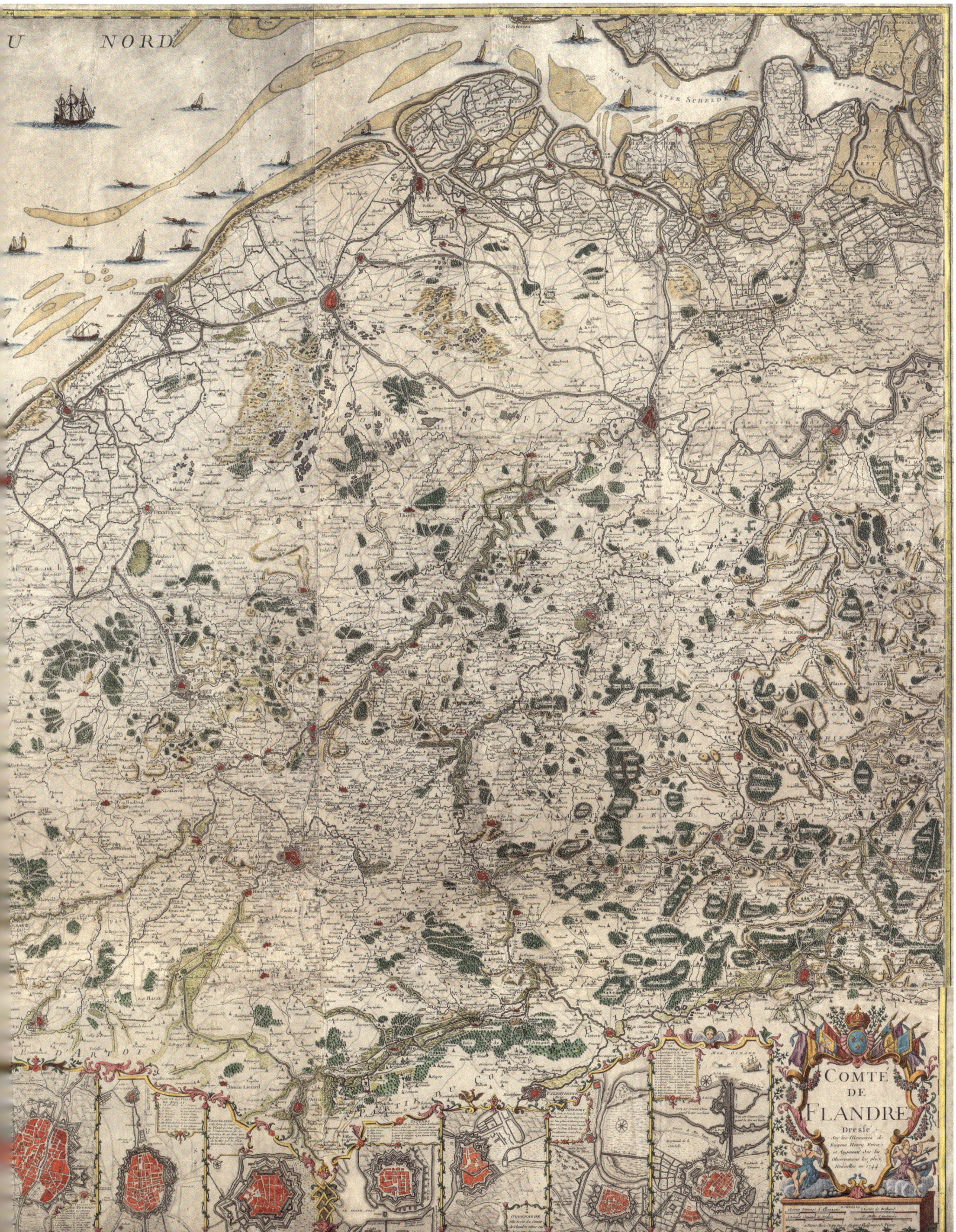
NORD
COMTE DE FLANDRE

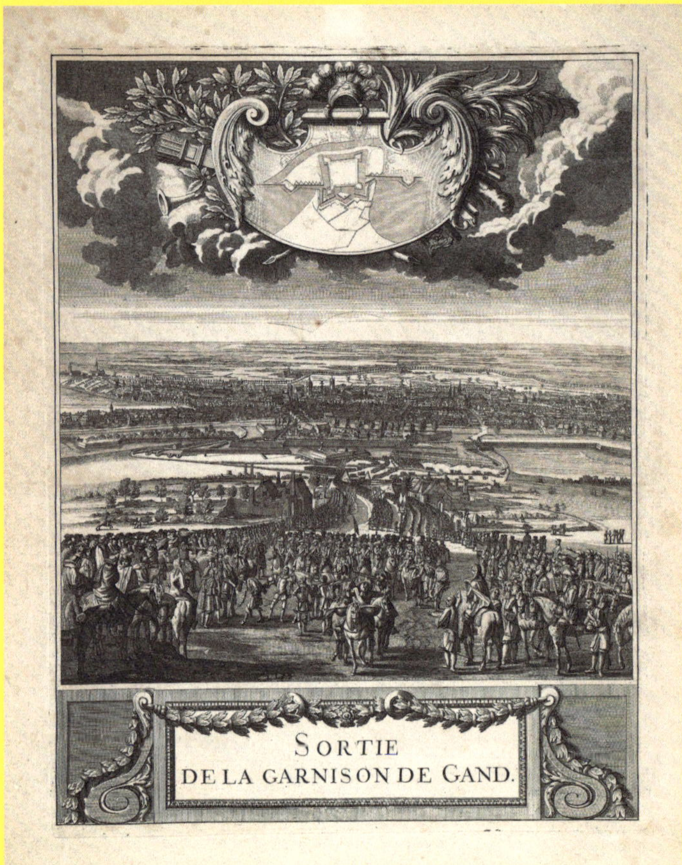

Louis XIV captures Ghent

These two prints show the capture of Ghent by Louis XIV (1638-1715) during his great European military campaign. They originally formed part of a monumental set devoted to his triumphs, *Les petites conquêtes de Louis XIV: plans et profils* (Paris, 1672-1693). It was a deluxe publication in folio format consisting of large prints depicting Louis XIV's campaign across European cities, most notably along the Rhine and in France and Belgium. This particular volume was number fifteen in a grandiose series called the *Cabinet du Roi*, which ran to 23 volumes and included some eight hundred prints celebrating the power and magnificence of Louis XIV, whose royal library oversaw the engraving, printing and distribution of the images. The best artists of the time, among them Charles Le Brun (1619-1690), Sébastien Le Clerc (1637-1714) and Jean Dolivar Cruyl (1641-1692), were commissioned to design prints of the royal palaces and gardens (Louvre and Versailles), the art collection and the king's military campaigns, and it is to these that the views of Ghent belong. The series positively drips with propaganda and glorification. A set, mostly bound in red or blue morocco and large in size (roughly 62.5 by 45 centimetres), was extremely expensive and not many were produced. The king obviously received a copy of the series, and his ambassadors presented a number of them as diplomatic gifts to foreign dignitaries and rulers.

Gand by Sébastien Le Clerc shows the siege of Ghent by Louis XIV. The print contains an inset map of the fortified walls of the captured city and an explanatory text. The propagandist intent is obvious straight away: the text lavishly praises Louis's clever tactics, the cunning with which he fooled his enemies and the speed (eight days) with which he captured the city and the fortress in March 1678. *Sortie de la garnison de Gand* by Jean Dolivar Cruyl shows the defeated garrison exiting the city after the surrender, with a view of the citadel located to the south. The defeated troops can be recognised by their crossed flags. The print features a cartouche with a map of the citadel, the fort abandoned by the defending troops as they march out of the city.

Jean Dolivar Cruyl
Sortie de la garnison de Gand, c.1672-1693
View of Ghent with the exit of the garrison after surrender
Copper engraving and etching, 522 × 452 mm
ANTWERP, THE PHOEBUS FOUNDATION

Sébastien Le Clerc
Gand, c.1672-1693
View of the siege of Ghent by Louis XIV
Copper engraving and etching, 523 × 452 mm
ANTWERP, THE PHOEBUS FOUNDATION

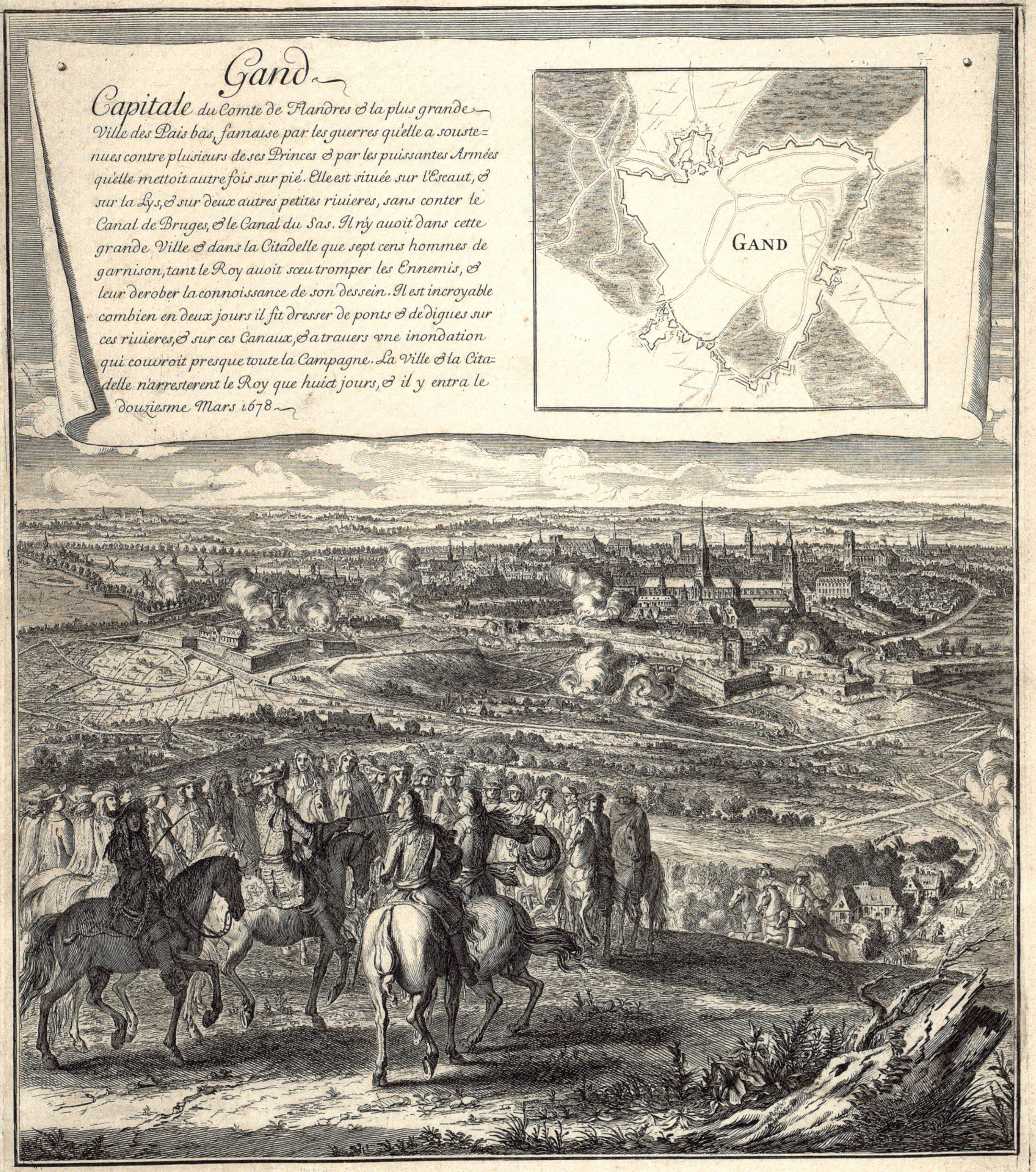
Gand
Capitale du Comte de Flandres & la plus grande
Ville des Pais bas, fameuse par les guerres qu'elle a souste=
nues contre plusieurs de ses Princes & par les puissantes Armées
qu'elle mettoit autrefois sur pié. Elle est située sur l'Escaut, &
sur la Lys, & sur deux autres petites riuieres, sans conter le
Canal de Bruges, & le Canal du Sas. Il n'y auoit dans cette
grande Ville & dans la Citadelle que sept cens hommes de
garnison, tant le Roy auoit sceu tromper les Ennemis, &
leur derober la connoissance de son dessein. Il est incroyable
combien en deux jours il fit dresser de ponts & de digues sur
ces riuieres, & sur ces Canaux, & a trauers vne inondation
qui couuroit presque toute la Campagne. La Ville & la Cita=
delle n'arresterent le Roy que huict jours, & il y entra le
douziesme Mars 1678.
GAND

At first sight, *Caerte van 't Scheldt ende Santvliet* appears to be a detailed and reliable map of the course of the river Scheldt from Dendermonde to Bergen op Zoom. The river winds horizontally through the cartographic image, the compass pointing east (north is left), while the words 'Brabant' and 'Vlaenderen' clearly identify the areas through which it flows. The title in full – 'Chart of the Scheldt and Zandvliet, showing the drowned, inundated land, newly formed marshes and creeks in and through the same between Bergen op Zoom and Antwerp' – suggests a river map in which the cartographer focused on meticulously mapping all the waterworks around the Scheldt, without omitting the smallest ditches, flooded zones or salt and freshwater marshes.

Yet nothing is what it seems: if we zoom in on the map a little closer, we see that there is much more going on in the river delta. A string of boats sails along the Western Scheldt towards Saftingergat (near Fort Sint Anna), a line of soldiers marches with horses and wagons along the road towards Zandvliet, and various dotted lines and texts run through the cartographic image. This is not so much a river map, therefore, as a news map dealing with the events of the Eighty Years War (1568-1648) or the Revolt of the Netherlands against Spanish rule. It shows the most recent developments in the conflict between the Dutch and the Habsburg Spanish for control of this strategic location between the territory of the Republic and the Spanish Netherlands. The words 'Improved from the news by Claes Jansz. Visscher, 1632' in the title cartouche should thus be taken very literally: this map was updated based on the latest reports. Visscher ensured in this way that his map remained newsworthy.

Mapmakers and publishers from the Southern Netherlands like Pieter Verbiest II (1607-1693) also updated their maps. This meant that publishers did not have to concern themselves unduly about inaccuracies, as they could correct the map later by issuing an improved version or a new 'state'. Re-engraving the copper plate was a convenient technique for cartographers, as it saved time and money. This was not only the case for news maps – atlases too were updated when the publisher became aware of a spatial, geographical change. Maps were and continue to be dynamic documents that move with the rhythm of news and historical and geographical realities.

CHAPTER IV

FRESH OFF THE PRESS

CARTOGRAPHIC UPDATES FOR THE BATTLE ON THE SCHELDT (1630s)

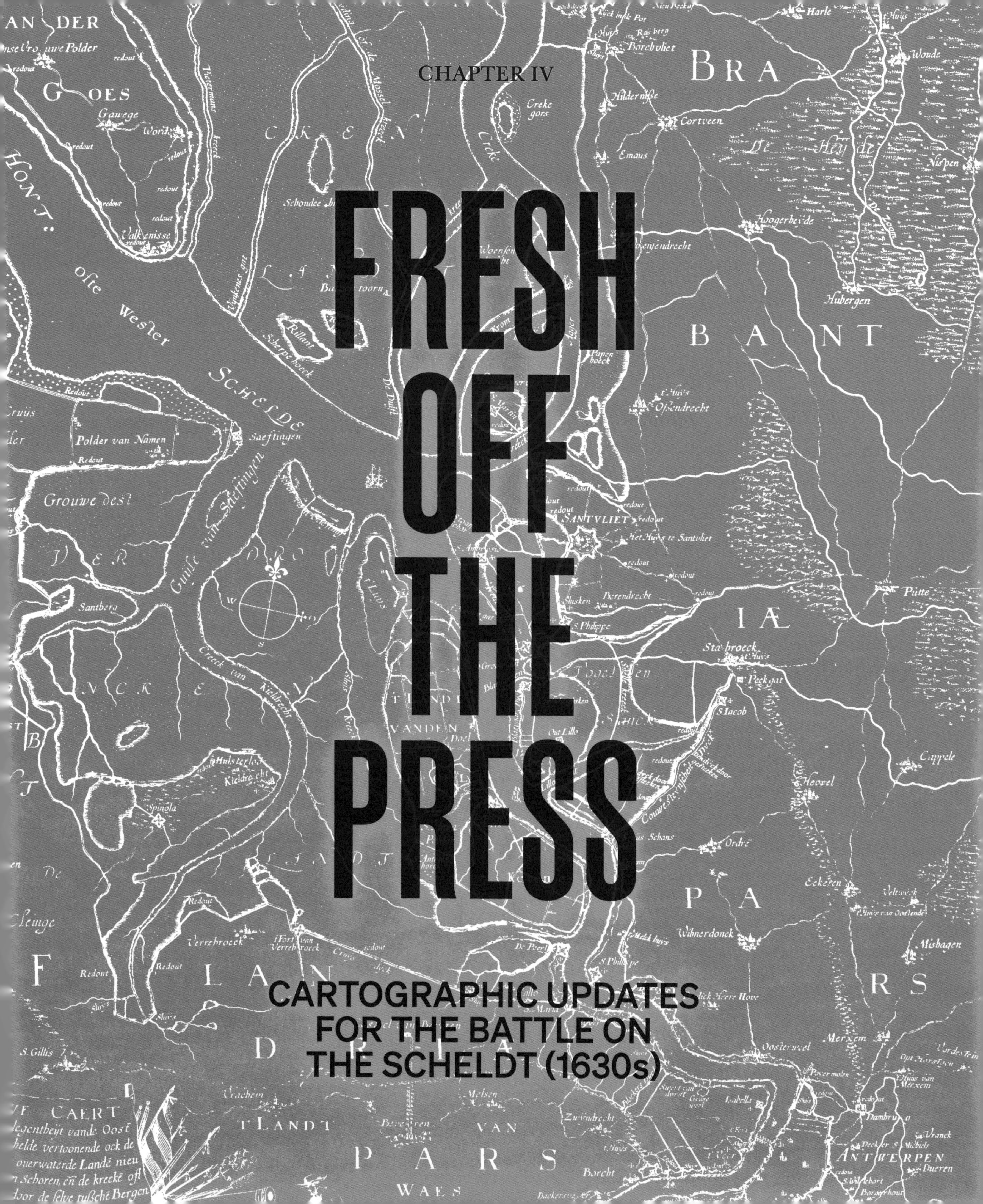

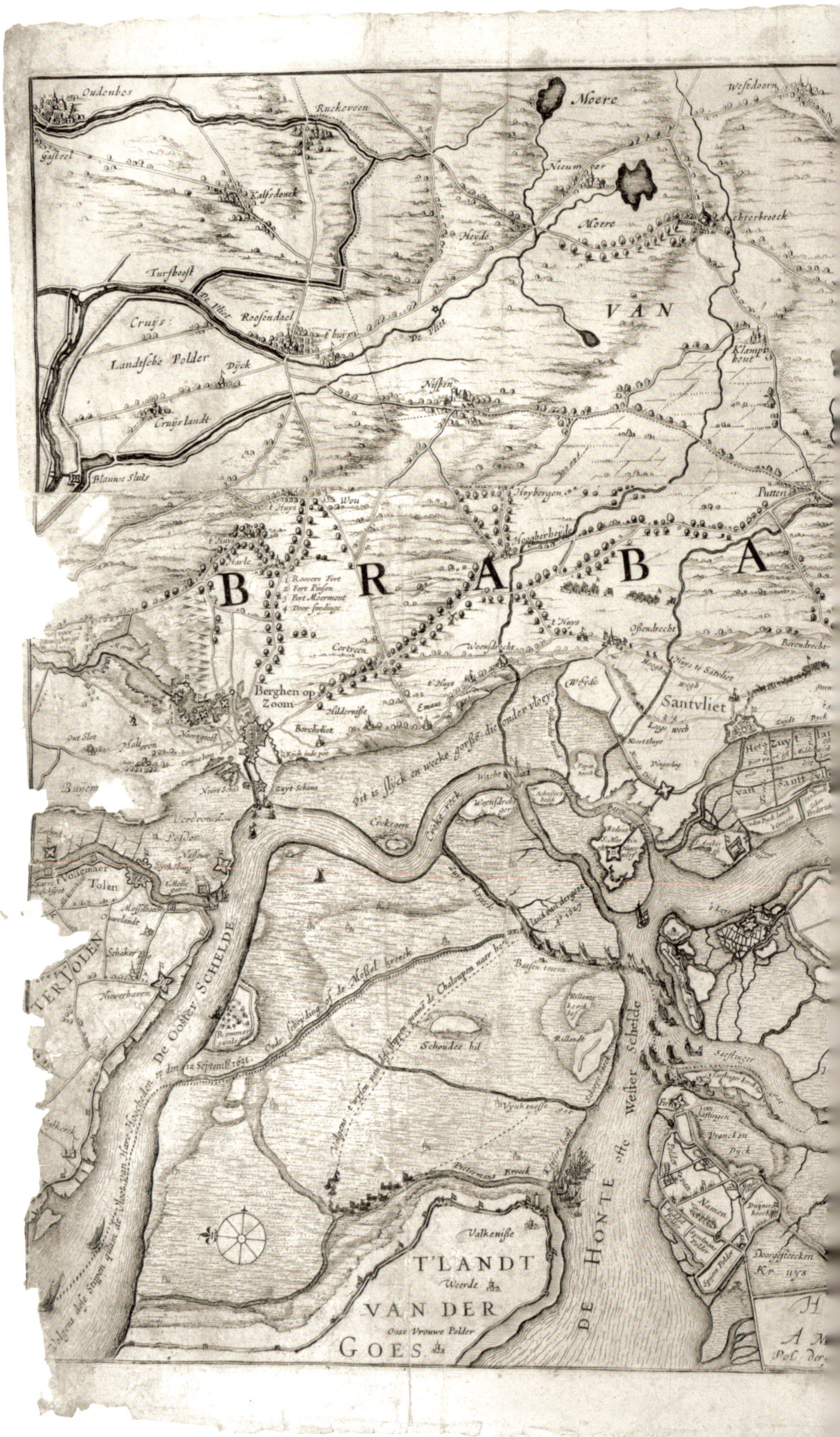

Claes Jansz. Visscher
Caerte van't Scheldt ende Sandvliet, 1632
Etching, 415 × 682 mm
ANTWERP, THE PHOEBUS FOUNDATION

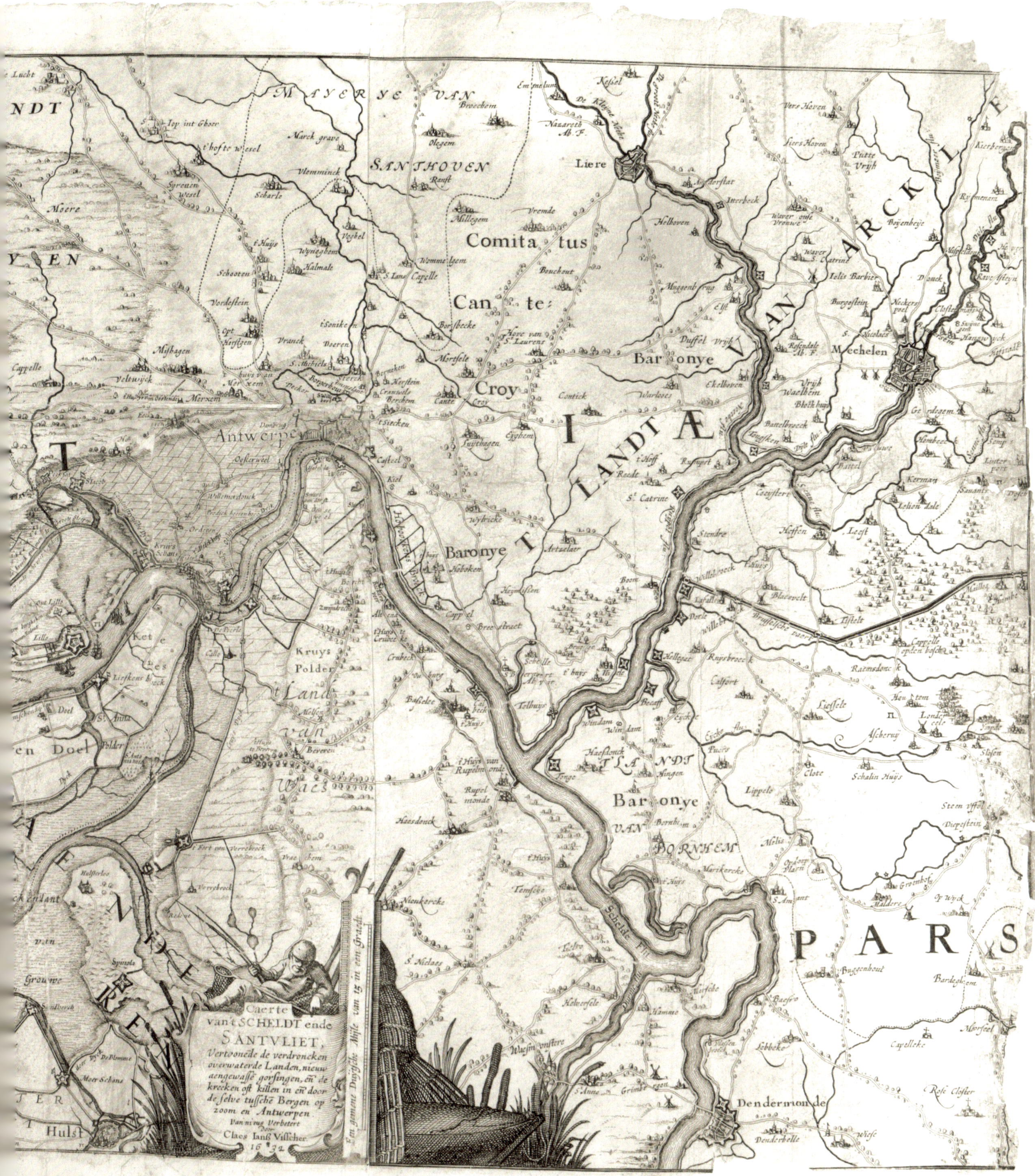
Caerte
van't SCHELDT ende
SANTVLIET,
Vertoonede de verdroncken
overwaterde Landen, nieuw
aengewassē gorsingen, en̄ de
kreecken oft killen in en̄ door
de selve tusschē Bergen op
zoom en Antwerpen
Van nieus Verbetert
Door
Claes Ianß Visscher.
1632
Een gemene Duytsche Myle van 15 in een Graedt
Comita tus
Can te
Baronye
Croy
Antwerpe
Liere
Mechelen
SANTHOVEN
Dendermonde
Baronye
DORNHEM
PARS
Hulst
Waes
Kruys Polder

BATTLE ON THE SCHELDT

The area around the river Scheldt shown on Visscher's map was the scene of fighting around 1630 between the armies of the Dutch Republic and the Habsburg Spanish, which vied for control over this strategic piece of land and water linking the northern and southern provinces. The conflict was part of the Revolt, in which the Republic of the Seven United Netherlands sought to overthrow Spanish rule (⤚→ p. 68). Ambrosio Spinola, commander of Spanish forces in the Low Countries (⤚→ p. 94), ordered the construction of fortifications around the strategic village of Zandvliet and garrisoned soldiers there, from where they could more easily attack the Northern Netherlands. The Spanish presence on the Scheldt created tension that eventually culminated in the Battle of the Slaak, fought on the night of 12 September 1631.

When the Twelve Years Truce (1609-1621) came to an end, the Spanish prepared to regain control of the Scheldt, over which the States army had held sway since the Fall of Antwerp in 1585. The Infanta Isabella – the Spanish governess of the Southern Netherlands – ordered the formation of a fleet with which to launch the assault, and on 4 September 1631, she sailed with the fugitive French queen Marie de' Medici and a grand retinue of ships from Brussels to Antwerp, where they received a festive welcome. This waterborne 'Joyous Entry' was a cover, however, for the deployment of naval vessels which, a week later, were put to an entirely different use.

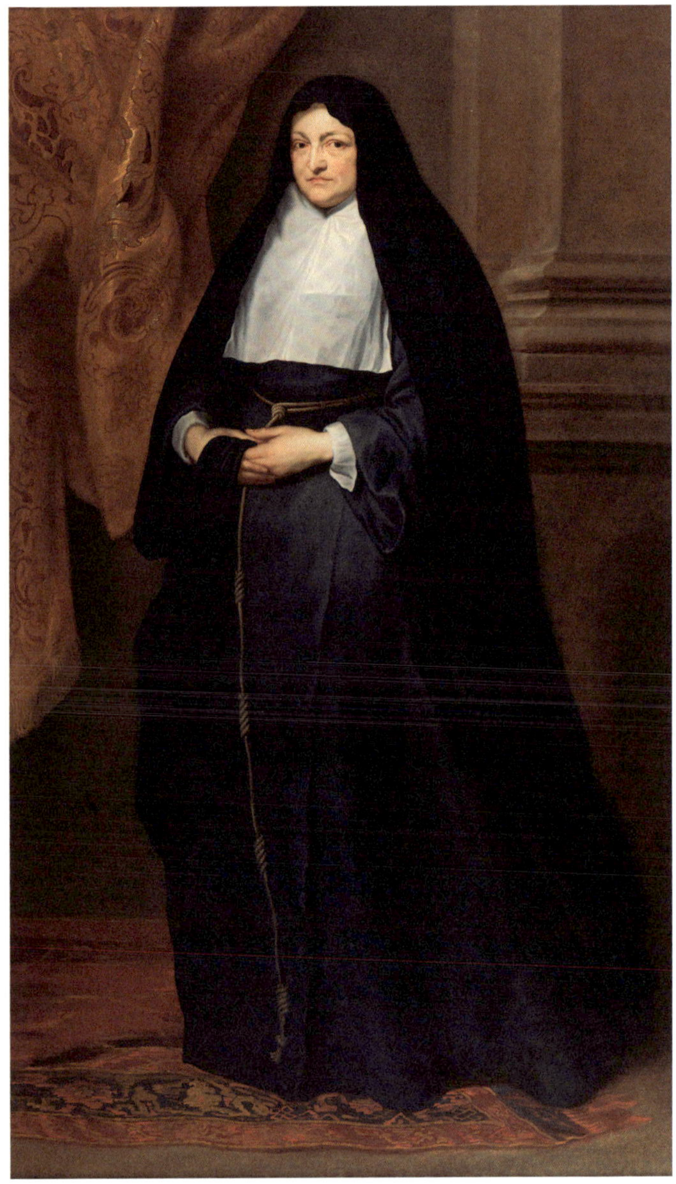

Anthony Van Dyck
Portrait of Infanta Isabella Clara Eugenia as a Poor Clare, c.1625
Oil on canvas, 197.5 × 114.5 cm
ANTWERP, THE PHOEBUS FOUNDATION

On 10 September, Isabella and Marie bade farewell to the Spanish fleet, which sailed from Antwerp along the Schans from Kallo towards Bergen op Zoom under the command of the Marquis of Aytona and John of Nassau-Siegen. The route is marked on Visscher's map with a dotted line labelled: 'these dots show the course sailed by the sloops of war to the land of the rebels, anno 1627'. During the night of 12 September, however, the fleet was ambushed by that of the Republic, which had got wind of the planned attack. The Spanish ships were driven into the Slaak, a narrow channel in which they were trapped ('these dots show the course sailed by Her Highness's fleet on 12 September 1631'). Things got even worse when a thick fog formed, triggering panic among those aboard. Leaping into the water, they swam to the nearest shore, where rebel troops were waiting for them. Most of the soldiers on the Spanish side – including a large number of mercenaries – were killed in this battle, in so far as they had not drowned already or been swallowed up in the tidal marshes. Their tragic fate is explicitly depicted in another of Visscher's news maps (see below), in which you can see figures wading through the mud up to their waists.

For a time after the drama of the Battle of the Slaak, the Spanish made no further attempt to conquer the area. The States army, however, saw an opportunity: in 1632 an army commanded by Count William of Nassau-Siegen (1592-1642) used captured weaponry and supplies to help take possession of the strategically located fortifications that the Spaniards had built on the banks of the Scheldt since 1627. The forts now flew the flag of the United Netherlands. Meanwhile, the dykes were breached to deliberately flood the entire area between Fort Sint-Jacob, the Kruisschans and Antwerp: the strategic flooding is indicated on the map by hatching (⟶ p. 107).

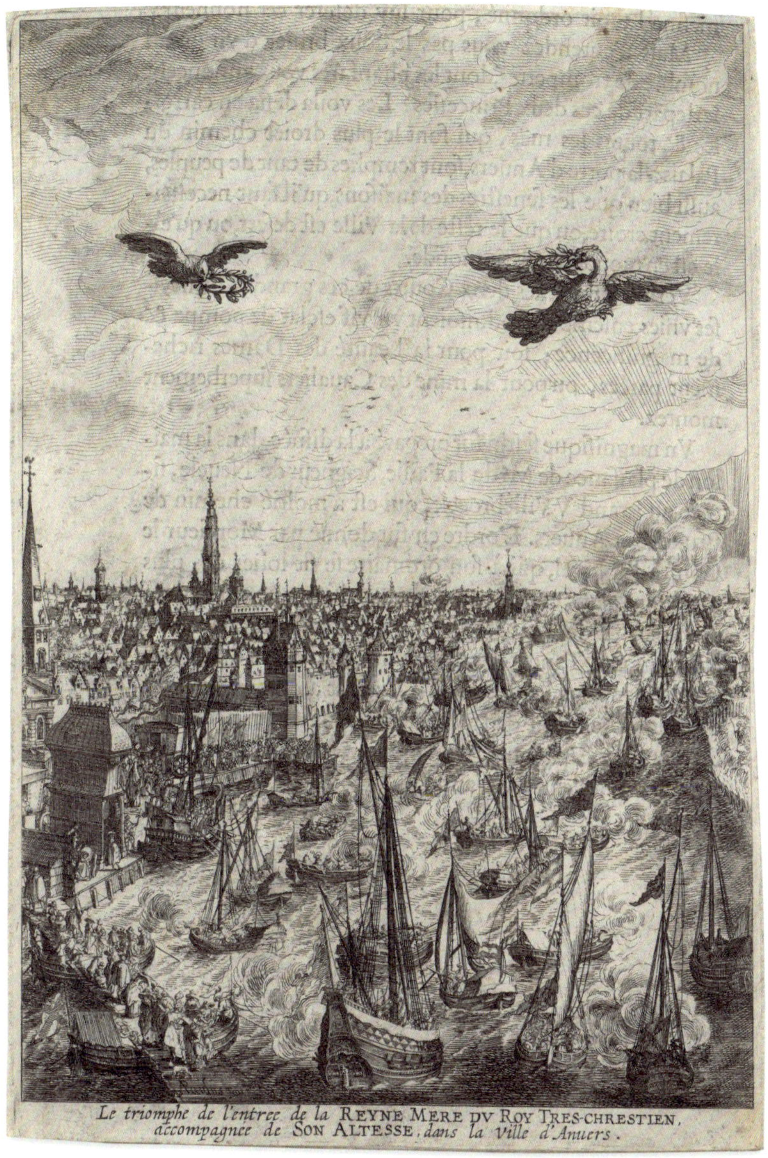

***Le triomphe de l'entree de la Reyne Mere [...] dans la ville d'Anvers*, 1632**
Bird's-eye view of the arrival of Marie de' Medici in Antwerp in 1631
Etching, 246 × 168 mm
ANTWERP, THE PHOEBUS FOUNDATION

CLAES JANSZ. VISSCHER ISSUES UPDATES

Developments on the Scheldt delta were closely followed by the Amsterdam engraver, printer and publisher Claes Jansz. Visscher, beginning with a map he published in 1627 to illustrate the run-up to the battle. As time went by, Visscher received more and more information, prompting him to scrape out some of the lines engraved in his original copper plate, which he then polished and re-engraved. A total of six different states of the map are known: two from 1627 and one each from 1631, 1632, 1638 and 1640 (↣ on 'states', see p. 20).

The Phoebus Foundation owns two copies from the series: the third state from 1631 (↣ p. 111) and the fourth, dated 1632 (↣ pp. 106-107). Visscher not only adjusted his copper plate several times; he also varied the map by adding extra sheets. He did so on the left side of the 1631 copy so that the Slaak would be included, while he placed the battle scene, in which the soldiers can be seen jumping into the water, in the upper right. A separately printed text in both Dutch and French explains the events depicted in the map under the title: 'Short and concise account of the Spanish armada that sailed from Antwerp on 10 September 1631.'

In 1632, a year later, Visscher took a very different approach: he removed the sheets on the left and the battle scene, and in their place he added extra sections of map at the top and on the right, which now included Rozendaal and Mechelen. Surprisingly, these do not depict any battles, figures or other events. Visscher might well have wanted to make these developments more visible by bringing them into the centre of the cartographic image. The composition is now symmetrical too, with the latest news events appearing more centrally on the map.

The different versions, intermediate forms and states gave Visscher the scope for endless variation, responding to current events and audiences and getting full use of the original 1627 copper plate. Where today we refresh our screens to keep up with the latest news developments, people in the seventeenth century could buy new, updated versions of Visscher's maps in his shop.
A separate map cost five to ten stuyvers (the price of a large rye loaf at the time was six and a half stuyvers, while a jug of beer cost half a stuyver). But each variant also had its own specific purpose and audience. The copies in The Phoebus Foundation's collection were seemingly intended as informative news maps for an international audience, given their large size with added map or text sheets.

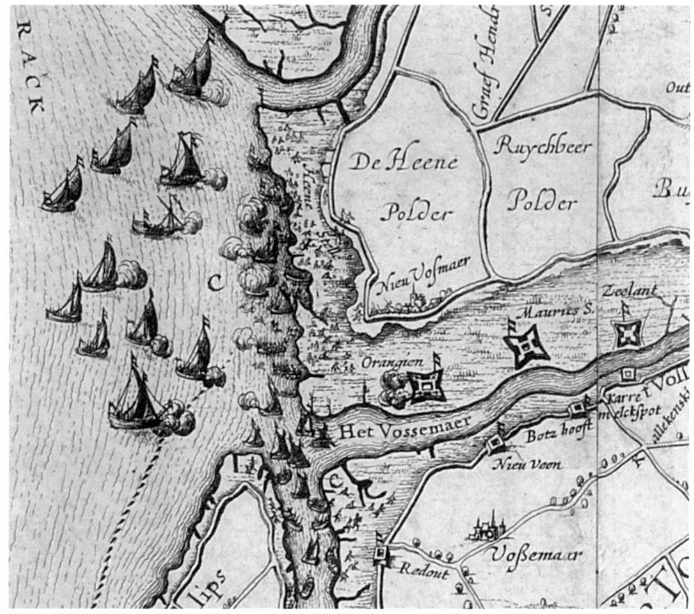

The Spanish ships are cornered in the Slaak. Panicking soldiers leap into the water, but States troops lie in wait for them.

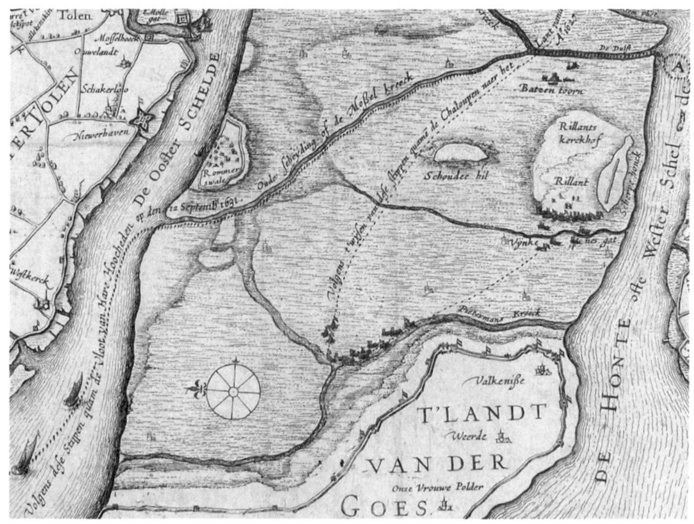

The dotted line shows the route taken by the Spanish fleet towards the Slaak channel.

Details from: Claes Jansz. Visscher, *Pas-caert van ghelege[n]theyt vande Schans te Santvliet*, 1631 (p. 111)

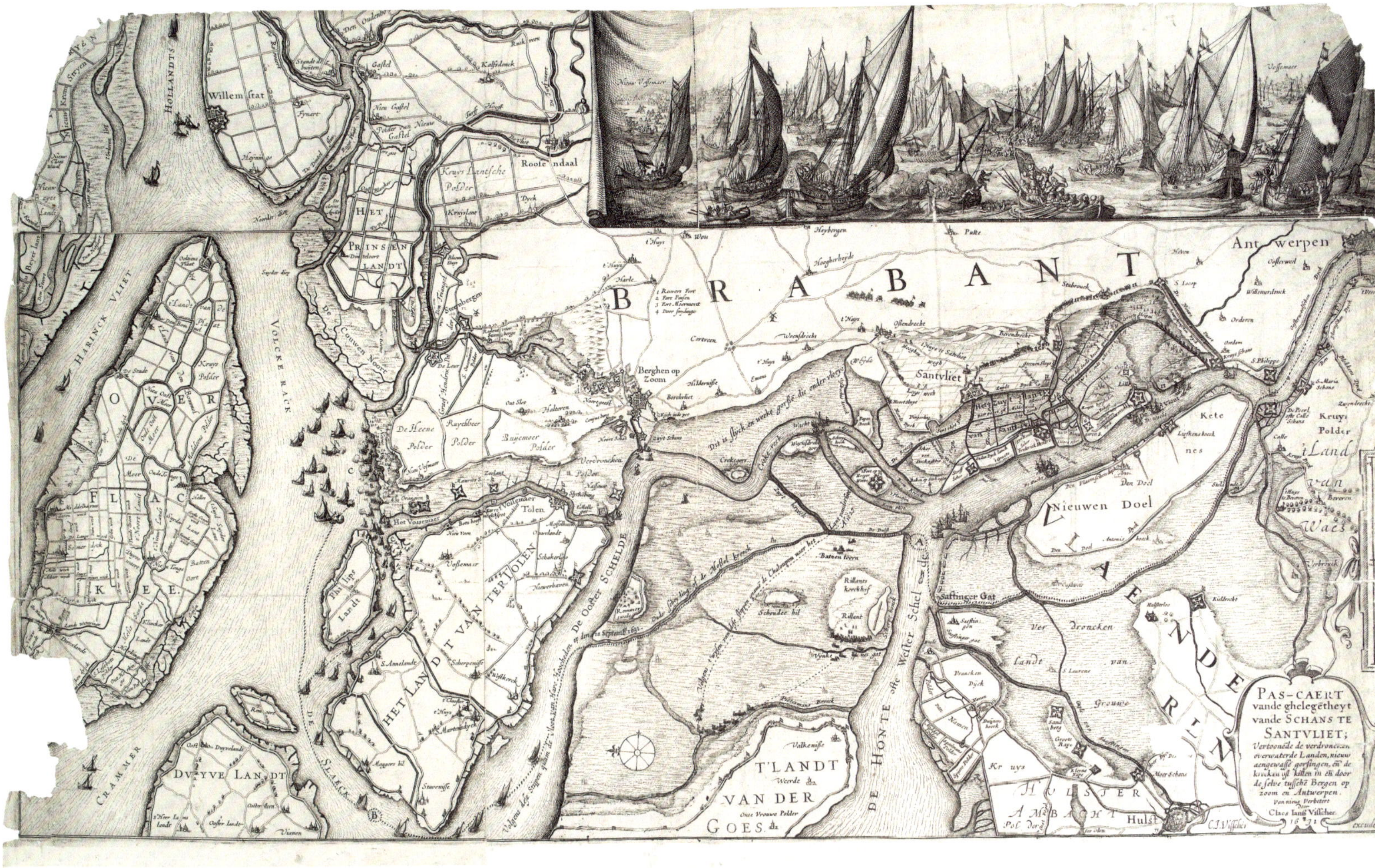

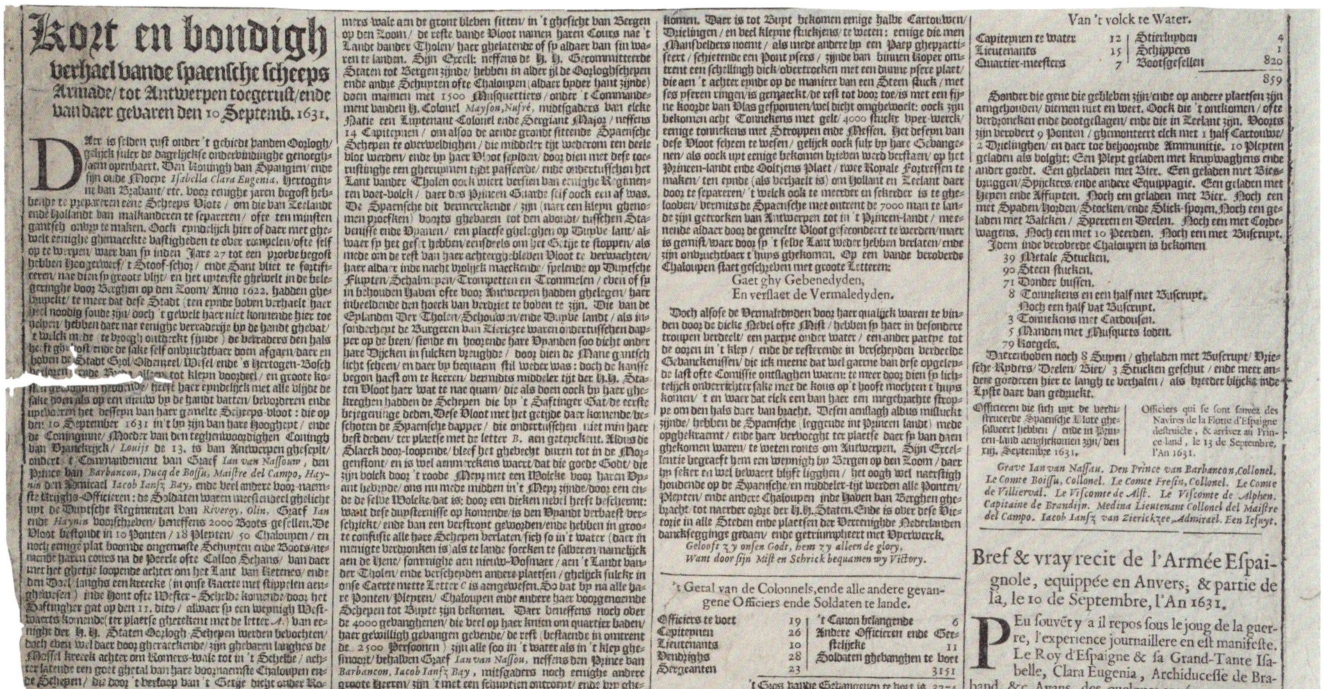

Kort en bondigh verhael vande Spaensche scheeps Armade, tot Antwerpen toegerust, ende vandaer gevaren den 10 Septemb. 1631.

Van 't volck te Water.

Bref & vray recit de l'Armée Espaignole, equippée en Anvers, & partie de là, le 10 de Septembre, l'An 1631.

Peu souvent y a il repos sous le joug de la guerre, l'experience journaillere en est manifeste. Le Roy d'Espaigne & sa Grand-Tante Isabelle, Clara Eugenia, Archiducesse de Braband, &c. Ayans, des quelques années en-ça com-

Claes Jansz. Visscher
Pas-caert van ghelege[n]theyt vande Schans te Santvliet, 1631
News map of the Battle of the Slaak
Etching, 380 × 640 mm
ANTWERP, THE PHOEBUS FOUNDATION

Text sheets
Part of: Claes Jansz. Visscher,
Pas-caert van ghelege[n]theyt vande Schans te Santvliet, 1631
ANTWERP, THE PHOEBUS FOUNDATION

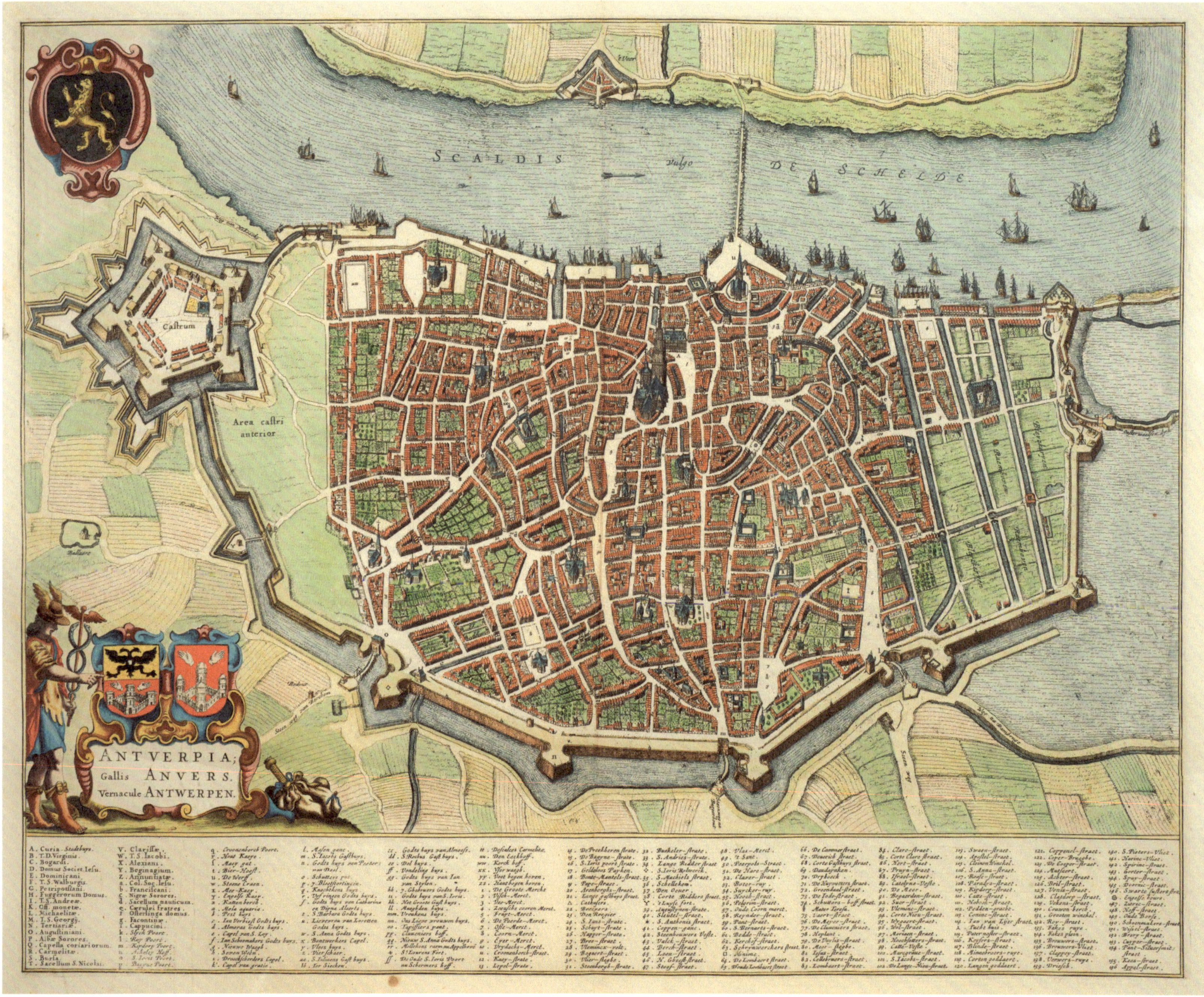

Joan Blaeu
Antverpia
From: Joan Blaeu, *Novum ac Magnum Theatrum Urbium Belgicae Regiae*, 1649
ANTWERP, THE PHOEBUS FOUNDATION

FROM NEWS MAP TO CITY ATLAS

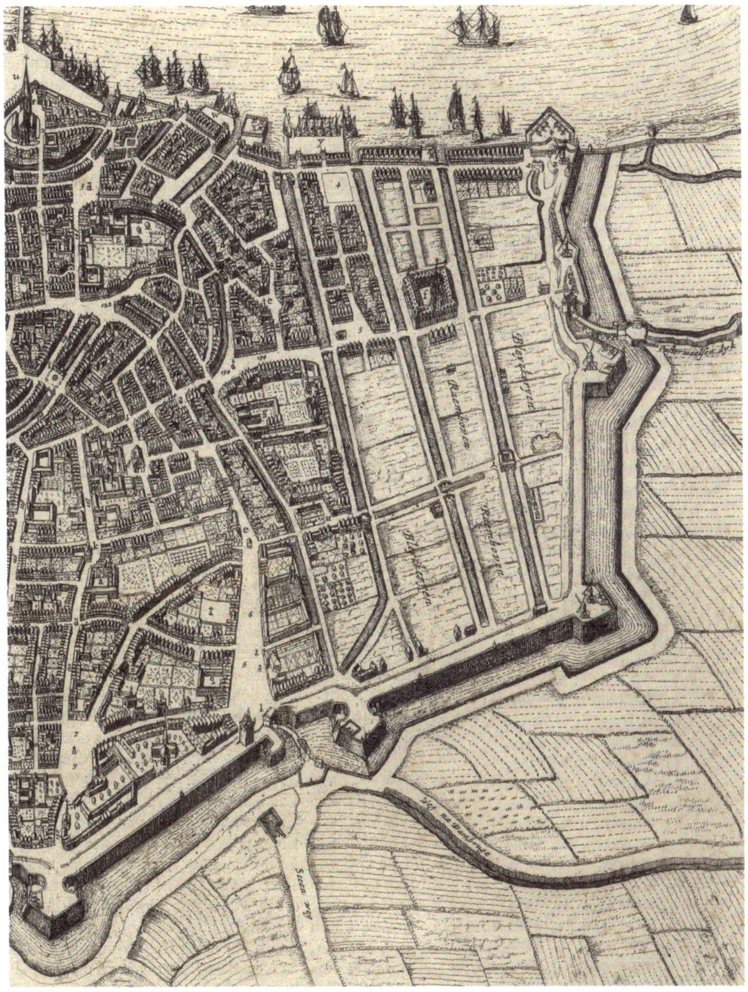

Visscher acted like a cartographic journalist, eagerly gathering news and not shying away from revising and correcting his mistakes. He even called on readers to point out any errors and to provide him with the latest, accurate information. He was not the only one, however, to respond adroitly to his customers and to current affairs: the technique of updating maps was applied not only in news cartography but also by other mapmakers and publishers. It is found, for instance, in the different editions and impressions of Joan Blaeu's atlases.

The impact of the Battle of the Slaak is apparent from the different states of the Antwerp city plans included in Blaeu's *Novum ac Magnum Theatrum Urbium Belgicae Regiae* (Amsterdam, 1649). The map of Antwerp from the first edition of this city atlas of the Low Countries shows the Oosterweel and Wilmarsdonk polders to the north of the city completely under water. William of Nassau had captured the Sint-Jacob and Kruisschans forts, as we saw in Visscher's 1632 map, and breached the Kauwenstein dyke to flood the entire polder next to Antwerp.

The water came right up to the city walls, and Blaeu's map shows the area to the right of them (the north of the city). The flooded polders were not drained again until 1651, three years after the Peace of Münster, when normality had returned once more. This development did not escape Blaeu's notice either: in his 1652 reissue of the city atlas, *Toonneel der steden van de Vereenigde Nederlanden* (Amsterdam, 1652), we find that he has adjusted the copper plate. The second state of the map of Antwerp shows the floods are gone and the fields and ditches neatly restored as if nothing had happened.

The polder near Antwerp redrained
Part of: Joan Blaeu, *Antverpia*, 1652 (pp. 118-119)

FROM THE OTHER SIDE

Mapmakers from the Southern, Spanish Netherlands are likely to have taken a different view of events than the pro-Republic Amsterdam-based cartographers Visscher and Blaeu. The publisher and mapmaker Pieter Verbiest II was born in Amsterdam, but his father Pieter Verbiest I (1571-1642/1643), who was a printmaker and map colourist, came from Antwerp and returned to the Southern Netherlands a few years after Pieter II's birth. From the city on the Scheldt, the younger Verbiest reported on developments in the surrounding area, just as Visscher did. His map too formed part of a series of updated documents dealing with events by the river. The copy in The Phoebus Foundation, the second state, does not show any drastic military events as yet. The difference in orientation is striking: Antwerp is pictured at the bottom, viewed from the south, whereas Visscher's map is viewed from the northwest.

The reclaimed area on the right bank of the Scheldt is quiet and still dry, with the exception of a few marshes along its banks. All the same, something gives away the presence of Republic troops to the attentive viewer: the dyke above the Oosterweel and Wilmarsdonk polder has been breached in two places. In a later state, Verbiest added hatching to indicate that the zone had been flooded in the meantime by William of Nassau. Verbiest simply forgot to adjust the year '1631' (below left) – an ironic slip, since the new state shows that he was doing his best to make his map more up-to-date and accurate.

Pieter Verbiest II
Nieuwe caert vande Ghelegentheijt vande Oost en Wester Schelde, 1631 [1632]
Map of the Eastern and Western Scheldt
Copper engraving, 538 × 403 mm
ANTWERP, THE PHOEBUS FOUNDATION

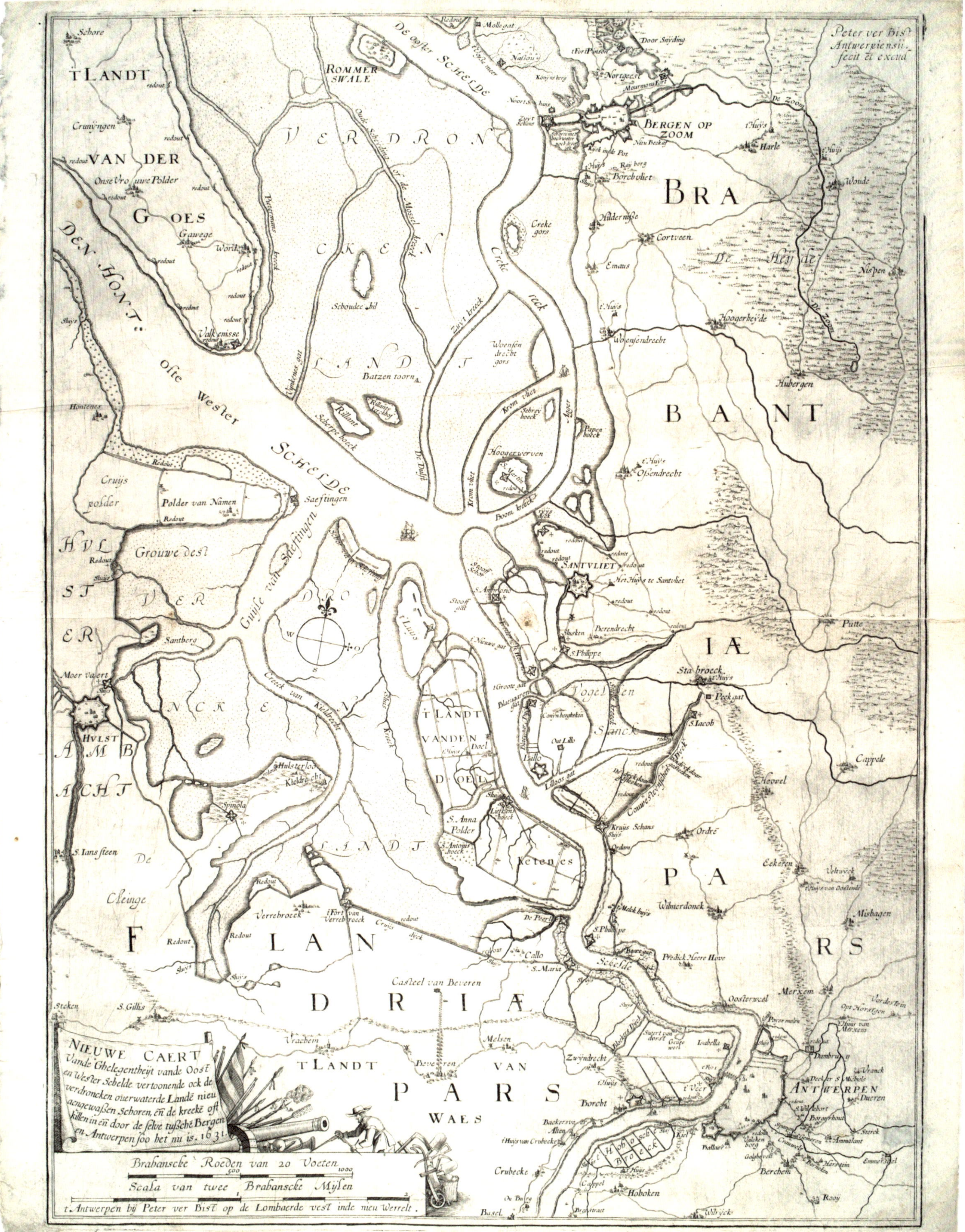
Peter ver Bist Antwerpiensis fecit et excud
NIEUWE CAERT
Vande Ghelegentheyt vande Oost en Wester Schelde vertoonende ock de verdronken ouerwaterde Landē nieu aengewaßen Schoren, ēn de kreekē oft killen in ēn door de selve tußchē Bergen en Antwerpen soo het nu is. 1631.
Brabanscke Roeden van 20 Voeten
500
1000
Scala van twee Brabanscke Mylen
t'Antwerpen by Peter ver Bist op de Lombaerde vest inde nieu Werrelt.
T'LANDT VAN DER GOES
DEN HONT: oste Wester SCHELDE
VERDRONCKEN LANDT
ROMMER SWALE
De ooster SCHELDE
BERGEN OP ZOOM
BRABANTIÆ PARS
FLANDRIÆ PARS
T'LANDT VAN PARS WAES
HVLST AMBACHT
T'LANDT VANDEN DOEL
SANTVLIET
LILLO
ANTWERPEN
Saeftingen
Hulst
Callo
Casteel van Beveren
Beveren
Zwyndrecht
Borcht
Hoboken
Berchem
Merxem
Eckeren
Wilmerdonck
Putte
Cappele
Stabroeck
Berendrecht
Ossendrecht
Woensdrecht
Hoogerheyde
Huybergen
Cortveen
Nispen
Wouwe
Harle
Verrebroeck
Kieldrecht
S. Gillis
Steken
Melsen
Vrachem
Crubecke
Schore
Cruyningen
Valkemisse
Hontenes
Santberg
Moer Vaert
Cruijs polder
Polder van Namen
Grouwe dest
Batzen toorn
Hoogerwerven
Papen hoeck
Ketenes
S. Anna Polder
Spinola
S. Iacob
S. Philippe
Kruijs Schans
Oosterweel
Dambrugh
Deurnen
Emaus
Hildernisse
Borchvliet
Nortgeest
Mollegat
Door Snijding
Krom vliet
Boom kreeck
Creke gors
Woensdrecht gors
Schoudee Jil
Rillant
Scherpe hoeck
Creeck van Kieldrecht
Guijse van Saeftingen
De Hey de
Predick Heere Hove
Isabella
Poyer molen
Sterck
Emmerbrgel
Rooy
Wilrijck
Berghstraet
Basel
De Cleinge
S. Ians steen
Mishagen
Veltwijck
Hoevel
Ordre
Peckgat
Vogel en Sanck
Out Lillo
Doel
De Poert

River map of the Scheldt

In 1585, the governor of the Low Countries, Alexander Farnese, laid siege to Antwerp. He barricaded the Scheldt with a line of ships running from bank to bank so that no outside aid or food could get through to the city (⟶ p. 68). The Calvinist rebels surrendered and Antwerp was taken by the Spanish Habsburgs. Following the city's fall, the Scheldt was cut off from direct overseas shipping for some two centuries, now by the Dutch. This, among other factors, robbed Antwerp of its position as an international trading metropolis.

The French period (1795-1815) breathed new life into the former centre of commerce after revolutionary France invaded the Southern Netherlands in 1792. To compensate for the military incursion, Paris reopened the Scheldt and declared that shipping could henceforth pass freely. Under Napoleon, the port was swiftly adapted to allow access by large ocean-going vessels and naval fleets. Hydrographic knowledge of the waterway was extremely outdated, however, as the Scheldt had been left high and dry, as it were, for hundreds of years. The French hydrographer Charles-François Beautemps-Beaupré (1766-1854) was, therefore, commissioned to survey the river up to the coast and beyond, to take soundings and to map its course accurately.

The result was a large, fold-out map of the Western Scheldt, consisting of forty-four sections pasted onto fabric complete with a special box in which to store them all. The numbers on the map are measurements of depth taken at low water, while the dotted lines indicate zones that were dry at that moment or remained so even at high tide. The only part of the coast to be mapped is the part directly along the riverbank, such as embankments and landmarks in several large towns. It was now possible once more for ships to sail from Antwerp without hindrance.

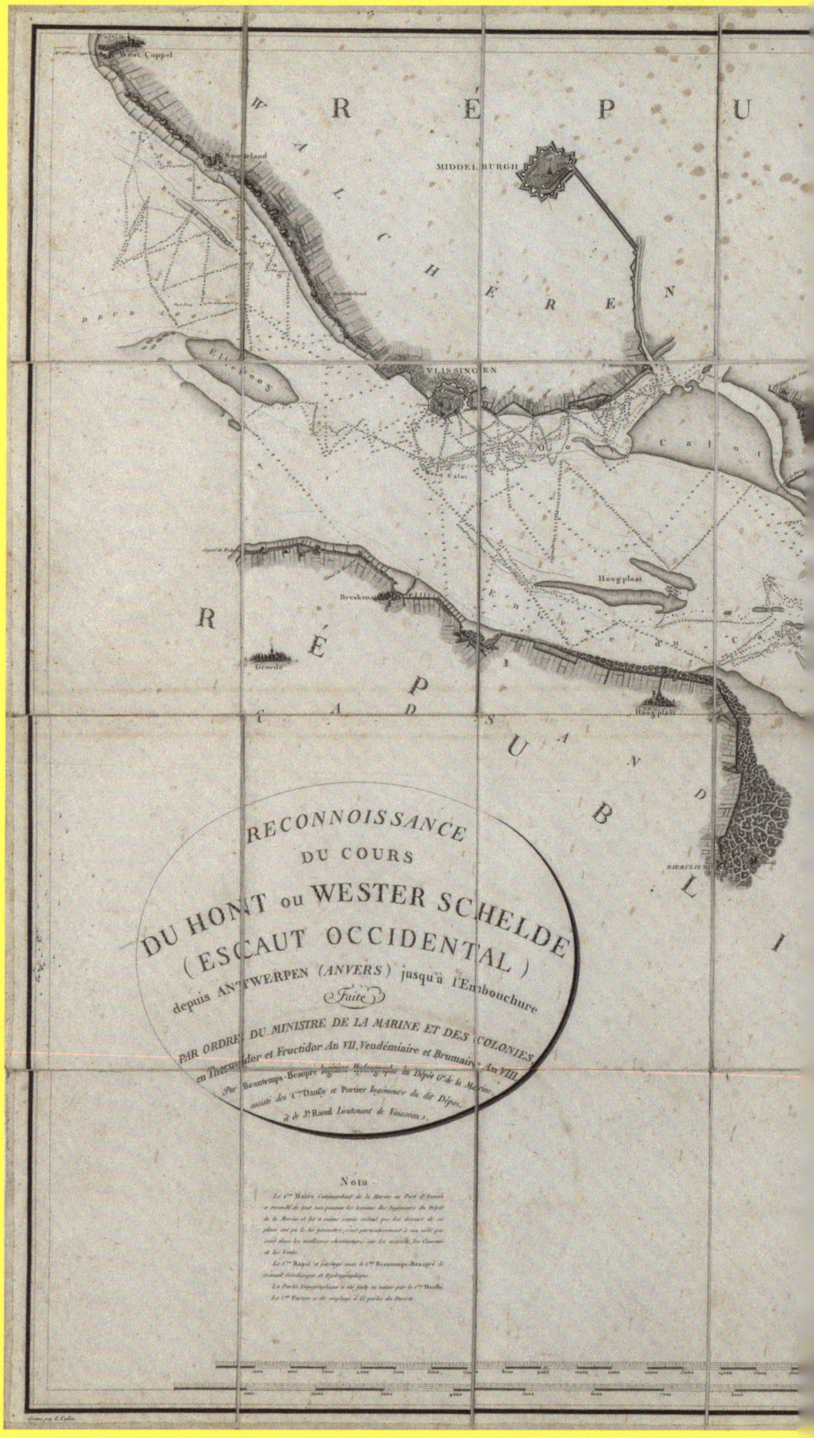

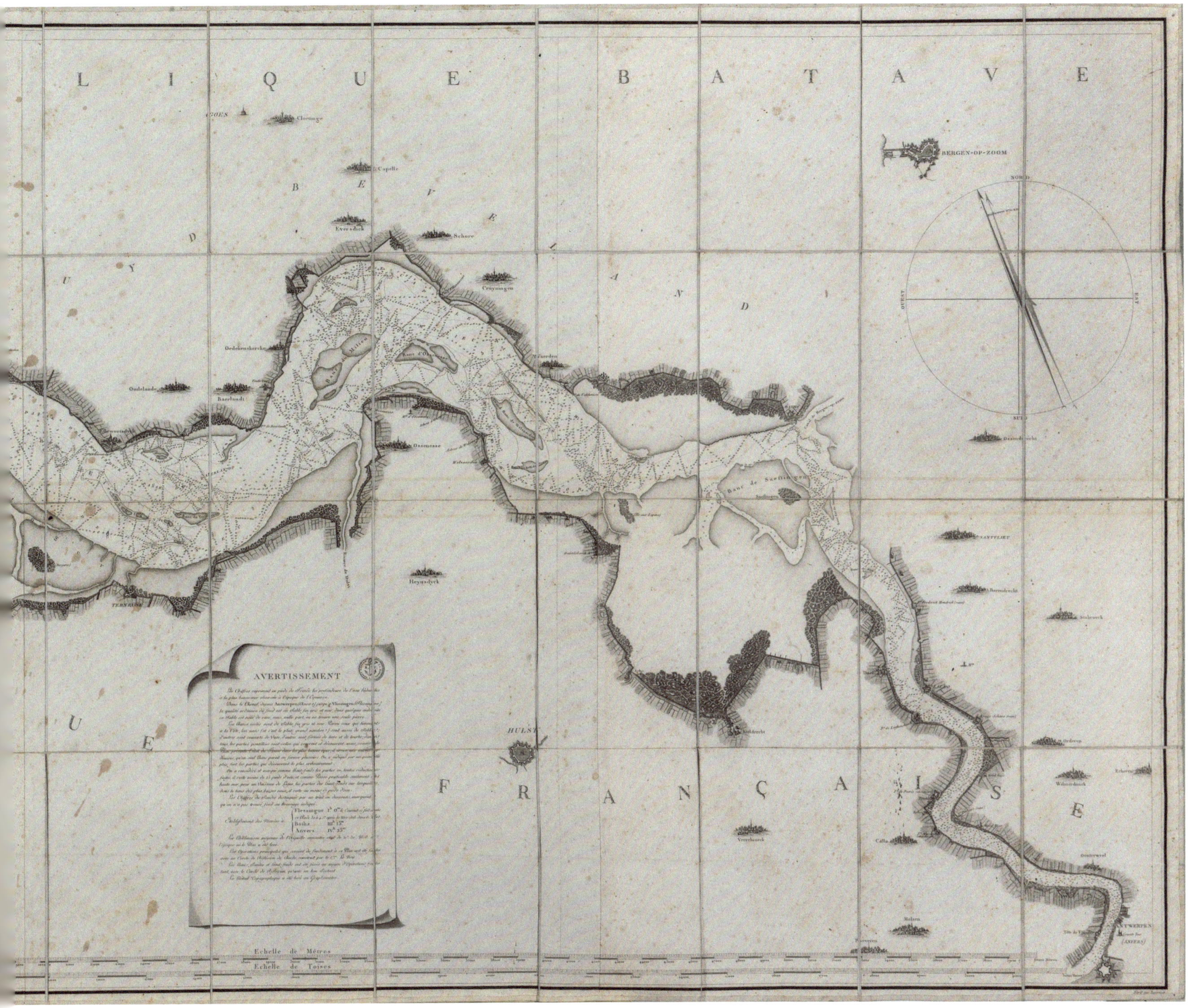

Charles-François Beautemps-Beaupré
Reconnoissance du cours du Hont ou Wester Schelde (Escaut Occidental) depuis Antwerpen (Anvers) jusqu'à l'Embouchure, 1799
Hydrographic map of the Western Scheldt
Copper engraving, 950 × 1778 mm
ANTWERP, THE PHOEBUS FOUNDATION

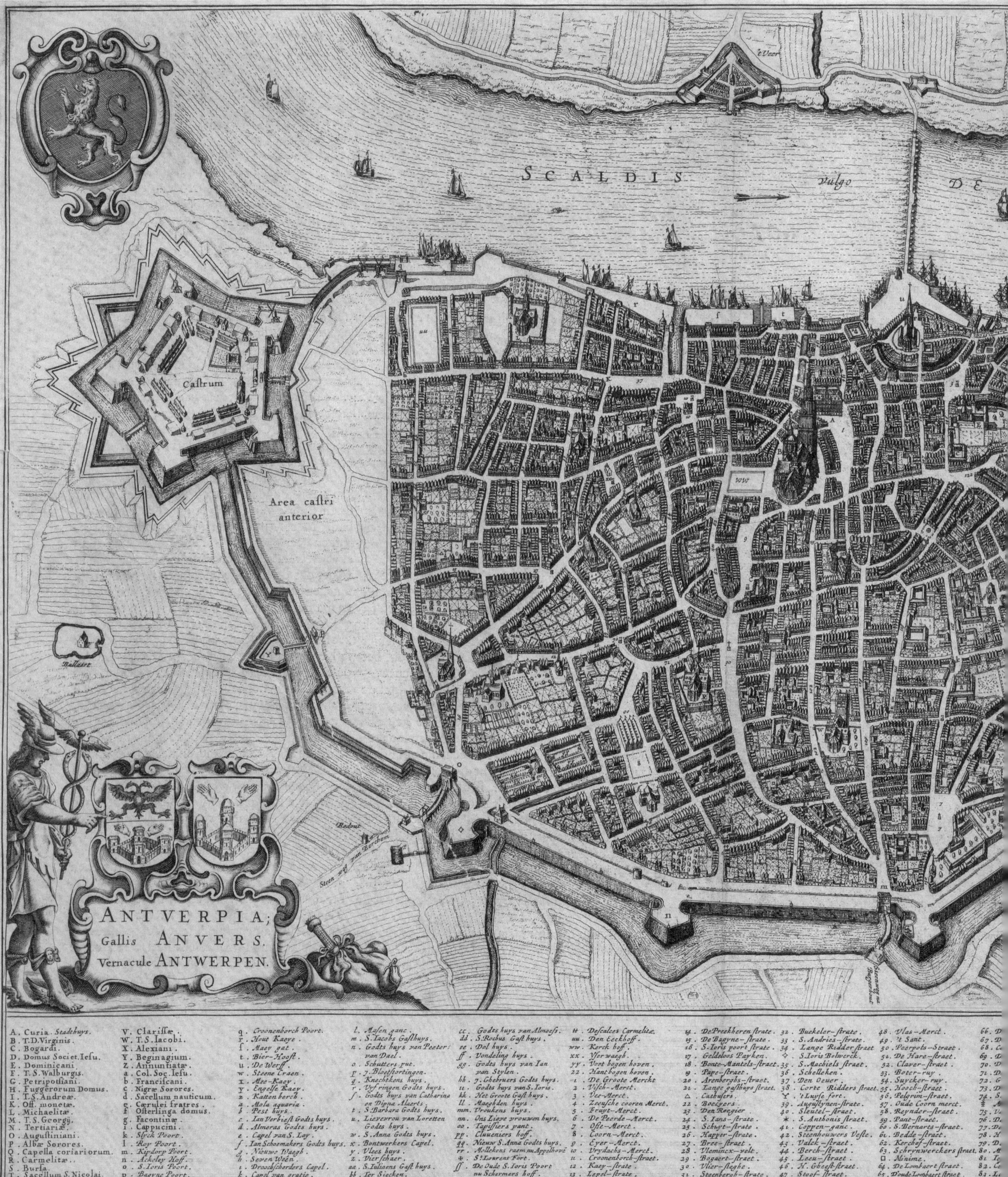

ANTVERPIA;
Gallis ANVERS.
Vernacule ANTWERPEN.
SCALDIS vulgo DE
Castrum
Area castri anterior
Ballaert
Redout
A. Curia. Stadthuys.
B. T.D.Virginis.
C. Bogardi.
D. Domus Societ.Iesu.
E. Dominicani
F. T.S.Walburgis.
G. Petripotsiani.
H. Fuggerorum Domus.
I. T.S.Andreæ.
K. Off. monetæ.
L. Michaelitæ.
M. T.S.Georgij.
N. Tertiariæ.
O. Augustiniani.
P. Albæ Sorores.
Q. Capella coriariorum.
R. Carmelitæ..
S. Bursa.
T. Sacellum S.Nicolai.
V. Clarissæ.
W. T.S.Iacobi.
X. Alexiani.
Y. Beginagium.
Z. Annuntiatæ.
a. Col.Soc.Iesu.
b. Franciscani.
c. Nigræ Sorores.
d. Sacellum nauticum.
e. Cærulei fratres.
f Osterlinga domus.
g. Facontinæ.
i. Cappucini.
k. Slyck Poort.
l. Roy Poort.
m. Kipdorp Poort.
n. Ackeley Hoff.
o. S.Ioris Poort.
p. Bagyne Poort.
q. Croonenborch Poort.
r. Hout Kaeye.
ſ. Maey gat.
t. Bier-Hooft.
u. De Werff.
w. Steene Craen.
x. Mee-Kaey.
y. Engelse Kaey.
z. Kattenberch.
a. Mola aquaria.
b. Pest huys.
c. Ian Verbiest Godts huys.
d. Almeras Godts huys.
e. Capel van S. Loy.
f. Ian Schoemakers Godts huys.
g. Nieuwe Waegh.
h. Seven Weën.
i. Droochscherders Capel.
k. Capel van gratie.
l. Mæsen ganc.
m. S.Iacobs Gasthuys.
n. Godts huys van Peeter van Dael.
o. Schutters put.
p. 7.Bloetstortingen.
q. Knechtkens huys.
r. Vyf ringen Godts huys.
ſ. Godts huys van Catharina en Dipna Alaerts.
t. S.Barbara Godts huys.
u. Lievevrou van Loretten Godts huys.
w. S.Anna Godts huys.
x. Bontwerkers Capel.
y. Vlees huys.
z. Vierschaer.
aa. S.Iuliaens Gast huys.
bb. Ter Siecken.
cc. Godts huys van Almoesſ.
dd. S.Rochus Gast huys.
ee. Dol huys.
ff. Vondeling huys.
gg. Godts huys van Ian van Stylen.
hh. 7.Ghebruers Godts huys.
ii. Godts huys van S.Ioris.
kk. Het Groote Gast huys.
ll. Maeghden huys.
mm. Vroukens huys.
nn. Ons Lieve vrouwen huys.
oo. Tapissiers pant.
pp. Cluueniers hoff.
qq. Nieuw S.Anna Godts huys.
rr. Mollekens raem nu Appelhovē
✠. S.t Lourens Fort.
ſſ. De Oude S.Ioris Poort nu Schermers hoff.
tt. Descalces Carmelitæ.
uu. Den Eeckhoff.
ww. Kerck hoff.
xx. Yserwaegh.
yy. Voet bogen hoven.
zz. Hant bogen hoven.
1. De Groote Merckt
2. Visch-Merct.
3. Vee-Merct.
4. Zeeusche cooren Merct.
5. Fruyt-Merct.
6. De Peerde-Merct.
7. Osse-Merct.
8. Coorn-Merct.
9. Eyer-Merct.
10. Vrydachs-Merct.
11. Croonenborch-straet.
12. Kaey-strate.
13. Lepel-strate.
14. De Preekheren strate.
15. De Bagyne-strate.
16. S.Ioris poort strate.
17. Geldeloos Payken.
18. Bonte-Mantels-straet.
19. Pape-straet.
20. Arenberghs-straet.
21. Lange gasthuys straet.
△. Cathusers.
22. Boecyers.
23. Den Roegier.
24. S.Ians-strate.
25. Schuyt-strate.
26. Happer-strate.
27. Bree-straet.
28. Vleminex-velt.
29. Bogaert-straet.
30. Vlier-steghe.
31. Steenbergh-strate.
32. Buekeler-strate.
33. S.Andries-strate.
34. Lange Ridder straet
◇. S.Ioris Bolwerck.
35. S.Machiels straet.
36. Schelleken.
37. Den Oeuer.
38. Corte Ridders straet.
Y. 't Luyse fort.
39. Augustynen-strate.
40. Sleutel-straet.
✶. S.Anthonis straet.
41. Coppen-ganc.
42. Steenhouwers Veste.
43. Valck-straet.
44. Berch-straet.
45. Leeu-straet.
46. H.Gheest-straet.
47. Stoof-straet.
48. Vlas-Merct.
49. 't Sant.
50. Peerpots-Straet.
51. De Hare-straet.
52. Claver-straet.
53. Boter-ruy.
54. Suycker-ruy.
55. Hooch-straet.
56. Pelgrim-straet.
57. Oude Coorn merct.
58. Reynder-straet.
59. Pant-straet.
60. S.Bernarts-straet.
61. Bedde-straet.
62. Kerchof-straet.
63. Schrynwerckers straet.
□. Minime.
64. De Lombaert straet.
65. D'oude Lombaert straet.

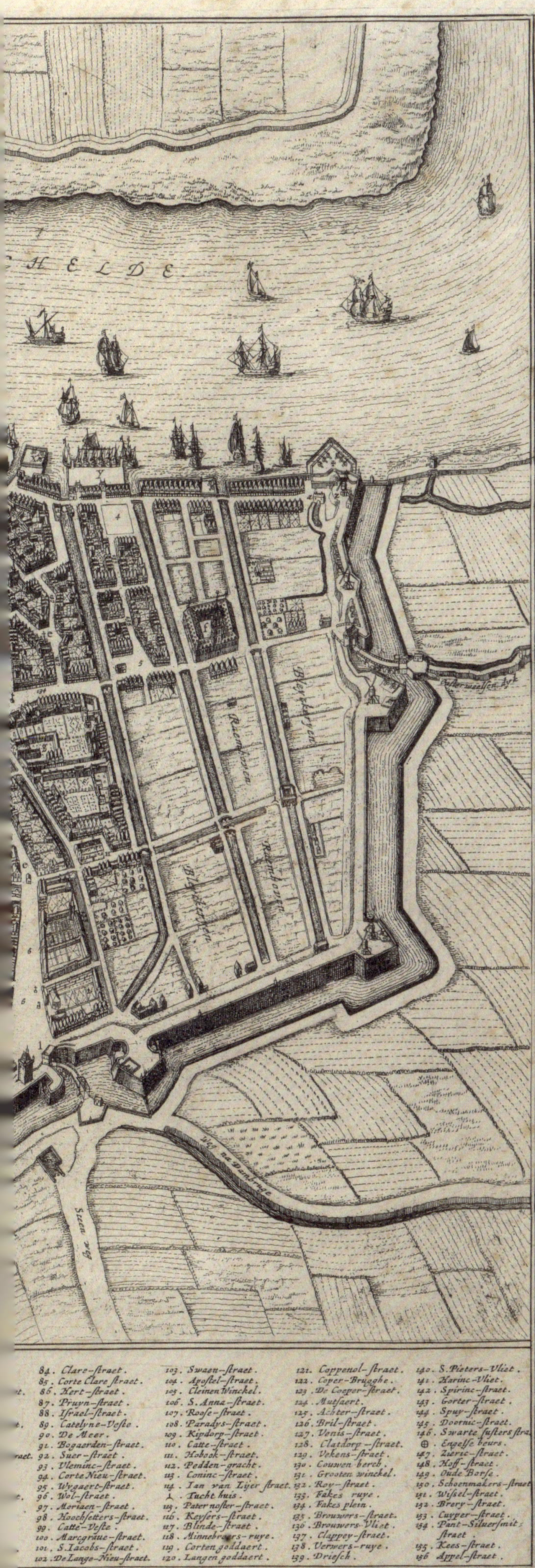

LITERATURE

L. Danckaert, 'Een kaart van het Antwerpse door Pieter Verbiest. Een voorbeeld van locale cartografie en historische actualiteit in de 17de eeuw', *De Gulden Passer*, 61-63, 1983-1985: 405-416.

P. Guns, *De Antwerpse noorderpolders in de 16de-17de eeuw, waterbouwkundig aboratorium 1933-2008*, Antwerp, 2008.

J. Leune, 'De Kruisschans aan de Schelde', in: M. Braem, *Poldergrond; bijdragen tot de geschiedenis van de Antwerpse Noorderpolders*, Hoevenen-Stabroek, 2013: 59-112.

J. Van Beylen, 'Een zeventiende-eeuwse "nieuwskaart" van de Slag op het Slaak: "Kort en bondigh verhael vande Spaensche Scheeps Armade tot Antwerpen toegerust ende van daer gevaren den 10 Septemb. 1631"', *Tijdschrift voor zeegeschiedenis*, 14, 1, 1995: 3-23.

A. van Schaik, 'A Delta of Time: Claes Jansz Visscher's Updated News Maps of Spanish-Dutch Battles in the Scheldt River Area, 1627-1640', *Early Modern Low Countries*, 7, 2, 2023: 154-189.

Joan Blaeu
Antverpia, 1652
Copper engraving and etching, 417 × 530 mm
ANTWERP, THE PHOEBUS FOUNDATION

The Northern and the Southern Netherlands enjoyed an ambivalent relationship in the period of the Eighty Years War, which dominated the late sixteenth and early seventeenth centuries. When the conflict began, the South was part of the revolt against Spanish rule. However, when Antwerp surrendered to the Spaniards in 1585, the Republic of the Seven United Netherlands found itself in opposition to the Southern, Spanish Netherlands. The two regions had much in common: a shared battle with the water, an urban culture and an overriding commercial mentality. Netherlanders had, moreover, already migrated from North to South and vice versa. After the Fall of Antwerp, however, especially large numbers of townspeople from the Southern Netherlands migrated north, although many stayed put or later returned. It was the first genuine split between North and South, although the two parts remained in close contact on a day-to-day level. There was also considerable interaction when it came to the making, printing and publishing of maps and books. Separation was only rendered permanent with the Peace of Münster in 1648, the end of the war, and the beginning of the Republic's sovereignty and independence.

The Northern and Southern Netherlands by no means always saw eye to eye economically, politically or religiously. So how was this conflict of identity reflected in cartography? From *Leo Belgicus* maps via Blaeu's city atlases to wall maps and the title prints of books, maps were part of society and held up a mirror to the Netherlanders. In a manner ranging from the subtle to the explicit, they show how people in the Low Countries of the turbulent sixteenth and seventeenth centuries sought a balance between unity and division.

CHAPTER V

UNITY AND DIVISION

THE *LEO BELGICUS* IN THE NORTHERN AND SOUTHERN NETHERLANDS

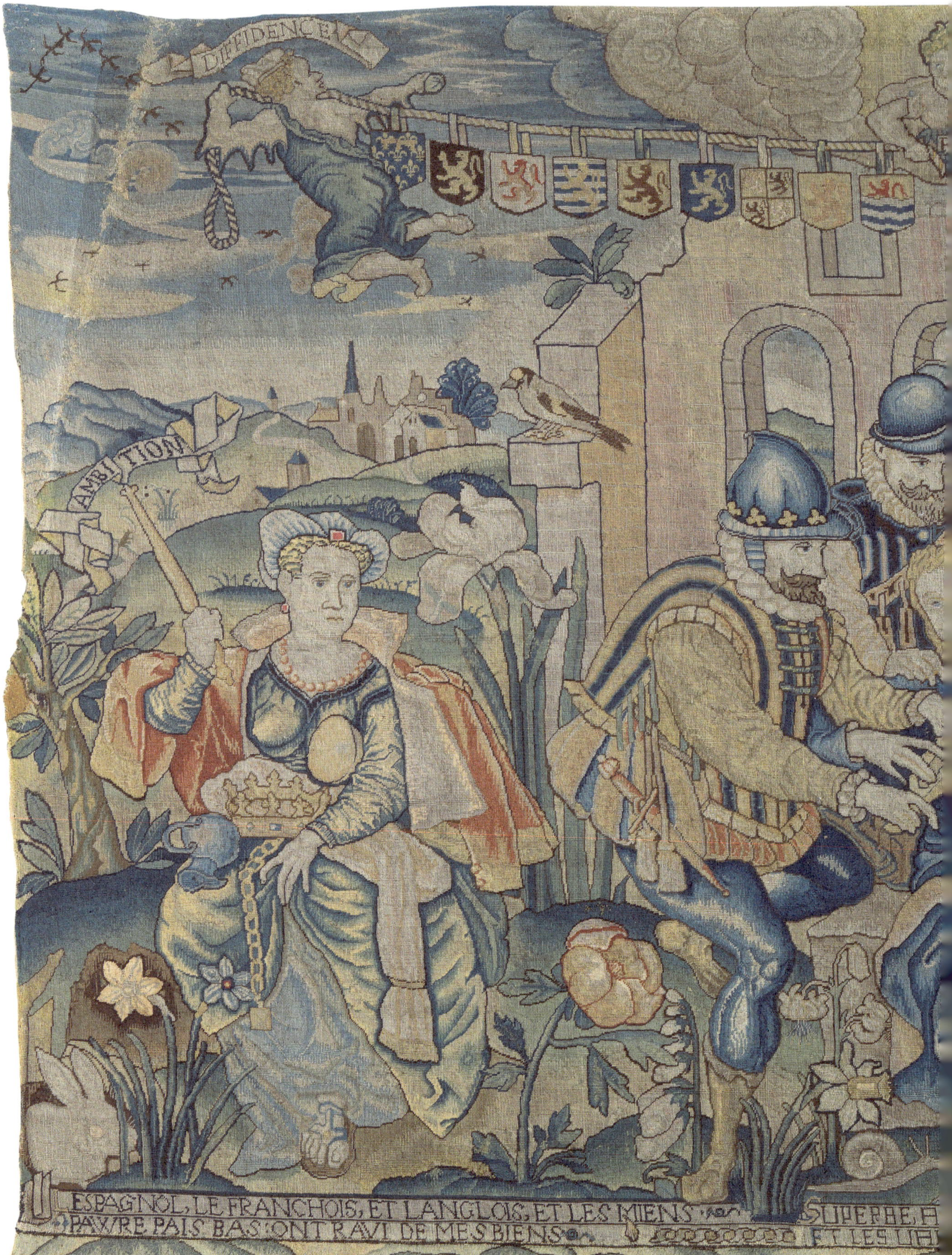

Brussels studio after Hans Collaert
Jammerklacht van de Nederlanden, 1597
Allegory of the plundering of the Low Countries by England, Spain and France
Embroidery in wool and silk, 222 × 162 cm
ANTWERP, THE PHOEBUS FOUNDATION

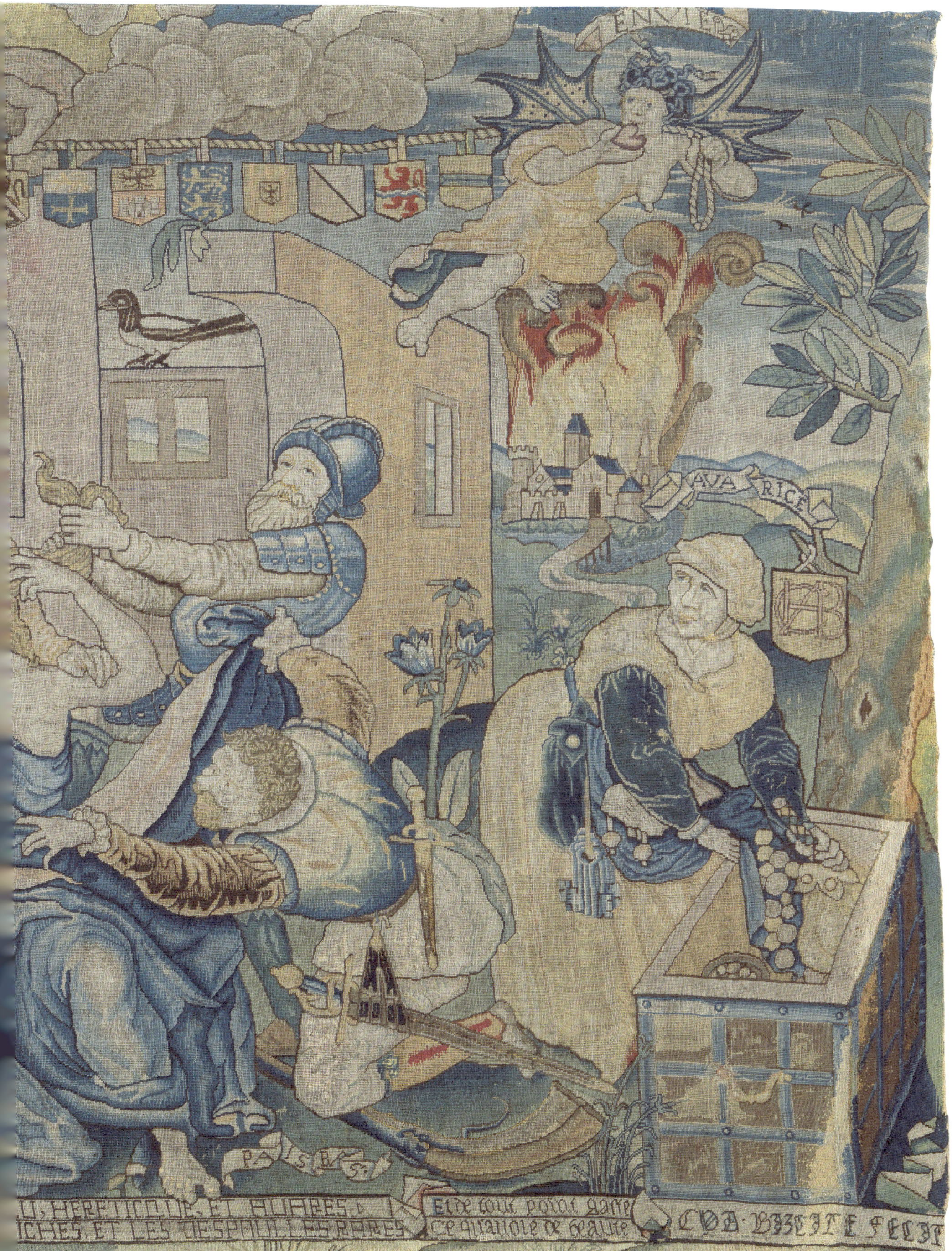
ENVIE
AVARICE

NEAT AND ORDERLY

The oldest map of the Low Countries in The Phoebus Foundation's collection comes from Abraham Ortelius's *Theatrum Orbis Terrarum* (Antwerp, 1570). This copy originated in the 1574 Latin edition of the atlas but has not been adjusted compared to the map in the first edition of 1570. It is oriented northwest, so that the curved coastline is centred symmetrically. The oval decorative frame is also curved so that it 'rhymes' with the coast in an aesthetically pleasing way. Together with the map of Flanders, it is the only map in Ortelius's world atlas with a striking decorative border like this. Although tensions were already running high around 1570 in the pictured area because of the Dutch Revolt (⟼ p. 68), Ortelius's map shows no trace of unrest, and the internal borders are indicated only by thin dotted lines, as if the decorative frame were meant to represent the actual border. The Low Countries form a seemingly peaceful unity here, as neat and orderly as can be.

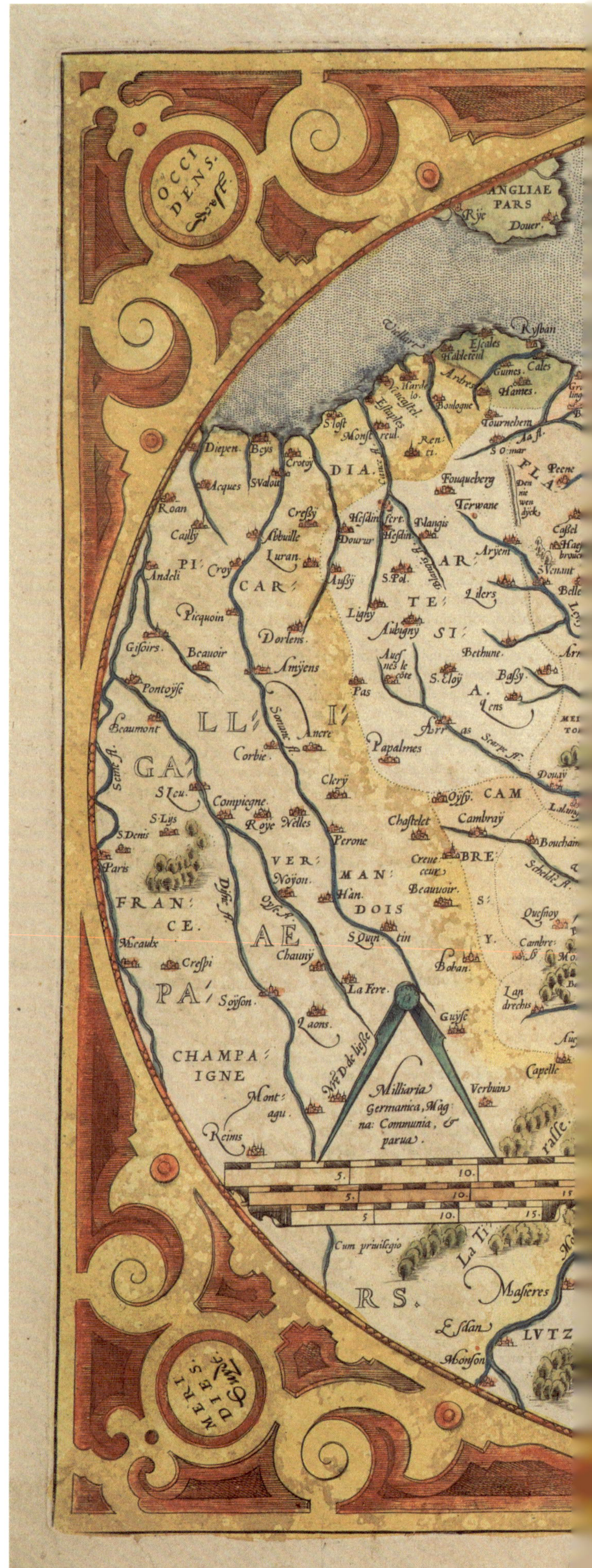

Descriptio Germaniae Inferioris
Map of the Low Countries
From: Abraham Ortelius, *Theatrum Orbis Terrarum*, 1574
ANTWERP, THE PHOEBUS FOUNDATION

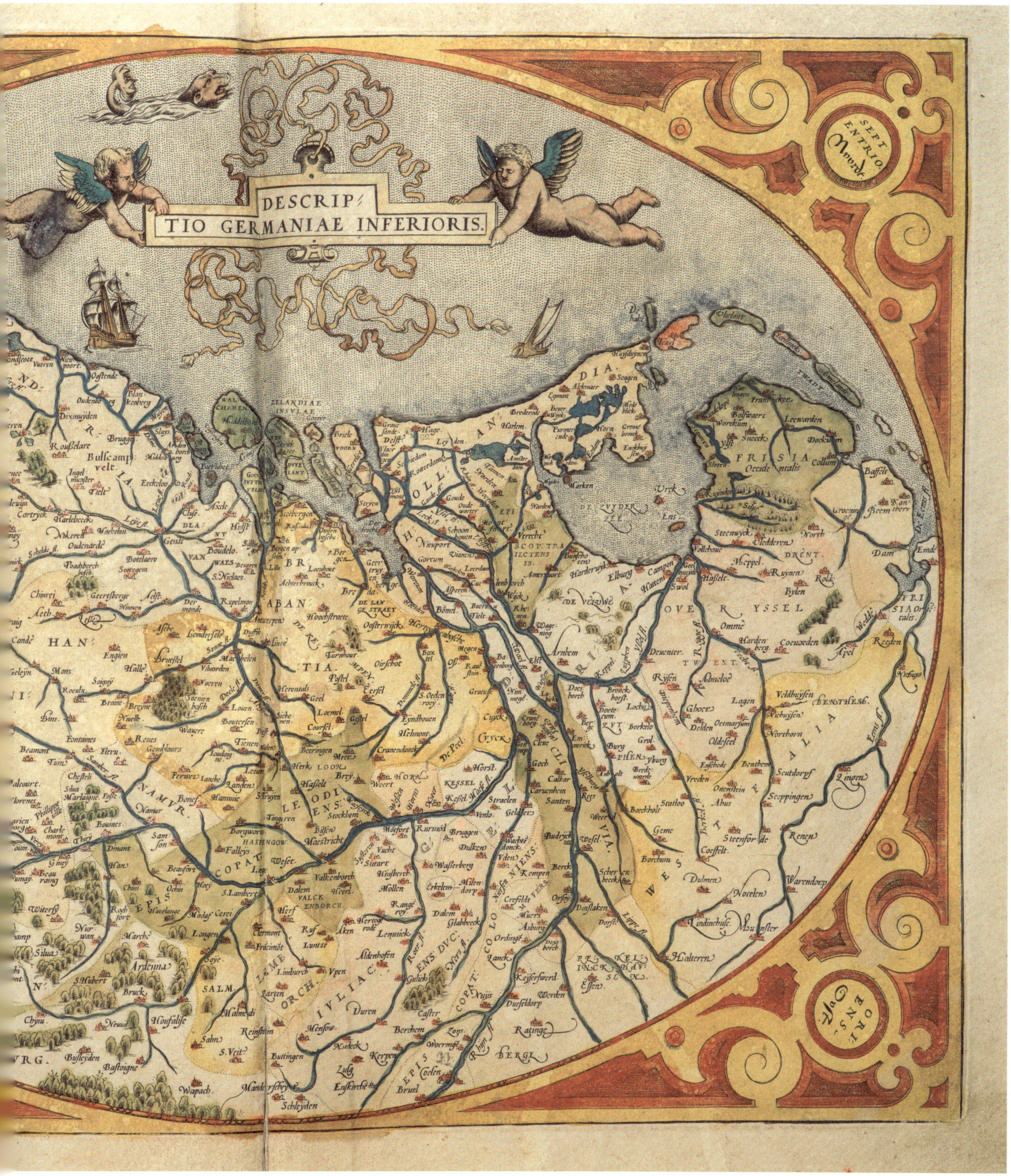
DESCRIP TIO GERMANIAE INFERIORIS
SEPTENTRIO Noordt
ORIENS Oost
ZELANDIAE INSULAE
HOLLANDIA
FRISIA Occidentalis
DE ZUYDER ZEE
OVER YSSEL
DRENT
BRABANTIA
NAMUR
LEODIENS EPISCOPAT
IULIACENS DUC
COLONIENS EPISCOPAT
CLIVIA
WESTFALIA
BENTHEM
ARDENNA
SALM
Arnhem
Maestricht
Antwerpen
Bruessel
Amsterdam
Harlem
Leyden
Utrecht
Deventer
Swol
Campen
Leewarden
Dockum
Emden
Dusseldorp
Munster
Halteren
Reenen
Warendorp
Lingen
Wesel
Cleve
Nymegen
Venlo
Ruremund
Gulick
Duren
Berchem
Zons
Ratinge

Abraham Ortelius

Abraham Ortelius or Wortels (1527-1598) was born and died in Antwerp, where he began his career as a map colourist. At least that is how he was registered from 1547 onwards in the city's Guild of St Luke. He began to deal in books in the 1550s and soon afterwards – at the urging of his friend and teacher Gerard Mercator (⟶ p. 202) – in maps too. He published his first map, an eight-pager showing the world, titled *Typus Orbis Terrarum*, in 1564. Ortelius travelled widely, including multiple trips to the Frankfurter Buchmesse (a major book and print fair) and several to Italy, as well as visits to France, England, Ireland and the Northern Netherlands in the company of Mercator.

Ortelius is considered to be the 'inventor' of the modern atlas thanks to his *Theatrum Orbis Terrarum* (Antwerp, 1570), the first commercial atlas of the world with maps arranged by continent, region and country. As a merchant and traveller, he acquired many of the maps in person and made grateful use of a wide international network. Rather than being a cartographer himself, Ortelius collected the best maps available and adapted them into a uniform whole. He also wrote texts on the back of them, with extensive references to classical authors. Nor did he do the engraving himself: he preferred to leave it to Frans Hogenberg.

Ortelius's atlases proved hugely successful, with numerous editions and translations into Dutch, French, German, English, Italian and Spanish, but they were less valuable in terms of scholarship than the atlases published by Mercator, for instance. As a cosmographer and cartographer himself, Mercator oversaw the mathematical and geographical content more strictly.

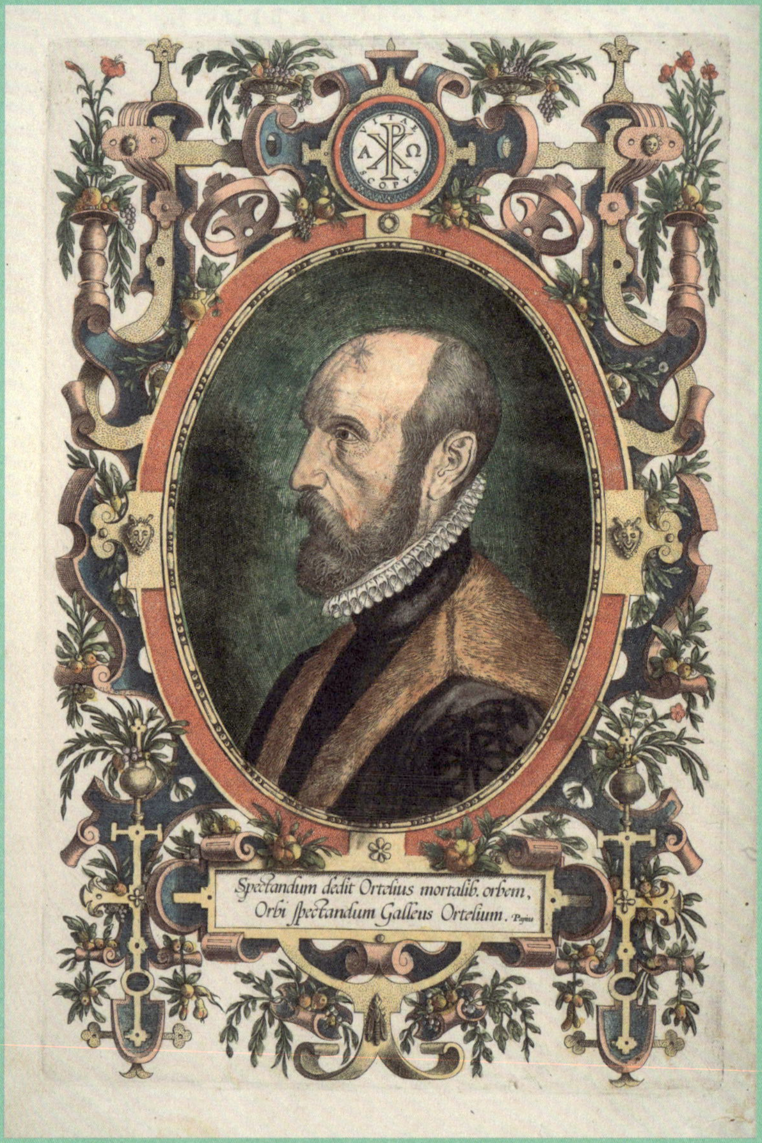

Portrait of Abraham Ortelius
From: Abraham Ortelius, *Theatrum Orbis Terrarum*, 1579
ANTWERP, THE PHOEBUS FOUNDATION

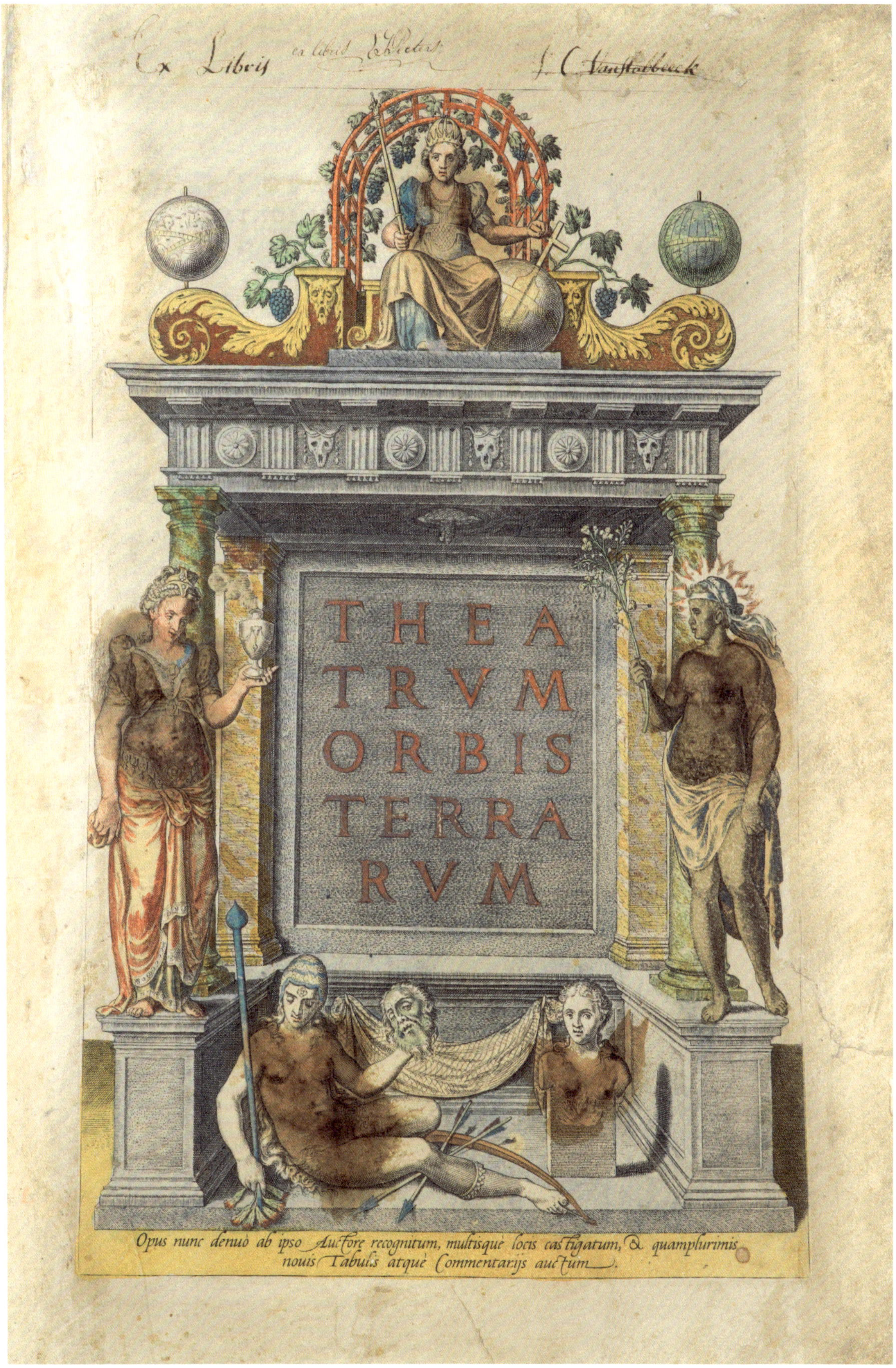

Title page
From: Abraham Ortelius, *Theatrum Orbis Terrarum*, 1579
ANTWERP, THE PHOEBUS FOUNDATION

THE ORIGINAL *LEO BELGICUS*

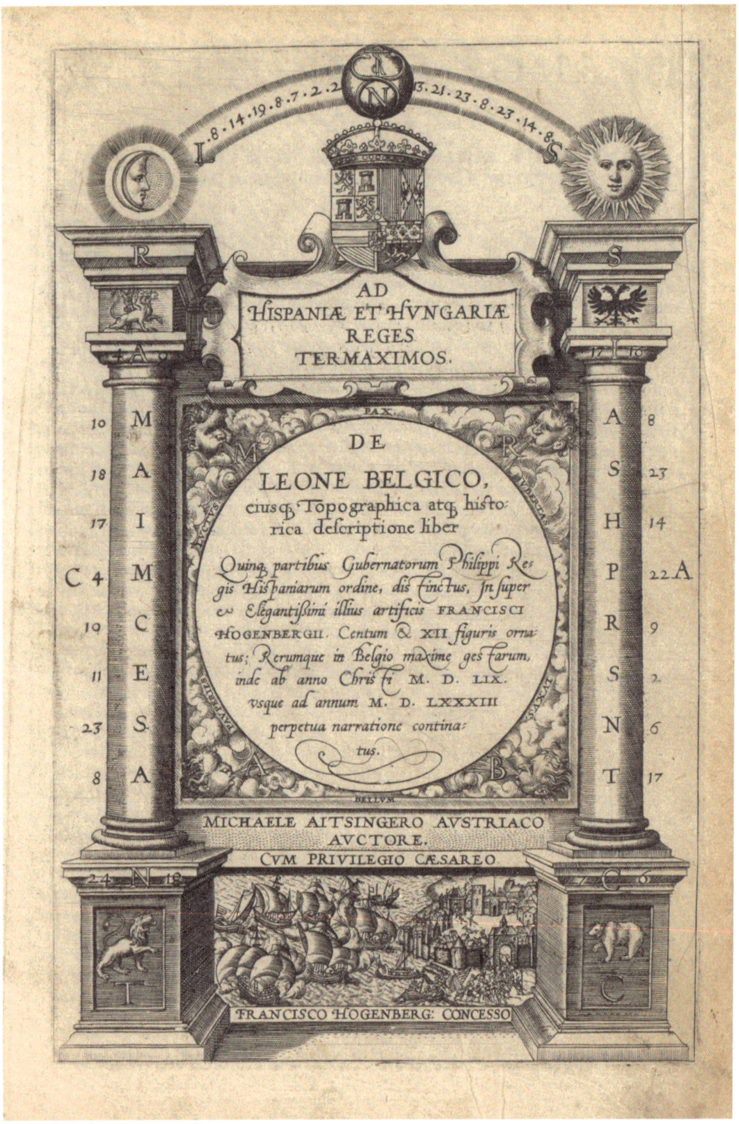

Title page
From: Michael Eytzinger, *De Leone Belgico*, 1583
ANTWERP, THE PHOEBUS FOUNDATION

It was not long before the simmering discontent in the Low Countries spilled over into civil war. The Revolt of the Netherlands against the Spanish erupted, and it was no longer possible to hide the divisions. The Austrian diplomat and historian Michael Eytzinger (1530-1598) wrote a book in Cologne entitled *De Leone Belgico* (Cologne, 1583), in which he described developments in neighbouring countries from the year 1559 onwards. The book was richly illustrated with news prints from the Cologne workshop of Frans Hogenberg, although Eytzinger had the captions at the bottom of the prints removed, on the grounds that they were unduly subjective and politically biased.

At the beginning of the book, Eytzinger placed a map of the conflict zone in the form of a lion: the *Leo Belgicus*, referring to *Gallia Belgica* – the Roman name for the northern part of Gaul between the rivers Seine and Rhine. 'Belgica' became the general Latin name for the Seventeen Provinces and the Low Countries. The choice of a lion is heraldic in origin: the animal appeared in numerous coats of arms in the Low Countries, meaning that it was already widely recognised and respected. The Union of Utrecht (1579) gave birth to the Republic of the Seven United Netherlands, for which the 'Generality Lion' was chosen as the coat of arms. While Eytzinger sought to remain as neutral as possible in the book, free of bias or prejudice, he had no control over the appropriation of his *Leo Belgicus* as the rebels' mascot. It was also a question of timing: the book appeared just as war was breaking out, when there was a need for catchy and powerful imagery. Perhaps this explains why the *Leo Belgicus* struck such a chord. The visually powerful image of the united Netherlands is likely to have instilled courage.

The copy of Eytzinger's *De Leone Belgico* owned by The Phoebus Foundation is missing its crucial *Leo Belgicus* map, further confirming how popular and sought-after the print was (and is). The Phoebus Foundation does, however, have a separate copy of the print taken at some point from a second edition of *De Leone Belgico* from 1586. The map became an icon which many others would continue to develop with variations and imitations. The *Leo Belgicus* took on a life of its own, becoming an element itself in the political debate and social climate in the Low Countries, as we will see from a number of other related maps from the collection.

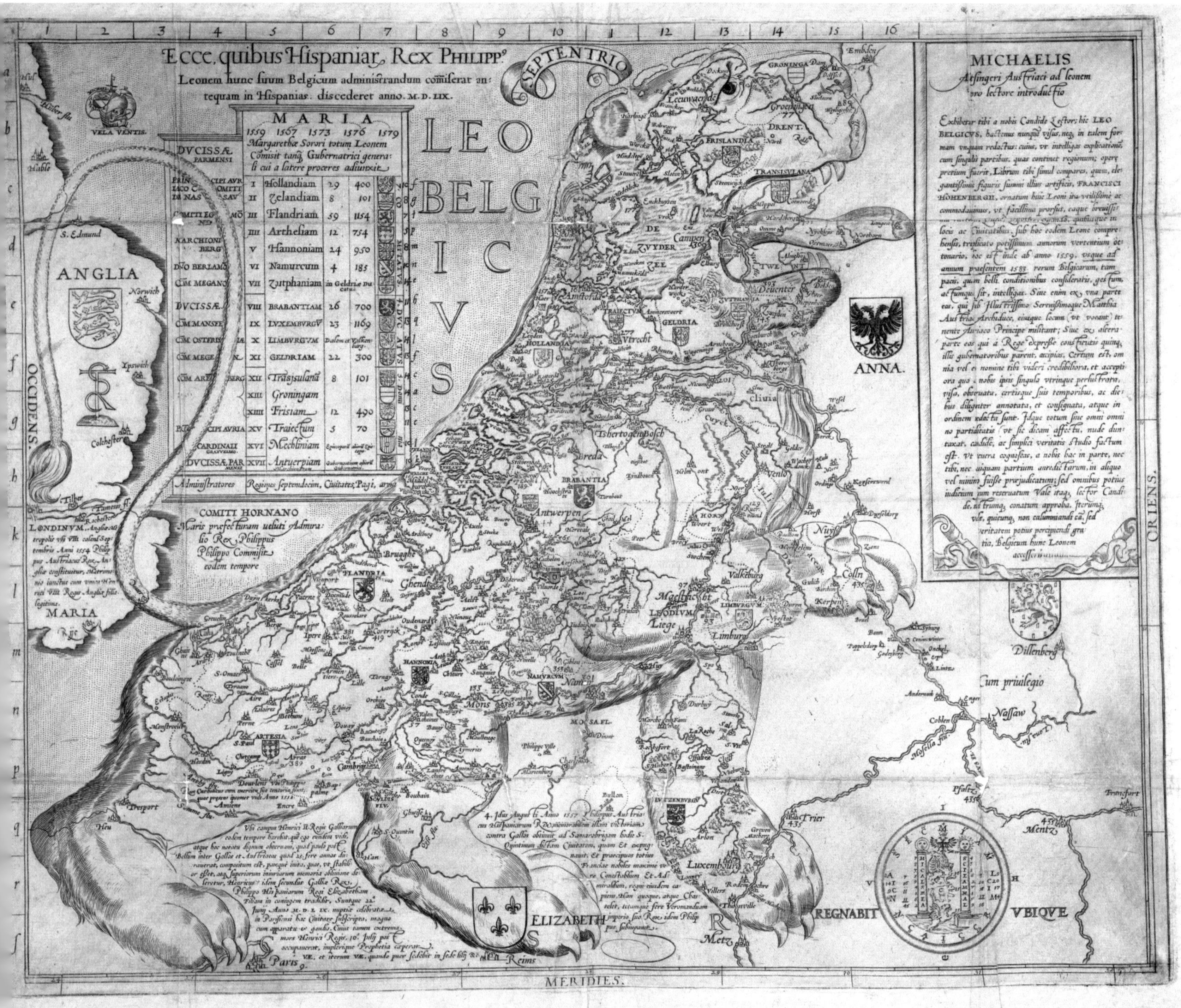

Frans Hogenberg
Leo Belgicus, 1586
Map of the Low Countries in the shape of a lion
Copper engraving, 365 × 495 mm
ANTWERP, THE PHOEBUS FOUNDATION

THE TRUCE LION BY CLAES JANSZ. VISSCHER

The Twelve Years Truce (1609-1621) allowed the United Provinces to catch their breath after years of suffering one battle and one siege after another. Although stadholder and general Maurice of Orange was opposed to the Truce, there was also a desperate need for economic recovery and social unity after years of warfare. Claes Jansz. Visscher belonged to the camp that fully embraced the Truce. His map, first issued around 1611, shows a lion sitting contentedly as it holds out the positive consequences of the imminent truce. The soldier snoozing on the right by the lion's front paw represents a break in the violence ('*slapende oorlogh*' – 'sleeping war'). The effects of the cessation of hostilities can be seen in the background: abundant livestock, peaceful agriculture, bustling commerce, expanding towns and safe travel all mean renewed well-being and prosperity. In this way, the 'Truce Lion' watches over not only the Seventeen Provinces but the welfare of its inhabitants too. North and South are presented here as separate: the decorative frame consists on the left of views of northern cities and on the right of cities from the south, while not one but two medallions hang from the sword held by the lion. The one on the left (with seven arrows for the seven northern provinces) represents the North, the one on the right (with the Burgundian cross) the South.

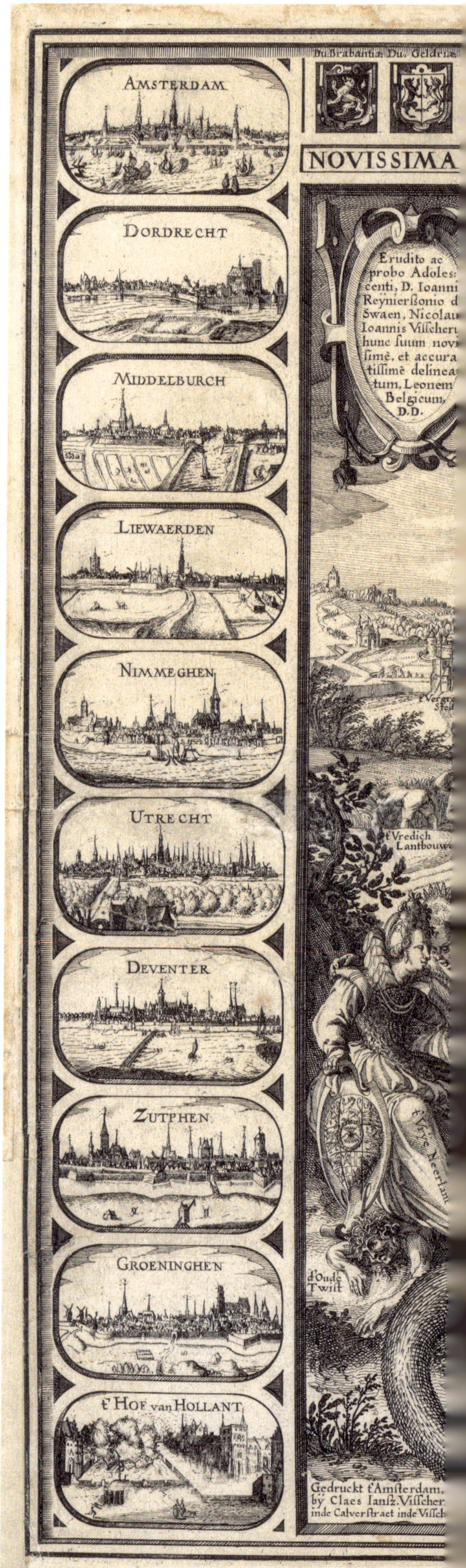

Claes Jansz. Visscher
***Novissima, et accuratissima leonis Belgici, seu septemdecim regionum descriptio*, c.1611**
Map of the Low Countries in the shape of a lion
Copper engraving and etching, 566 × 690 mm
ANTWERP, THE PHOEBUS FOUNDATION

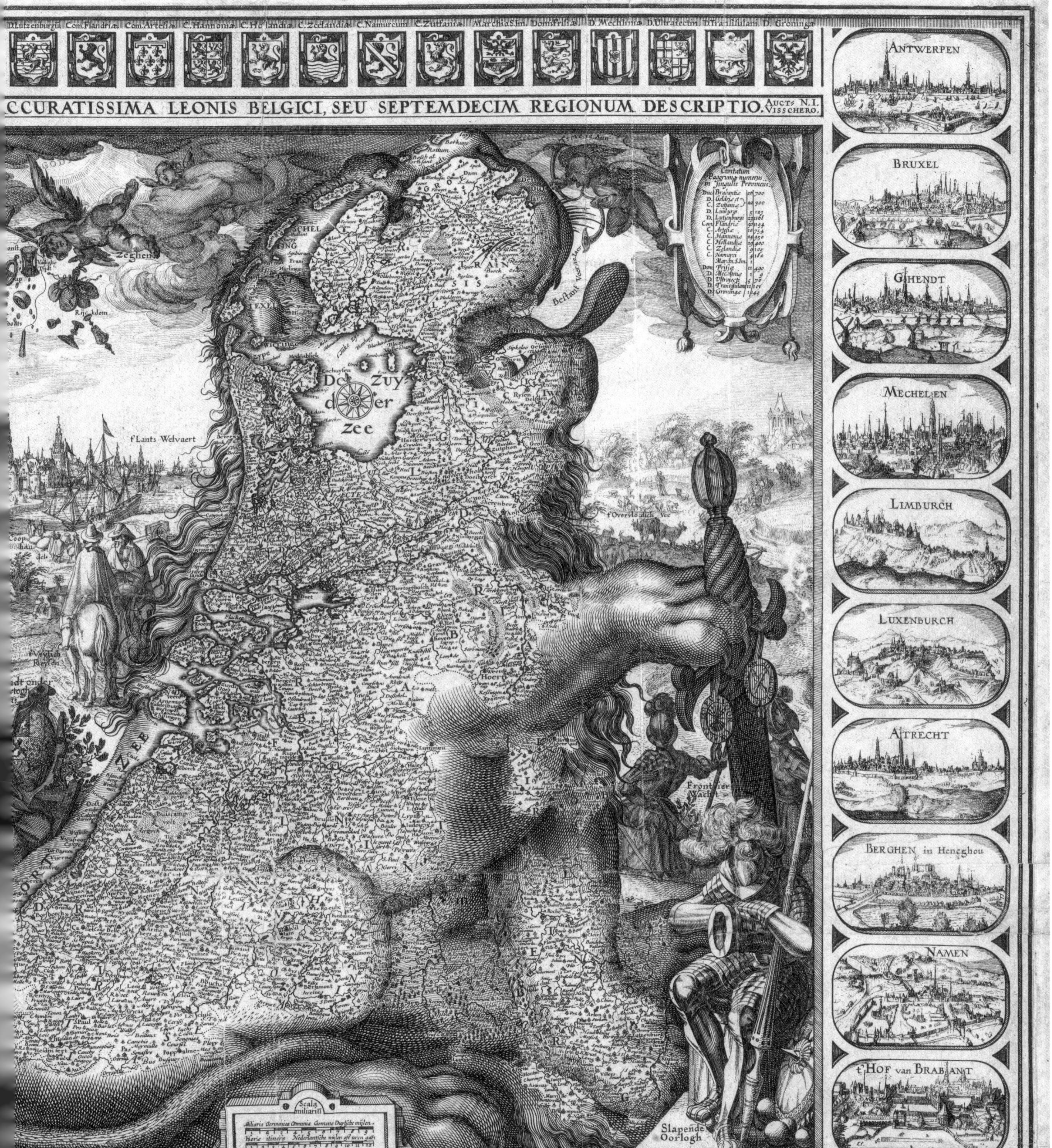
CCURATISSIMA LEONIS BELGICI, SEU SEPTEMDECIM REGIONUM DESCRIPTIO.
ANTWERPEN
BRUXEL
GHENDT
MECHELEN
LIMBURCH
LUXENBURCH
ATRECHT
BERGHEN in Henegou
NAMEN
t'HOF van BRABANT
De Zuyder zee
t'Lants Welvaert
Slapende Oorlogh
Scala miliarii

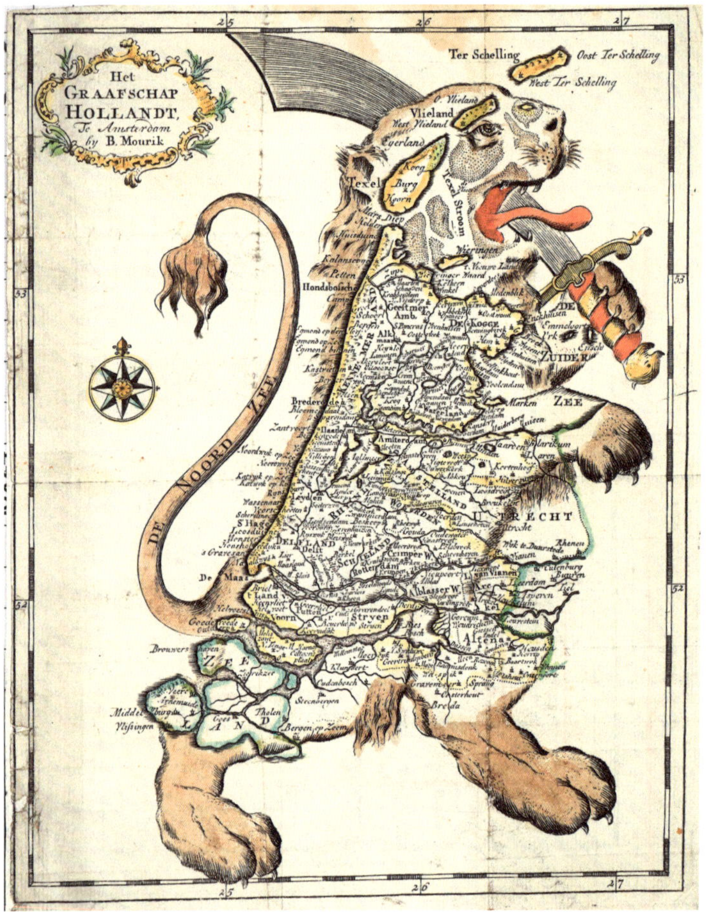

THE LION OF HOLLAND

The Twelve Years Truce was followed in 1621 by a fragile period in which the war-torn North and South grew increasingly apart. Many maps now showed just the northern or the southern part of the Low Countries. The northern provinces, with Holland taking a leading role, appropriated the *Leo Belgicus* in opposition to the South. Claes Jansz. Visscher responded to this societal trend once more, with his map *Comitatus Hollandiae denuo forma Leonis*: the County of Holland (still depicted here with Zeeland) in the shape of a lion. Unlike the Truce Lion, this beast is no longer relaxed, sitting on its behind, but instead roars angrily. Its raised sword bears the menacing, pugnacious words '*Patriae defensio*' – 'I defend the fatherland'. Holland was ready to hurl itself back into the fray. The Lion of Holland remained a symbol of the province even after the end of the Eighty Years War in 1648. As late as 1780, *Leo Hollandicus* was still flaunting himself in a title print for the *Naamregister* for Holland – an administrative list of people serving on city councils. Tensions between North and South might have eased by the end of the eighteenth century, but patriotic pride in Holland persisted.

***Het Graafschap Hollandt*, 1780**
The Leo Belgicus as title print for Bernardus Mourik's Naamregister
Hand-coloured copper engraving, 177 × 231 mm
ANTWERP, THE PHOEBUS FOUNDATION

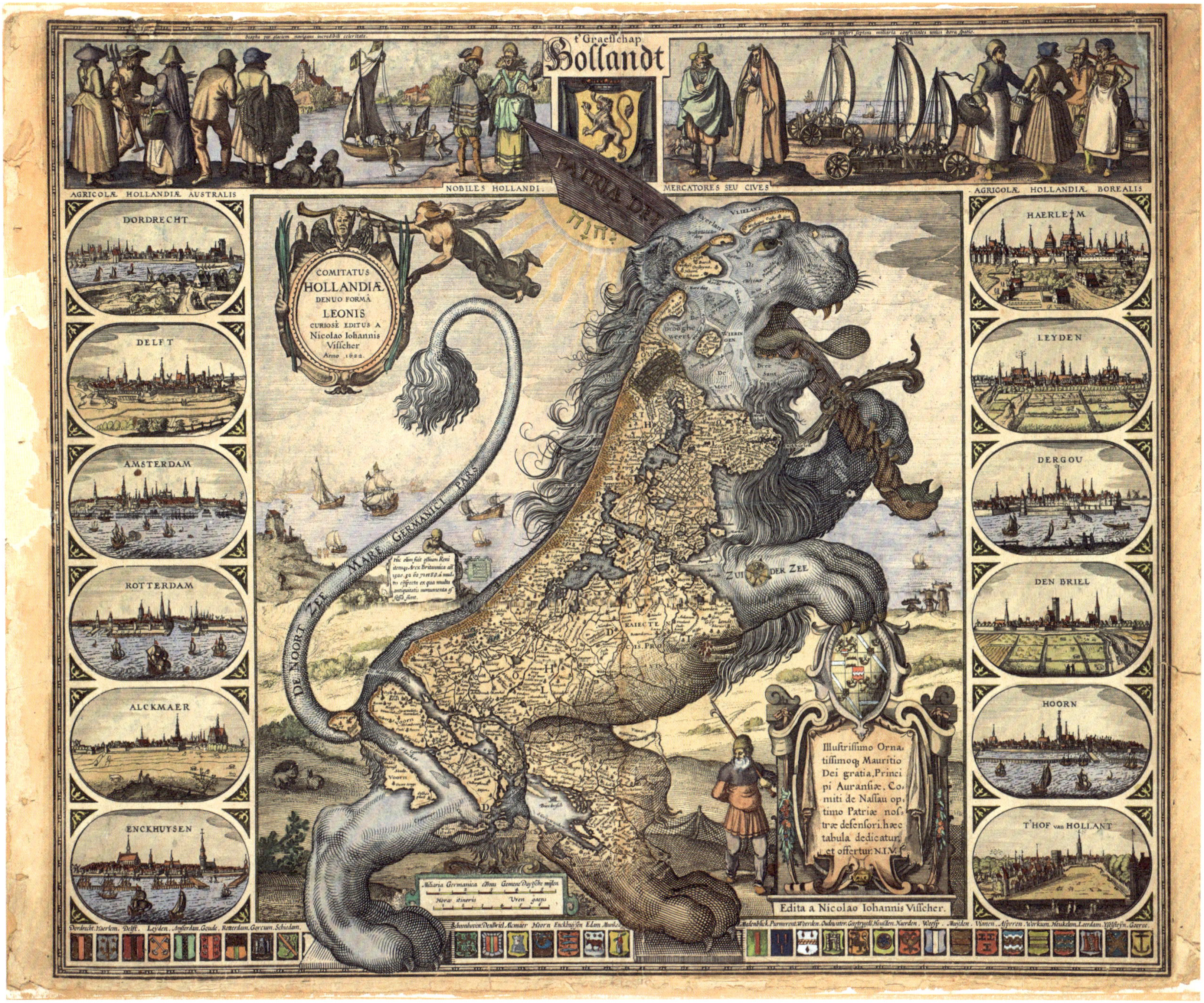

Claes Jansz. Visscher
Comitatus Hollandiae denuo forma Leonis, 1622
Map of the province of Holland in the shape of a lion
Hand-coloured copper engraving and etching, 503 × 611 mm
ANTWERP, THE PHOEBUS FOUNDATION

CITY ATLAS OF THE SEPARATED NETHERLANDS

'From my print workshop, on the very day that peace was declared between the King of Spain and the United Dutch Provinces.' This was the final sentence of the foreword penned by Joan Blaeu for the Dutch translation of his great city atlas of the world. The first volume contained the cities of the Northern Netherlands. The year 1648, which saw the signing of the Peace of Münster, was the ideal time for him to market the atlas. It marked the end of the Eighty Years War and the beginning of the definitive separation of the Southern, Habsburg Netherlands and the Northern, 'United' Netherlands. The first volume of his universal city atlas came in two tomes, therefore, one devoted the cities of the United Provinces, the second to the Spanish Netherlands. The former was dedicated to the States General and the stadholder, the latter to the King of Spain. Blaeu did not intend the two tomes to be viewed in a particular order. In the South, therefore, people will have treated the tome devoted to the Spanish Netherlands as the first, while those in the North will have seen 'their' section as tome one.

The Phoebus Foundation has a set of five atlases by Joan Blaeu that are bound in a uniform manner in parchment and probably remained together that way: the first volume of the *Atlas Novus* (1644-1645), the second volume of the *Atlas Novus* (1645), the *Novus atlas Sinensis* (Atlas of China, 1655) and, lastly, the atlas of the cities of the Low Countries (1649) in two volumes, the very first edition of the city atlas in Latin. The volume devoted to the United Netherlands, *Novum ac Magnum Theatrum Urbium Belgicae Liberae et Foederatae* ('New and Great Theatre of the Cities of the Free and United Netherlands'), is the first in this specific set, while the *Novum ac Magnum Theatrum Urbium Belgicae Regiae* ('New and Great Theatre of the Cities of the Royal Netherlands') is the second. The compiler or previous owner is thus likely to have been someone from the Northern Netherlands. Shortly after the Latin edition, Blaeu also published an edition in Dutch, titled *Toonneel der steden van de [...] Nederlanden*. His ambition was to expand the atlas to include cities from all over the world, but in the end he got no further than a few additional volumes on Italian cities.

Joan Blaeu's celebrated city atlas of the Low Countries was much more, however, than simply a collection of urban maps. There were, of course, practical reasons for him to begin with the Low Countries, as this was the region for which he had the most material available, but there was also a commemorative motive. In the Dutch translation, he wrote about the bravery and pugnacity shown in the course of eighty years in the Low Countries, 'in which so many armies and riches melted away, lands and cities spoiled' – and gave this recent history as the reason for the innovation. The atlas was also a symbol, therefore, of the peace that had finally returned and a monument to all the Netherlandish cities and citizens who had been involved in the conflict. The fact that he divided the Low Countries into two volumes was entirely in keeping with the political situation: 'I offer you the Netherlands in two volumes, [...] which division will not, undoubtedly, displease them, since it was done by themselves, and ought, therefore, to be distasteful to no one.'

Blaeu responded to current events and to the political mood, therefore, but did so with considerable historical awareness and respect for the Revolt, all of which is reflected in how the atlas is compiled. Present and past are united. The two volumes consist of over two hundred maps, but a substantial part – found on the reverse of the maps in the atlas – consists of text recounting the history of the various cities and locations, small and large. The written sections also address specific events during the Revolt, while the maps themselves are far from static, timeless city plans. Take the map of Ostend, which is also simultaneously a narrative of the siege of that town (1601-1604) (↣ p. 94). The crossed (Spanish) and striped flags (Republic), the half-moons, the encirclement of the city, and Spinola's camp in the lower left tell us that it is the siege that we see depicted here. The accompanying text lists the numbers of casualties, while the map tells the story of the event and immortalises the act of war through its inclusion in Blaeu's atlas of the world's cities.

Besides the narrative siege maps and city plans, the work also includes bird's eye views, panoramas and maps of fortifications. Blaeu consulted cartographic material but will also have drawn on history books and chronicles for the historical descriptions on the reverse of the maps. His atlas not only depicts and maps locations in the Low Countries; it also narrates and commemorates the experience of the region's cities during the Eighty Years War.

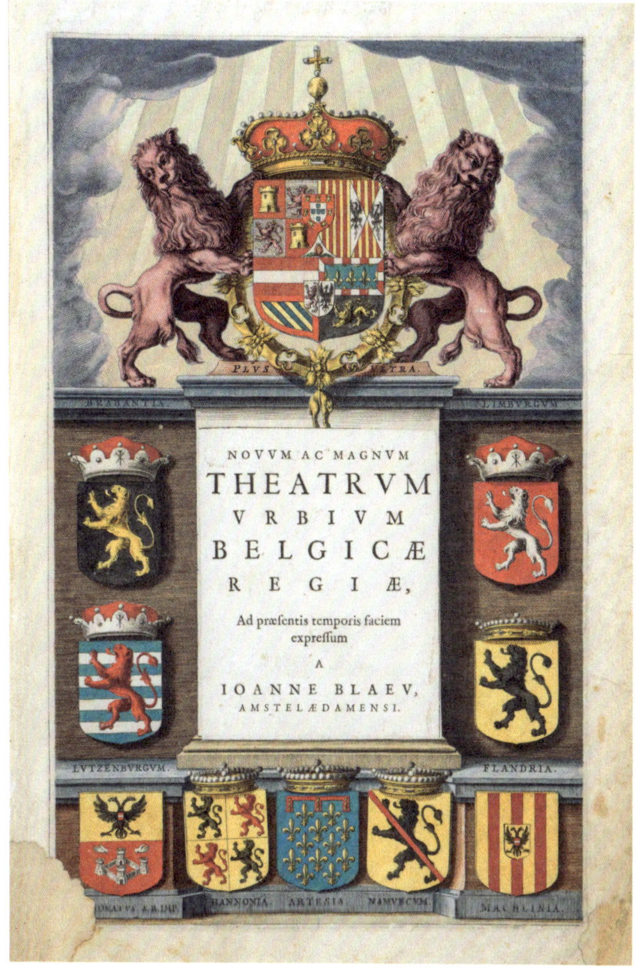

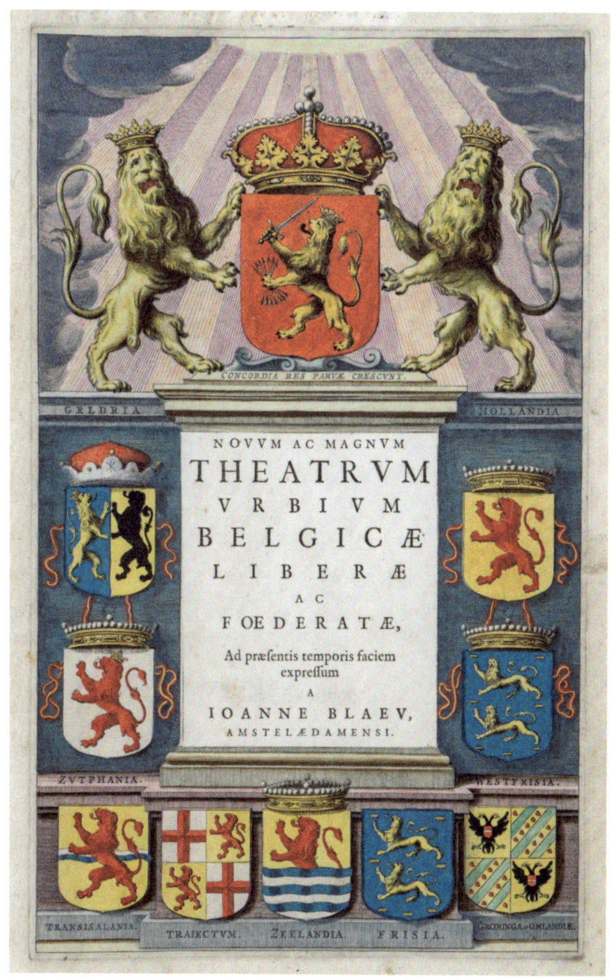

Title page
From: Joan Blaeu, *Novum ac Magnum Theatrum Urbium Belgicae Regiae*, 1649
ANTWERP, THE PHOEBUS FOUNDATION

Title page
From: Joan Blaeu, *Novum ac Magnum Theatrum Urbium Belgicae Liberae et Foederatae*, 1649
ANTWERP, THE PHOEBUS FOUNDATION

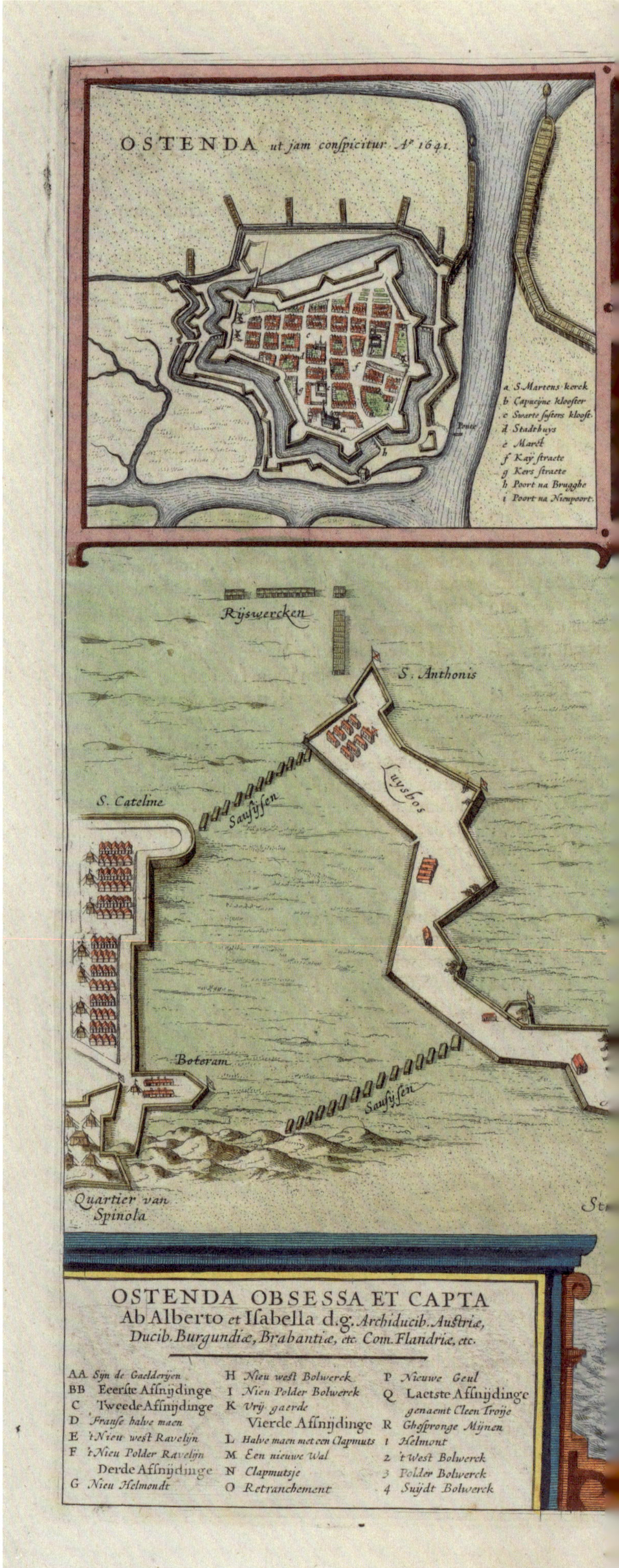

Ostenda Obsessa et Capta
Map of the Siege of Ostend from 1601 to 1604
From: Joan Blaeu, *Novum ac Magnum Theatrum Urbium Belgicae Regiae*, 1649
ANTWERP, THE PHOEBUS FOUNDATION

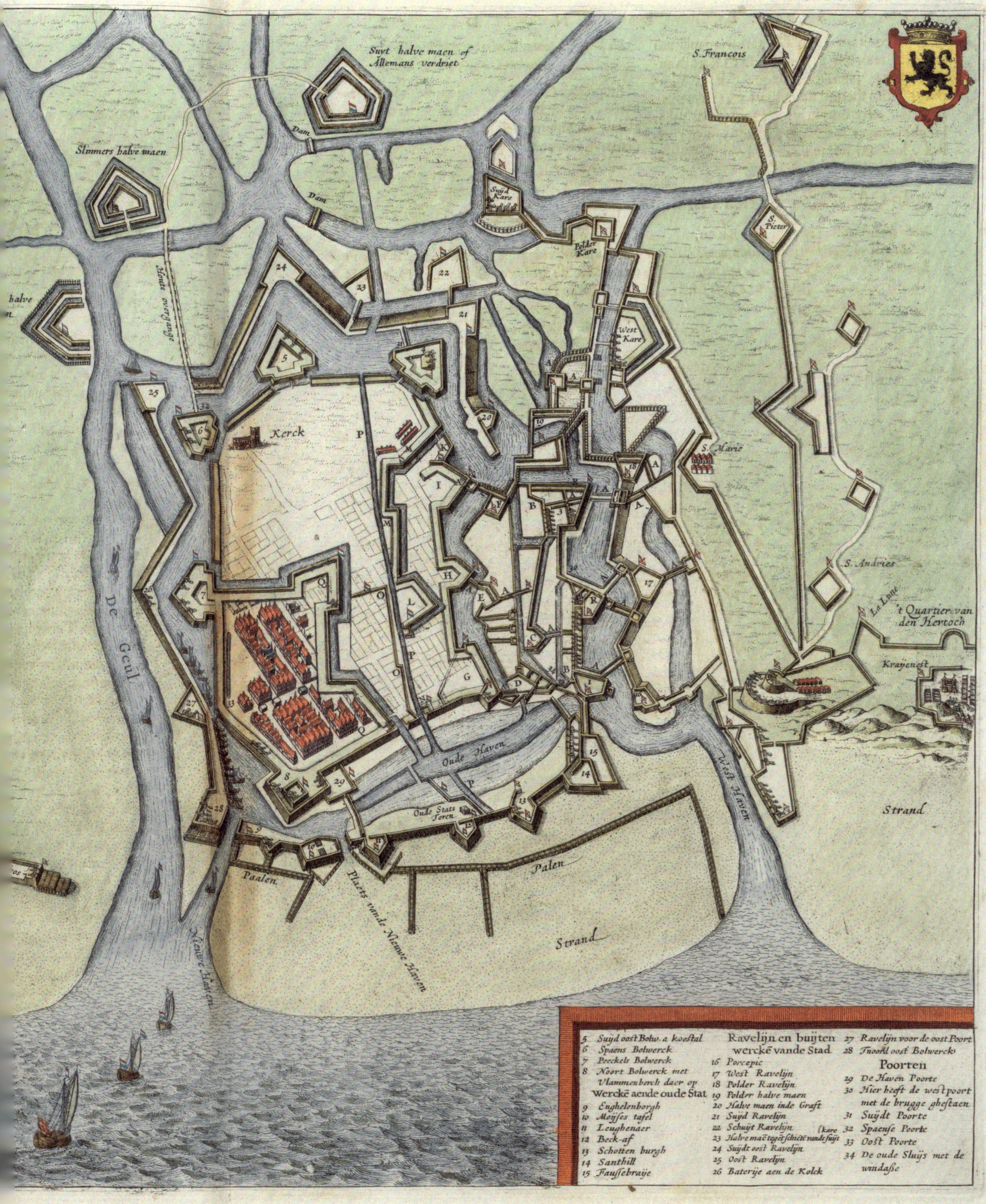
Suyt halve maen of Allemans verdriet
S. Francois
Slimmers halve maen
Dam
Dam
Suyd Kare
Polder Kare
S. Pieter
West Kare
Kerck
S. Marie
S. Andries
La Lonne
't Quartier van den Hertoch
Krayenest
De Geul
Oude Haven
Oude Stats Toren
West Haven
Strand
Paalen
Palen
Strand
Plaets vande Nieuwe Haven
Nieuwe Haven
5 Suijd oost Bolw. a koestal
6 Spaens Bolwerck
7 Peeckels Bolwerck
8 Noort Bolwerck met Vlammenberch daer op
Werckē aende oude Stat
9 Enghelenborgh
10 Moijses tafel
11 Leughenaer
12 Beck-af
13 Schotten burgh
14 Santhill
15 Faussebraije
Ravelijn en buijten werckē vande Stad
16 Porcepic
17 West Ravelijn
18 Polder Ravelijn
19 Polder halve maen
20 Halve maen inde Graft
21 Suijd Ravelijn
22 Schuijt Ravelijn
23 Halve maē tegē schietē vande suijt kare
24 Suijdt oost Ravelijn
25 Oost Ravelijn
26 Baterije aen de Kolck
27 Ravelijn voor de oost Poort
28 Tnoord oost Bolwerck
Poorten
29 De Haven Poorte
30 Hier heeft de west poort met de brugge ghestaen
31 Suijdt Poorte
32 Spaense Poorte
33 Oost Poorte
34 De oude Sluijs met de windaße

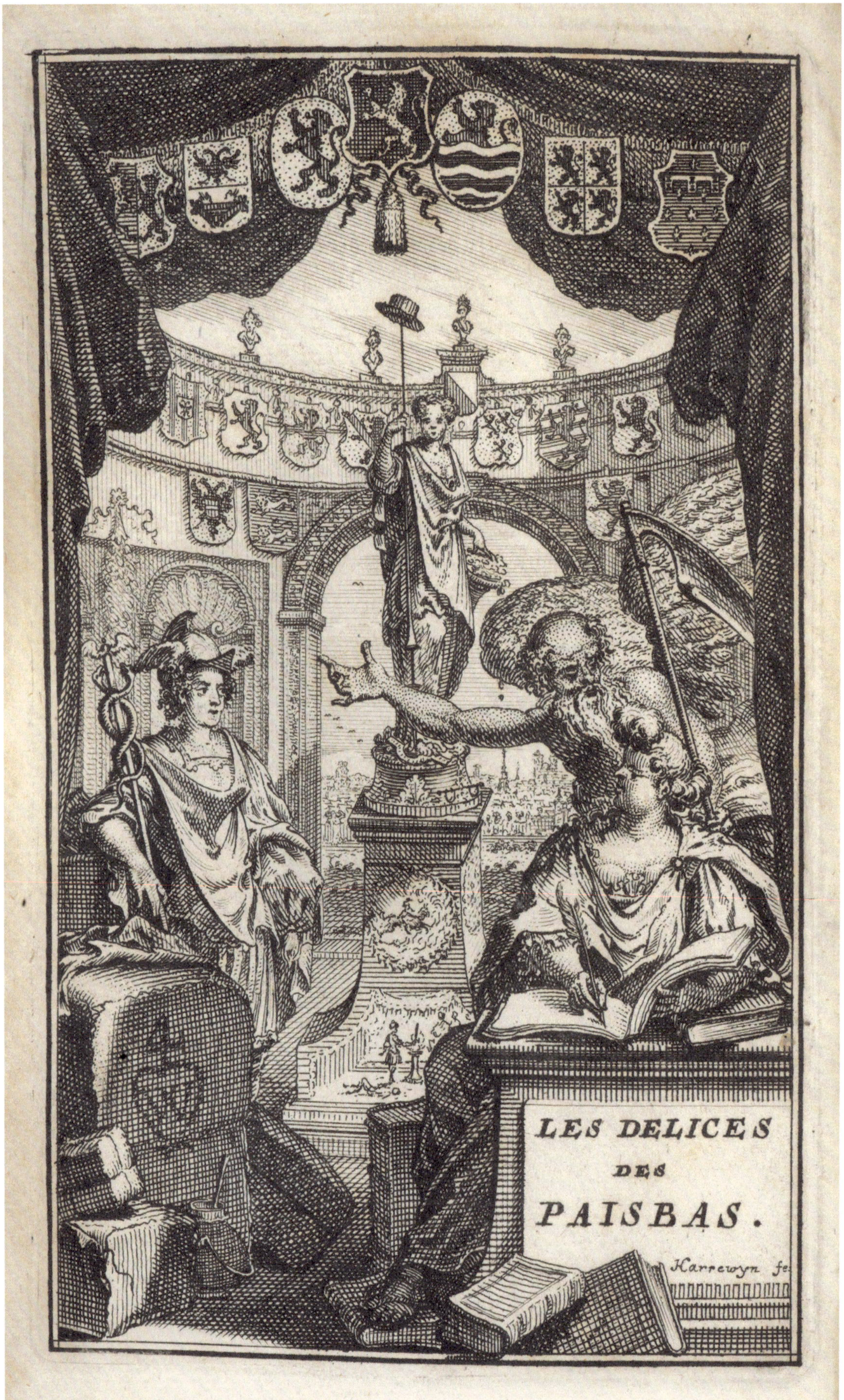

Title page
From: Jean-Baptiste Christyn, *Les Delices des Pais-Bas*, 1711
ANTWERP, THE PHOEBUS FOUNDATION

THOSE GLORIOUS, REUNITED NETHERLANDS

The tendency to represent the Northern and Southern Netherlands as separate entities was by no means always observed. Maps showing all Seventeen Provinces continued to be produced until well into the eighteenth century. *Les Delices des Pais-Bas* ('Delights of the Low Countries'; Brussels, 1711) describes the region as a united whole, despite all the annexations and conflict. The multi-volume illustrated book is a chorography (⟼ p. 60) by Jean-Baptiste Christyn (1630-1690), containing both geographical and historical descriptions, with illustrations and maps etched by the Brussels printmaker Jacob Harrewijn (1660-1727).

The first edition in 1697 ran to three volumes, but the work was expanded in the eighteenth century (until 1793) to five volumes. Volume one is introduced by a fold-out map of the Seventeen Provinces, engraved by Harrewijn, with no clear indication of the border between North and South. Nor does the map inform the reader that the union of the Seventeen Provinces has not existed since the Fall of Antwerp in 1585. The coat of arms of Flanders appears next to that of Friesland in the frame on the left.

Why might such a harmonious presentation have been chosen? The answer is that this was in keeping with the tradition of chorography and of Guicciardini, in which a region or country was presented as a whole. There might have been commercial considerations too: a larger territory meant a bigger market. The text also presents a narrative, however, in which the shared values, characteristics and history of the Low Countries are stressed. The book looks back wistfully at the early history of the region's provinces, counties, duchies and aristocratic estates, from the *Bello Gallico* in classical antiquity to the dukes of Burgundy and the revolt against Philip II. These were the shared roots that connected the Northern and Southern Netherlands, and the focus is on the traits and experiences that united Netherlanders rather than those that divided them. North and South shared a highly urbanised landscape, independent cities and regions, and relative freedom and tolerance.

IN THE LION'S SHADOW

The *Tafel van de XVII Nederlandze Provincien* also offers a survey of the Seventeen Provinces as if they still belonged together, with a key to the cities, coats of arms, the principal towns and villages, political and religious diversity, trades, inventions, historical information and chambers of the Dutch East India Company. The publisher of this print, Pieter Schenk I (1660-1719), added a map of the Low Countries in the upper left, but not just any map – we recognise the sketchy contours as those of the *Leo Belgicus*. Schenk reused Caspar Specht's original map from 1705, the title of which states: 'The Seventeen Provinces, being 4,928 square miles in extent [...] the upper part of the lion contains the United Netherlands' (although it takes a degree of imagination to make out the *Leo Belgicus* in the map). Schenk reissued the *Tafel* in 1707, accentuating the lion's contours. The beast's tail now curls into the North Sea, its velvety mane hangs along the shore, and its teeth bite mercilessly through the frame of the map. Despite the separation of North and South, the lion's shadow thus lived on as a reminder of the Netherlandish provinces that once bunched together as a fist against their common enemy.

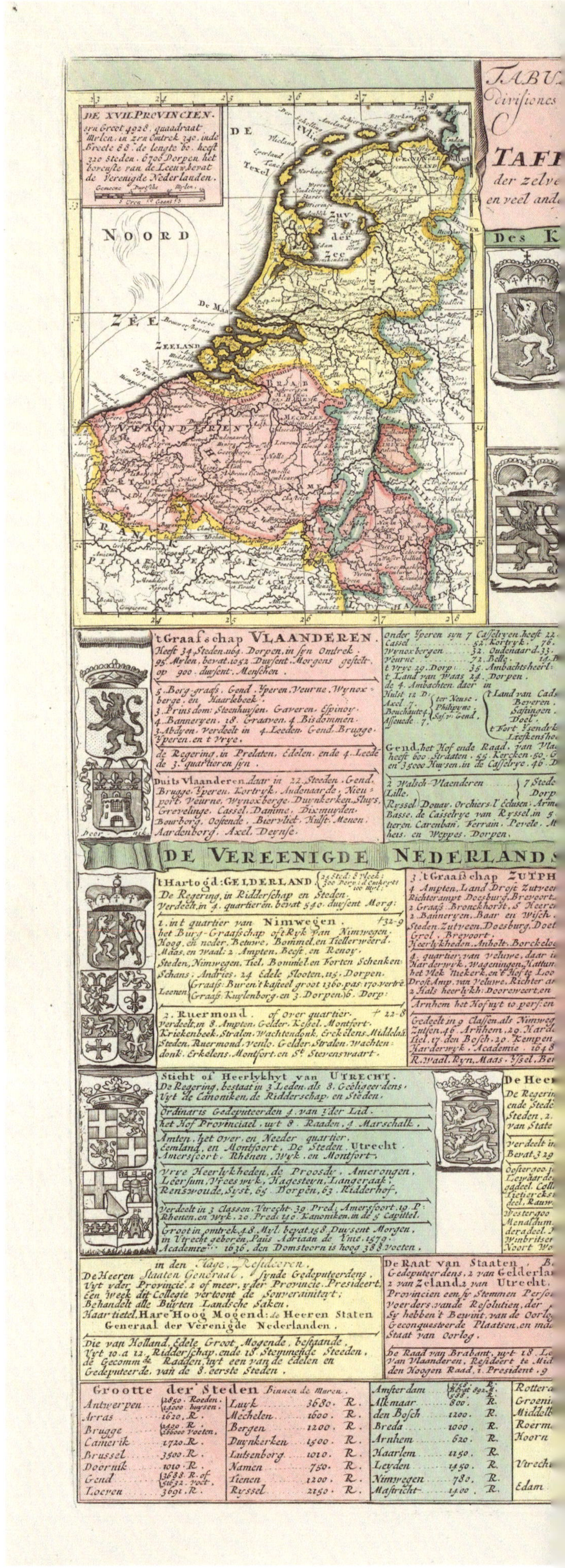

Pieter Schenk I and Caspar Specht
Tafel van de XVII Nederlandze Provincien, 1707
Table of the Seventeen Provinces with descriptions
Hand-coloured copper engraving, 485 × 546 mm
ANTWERP, THE PHOEBUS FOUNDATION

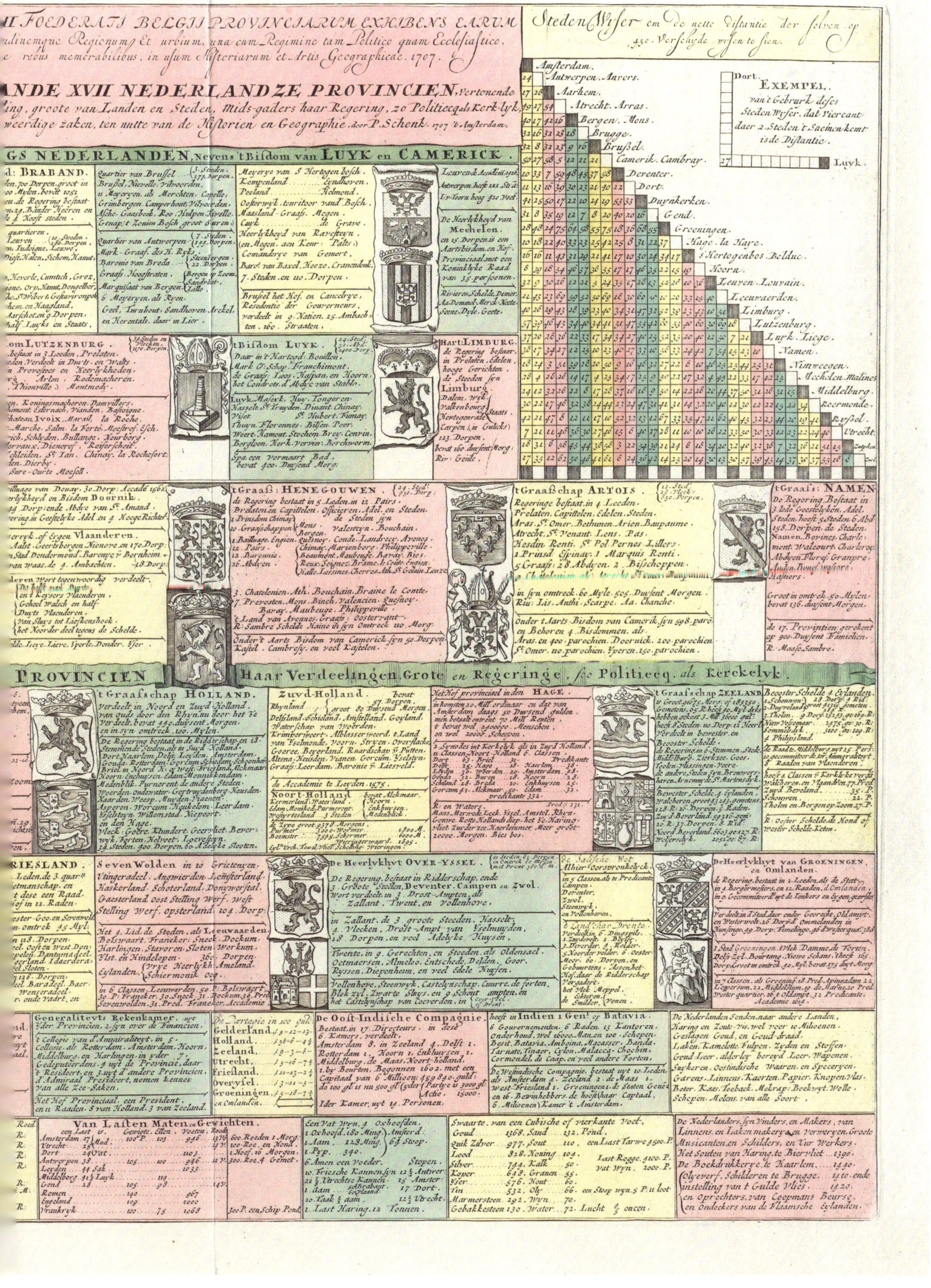
II FOEDERATI BELGII PROVINCIARUM EXHIBENS EARUM
Steden Wyser om de nette distantie der selven op 450 verschyde wysen te sien.
ANDE XVII NEDERLANDZE PROVINCIEN
GS NEDERLANDEN, Nevens 't Bisdom van LUYK en CAMERICK
EXEMPEL
PROVINCIEN
Haar Verdeelingen, Grote en Regeringe, so Politiecq als Kerckelyk.
Van Laiten Maten en Gewichten

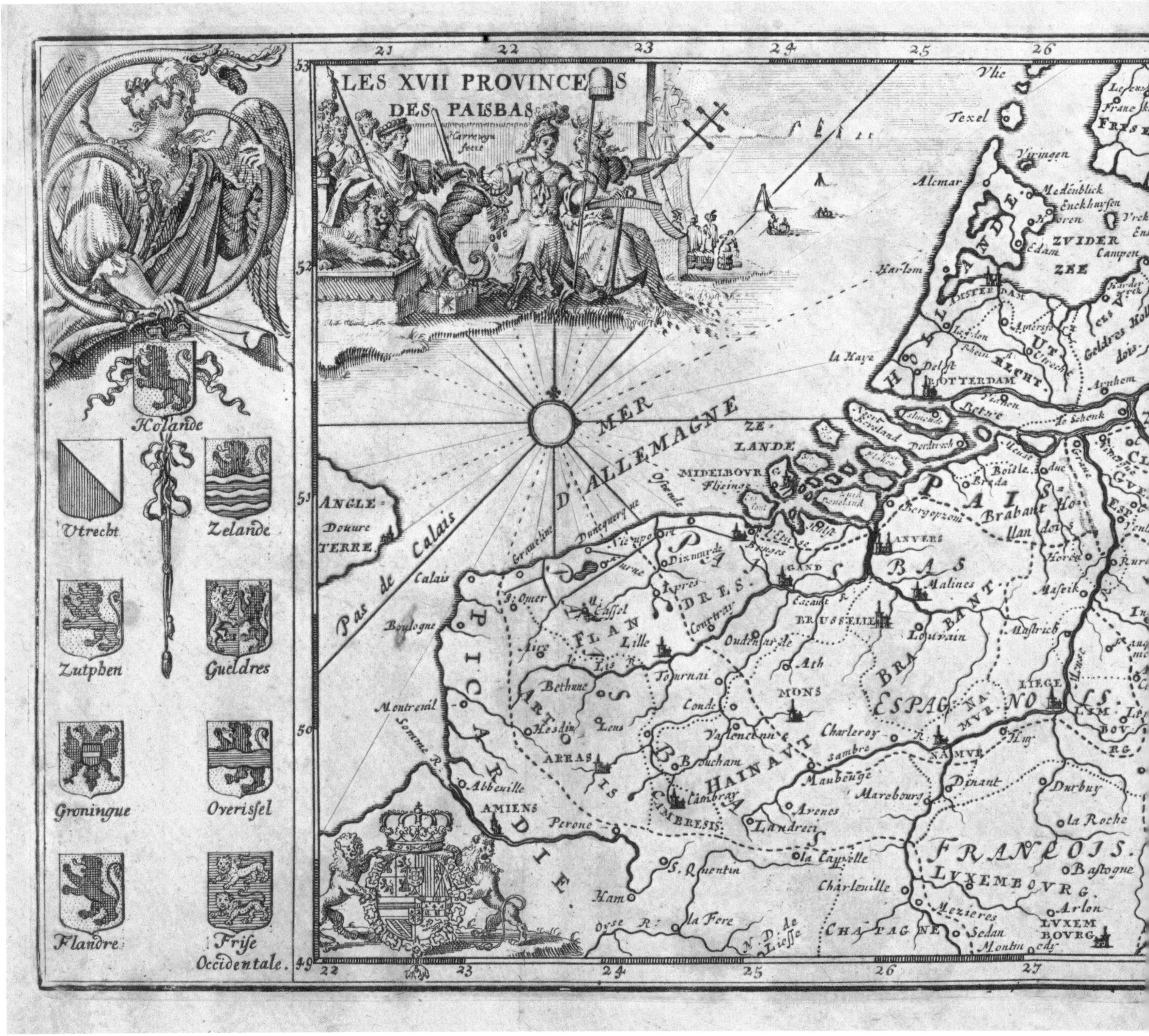

Jacob Harrewijn
Les XII Provinces des Paisbas
Map of the twelve Dutch provinces, flanked by their coats of arms
From: Jean-Baptiste Christyn, *Les Delices des Pais-Bas*, 1711
ANTWERP, THE PHOEBUS FOUNDATION

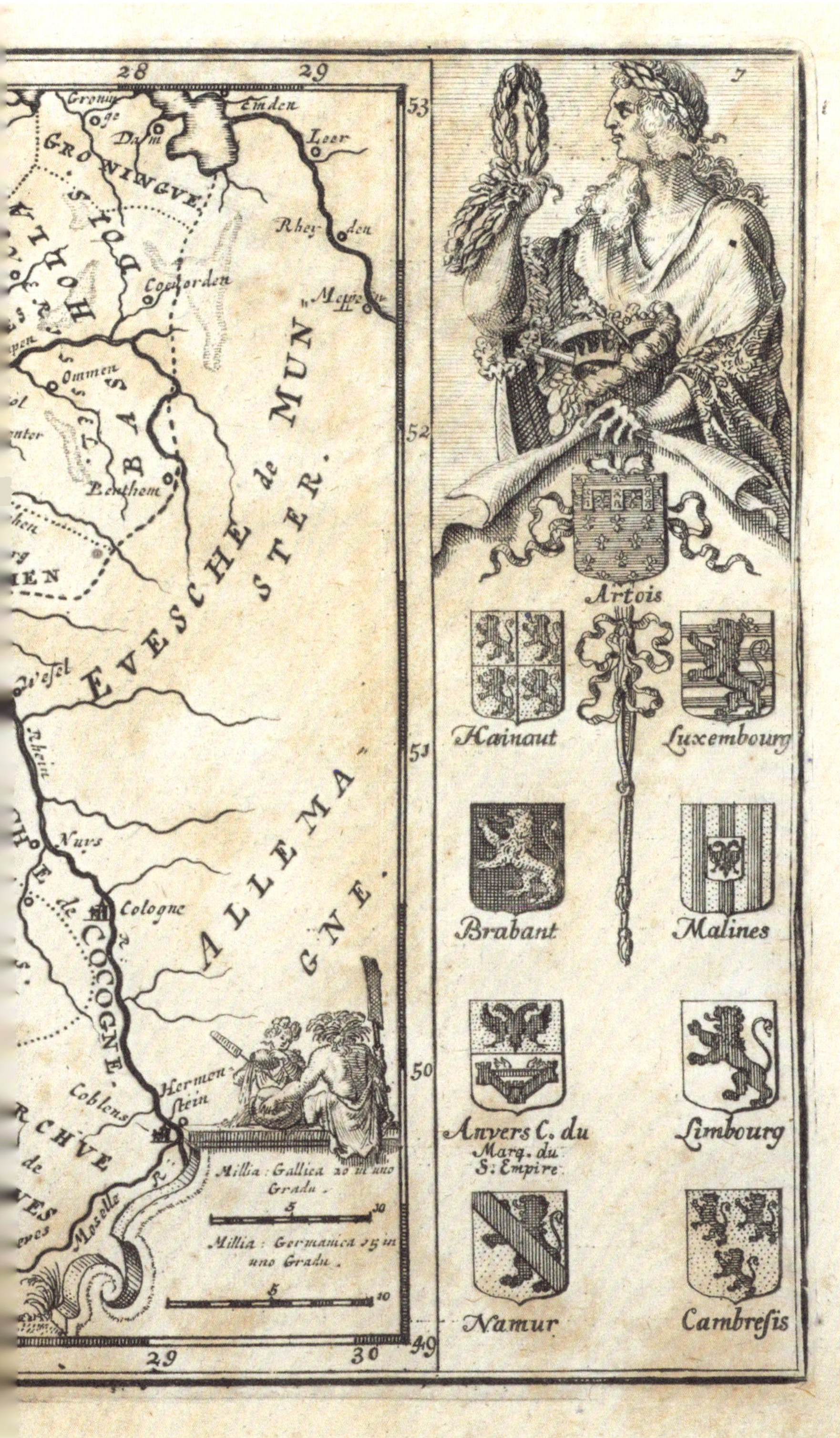

LITERATURE

M. Donkersloot-de Vrij, *Drie generaties Blaeu: Amsterdamse cartografie en boekdrukkunst in de zeventiende eeuw*, Amsterdam, 1992.

C. Engrand, 'Les Délices des Pays-Bas, miroir des Dix-sept provinces?', *Revue du Nord*, 2-3, 2005: 487-511.

G. Schilder, *Monumenta Cartographica Neerlandica, Vol. VI: Nederlandse foliokaarten met decoratieve randen, 1604-60*, Alphen aan den Rijn, 2000.

R. Tooley, *Leo Belgicus: an Illustrated List of Variants*, London, 1963.

H. van der Heijden, *Leo Belgicus: an Illustrated and Annotated Carto-Bibliography*, Alphen aan den Rijn, 2006.

H. van der Heijden, *Oude kaarten der Nederlanden, 1548-1794: historische beschouwing, kaartbeschrijving, afbeelding, commentaar*, Alphen aan den Rijn, 1998.

M. van den Broecke, *Ortelius Atlas Maps: an Illustrated Guide*, 't Goy, 1996.

B. van 't Hoff, *Toonneel der steden van de Vereenighde Nederlanden, met hare beschrijvingen (Holland en West-Vriesland / Utrecht)*, Amsterdam / Brussels, 1966.

The bombardment of Brussels

The colourful map on the right-hand page was engraved by the Brussels artist and engraver Jacob Harrewijn (1660-1727). The design and possible measurements are attributable to the *mathesios studiosi* (mathematicians) J. Laboureur and J. Van der Baren, about whom little else is known. What is striking about this map is that it mentions the bombardment of 1695 in the lower left. On 13, 14 and 15 August, the army of Louis XIV bombarded the city centre as part of the Nine Years War (1688-1697) between France and an alliance of several European powers, including the Spanish Netherlands. The shelling and resulting fires were among the greatest disasters in the history of the Southern Netherlands. There is little trace of this devastation in the map, however, merely a dotted line in the middle marking the affected area plus a line of infantry on the right, outside the ramparts. The work is dedicated to the target of the attack, Maximilian Emmanuel (1662-1726), elector of Bavaria and governor of the Spanish Netherlands between 1691 and 1706.

With the allegory of victory in the upper right and the personification of the city's wealth lower right, the map presents this figure as if he were the victor, despite the immense damage and loss of life inflicted on Brussels. This is in stark contrast to a print by Nicolas de Fer, also made after the bombardment, which clearly depicts the batteries and the shelling, and in which it is the inside of the walls that has been left conspicuously blank. Harrewijn's large map, including the original city centre, might have been prompted by his desire to restore its past glory.

The plan of Brussels, including the street names in Dutch, is clearly visible because of the stylized blocks of buildings. Only the principal edifices, explained in the lower right-hand corner of the legend, are depicted three-dimensionally – a typically eighteenth-century form of representation. Not many copies of this map have been traced.

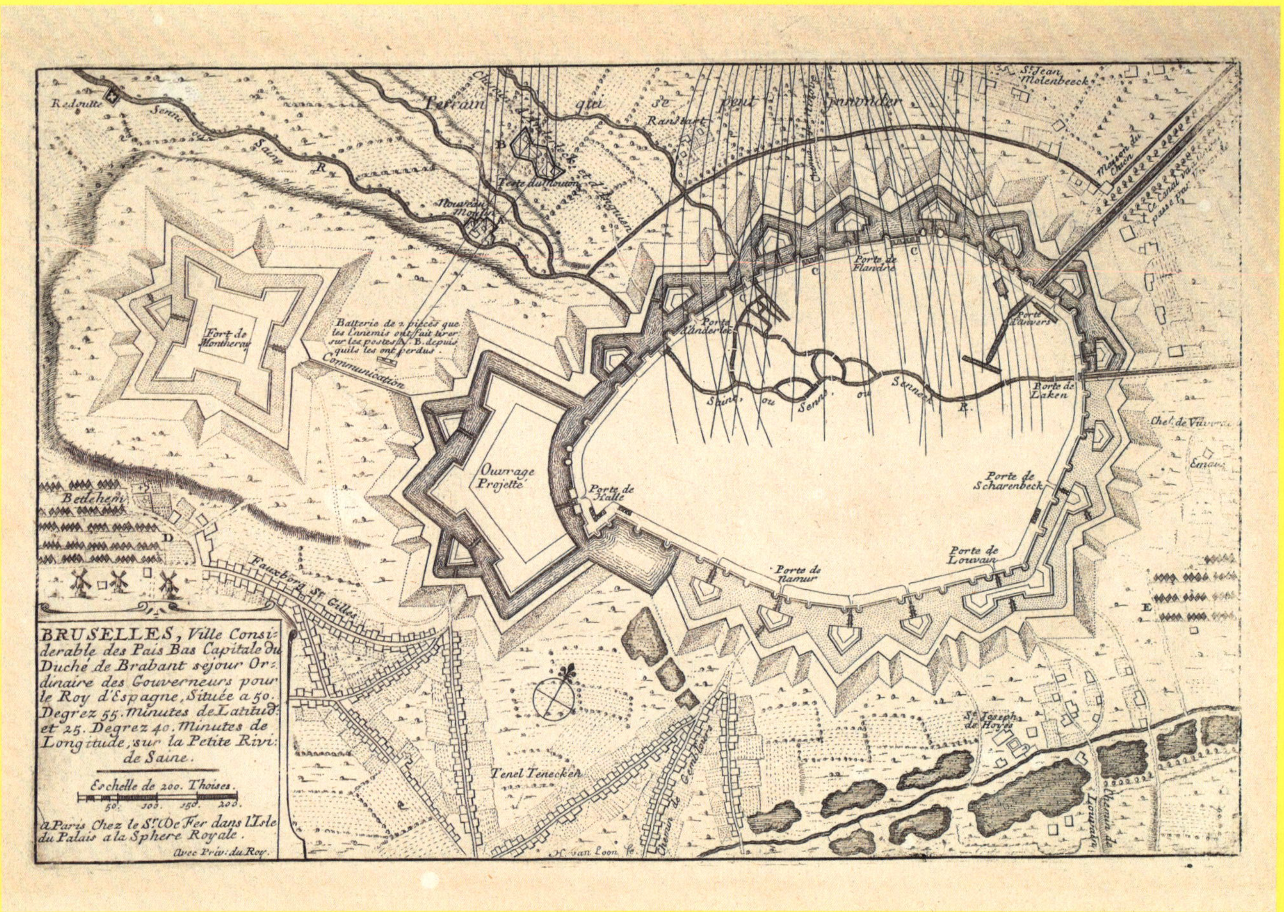

Harmanus Van Loon after Nicolas de Fer
Bruselles, Ville Considerable des Pais Bas Capitale du Duché de Brabant, c.1695
Map of the bombardment of Brussels by the French in 1695
Etching, 190 × 300 mm
ANTWERP, THE PHOEBUS FOUNDATION

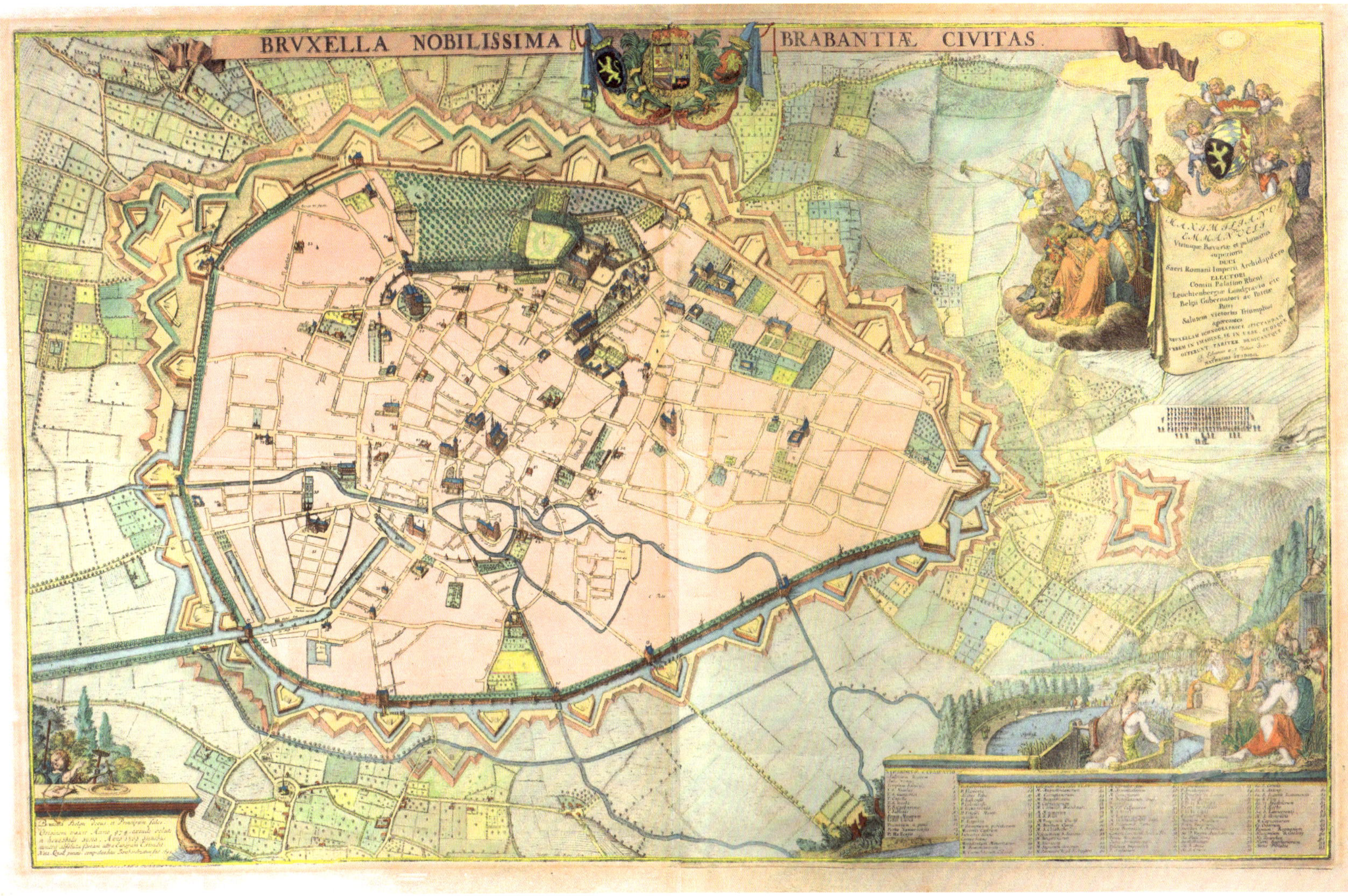

Jacob Harrewijn after J. Laboureur and J. Van der Baren
Bruxella nobilissima Brabantia civitas, c.1695
Hand-coloured copper engraving and etching, 638 × 1028 mm
ANTWERP, THE PHOEBUS FOUNDATION

It was midway through the sixteenth century that the Italian Lodovico Guicciardini (1521-1589) came to Antwerp to seek his fortune. Feasting his eyes on what he found, the Florentine merchant penned his *Descrittione di tutti i Paesi Bassi, altrimenti detti Germania inferiore* ('Description of all the Low Countries, also known as Lower Germany'; Antwerp, 1567), a book that would cause a sensation. Not that Guicciardini was the only Italian to venture onto Southern Netherlandish soil: in an earlier chapter we already mentioned Carolus Scribani, whose father was an Italian nobleman dispatched to Brussels to work as advisor to Margaret of Parma, and there were plenty of other migrants from Italy and the Iberian Peninsula in the region too. The Revolt of the Netherlands against Spanish rule brought a fresh influx of Italians, who accounted for many of the soldiers deployed by the Spanish-Habsburg King Philip II in his conflict with the rebels. Many were relatively educated men who recorded and published their experiences and observations. And while Netherlanders did not always agree with the Italian perspective on the war itself, they happily devoured the work of southern-European historians. Books by Italian authors thus spread throughout the Low Countries in the form of translations and reprints. They also included a good deal of cartographic material, another respect in which Guicciardini set the trend with his amply illustrated city atlas.

We saw in previous chapters how people in the Low Countries themselves viewed their own surroundings and cities, the events of the Eighty Years War, and the unity and separation of the Northern and Southern Netherlands, but how did foreigners view the region's people, cities and recent history? The Phoebus Foundation has numerous examples in its collection of Italian books and maps that focus on the Low Countries. Their authors were not shy about expressing their view of events, concentrating on their own victories while portraying what they saw as the godless, rebellious Protestant Netherlanders in a very different light. Antwerp – still under Spanish dominion – was a safe haven in which to print contentious ideas like this.

CHAPTER VI

THROUGH OTHERS' EYES

ITALIANS IN THE SOUTHERN NETHERLANDS

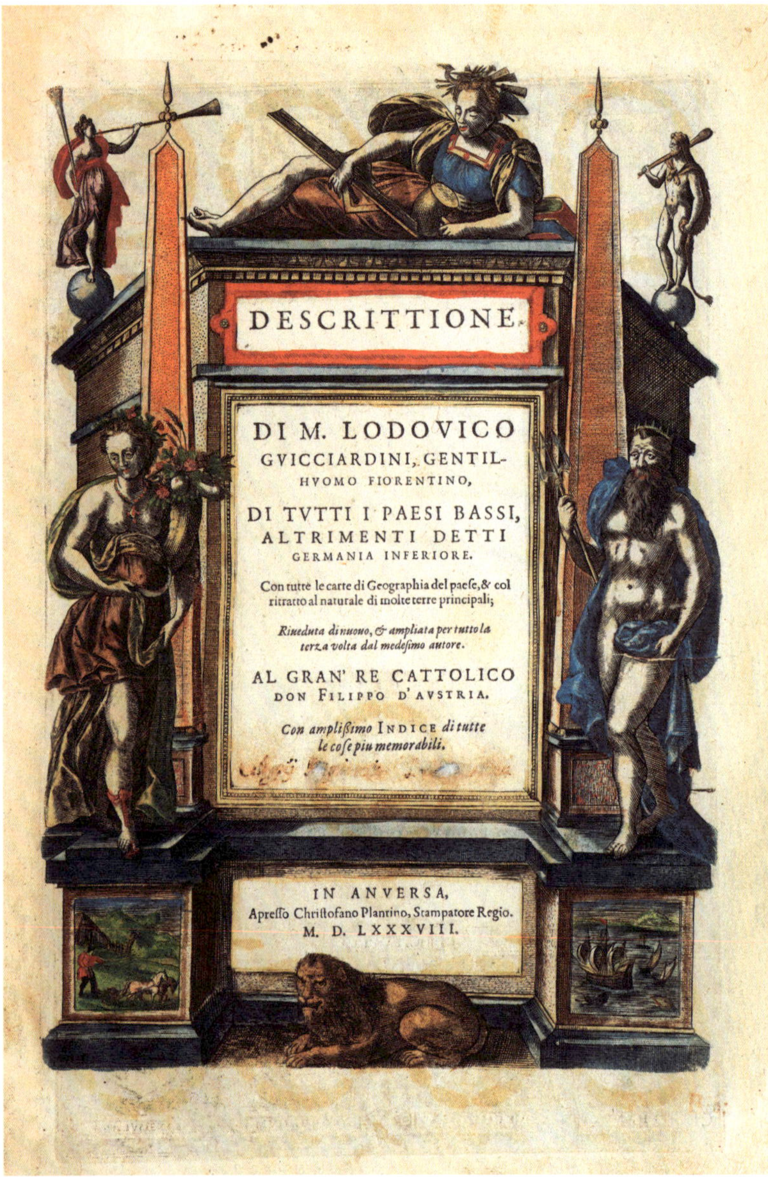

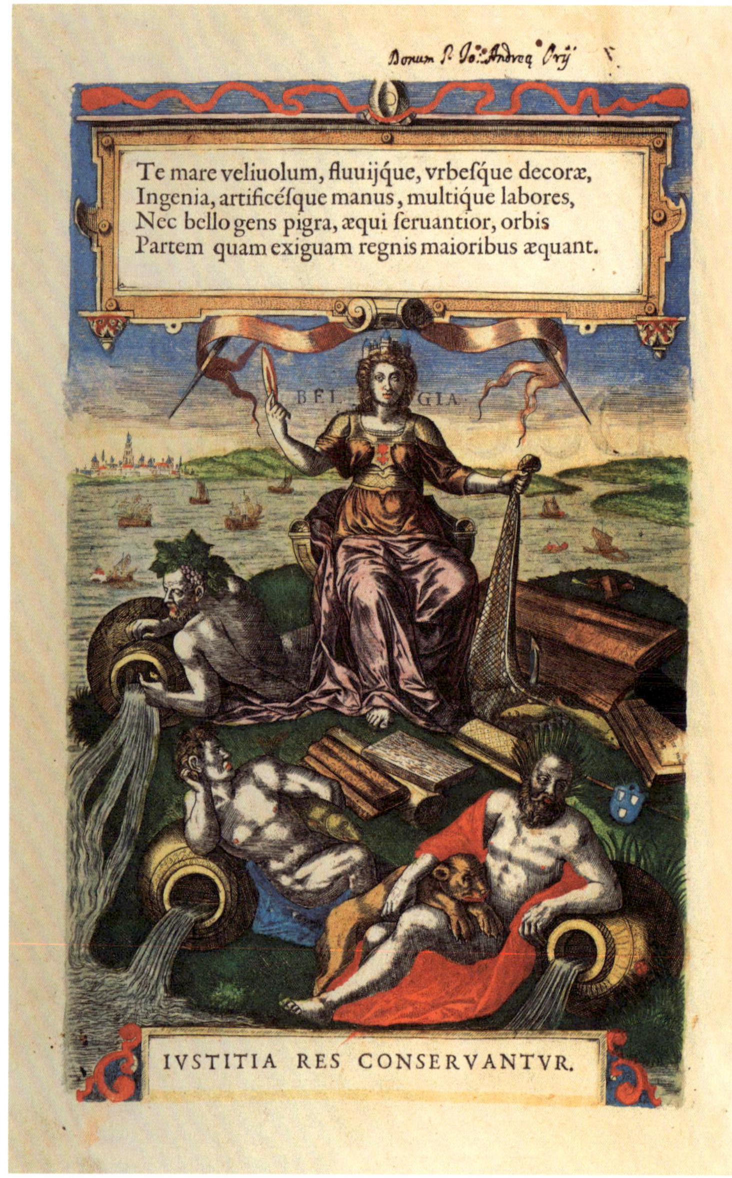

Title page and title print
From: Lodovico Guicciardini, *Descrittione di tutti i Paesi Bassi*, 1588
ANTWERP, THE PHOEBUS FOUNDATION

THE LOW COUNTRIES IN EVOCATIVE DETAIL

Lodovico Guicciardini was born into an old and wealthy line of Florentine merchants. He was dispatched to Antwerp in 1542 to look after the family's business interests in the city, but he promptly fell in love with both a local woman and his new home, and ended up staying there until his death. Before long, Guicciardini turned his back on commerce and devoted himself to writing instead. As he travelled around the Low Countries, he described in detail what he found: the people, their customs, the landscape, local costume, flora and fauna. Willem Silvius published the first edition of the *Descrittione di tutti i Paesi Bassi* in Antwerp in 1567. Guicciardini's magnum opus was a great success, with more than sixty reprints in different languages ultimately published in cities all over Europe.

The Phoebus Foundation has a copy of the third edition of the *Descrittione*, printed in 1588 by none other than the book magnate Christophe Plantin at his *Officina Plantiniana* on Antwerp's Vrijdagmarkt. This was also the final, and hence most complete, Italian edition. It is a unique copy, in which all 78 copper engravings have been richly coloured in, one by one, and inserted in the appropriate spot (with the corresponding texts in the book), rather than merely collected at the back, as is the case in many other copies. The book is bound in contemporary gold-stamped black calfskin, and Plantin ensured that it was extensively illustrated with regional maps as well as city plans and views. He also offered the prints for sale separately.

Guicciardini went on to be published by other houses too. These did not have new copper plates made for each edition but drew as fully as possible on the existing maps and other illustrations. Georg Braun and Frans Hogenberg's city atlas, *Civitates Orbis Terrarum* (⟼ Chapter II) and Abraham Ortelius's *Theatrum Orbis Terrarum* (⟼ p. 126) served as important examples for many of the maps and prints, while the bird's-eye map of Antwerp from the first volume of their six-volume city atlas formed the basis for Plantin's edition of Guicciardini several years later. This is especially clear when we compare the two prints (*Antwerpiae* on pp. 150-151 and *Anverpia* on p. 66 respectively).

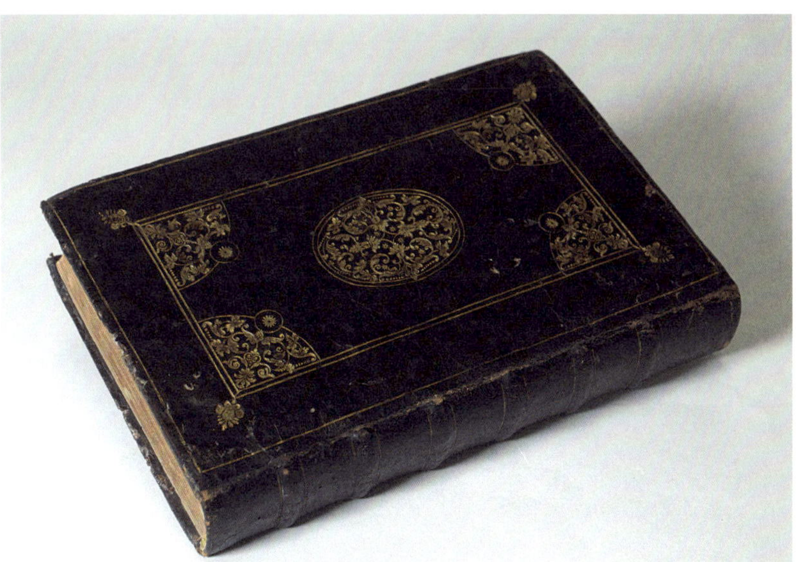

Lodovico Guicciardini
Descrittione di tutti i Paesi Bassi, Antwerp, 1588
ANTWERP, THE PHOEBUS FOUNDATION

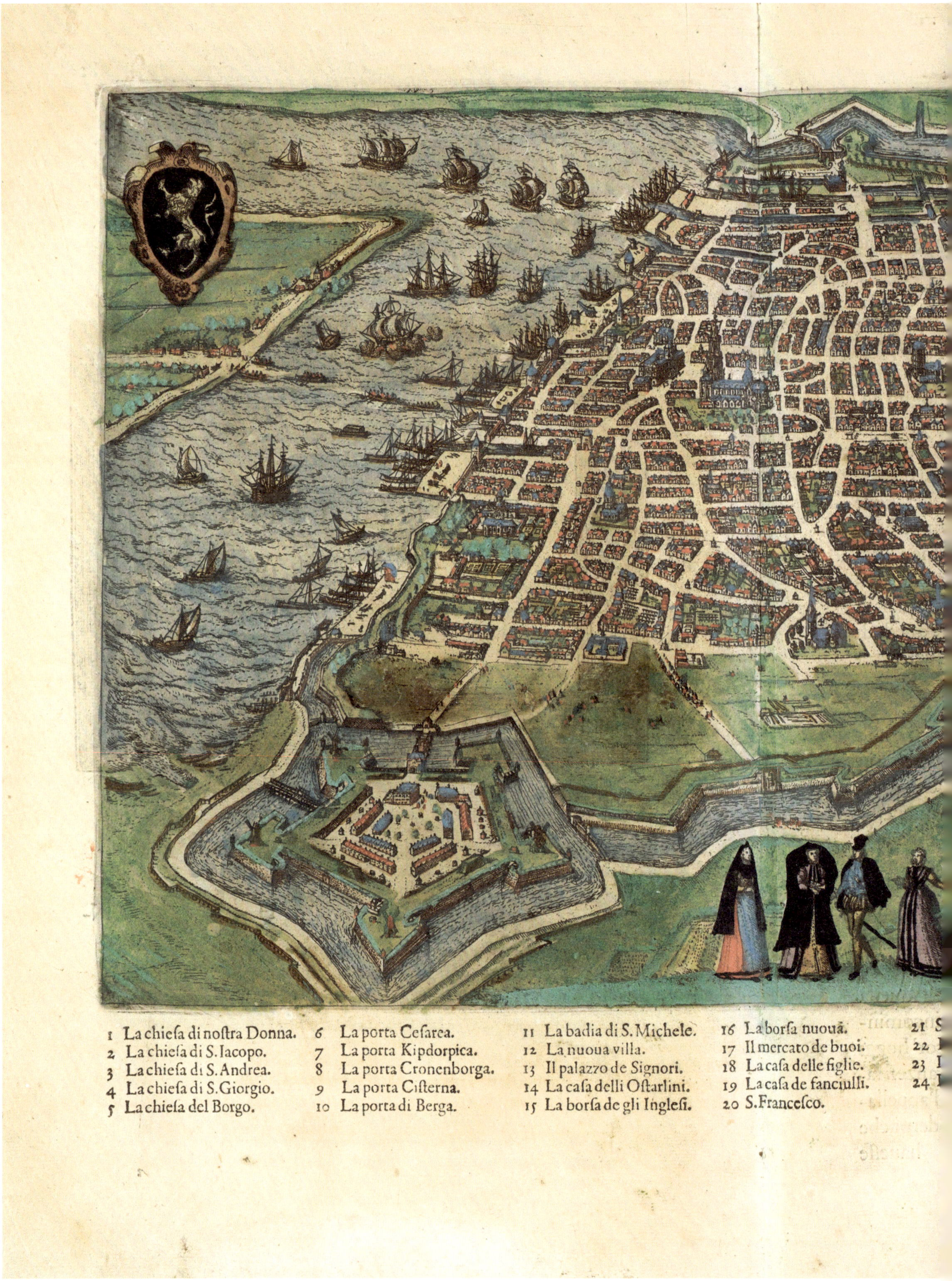
1 La chiesa di nostra Donna.
2 La chiesa di S. Iacopo.
3 La chiesa di S. Andrea.
4 La chiesa di S. Giorgio.
5 La chiesa del Borgo.
6 La porta Cesarea.
7 La porta Kipdorpica.
8 La porta Cronenborga.
9 La porta Cisterna.
10 La porta di Berga.
11 La badia di S. Michele.
12 La nuoua villa.
13 Il palazzo de Signori.
14 La casa delli Ostarlini.
15 La borsa de gli Inglesi.
16 La borsa nuoua.
17 Il mercato de buoi.
18 La casa delle figlie.
19 La casa de fanciulli.
20 S. Francesco.
21
22
23
24

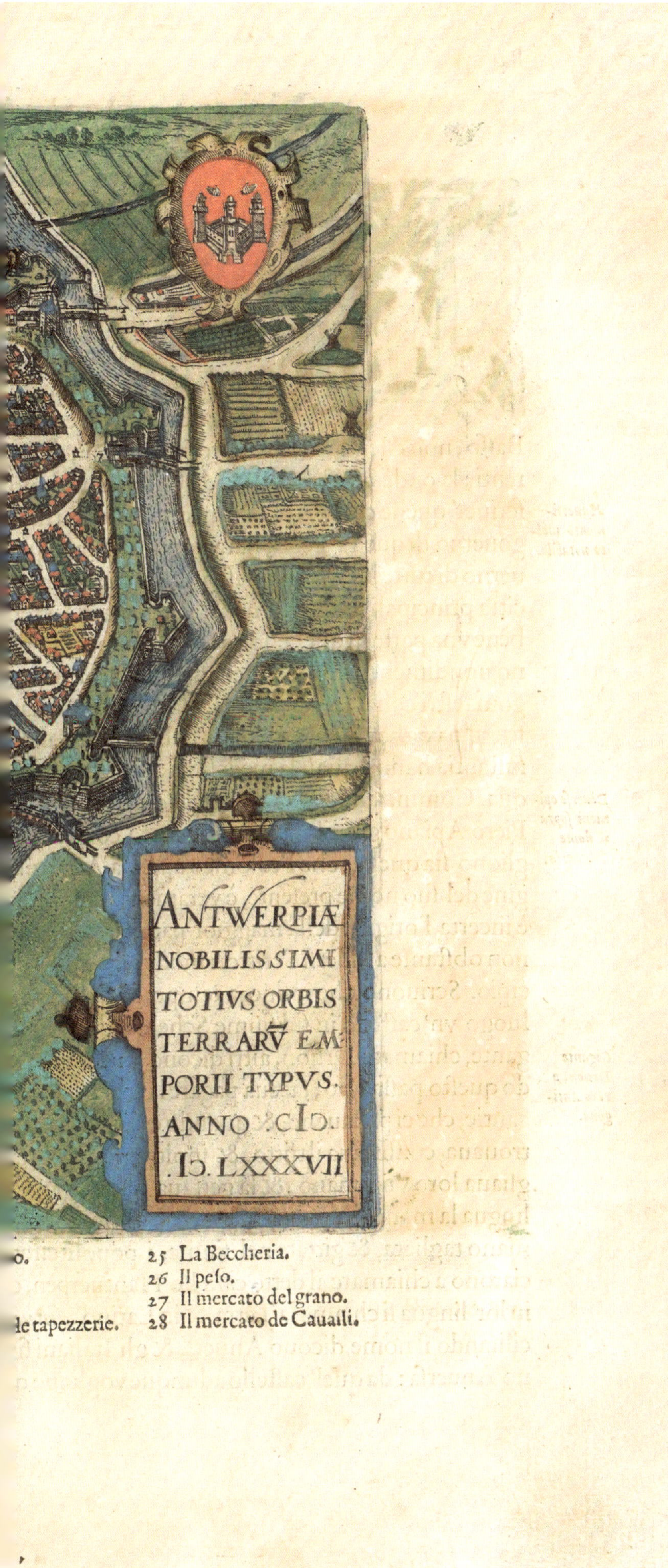

Antwerpiae nobilissimi totius orbis terraru[m] emporii typus anno 1587
Bird's-eye map of Antwerp
From: Lodovico Guicciardini, *Descrittione di tutti i Paesi Bassi*, 1588
ANTWERP, THE PHOEBUS FOUNDATION

ITALIANS AT THE FRONT

While Guicciardini was busy describing the Low Countries in evocative detail, the situation in the region grew unstable. Around the time that the first edition of his work appeared in 1567, the Protestant provinces of Holland and Zeeland rose up against their Catholic, Spanish-Habsburg rulers (⟶ p. 68). Guicciardini condemned the war, which he saw as a blot on his adopted and much-loved home, yet there was no shortage of interest on the part of Italians in the course of the northern conflict. And not only Italians: numerous studies were penned all over Europe concerning the wars in the Low Countries and France. They included international histories by the Englishman William Camden (1551-1623) and the Frenchman Jacques-Auguste de Thou (1553-1617), and the *Histoire generale de la guerre de Flandre* (Paris, 1611) by Gabriel Chappuys (1550-*c*.1611), historian to the king of France. The latter features a large number of maps and prints depicting episodes in the conflict, such as the line of ships bridging the Scheldt.

The uprising and the tensions it sparked spilled across borders, impacting neighbouring countries: if the Protestant rebels were to succeed in toppling the Spanish-Habsburg Empire, the political and economic reverberations would be felt as far away as the Italian peninsula, where the Spaniards had long ruled peacefully. Italians also fretted about the danger posed to their Catholic faith, prompting various well-to-do gentlemen to volunteer to fight for the Habsburg cause. It was exceedingly important at such a moment to seize and maintain control of the flow of information. There was no shortage of educated Italians at the front, and these took it upon themselves to keep their homeland informed about what was going on, albeit from their own perspective.

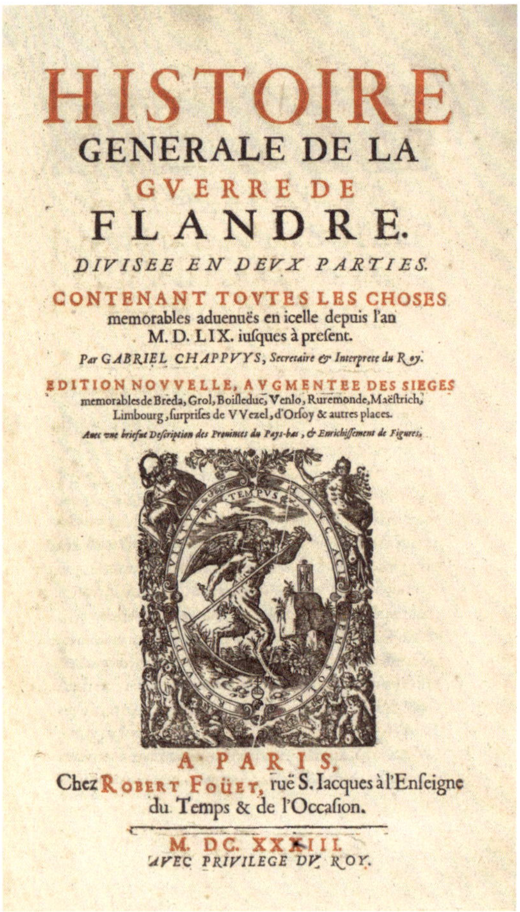
HISTOIRE
GENERALE DE LA
GVERRE DE
FLANDRE.
DIVISEE EN DEVX PARTIES.
CONTENANT TOVTES LES CHOSES
memorables aduenuës en icelle depuis l'an
M. D. LIX. iusques à present.
Par GABRIEL CHAPPVYS, Secretaire & Interprete du Roy.
EDITION NOVVELLE, AVGMENTEE DES SIEGES
memorables de Breda, Grol, Boisleduc, Venlo, Ruremonde, Maëstrich, Limbourg, surprises de VVezel, d'Orsoy & autres places.
Avec vne briefue Description des Provinces du Pays-bas, & Enrichissement de Figures.

A PARIS,
Chez ROBERT FOÜET, ruë S. Iacques à l'Enseigne
du Temps & de l'Occasion.
M. DC. XXXIII.
AVEC PRIVILEGE DV ROY.

The military treatise *Delle guerre di Fiandra* ('On the Wars in Flanders'; Antwerp, 1609) recounts the events of the Revolt in the Southern Netherlands as seen through the eyes of the Italian general Pompeo Giustiniano (1569-1616). Born in Corsica, the soldier had travelled to the Low Countries to fight for the Spanish king in the international 'Army of Flanders' between 1601 and 1609. He was involved most notably in the three-year Siege of Ostend (⟶ p. 94), a gruelling campaign that saw him rise to the rank of *maestro de campo*. Giustiniano's experiences were chronicled by his Tuscan compatriot Giuseppe Gamurini, of whom little is known other than that he was a military engineer who also fought in the campaign in Flanders. Gamurini adapted Giustiniano's military reports, so that the resulting treatise reads more like an army manual than a history of the war, complete with detailed explanations of siege tactics and troop movements. He added around thirty large engravings to illustrate the text.

Title page
From: Gabriel Chappuys, *Histoire generale de la guerre de Flandre*, 1633
ANTWERP, THE PHOEBUS FOUNDATION

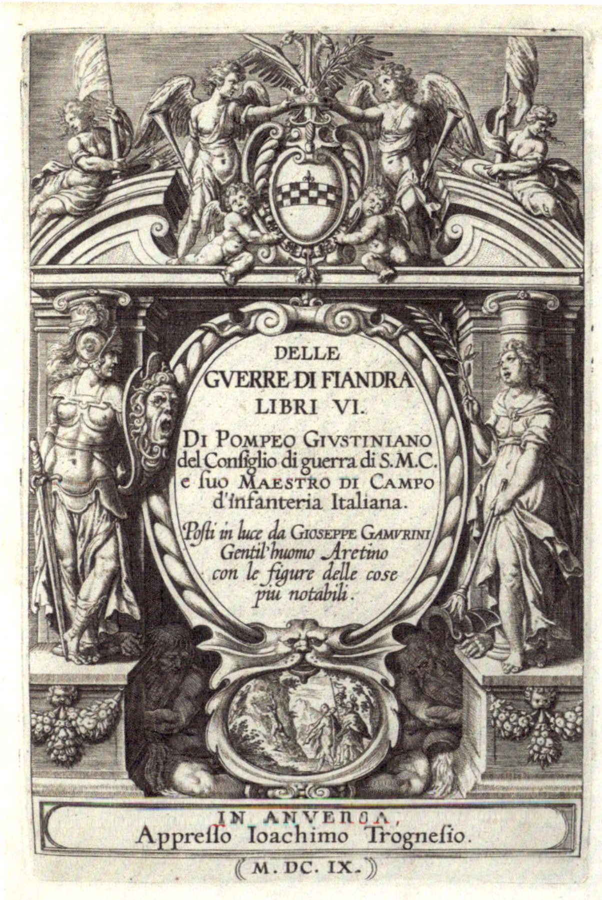

Gamurini compiled *Delle guerre di Fiandra* and adapted it into a coherent narrative, which the Antwerp publisher Joachim Trognesius (active until 1624) printed in 1609 – the very year in which the Twelve Years Truce (1609-1621) began. The book was distributed in Italy too, with editions appearing in Venice and Milan from 1615 onwards. Initially available only in Italian, a Latin translation was published two years later in Cologne, gaining numerous non-Italian readers and enabling the account to cross borders. Copies of the book are today found all over the world.

Galeazzo Gualdo Priorato (1606-1678), born into a family of Italian counts in Vicenza, also volunteered for the front, where he served as a mercenary in all sorts of places and for different sides. In the course of the Thirty Years War he fought for the Habsburgs and the Dutch alike. He recorded his adventures in the book *Teatro del Belgio* (Frankfurt, 1683), which includes maps of the fortresses at Lillo and Liefkenshock as well as one of the city of Antwerp. These are schematic, abstracted images with little additional information aside from a few figures in the landscape, vessels on the Scheldt and important buildings like the Cathedral of Our Lady in Antwerp. Although Priorato praised the '*Gran Monarca della Spagne*', his intention was to recount the events without embellishment and as simply as possible. The maps in his work are sober and neutral – the image he wished to present of himself too – and so lack vivid details and omit references to violence and action.

Almost all of these are 'story maps' in which spatial and textual information was combined to help communicate the story of a particular event. In this instance, the maps narrate the war in Flanders: the attempts to lay siege to 's-Hertogenbosch (1601 and 1603), the loss of the town of Grave (1602) and Ambrosio Spinola's successful campaign in the east, which saw the capture of Groenlo (1606) and Rijnberk (1606), but first and foremost the grinding Siege of Ostend (1601-1604), which eventually fell to the Spanish Habsburgs after colossal effort and suffering. Giustiniano and Spinola are the main protagonists in these successive feats of arms at a stage of the Revolt in which the Spanish army – bolstered by Italians and other international mercenaries – gained the upper hand over Flemish cities and fortifications. The reader is struck by the emphasis that Giustiniano's work places on these victories and how little attention is paid to Spanish losses.

Title page
From: Pompeo Giustiniano and Giuseppe Gamurini, *Delle guerre di Fiandra*, 1609
ANTWERP, THE PHOEBUS FOUNDATION

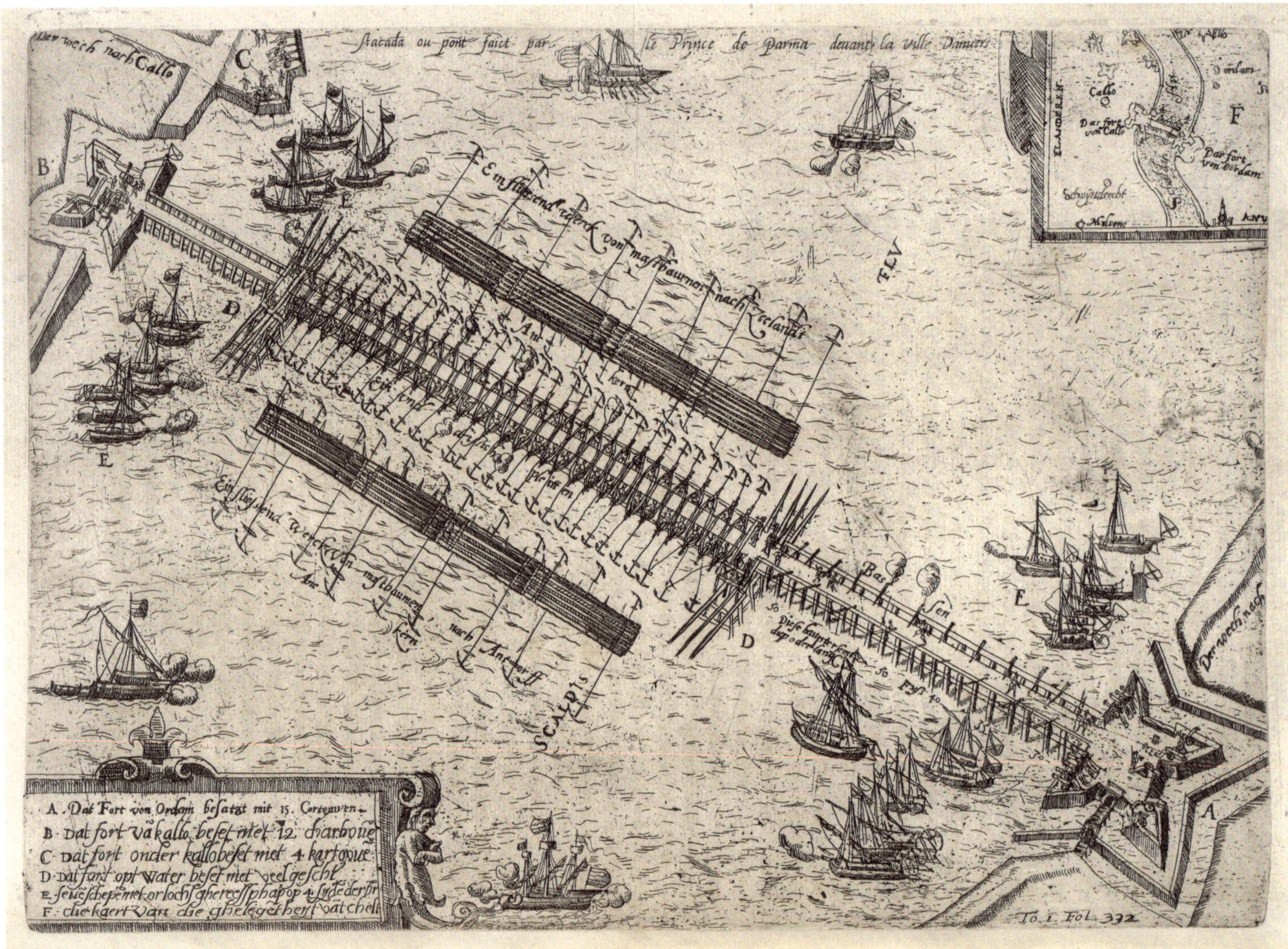

Stacada ou pont faict par le Prince de Parma devant la ville Damiers
Line of ships across the Scheldt between Fort Kallo and Fort Oordam
From: Gabriel Chappuys, *Histoire generale de la guerre de Flandre*, 1633
ANTWERP, THE PHOEBUS FOUNDATION

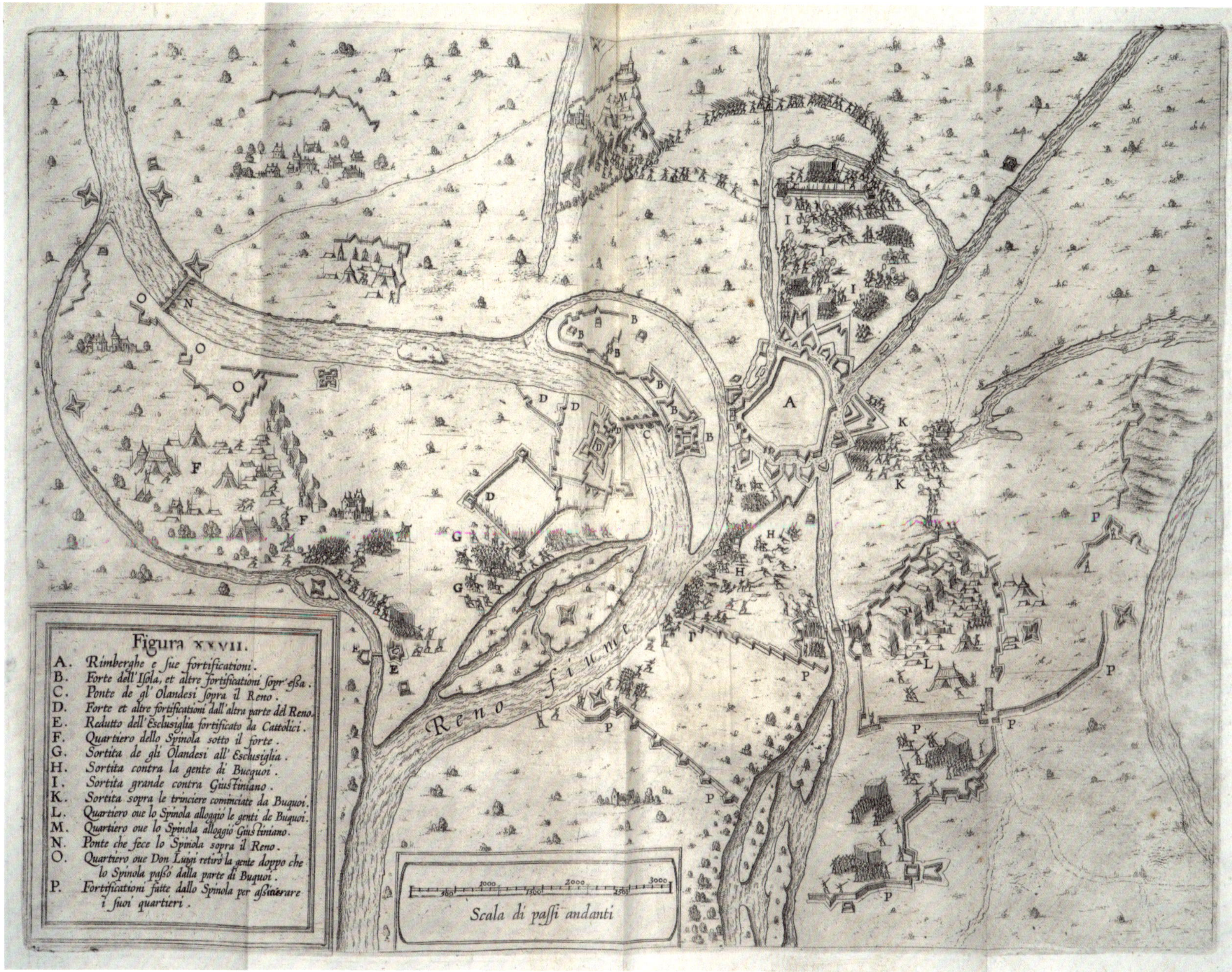

Giuseppe Gamurini
Figura XXVII
Map of the Siege of Rijnberk in 1606
From: Pompeo Giustiniano and Giuseppe Gamurini, *Delle guerre di Fiandra*, 1609
ANTWERP, THE PHOEBUS FOUNDATION

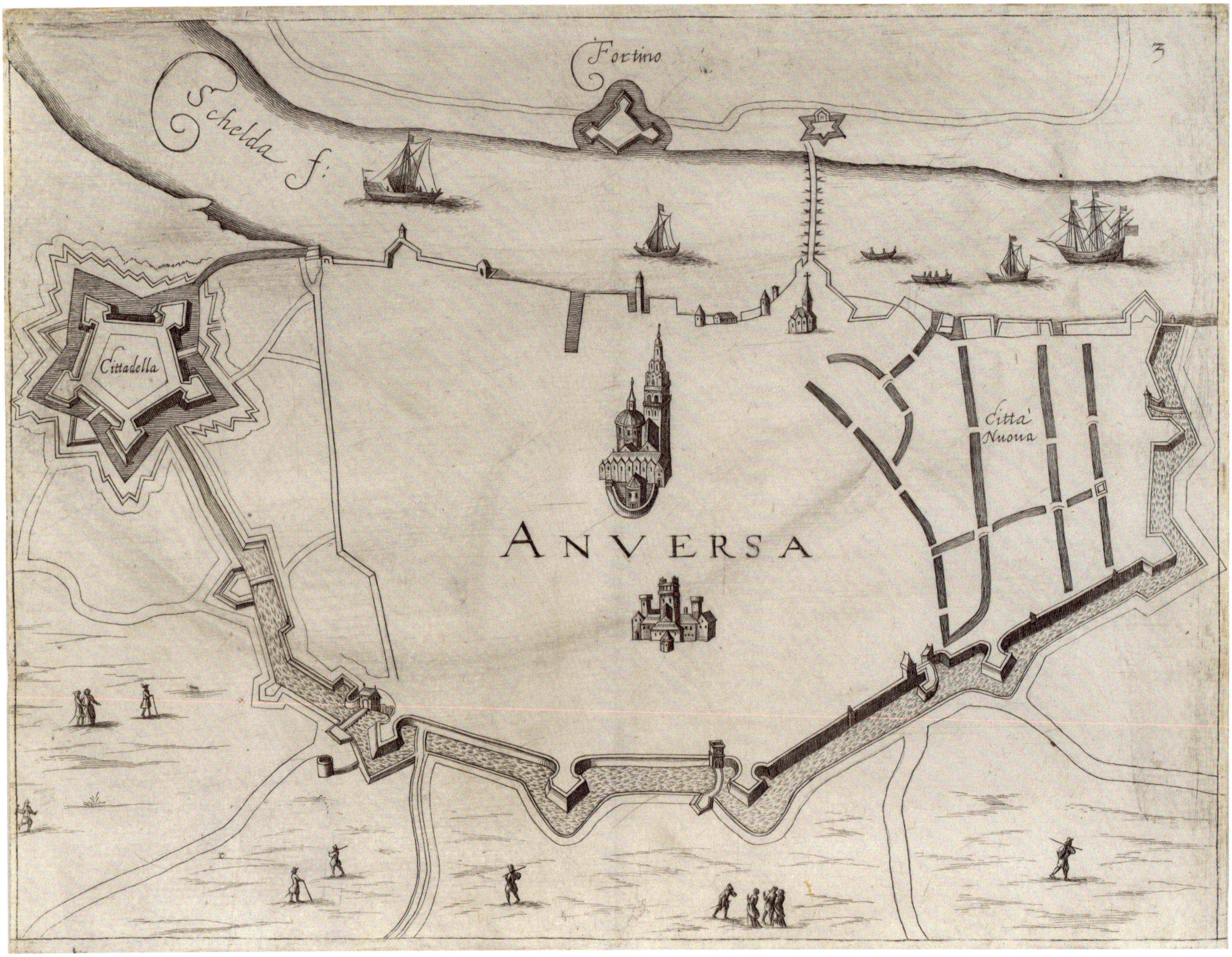

Plan of the ramparts and fortifications of Antwerp
From: Galeazzo Gualdo Priorato, *Teatro del Belgio*, 1673
ANTWERP, THE PHOEBUS FOUNDATION

Maps of the forts of Lillo and Liefkenshoek
From: Galeazzo Gualdo Priorato, *Teatro del Belgio*, 1673
ANTWERP, THE PHOEBUS FOUNDATION

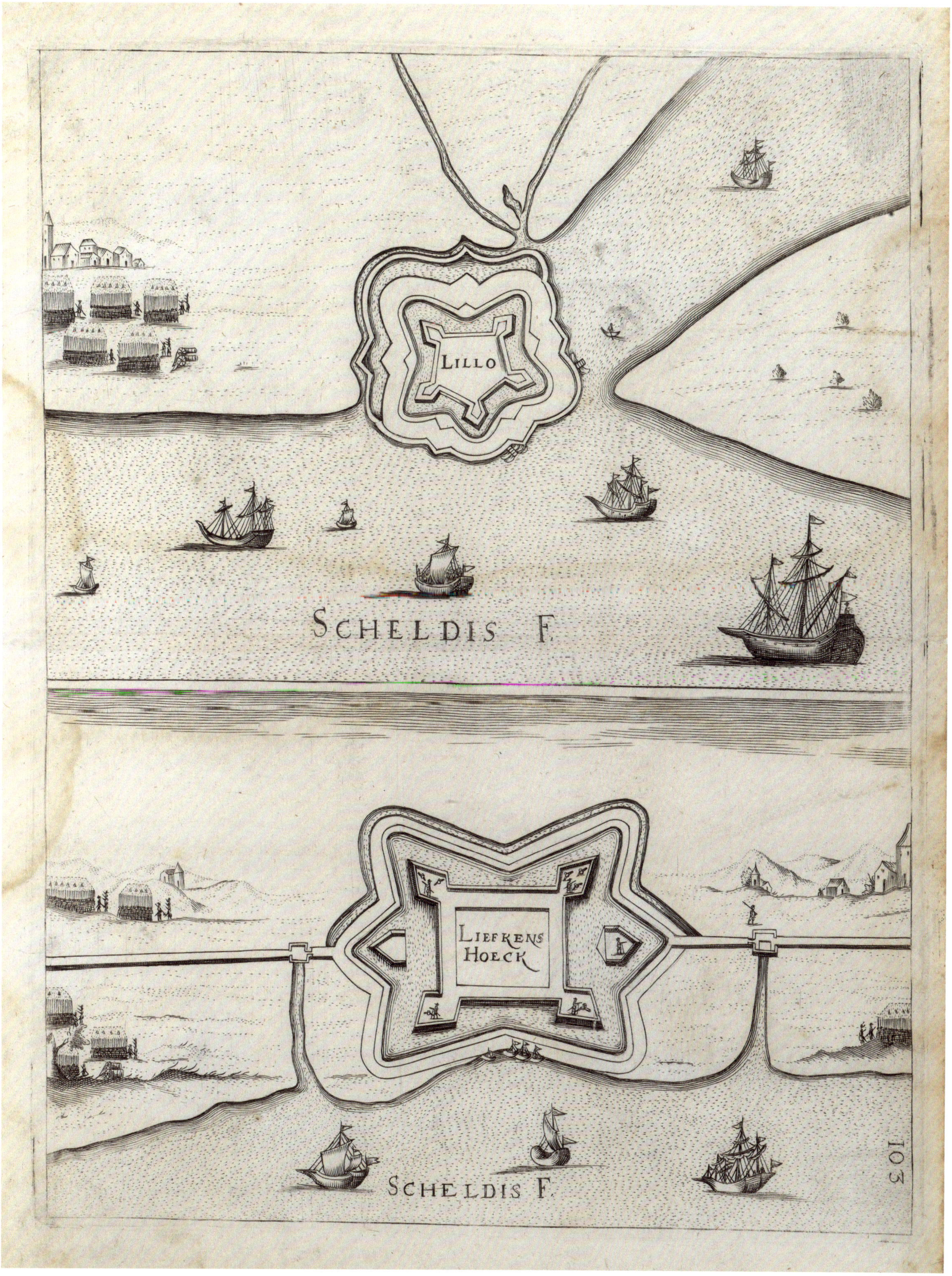
LILLO
SCHELDIS F.
LIEFKENS HOECK
SCHELDIS F.
103

RELIGIOUS CHRONICLERS FROM ROME

While Italians were mostly supporters of Spain and did not shrink from expressing their Catholic, Habsburg sympathies in their texts and maps, the works they produced were still viewed as important sources by readers in the Low Countries. Italian history books were sought after by the intellectual elite in the Southern and the Northern Netherlands alike. Peter Paul Rubens (1577-1640) owned or had perused Giustiniano, as had the Dutch historians Hugo de Groot (1583-1645) and Pieter Cornelisz. Hooft (1581-1647), the latter drawing on it for his great history of the Low Countries, *Nederlandsche historien* (Amsterdam, 1642). The most frequently read and cited Italian authors were, however, Famiano Strada (1572-1649) and Guido Bentivoglio (1579-1644).

Strada was a Jesuit and teacher of rhetoric and Latin from Rome, which is where he wrote his famous history of the Revolt of the Netherlands between 1555 and 1590, commissioned by Alexander Farnese. Each of the two volumes of *De Bello Belgico* (vol. 1: Rome, 1632, vol. 2: Rome, 1647) ran to ten parts. He almost never left his native country and chiefly relied on standard works like Guicciardini's *Descrittione*. With Farnese as his patron, however, Strada also had access to the Farnese family archives in Rome and could thus publish material that had not previously been available. First-hand information of this kind, ranging from personal correspondence to the reports of spies and messages to ambassadors, made his book extremely valuable.

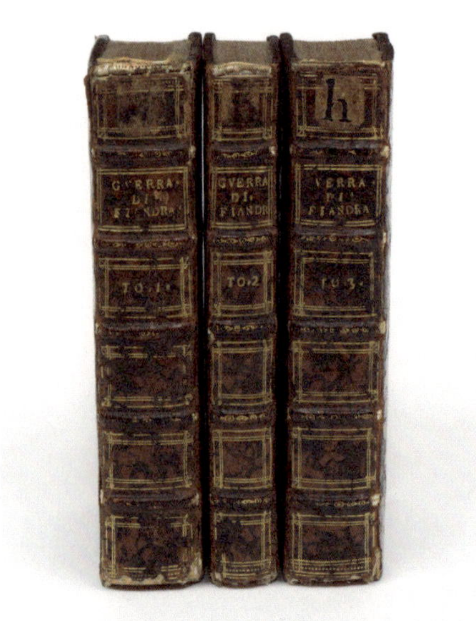

Guido Bentivoglio
Della Guerra di Fiandra, 1635-1640
ANTWERP, THE PHOEBUS FOUNDATION

The result was a popular and widely disseminated tome published in six languages, although it also drew criticism for its highly partisan account, in which the Revolt is narrated from a firmly pro-Spanish perspective. One of Strada's fiercest critics was his fellow historian Cardinal Guido Bentivoglio, who felt the book deviated too far from historical reality. For his own part, Bentivoglio took a more nuanced approach and – unlike Strada – did actually visit the Low Countries in person. He took part in Brussels in the negotiations between delegates of Spain and the Republic to bring about the Twelve Years Truce and witnessed the Revolt more or less at first hand, describing his experiences in *Della Guerra di Fiandra* (Cologne, 1633-1639).

Strada's first volume, *decas prima*, covers the years 1555-1578 and was published in Cologne in 1632, with the second – *decas secunda*, on the years 1578-1590 – following in 1647. The two-volume copy in The Phoebus Foundation's collection is the second edition in Latin. The place of publication is falsely stated as Cologne on the title page, to divert attention from the fact that it was actually published by Abraham Elzevier in Leiden. This first edition published and distributed in the Low Countries includes illustrations with portraits of important figures; only in later editions were large fold-out illustrations and maps of the events incorporated. This book too was reprinted and translated many times and even included a Polish edition.

The Dutch and Latin editions had a title page illustrated with a *Leo Belgicus* map – a motif in which the Low Countries were symbolically depicted in the form of a lion. It was first introduced by Michael Eytzinger in his book *De Leone Belgico* (Cologne, 1583) (⟼ Chapter V). The Dutch publishers of Famiano Strada's *De Bello Belgico* were quick to incorporate the title page in the editions they produced during the Revolt, as seen, for instance, in the title print of the Amsterdam edition by Johannes Janssonius (1588-1664) of around 1648. The popularity of the lion map grew immensely in the period in question as a symbol of civic pride and confidence – a statement of the unity of the Low Countries in the face of Spanish rule and a symbol that was plainly shared both within the region and abroad.

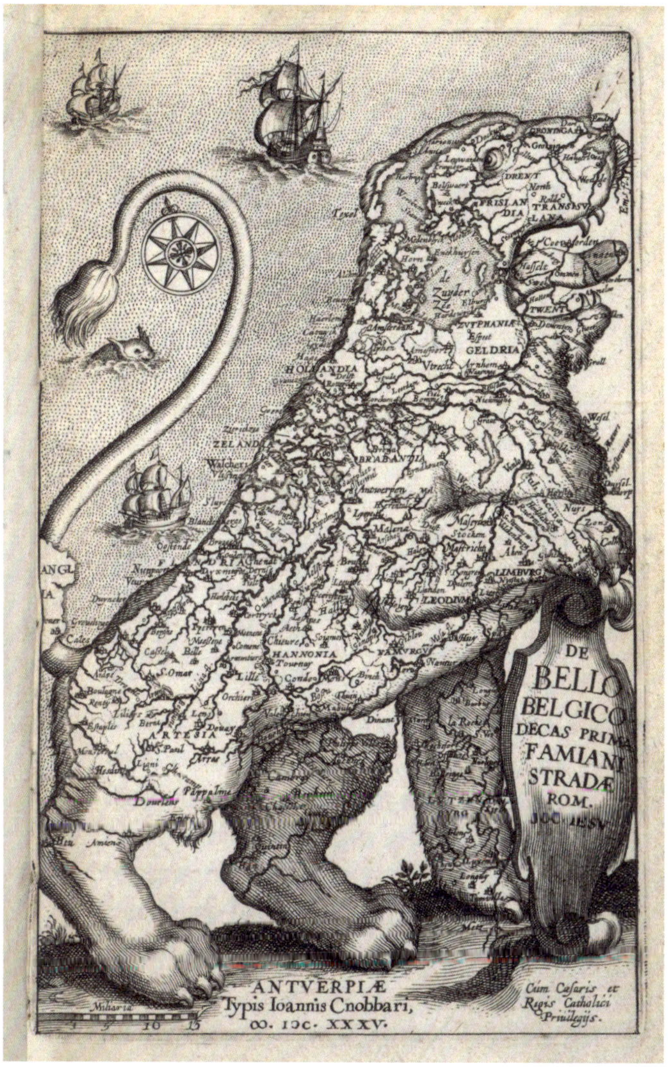

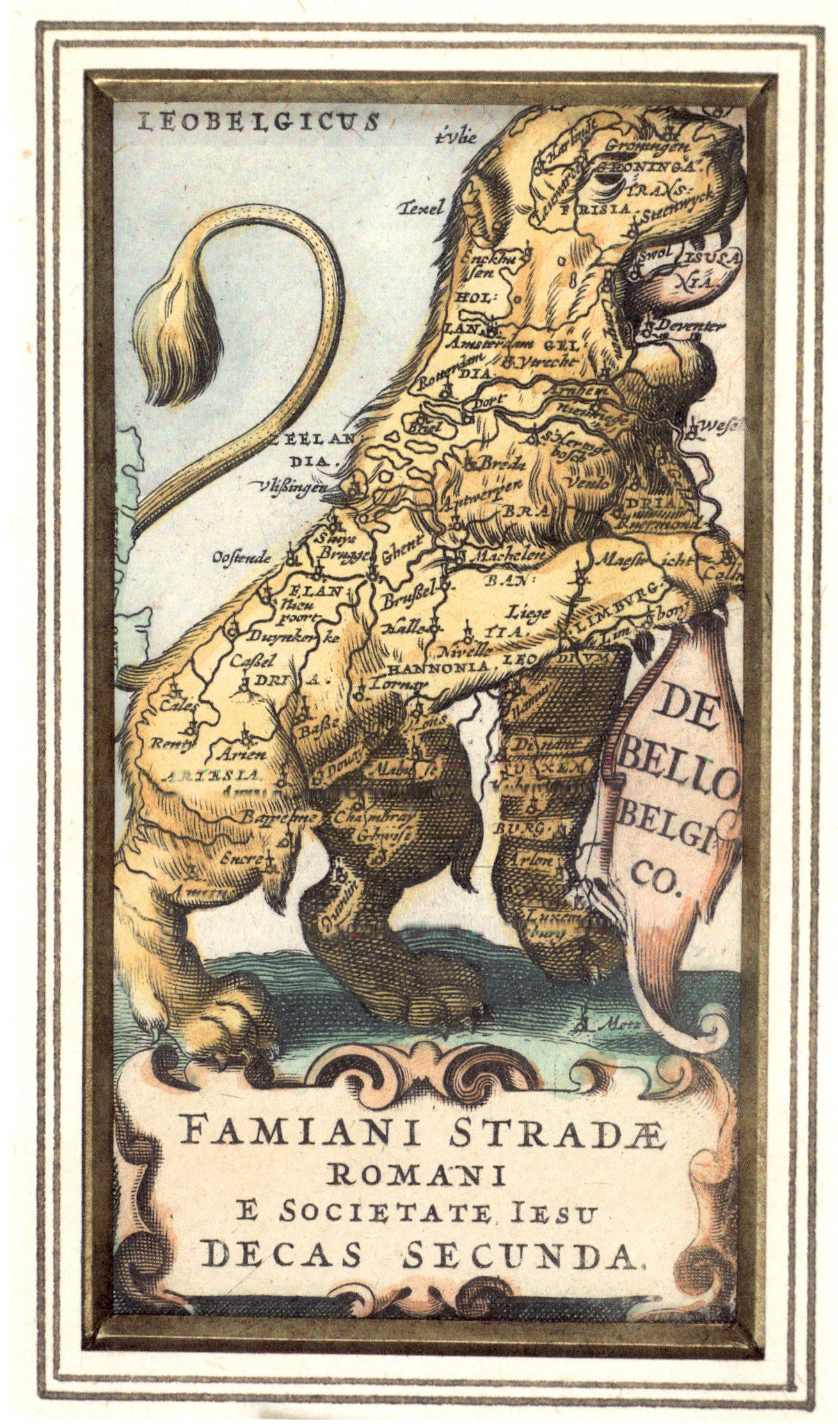

Leo Belgicus as title print
From: Famiano Strada, *De Bello Belgico. Decas prima* [volume 1], 1635
ANTWERP, THE PHOEBUS FOUNDATION

Leo Belgicus as title print
From: Famiano Strada, *De Bello Belgico. Decas secunda* [volume 2], 1648
ANTWERP, THE PHOEBUS FOUNDATION

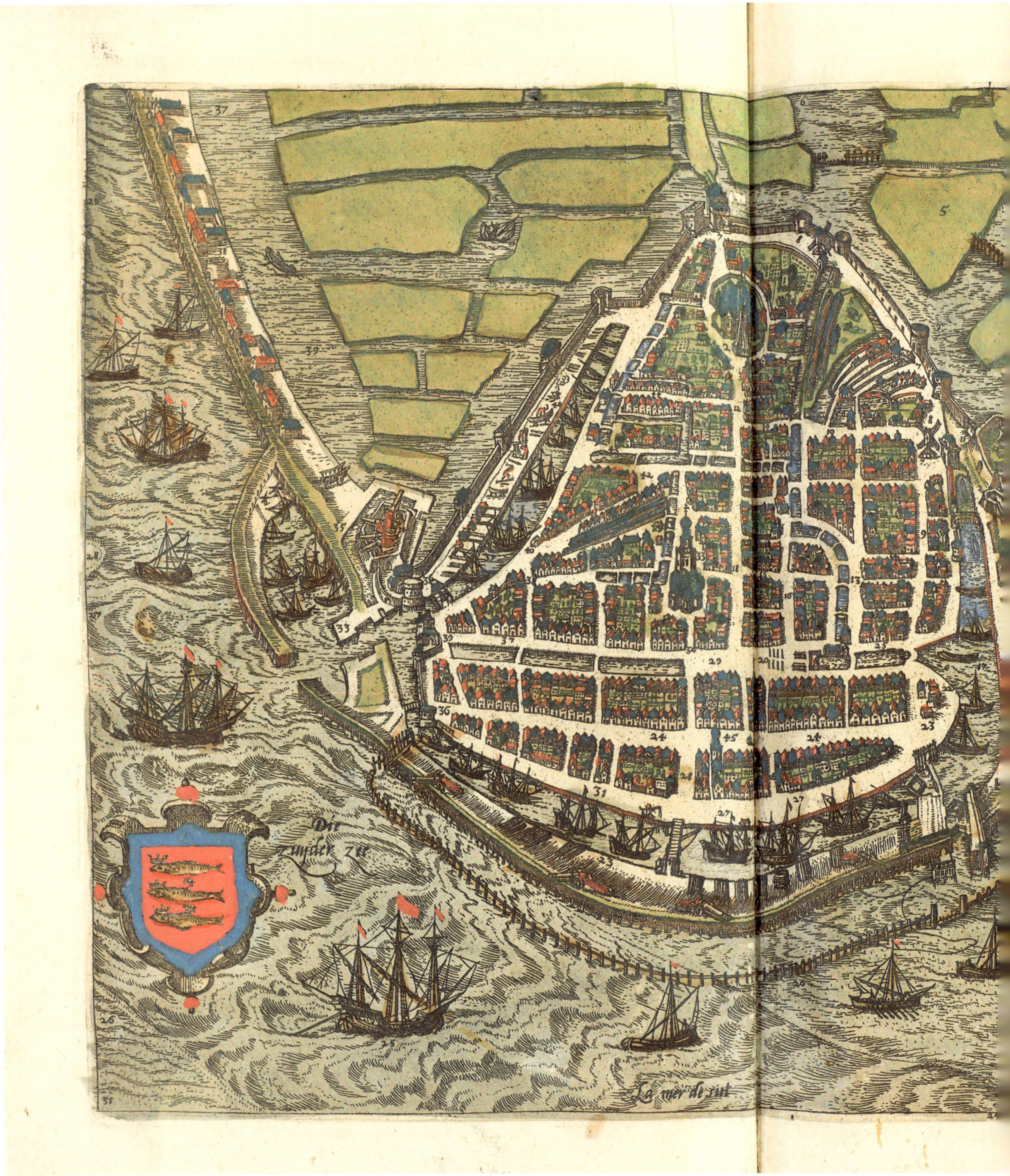
Die zuyder zee
La mer de zut

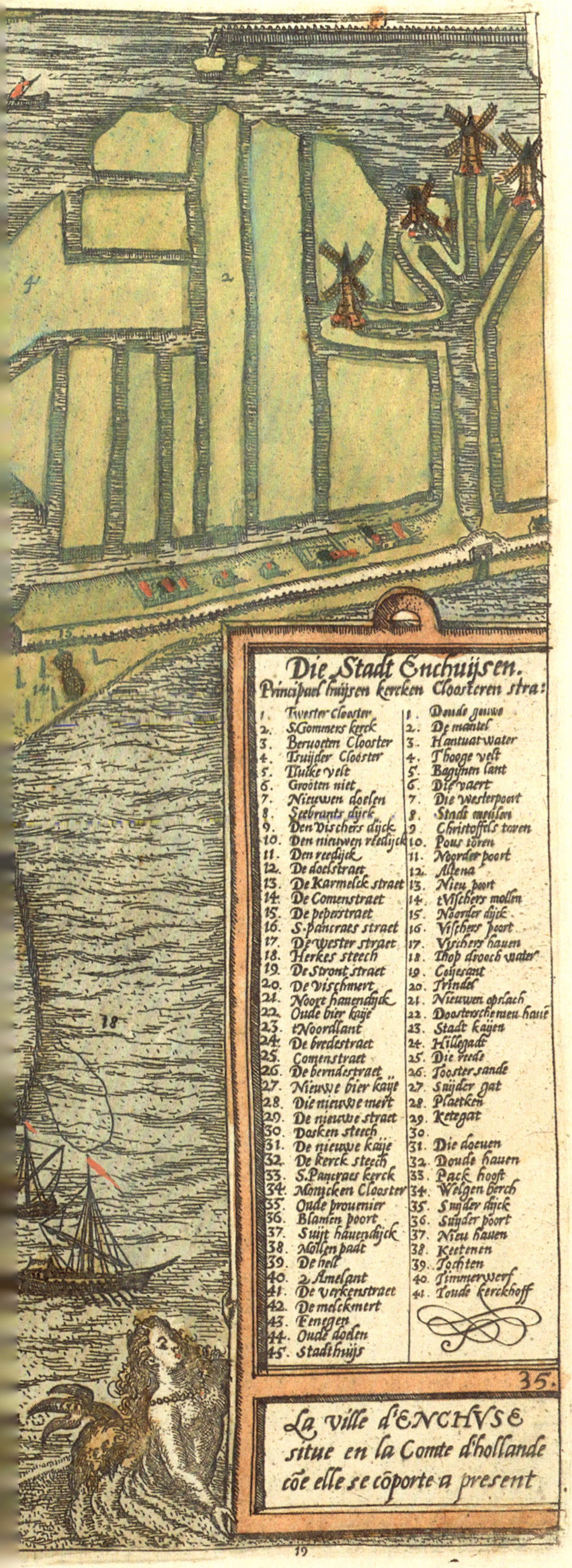

LITERATURE

P. Arblaster, *From Ghent to Aix: How They Brought the News in the Habsburg Netherlands, 1550-1700*, Leiden, 2014.

D. Aristodemo, *Descrittione di tutti i Paesi Bassi: Lodovico Guicciardini (1521-1589) in de ban van de Lage Landen*, Antwerp, 2020.

H. Deys, M. Franssen, V. Van Hezik et al., *Guicciardini Illustratus. De kaarten en plannen in Lodovico Guicciardini's Beschrijving van de Nederlanden*, 't Goy-Houten, 2001.

F. Donnet, *Les imprimeurs Trognaesius et leur famille*, Antwerp, 1919.

N. Lamal, 'Publishing Military Books in the Low Countries and in Italy in the Early Seventeenth Century', *Specialist Markets in the Early Modern Book World*, 40, 2015: 223-240.

N. Lamal, *Italian Communication on the Revolt in the Low Countries (1566-1648)*, Leiden, 2023.

C. Reijner, 'Italiaanse geschiedschrijvers over de Nederlandse Opstand, 1585-1650: Een transnationale geschiedenis', PhD thesis, Leiden University, 2020.

A. Simoni, 'Soldier's Tales. Observations on Italian military books published at Antwerp in the early 17th century', in: D. Rhodes & D. Reidy (ed.), *The Italian book 1465-1800*, London, 1993: 255-290.

A. Simoni, 'Sans Frontières. Italo-Dutch books 1565-1629', *La Bibliofilia*, 104, 2002: 57-82.

H. van der Heijden, *Leo Belgicus: an Illustrated and Annotated Carto-Bibliography*, Alphen aan den Rijn, 2006.

Die Stadt Enchuijsen
Map of the town of Enkhuizen
From: Lodovico Guicciardini, *Descrittione di tutti i Paesi Bassi*, 1588
ANTWERP, THE PHOEBUS FOUNDATION

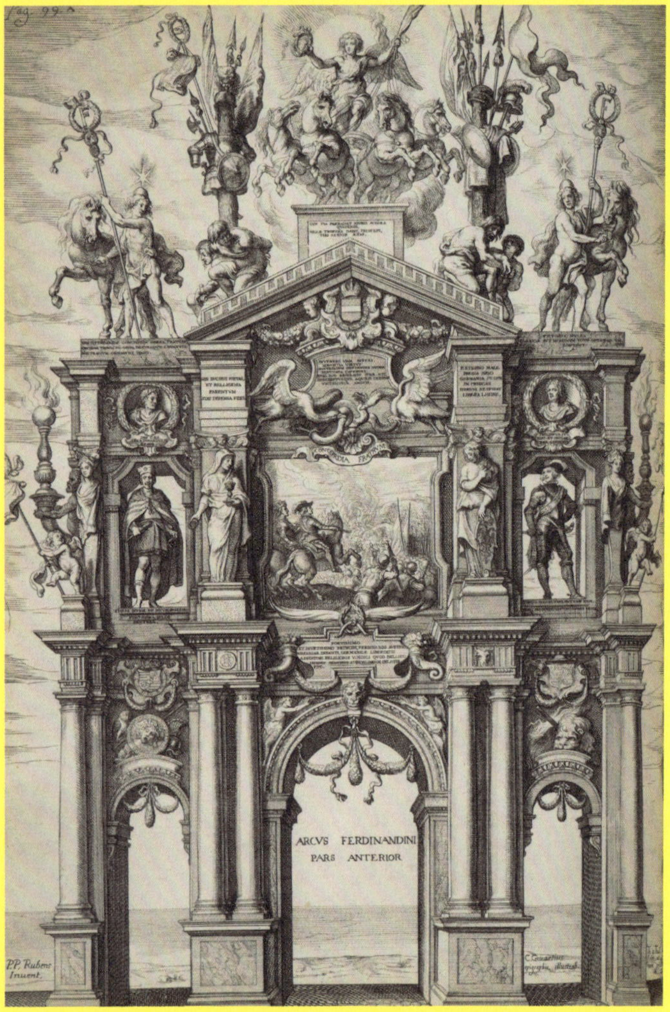

Joyous entry in Antwerp

This bird's-eye map of Antwerp shows the ceremonial reception or 'joyous entry' of Cardinal Infante Ferdinand of Austria (1609-1641), son of Philip III of Spain (1578-1621) and governor of the Southern Netherlands, into the city on 15 May 1635. The event celebrated his victory with the Catholic Imperial Army in the Battle of Nördlingen (5-6 September 1634) over the combined Protestant armies of Sweden and its German allies. This battle was part of the Thirty Years War (1618-1648), which involved several European powers. The city pulled out all the stops, commissioning Rubens to design triumphal arches, stages, tableaux and floats to welcome the great general. Numerous Antwerp artists worked on the execution of the master's designs under his supervision.

As frequently occurred on these occasions, a book was produced to commemorate the event. *Pompa introitus Ferdinandi Austriaci Hispaniarum infantis* (Antwerp, 1641), written by Gaspar Gevartius (1593-1666), featured this map of the entry in Antwerp made by Theodoor Van Thulden (1606-1669), who took an existing map by Joris Hoefnagel and added the procession passing the fortress, crossing the bridge and entering the city streets via the main gate. Cannons fire a salute and flags are raised: Antwerp is ready to celebrate. The book also contains illustrations of the spectacular artworks that were installed for the occasion, engraved after the original oil sketches made by Rubens.

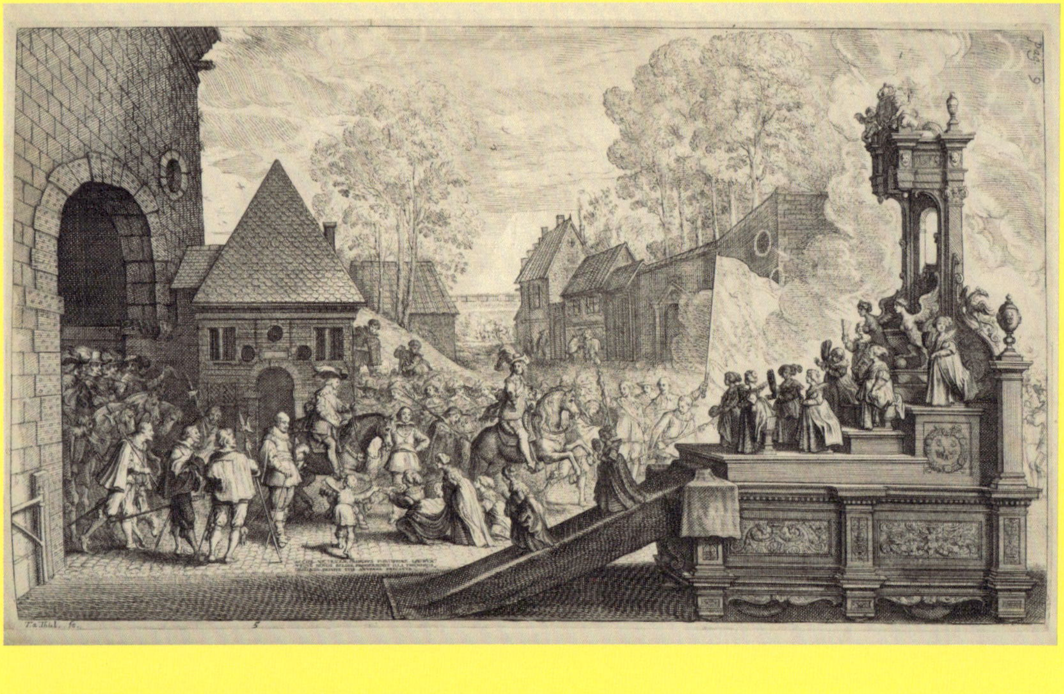

Triumphal arch and view of the procession
From: Gaspar Gevartius, *Pompa introitus Ferdinandi Austriaci Hispaniarum infantis*, 1641
ANTWERP, THE PHOEBUS FOUNDATION

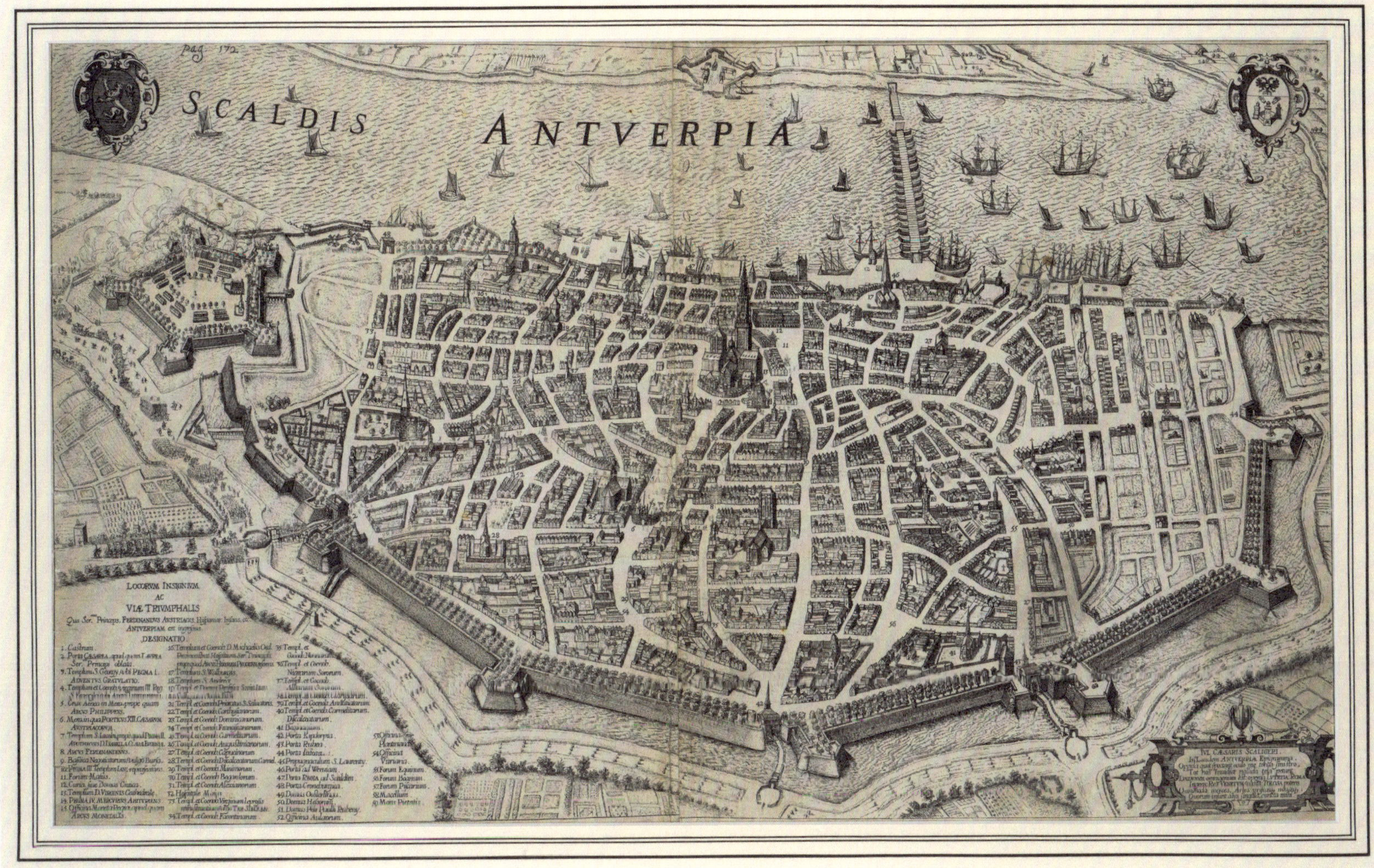

Theodoor Van Thulden
Locorum Insignium ac Viae Triumphalis, c.1639-1641
Bird's-eye map of Antwerp with the entry of Ferdinand of Austria in 1635
Etching, 545 × 760 mm
ANTWERP, THE PHOEBUS FOUNDATION

An unfinished military map for Louis XIV

Somewhat puzzlingly, this anonymous manuscript map has not been finished. Two rectangles, in the bottom left and the upper right corner, were left blank to allow a title, legend or similar to be added, but this never happened. Is it a design for a map? Did the cartographer simply not get round to finishing it? Or was there something wrong here? There is little in the handmade map itself to suggest its purpose other than the area it depicts between Dunkirk, Ghent, Dinant, Namur, Leuven and Brussels. Following some detective work, it was found to be an unfinished version of a map included in a series of military atlases from the years 1694-1703 entitled *Camps et ordres de marches de l'armée du Roy*.

The atlases were created by an engineer and cartographer called 'Pennier'. Employed by King Louis XIV, he followed the French army on its campaign in the Southern Netherlands, surveying and mapping the territory for military ends. The map in question is found in the volume for 1692, where it is titled: 'Camps and marching orders of the Royal Army in Flanders commanded by my lord Marshal the Duke of Luxembourg in the year 1692. Recorded at the locations and drawn by Mr Pennier, Engineer and Geographer to the King.' Copies of this series of atlases can be found in the Bibliothèque Nationale in Paris and the Royal Library in Brussels.

The maps offer an account of the course of the campaign. Accompanying texts provide further explanation of the symbols, such as the red crosses representing defensive features. Small plans of the defences have also been added to the larger fortified towns (such as *Tournay* or Tournai). *Camps et ordres de marches* is the first map in this volume of the atlas and served as an overview of the military advance. It is unusual for a preparatory design to have survived, and it offers an insight into the process by which the maps were made. This version simply provides the geographical basis, in which we make out the outlines of each town's defences. In the final, completed version, troop positions and lines of advance were drawn in and coloured. The empty frames are filled there with text describing the course of the campaign, with the dates in chronological order. Military maps like this contained sensitive information and were not intended for wider distribution, which is why they were also done by hand and are so particularly rare. Pennier made just a few copies, for the king himself and certain advisers, but no more. It was important, after all, that they did not fall into enemy hands.

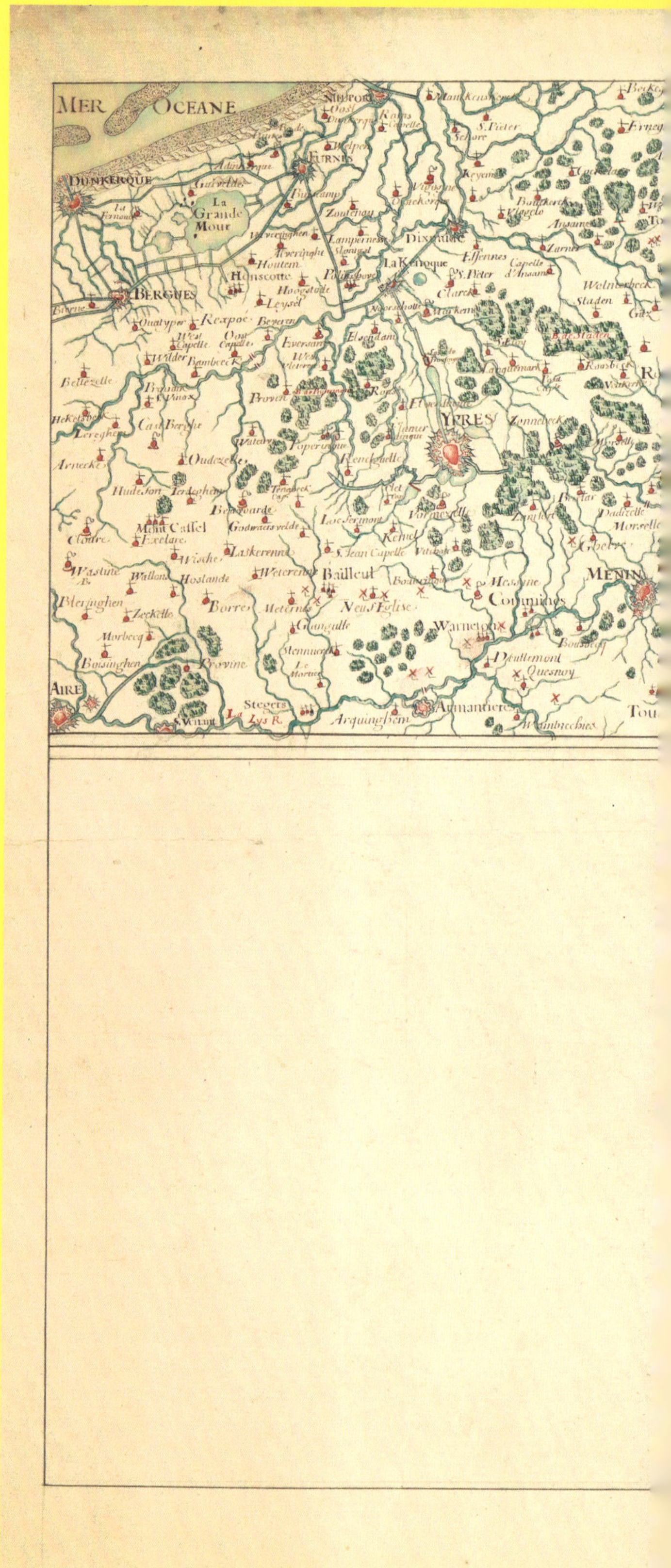

Pennier
Manuscript map of Louis XIV's campaign in Flanders in 1692, c.1694-1703
Pen and ink, 505 × 808 mm
ANTWERP, THE PHOEBUS FOUNDATION

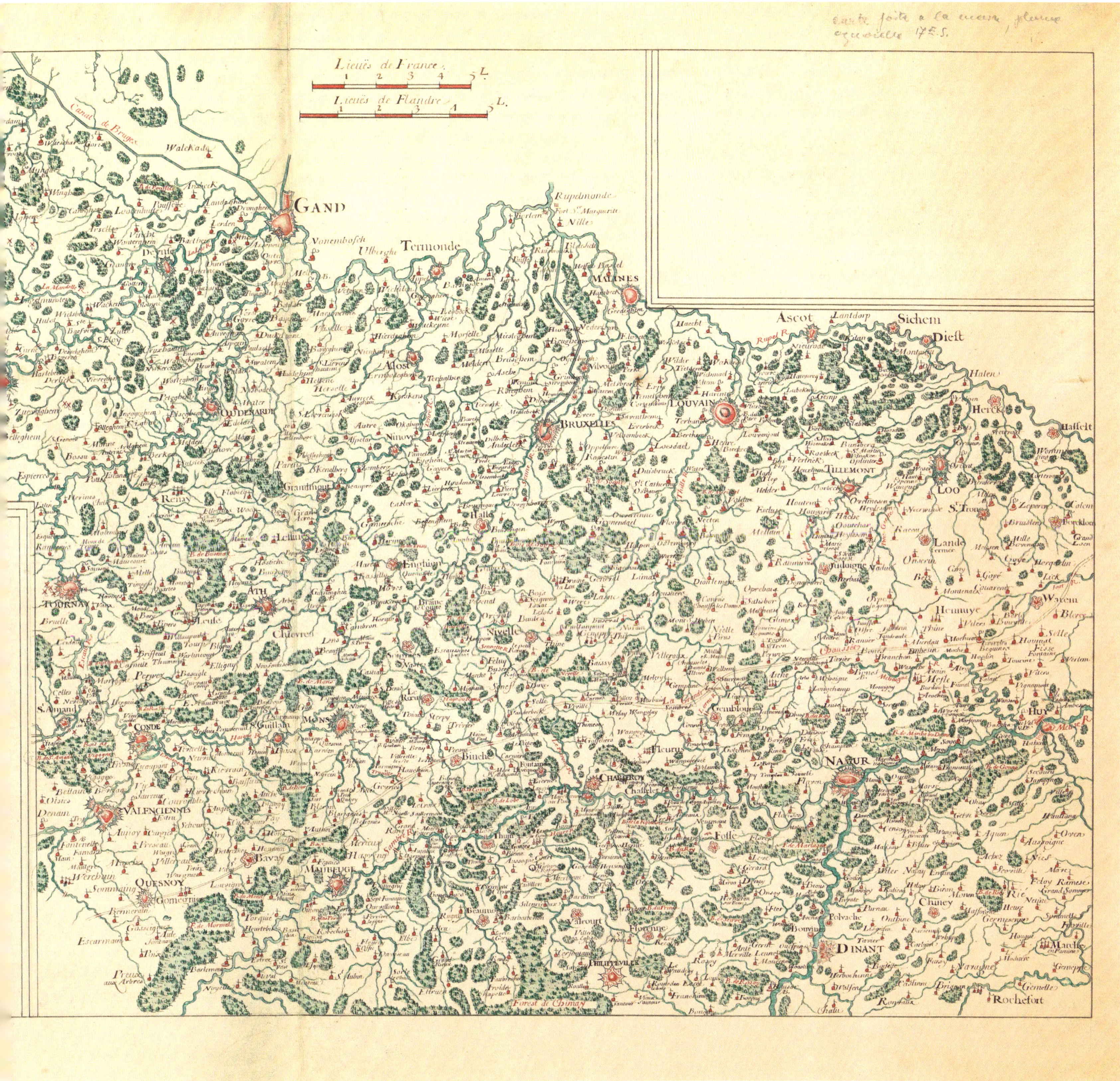
Lieuës de France
Lieuës de Flandre
Canal de Bruges
GAND
Termonde
Rupelmonde
MALINES
Ascot
Sichem
Dieſt
Aloſt
Ninove
OUDENARDE
BRUXELLES
LOUVAIN
TILLEMONT
Herck
LOO
S. Trons
Lande
Grammont
Renay
Halle
Enghien
TOURNAY
ATH
Leuze
Chievres
Nivelle
Hennuye
Warem
S. Amand
CONDE
MONS
S. Guillain
Binch
Fleurus
Gemblours
NAMUR
HUY
VALENCIENNES
QUESNOY
Bavay
MAUBEUGE
CHARLEROY
Philippeville
Florenne
DINANT
Ciney
Rochefort
Forest de Chimay

THE PAINTING *SHIPS IN A STORM* by Bonaventura Peeters I shows vessels with taut sails making their way towards the open sea beneath banks of menacing storm clouds. How did people use to navigate at sea? As we saw in previous chapters, Netherlanders and other Europeans travelled widely by land in the early modern era, but what about their journeys by water? As the horizons expanded for Western European shipping, demand for accurate charts and serviceable navigational instruments grew. New knowledge of coastlines and the steady development of navigation as a science spread relatively quickly thanks to printed books and maps.

Many mapmakers continued to draw on medieval knowledge of seafaring, but recent new shipping routes and 'discoveries' created fresh impetus for all manner of new sea charts. The end of the seventeenth century saw not only the appearance of practical pilot guides and rutters or *leeskaarten* ('a book to be read as a chart'), but also deluxe maritime atlases, coloured in, heightened in gold and stylishly bound. So which charts were actually used at sea, and which were intended more to be collected and admired? And how did nautical practice combine with commerce in maritime cartography?

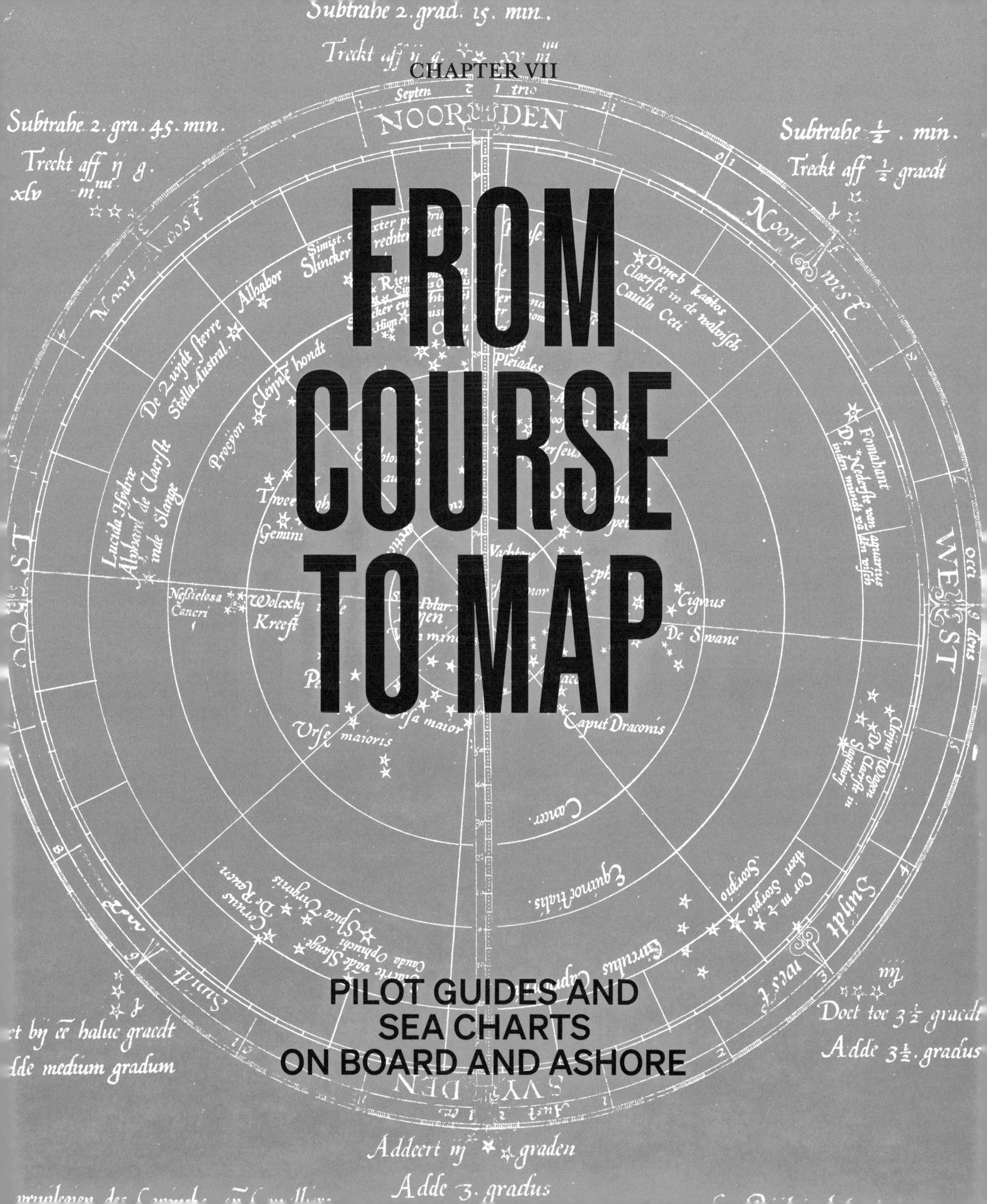

CHAPTER VII

FROM COURSE TO MAP

PILOT GUIDES AND SEA CHARTS ON BOARD AND ASHORE

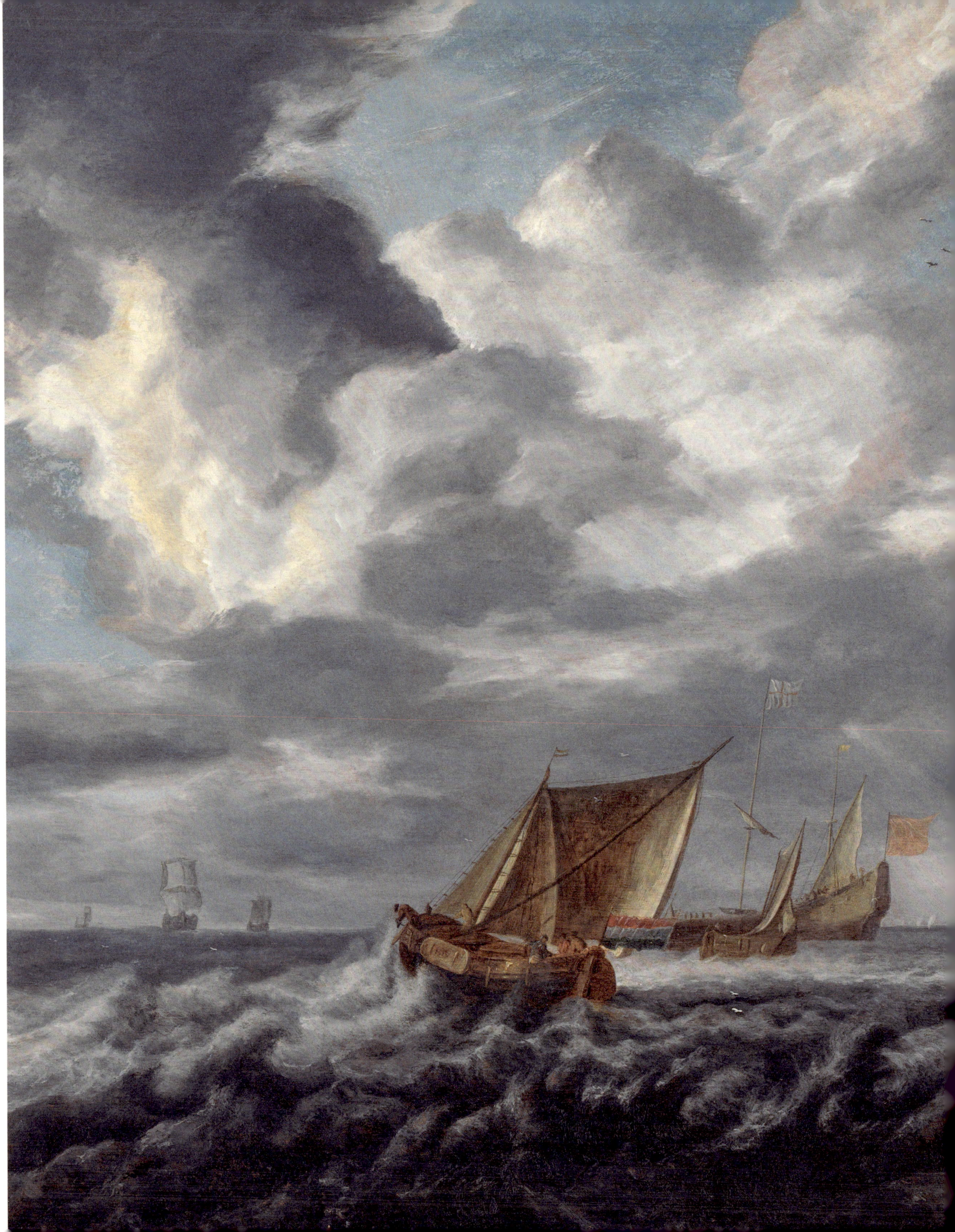

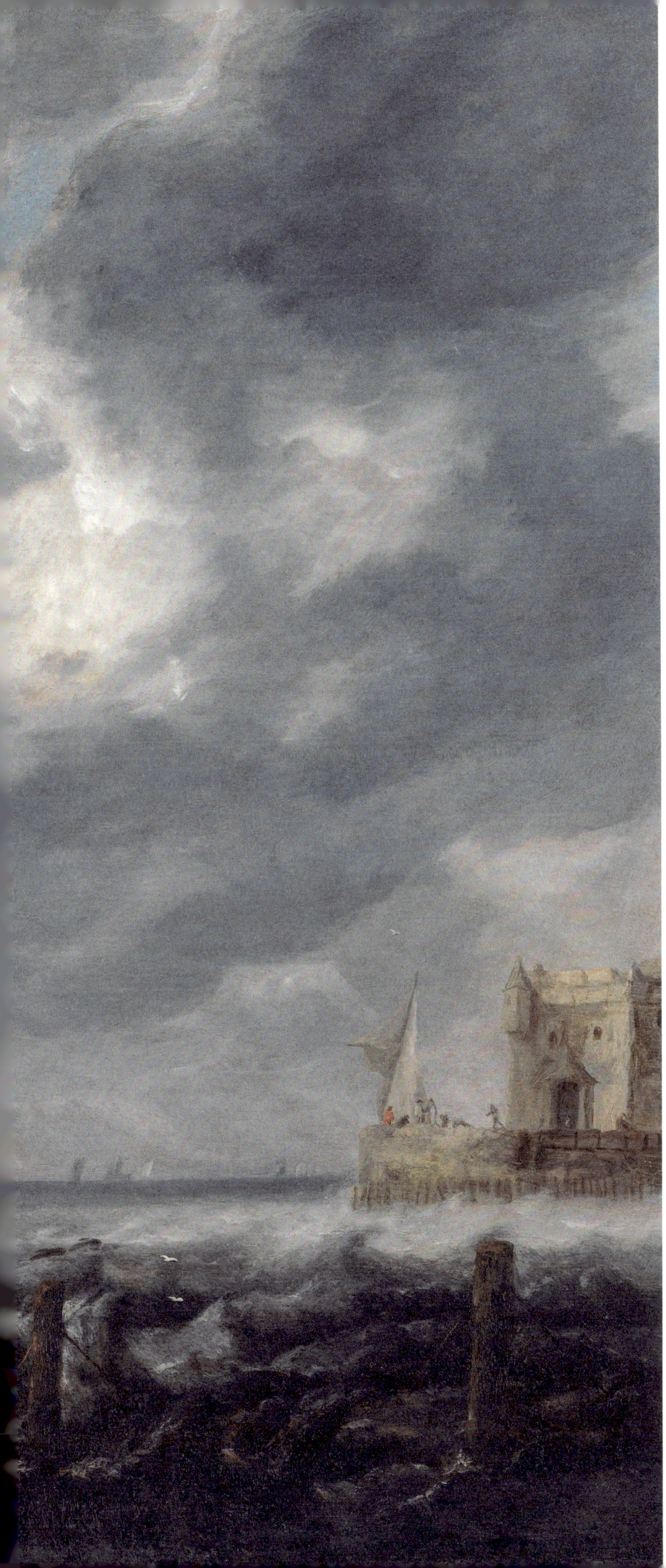

Bonaventura Peeters I
Ships in a Storm, c.1640
Oil on panel, 58.4 × 74.2 cm
ANTWERP, THE PHOEBUS FOUNDATION

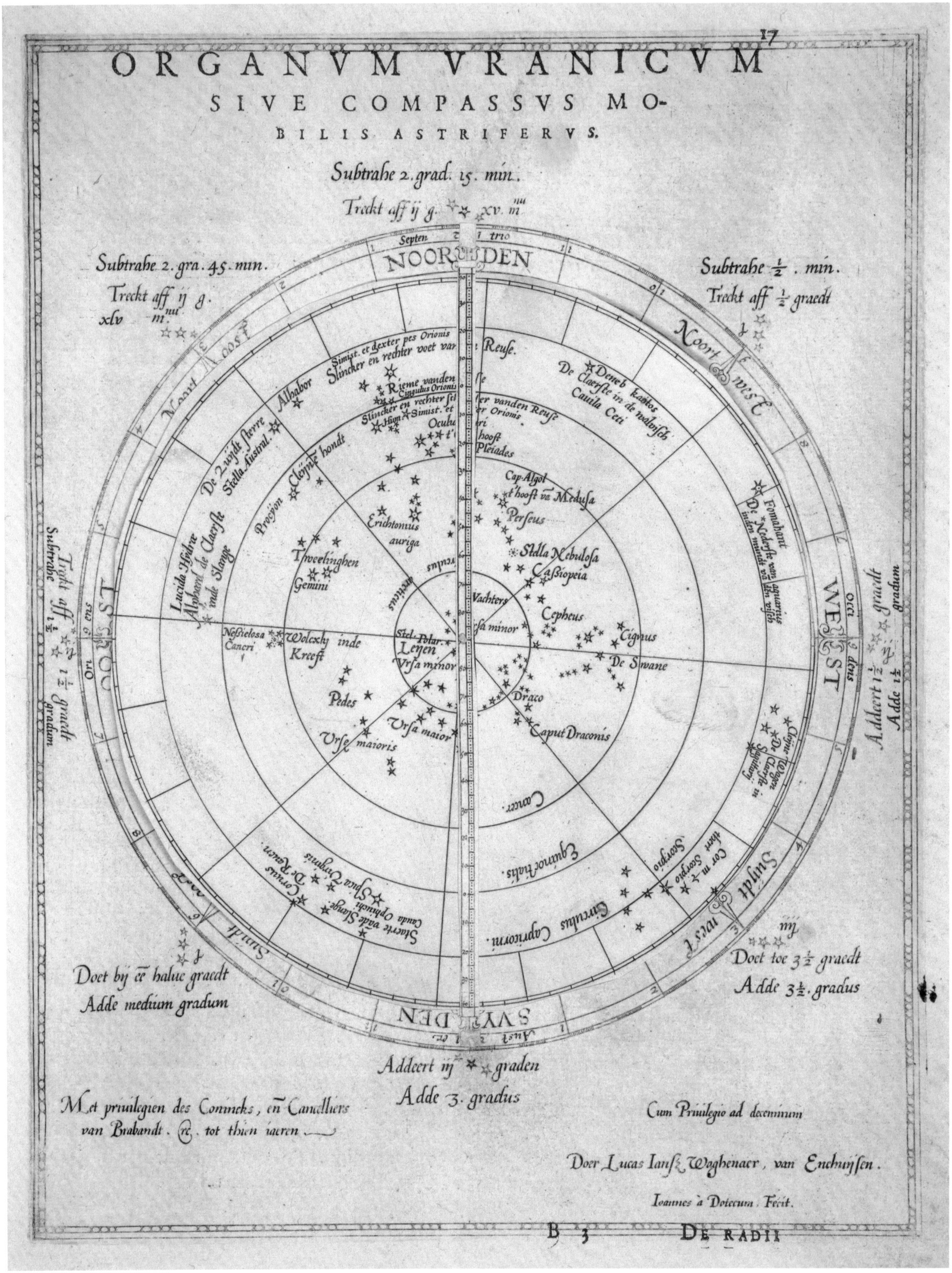
17
ORGANVM VRANICVM
SIVE COMPASSVS MO-
BILIS ASTRIFERVS.
Subtrahe 2. grad. 15. min.
Subtrahe 2. gra. 45. min.
Subtrahe ½. min.
Trecht aff ½ graedt
NOORDEN
WEST
SVYDEN
Addeert iij graden
Adde 3. gradus
Doet bij ee halue graedt
Adde medium gradum
Doet toe 3½ graedt
Adde 3½. gradus
Addeert 1½ graedt
Adde 1½ gradum
Met priuilegien des Coninckx, en Cancelliers
van Brabandt. &c. tot thien iaeren
Cum Priuilegio ad decennium
Doer Lucas Ians; Waghenaer, van Enchuijsen.
Ioannes à Dotecum. Fecit.
B 3
DE RADII

WAGHENAER'S *SPIEGHEL*

Shipping and overseas trade boomed in the Low Countries in the sixteenth century, with improving knowledge of trade routes a major key to success. The introduction of the printing and plate press meant that information could be disseminated more widely and that an attractive market arose for it. All the same, the first 'charts' to be produced were far from graphic, visual representations of the earth's surface but rather written sailing instructions and descriptions of routes and ports. These are also referred to nowadays as rutters or *leeskaarten*, given their textual nature. At most, pilot guides of this kind featured a few woodcuts of coastal landmarks.

Lucas Jansz. Waghenaer (1533/1534-1606) was the first to have the bright idea of combining *leeskaarten* with maritime maps, and his *Spieghel der zeevaerdt* ('Mirror of Seafaring') became the first printed pilot guide to feature large maps in folio format. The navigator from Enkhuizen set about spreading his knowledge, drawing on his years of experience and his contacts with Italian, Portuguese and Spanish mariners. He started work on his book in 1579, focusing on Europe's coasts (it was not intended for ocean-going navigation). The first part comprised charts of the 'western navigation' (as far as Cádiz in Spain), while the second part featured the 'eastern navigation', towards the Baltic Sea and Finland. Waghenaer drew up the maps for the first part by himself. Joannes van Doetecum (1530-1605) then engraved the drawings, producing artistically sophisticated products decorated lavishly with ships and sea monsters. Because of this and in spite of the book's practical approach, it immediately became an object to admire and collect.

The first volume of the *Spieghel* in Dutch was printed in Leiden in 1584 by the Antwerp publisher Christophe Plantin, who had temporarily fled the war to the Northern Netherlands. A second volume followed in 1585, with editions in Latin, English, French and German coming soon after. The Phoebus Foundation has a copy of the rare first Latin translation from 1586, printed by Plantin's son-in-law Franciscus Raphelengius (Frans van Ravelingen; 1539-1597). There is also a copy in the collection of the first French edition from 1590, the text of which was printed in Antwerp by Jean Bellère (1526-1595). Cornelis Claesz. (*c.*1551-1609), who emigrated from the Southern Netherlands to Amsterdam, printed the illustrations.

A striking feature of both copies is that they still contain their astronomical *volvelles* – rotating paper instruments that were inserted in the book (⟶ Chapter XI) as an aid to determining the declinations (deviations) of the stars. They consist of a page with a separate disc on it showing the constellations and a separate point on top. The elements are attached to one another to allow them to rotate. Sailors would turn the disc to match their own view of the heavens and then use a maritime compass to set the desired course. The *Spieghel* was a milestone in European maritime cartography, preparing the way for the modern marine atlas. So famous did it become that the word 'Waggoner' (derived from 'Waghenaer', its creator) became a generic name in English for pilot guides.

Volvelle
From: Lucas Jansz. Waghenaer, *Speculum Nauticum*, 1586
ANTWERP, THE PHOEBUS FOUNDATION

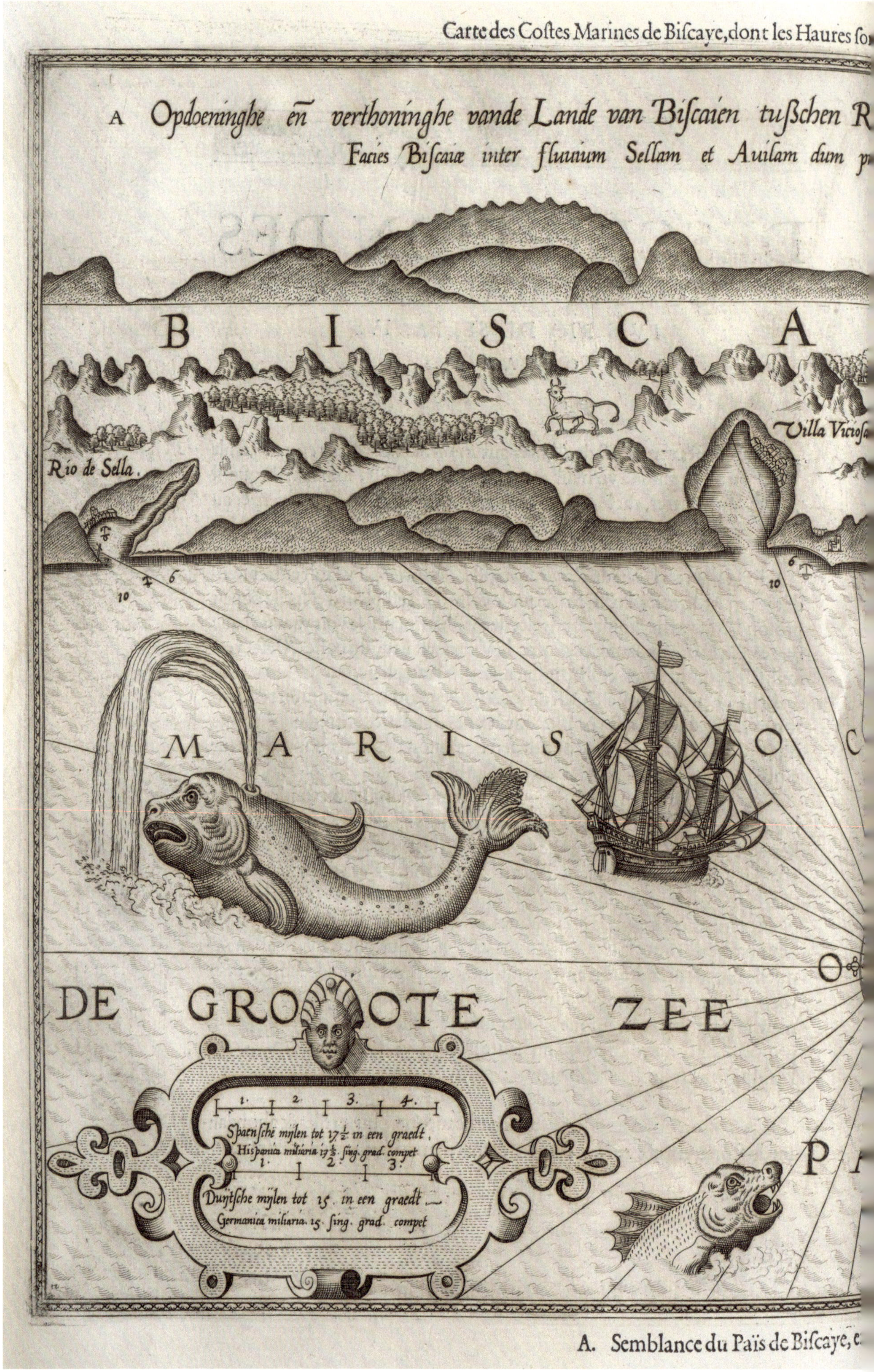

Joannes van Doetecum
Die Caerte vande Zeekusten van Biscaien, Zeer quade havens voer groote schepen, tussche[n] Rio de Sella e[nde] Aviles
Sea chart of the Spanish coastline between Rio de Sella and Aviles
From: Lucas Jansz. Waghenaer, *Miroir de la navigation*, 1590
ANTWERP, THE PHOEBUS FOUNDATION

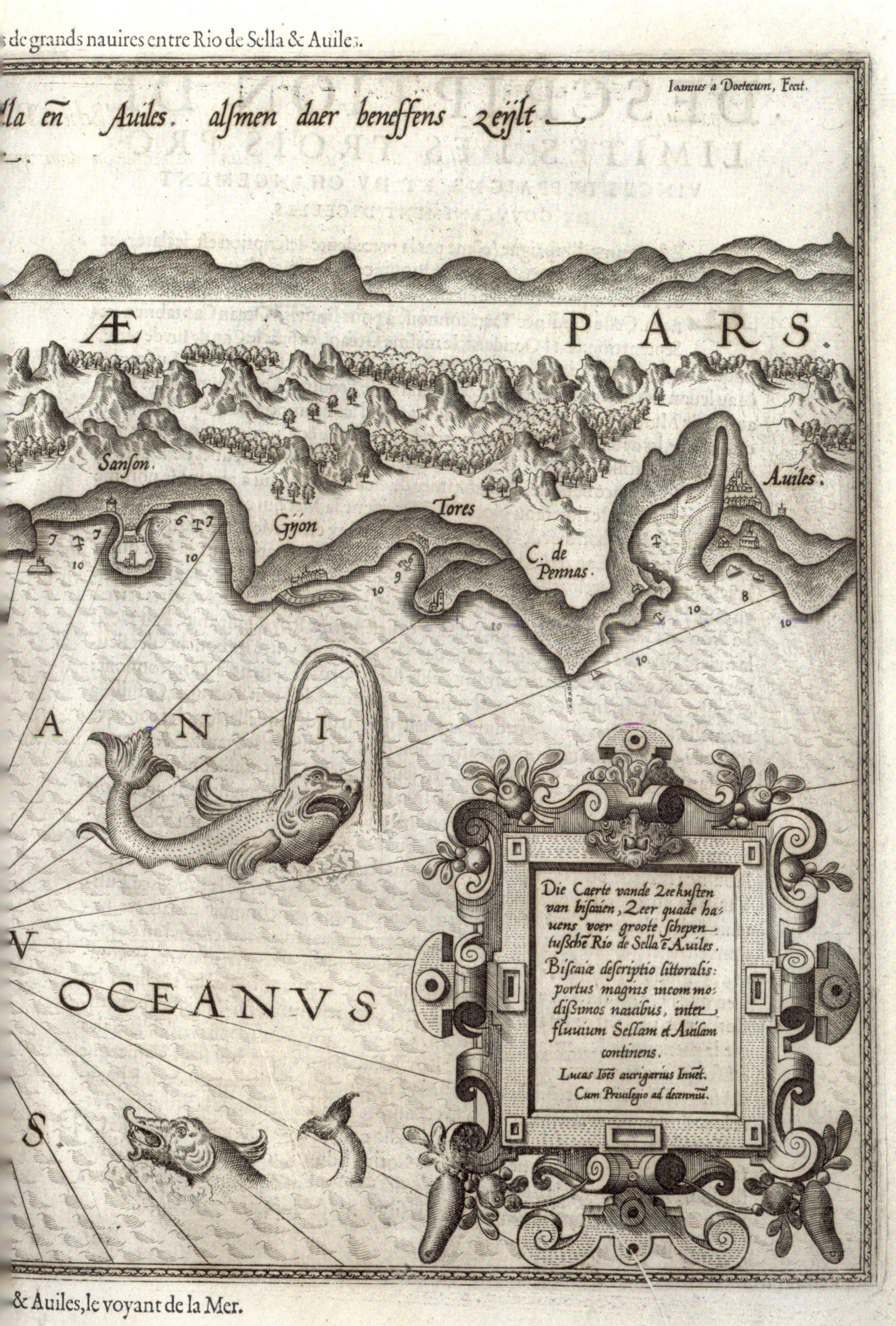
s de grands nauires entre Rio de Sella & Auiles.
Ioannes a Doetecum, Fecit.
lla en Auiles. alsmen daer beneffens zeylt
Æ PARS.
Sanson.
Gijon
Tores
C. de Pennas.
Auiles.
A N I
V
OCEANVS
S.
Die Caerte vande zeekusten van biscaien, zeer quade hauens voer groote schepen tusschē Rio de Sella ē Auiles.
Biscaiæ descriptio littoralis: portus magnis incommodissimos nauibus, inter fluuium Sellam et Auilam continens.
Lucas Ioēs aurigarius Inuēt.
Cum Priuilegio ad decennīū.
& Auiles, le voyant de la Mer.

Sea charts and shipboard instruments

Sailing instructions and detailed maps of coasts and ports were not enough for mariners when it came to longer voyages and crossing seas and oceans. The network of lines in the shape of a star that we find on Waghenaer's charts were not merely ornamental. Sea charts are identifiable by lines in differen compass directions: they enabled navigators to calculate direction and estimate distances. They developed from earlier 'portolan' charts from Southern Europe – hand-drawn parchment maps of coastlines and ports, the earliest examples of which date back as early as the thirteenth century. In the late sixteenth century, Gerard Mercator introduced the system of increasing latitude lines known after him as the 'Mercator projection' (⟶ p. 202), in which the curved, three-dimensional earth is represented in rectangular, two-dimensional form allowing a course to be calculated. Sea charts on which a course was actually plotted were large in size and loose-leaf for ease of use.

The navigator at sea needed more, however, than a sea chart alone: having the right instruments on board was important too. A compass and ruler already took the mariner a long way, but an armillary sphere, quadrant, terrestrial and celestial globes, and protractor could also help with measurements and positioning. The crucial role played by navigational and maritime instruments is illustrated by the title page of Waghenaer's *Miroir de la navigation* (the French translation of *Spieghel der zeevaerdt*), engraved by Van Doetecum, which depicts contemporary instruments. Hanging in the two upper corners we find (from top to bottom) a quadrant, a nautical astrolabe and a cross-staff or Jacob's staff. They appear on either side of a pair of globes, celestial (left) and terrestrial (right), and two hourglasses at the top of the title print's architectural frame. We also see sounding lines, held by two figures, and two large dividers in the bottom corners with nautical compasses between their legs.

Waghenaer's book also contains instructions for making charts. This too required special instruments, including a reduction compass like the one by an anonymous maker from the Leuven school, which is used to resize sketches to a specific scale. The instrument is placed on a piece of paper and has adjustable points that can be used to transfer and compare measurements and geometric proportions.

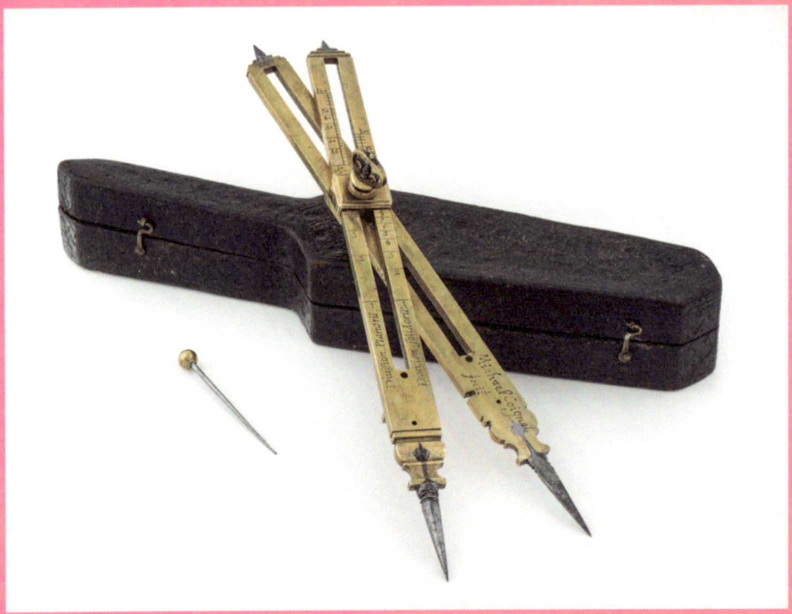

Reduction compass, c.1600
Brass, 26 × 6 × 4 cm
ANTWERP, THE PHOEBUS FOUNDATION

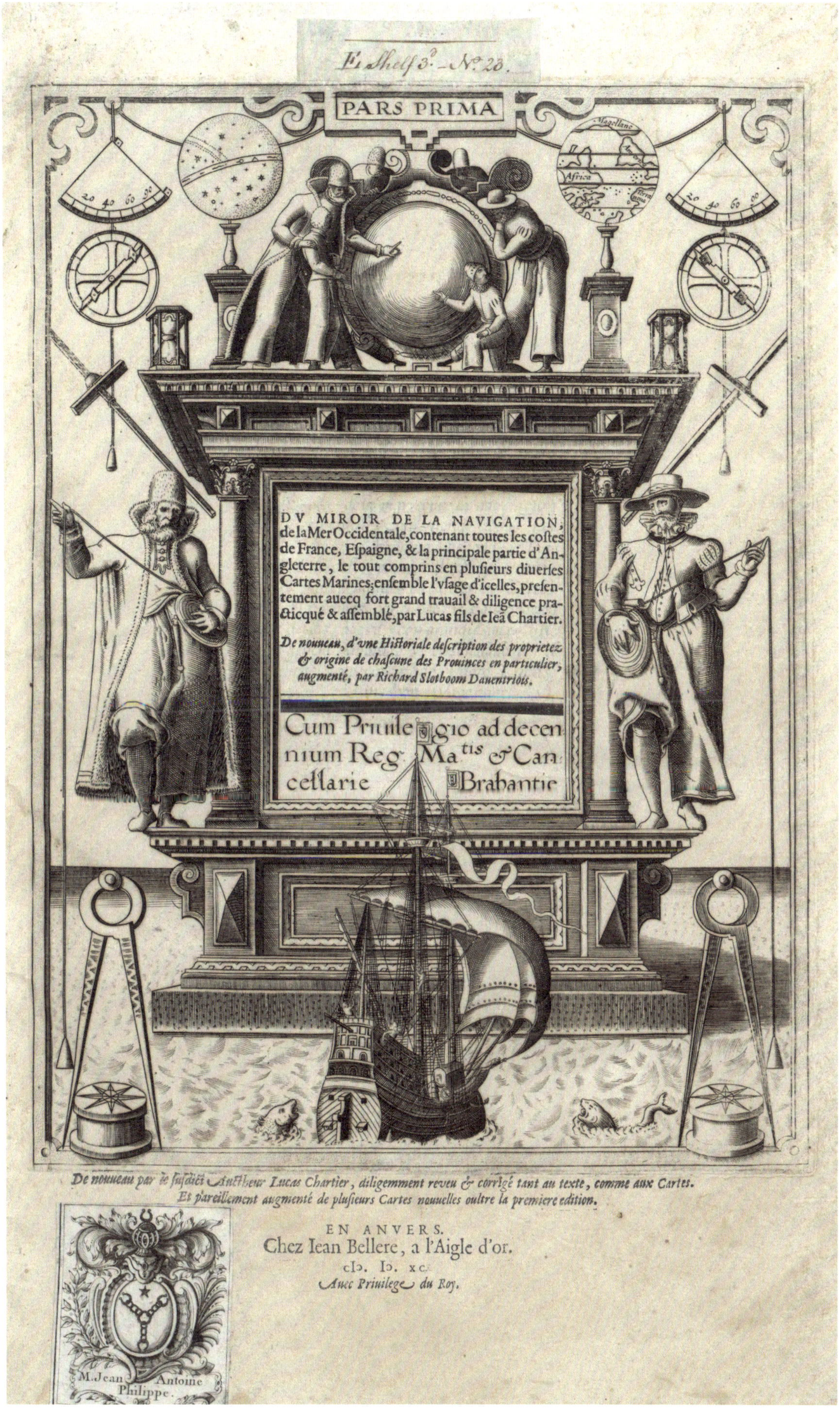

Title page
From: Lucas Jansz. Waghenaer, *Miroir de la navigation*, 1590
ANTWERP, THE PHOEBUS FOUNDATION

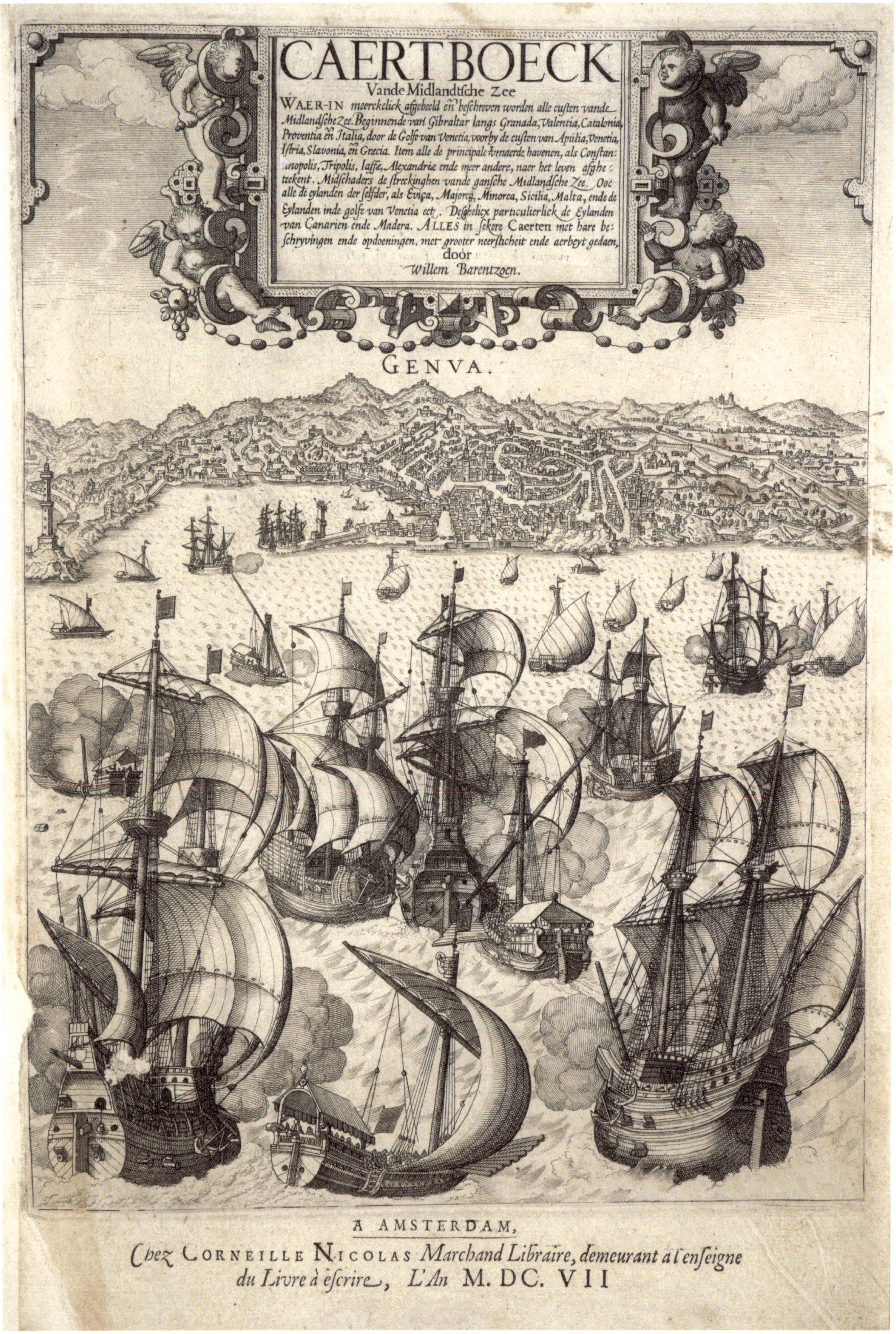
CAERTBOECK
Vande Midlandtsche Zee
WAER-IN meerckelick afgebeeld en beschreven worden alle custen vande
Midlandsche Zee. Beginnende van Gibraltar langs Granada, Valentia, Catalonia,
Proventia en Italia, door de Golfe van Venetia, voorby de custen van Apulia, Venetia,
Istria, Slavonia, en Grecia. Item alle de principale vermaerde havenen, als Constan:
tinopolis, Tripolis, Iaffe, Alexandriæ ende meer andere, naer het leven afghe:
teekent. Midschaders de streckinghen vande gansche Midlandsche Zee. Ooc
alle de eylanden der selfder, als Eviça, Majorca, Minorca, Sicilia, Malta, ende de
Eylanden inde golfe van Venetia ect. Desghelicx particulierlick de Eylanden
van Canarien ende Madera. ALLES in sekere Caerten met hare be:
schryvingen ende opdoeningen, met grooter neersticheit ende aerbeyt gedaen,
door
Willem Barentzoen.
GENVA.
A AMSTERDAM,
Chez CORNEILLE NICOLAS Marchand Libraire, demeurant a l'enseigne
du Livre à escrire, L'An M. DC. VII

WILLEM BARENTSZ. AND HIS *CAERTBOECK*

Amsterdam was a bustling port in the seventeenth century where nautical knowledge from every quarter converged. Publishers quickly spotted the money to be made from the burgeoning field of nautical cartography. Cornelis Claesz. kept a close eye on voyages of exploration, from which he learned that Dutch ships had sailed through the Strait of Gibraltar towards Italy. He had already acquired the rights to Waghenaer's *Spieghel* (which he printed from 1589 onwards), and so all he needed to do was to add charts of the Mediterranean. He commissioned the mariner and cartographer Willem Barentsz. (*c.*1550-1597) to draw them up for him.

To create the maps for the *Caertboeck Vande Midlandtsche Zee* ('Map Book of the Mediterranean Sea'; 1609), Barentsz. relied heavily on existing Italian pilot guides and charts, which were important sources for mariners, as the Italians had already charted the region in great detail. This is reflected at the back of the book with a translation of Paolo Gerardo's *Il Portolano del Mare* (Venice, 1584), which contains sailing instructions and descriptions of the coast for sailing the Mediterranean. The texts and coastal profiles were original additions by Barentsz. himself. Claesz. compiled the volume into a proper ensemble and had the maps skilfully engraved by Jodocus Hondius I (1563-1612) and Pieter Van den Keere. The navigation expert Petrus Plancius (1552-1622) also oversaw the quality of detailed charts (⟶ Chapter VIII). The approach was similar to that of Waghenaer's *Spieghel*, with the focus on accurate navigation from port to port along the coast. The copy in The Phoebus Foundation's collection is in French. Demand for the book in this lingua franca was clearly high, as similar translations were published in 1607, 1608 and 1609. Only two complete copies of this 1609 edition are known.

Title page
From: Willem Barentsz., *Caertboeck Vande Midlandtsche Zee*, 1609
ANTWERP, THE PHOEBUS FOUNDATION

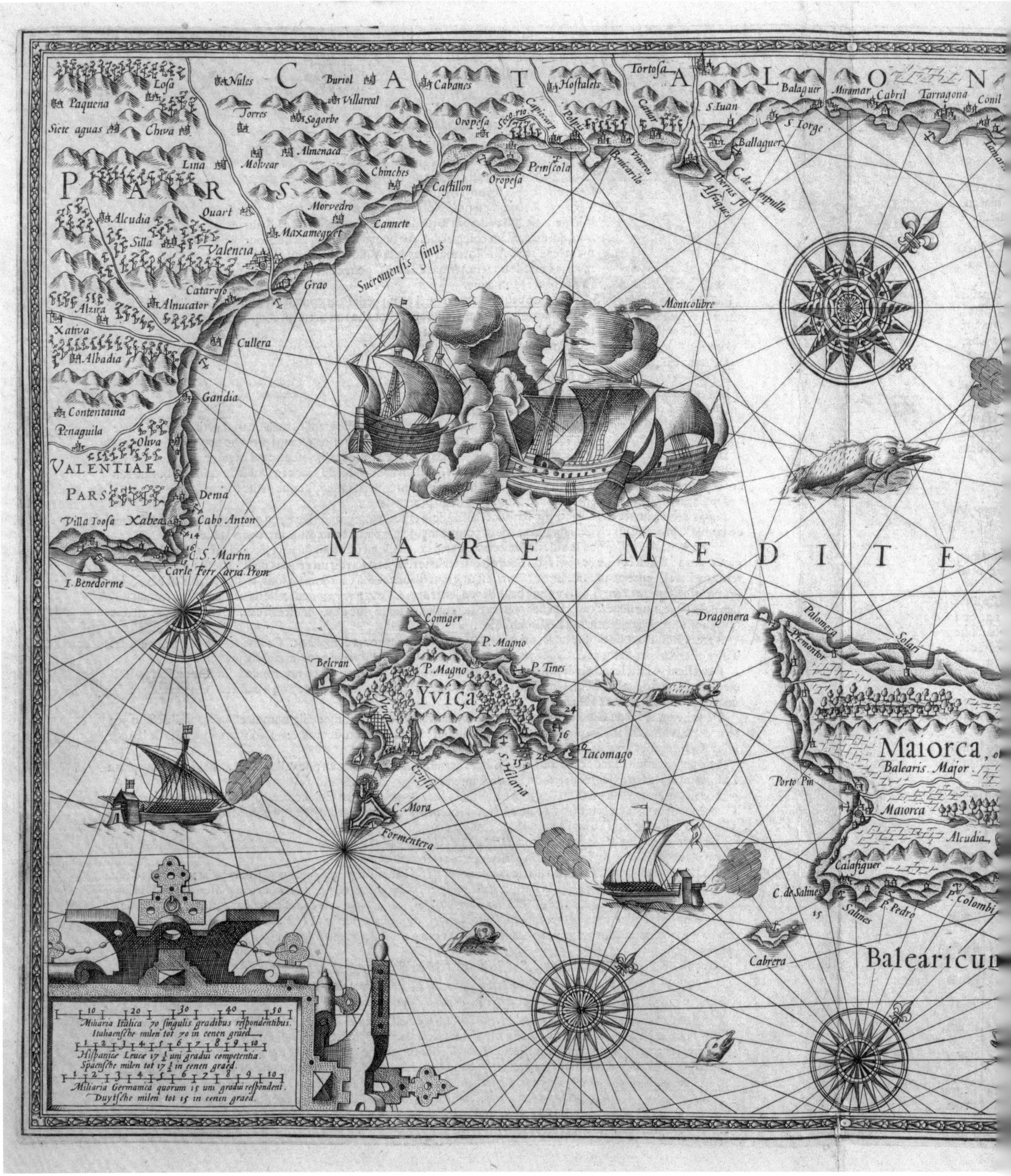
CATALONIA
VALENTIAE PARS
MARE MEDITE
Yviça
Maiorca, o
Balearis Major
Balearicum
Formentera
Tortosa
Burriol
Villareal
Sogorbe
Almenaca
Chinches
Castillon
Oropesa
Peniscola
Benicarlo
Vinaros
Iberus fl
Alfaques
C. de Ampulla
S. Iuan
S. Iorge
Ballaguer
Miramar
Cabril
Tarragona
Conil
Morvedro
Valencia
Grao
Cullera
Gandia
Xativa
Denia
Cabo Anton
Xabea
Villa Ioosa
C. S. Martin
Carle Ferr aria Prom
I. Benedorme
Sucronensis sinus
Montcolibre
Coniger
P. Magno
P. Tines
Belcran
Tacomago
S. Hilaria
C. Mora
Dragonera
Palomera
Solari
Promontor
Porto Pin
Alcudia
Calafiguer
C. de Salines
Salines
P. Pedro
P. Colombi
Cabrera
Miliaria Italica 70 singulis gradibus respondentibus.
Italiaensche milen tot 70 in eenen graed
Hispanicae Leucae 17 ½ uni gradui competentia.
Spaensche milen tot 17 ½ in eenen graed.
Miliaria Germanica quorum 15 uni gradui respondent.
Duytsche milen tot 15 in eenen graed.

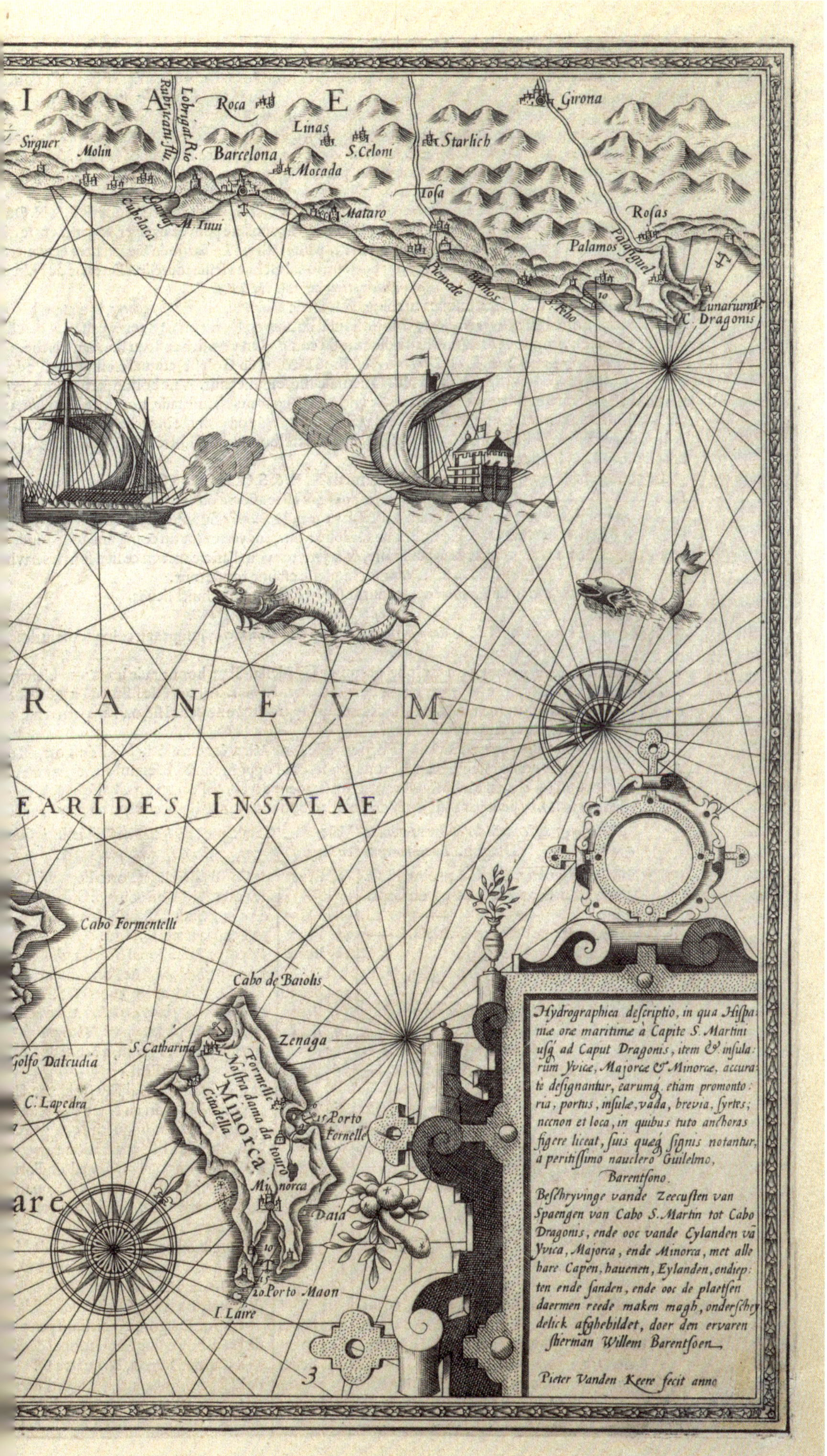

Pieter Van den Keere
Hydrographica descriptio, in qua Hispaniae orae maritimae à Capite S. Martini usq'ad Caput Dragonis / Beschryvinge vande zeecusten van Spaengen van Cabo S. Martin tot Cabo Dragonis
Sea chart of the Spanish coast and the islands of Mallorca and Ibiza
From: Willem Barentsz., *Caertboeck Vande Midlandtsche Zee*, 1609
ANTWERP, THE PHOEBUS FOUNDATION

WILLEM JANSZ. BLAEU'S *LICHT DER ZEE-VAERT*

A key figure in Netherlandish nautical cartography was the Amsterdam mapmaker and publisher Willem Jansz. Blaeu (⟼ Chapter X), who also made instruments and was an expert on navigation. His chart of the coasts of eastern England comes from the first volume of his pilot guide *Het Licht der Zee-vaert* ('The light of seafaring'; Amsterdam, 1608) and shows the coast from Dover to Great Yarmouth. An inset map at the top indicates the course of the Thames. Blaeu adopted the pilot guides produced by Waghenaer as his starting point, improving the material wherever possible. In his characteristic way, he made each map a genuine work of art, richly decorated with title cartouches and highly ornamental scale indicators, and often coloured. In 1633, he was appointed cartographer of the Dutch East India Company, for which he trained new navigators in Amsterdam to read charts and plot courses at sea. He was also responsible for providing ships with the right charts and instruments.

Next fold-out map:
Willem Jansz. Blaeu
Ware afbeeldinge der zeecusten van Engelant gelegen tuschen Doveren en[de] Iarmuijen, c.1608
Sea chart of the coast of eastern England
Hand-coloured copper engraving and etching, 257 × 560 mm
ANTWERP, THE PHOEBUS FOUNDATION

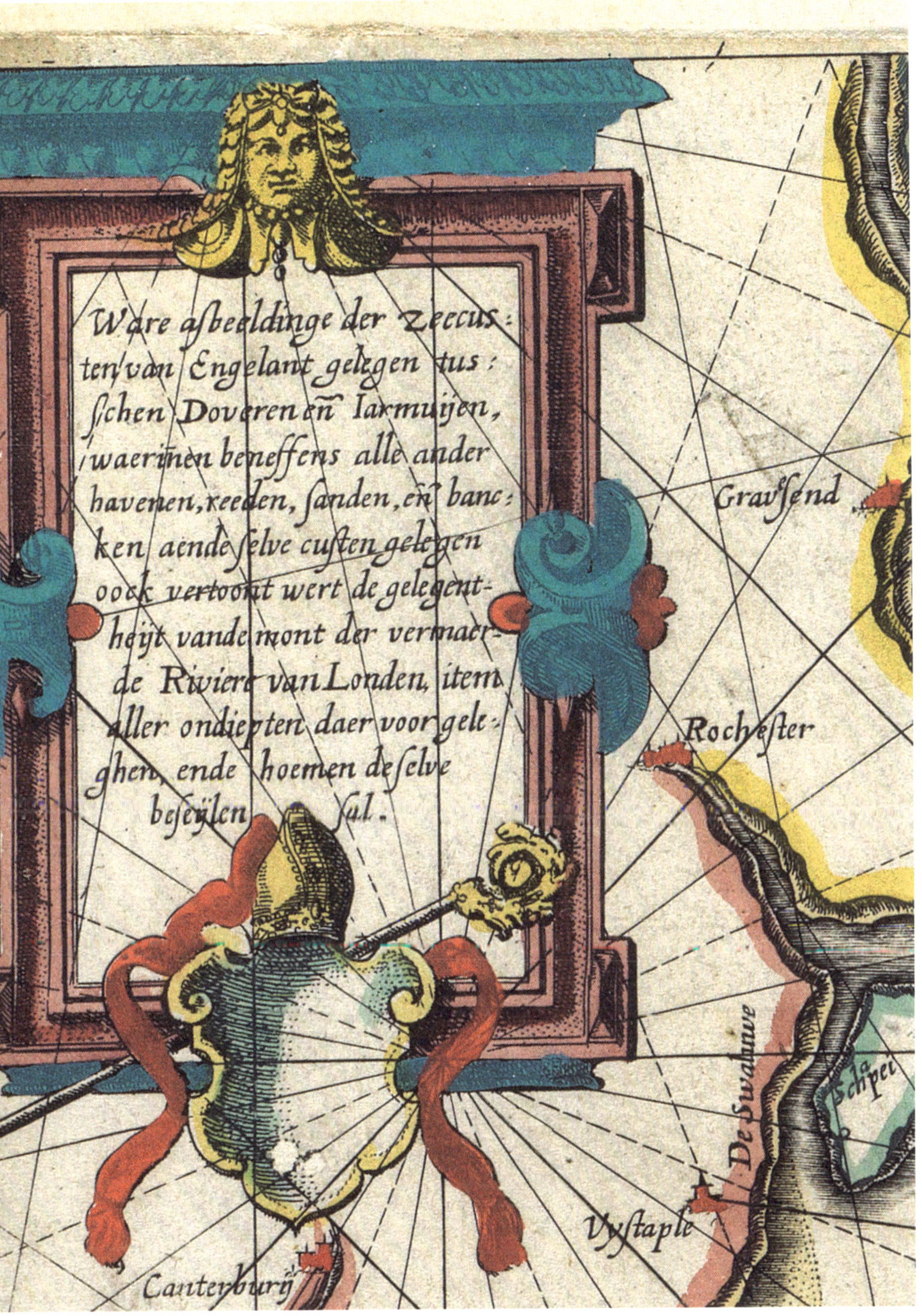
Ware afbeeldinge der Zeecus:
ten van Engelant gelegen tus:
schen Doveren en̄ Iarmuijen,
waerinen beneffens alle ander
havenen, reeden, sanden, en̄ banc:
ken aende selve custen gelegen
oock vertoont wert de gelegent:
heyt vande mont der vermaer:
de Riviere van Londen, item
aller ondiepten daer voor gele:
ghen, ende hoemen deselve
beseijlen sal.
Gravsend
Rochester
Vystaple
Canterbury

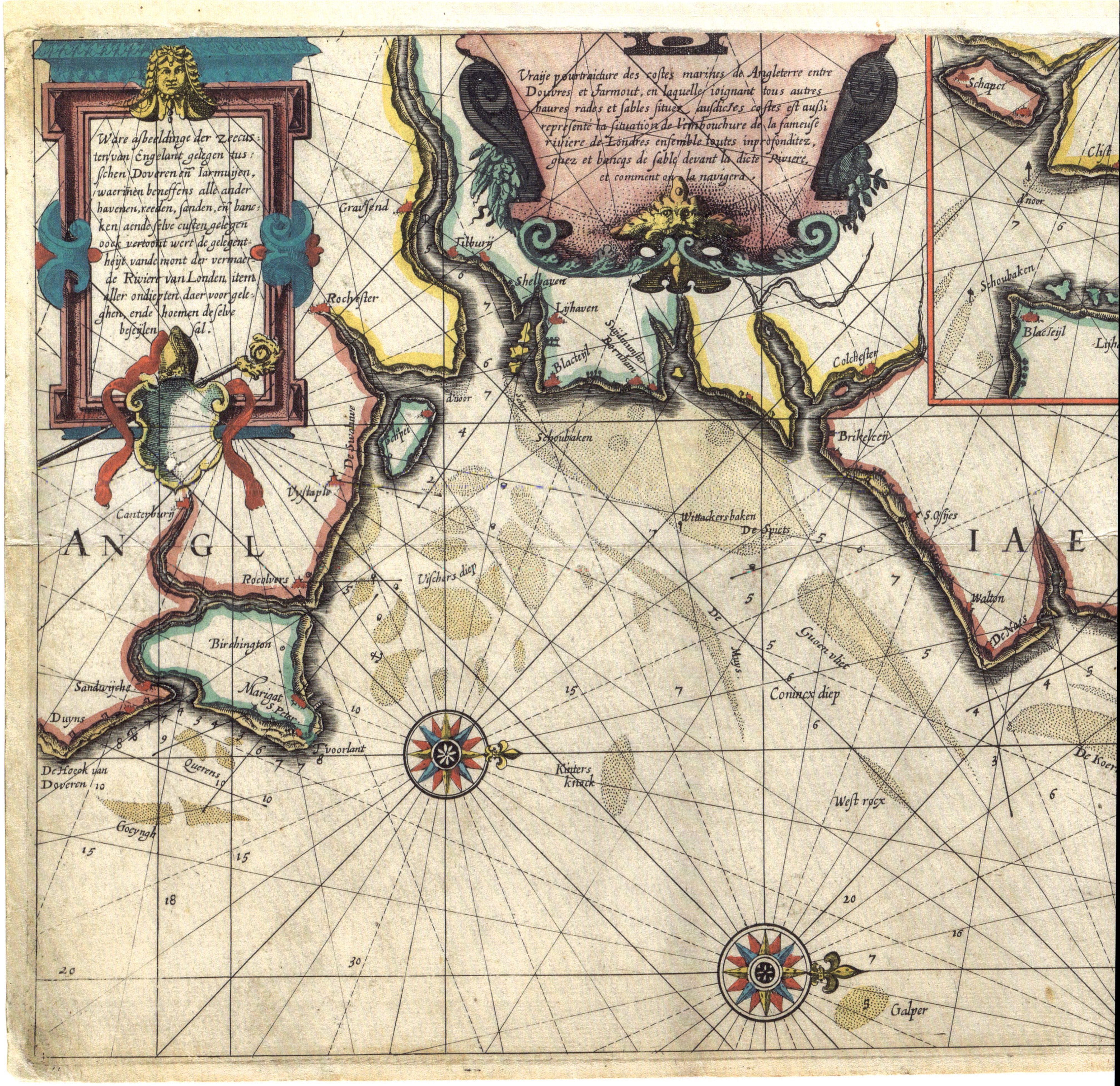
Vraije pourtraicture des costes marines de Angleterre entre
Doubres et Iarmout, en laquelle ioignant tous autres
havres rades et sables situez audictes costes est aussi
represente la situation de l'embouchure de la fameuse
riviere de Londres ensemble toutes in profonditez,
guez et bancqs de sable devant la dicte Riviere,
et comment on la navigera.
Ware afbeeldinge der Zeecus:
ten van Engelant gelegen tus:
schen Doveren en̄ Iarmuijen,
waerinen beneffens alle ander
havenen, reeden, sanden, en̄ banc:
ken aende selve custen gelegen
oock vertoont wert de gelegent:
heyt vande mont der vermaer:
de Riviere van Londen, item
aller ondiepten daer voor gele:
ghen, ende hoemen deselve
beseijlen sal.
Gravsend
Tilbury
Rochester
Shelhaven
Lyhaven
Blaceyil
Colchester
Brikelcey
Schoubaken
Canterbury
Vystaple
A N G L
I A E
Rocolvers
Visschers diep
Wittackersbaken
De Spits
S. Ossjes
Walton
De Nats
Birchington
Marigat
Sandwijche
Duyns
De Hoeck van
Doveren
Querens
Goeyngh
Conincx diep
Guern vliet
West rocx
Kinters
Galper
Schaper

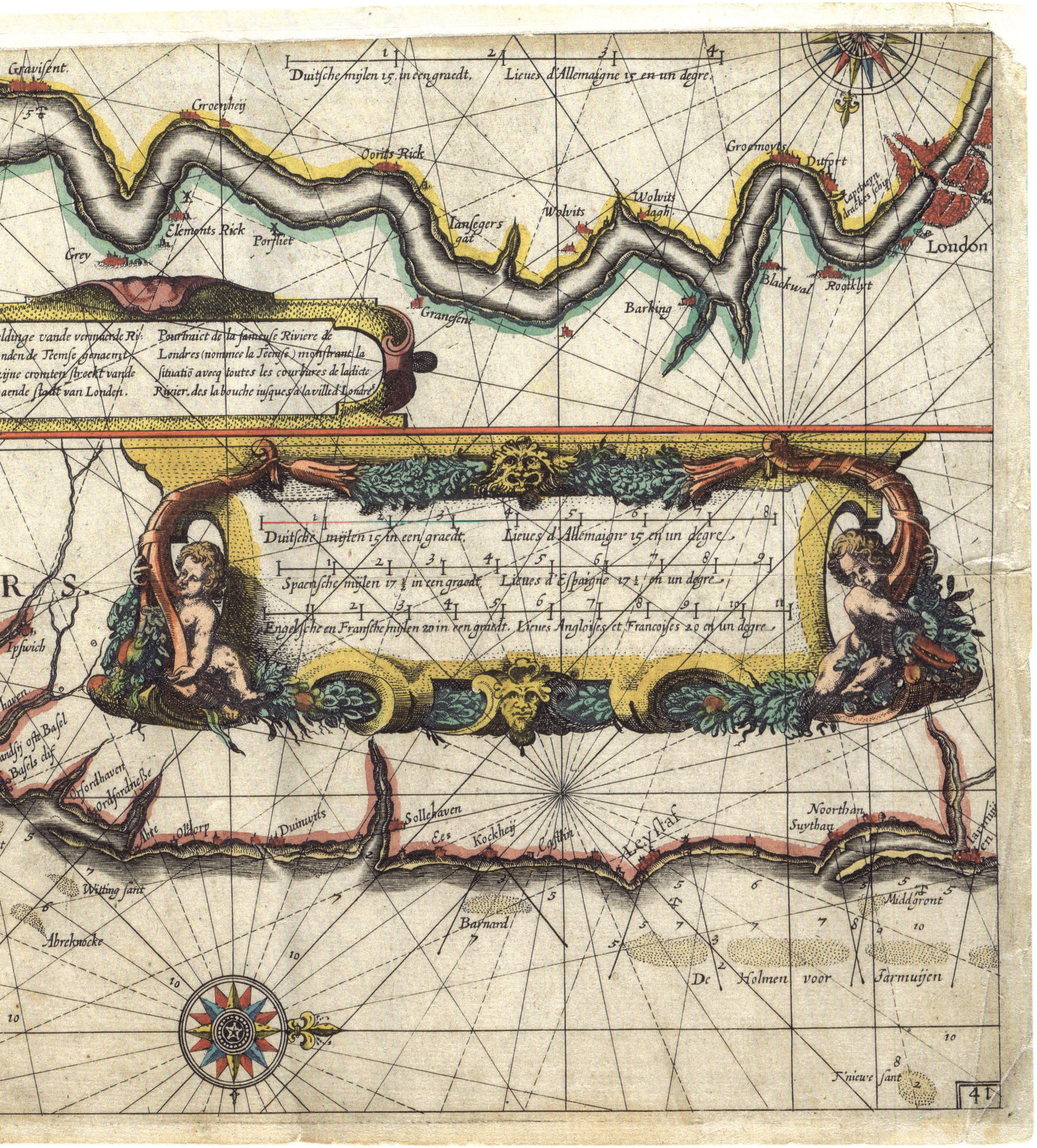
Duitsche mijlen 15 in een graedt. Lieues d'Allemaigne 15 en un degre.
Gravisent.
Groenheij
Dorits Rick
Elements Rick
Porsfliet
Grey
Tansegers gat
Wolvits
Wolvits dagh
Groenwits
Dutfort
London
Blackwal
Rooklyt
Barking
Gransent
Pourtraict de la fameuse Riviere de Londres (nommee la Teemse) monstrant la situatiō aveq toutes les courbures de la dicte Rivier. des la bouche iusques a la ville de Londre.
Duitsche mijlen 15 in een graedt. Lieues d'Allemaign 15 en un degre.
Spaensche mijlen 17½ in een graedt. Lieues d'Espaigne 17½ en un degre.
Engelsche en Fransche mijlen 20 in een graedt. Lieues Angloises et Francoises 20 en un degre.
Ipswich
Ortfordhaven
Ortfordnesse
Abre
Oltorp
Duinwits
Sollehaven
Ees
Kockheij
Cosstin
Leystaf
Noorthan
Suythan
Witting sant
Abreknocke
Barnard
Middgront
De Holmen voor Jarmuijen
Nieuwe sant
41

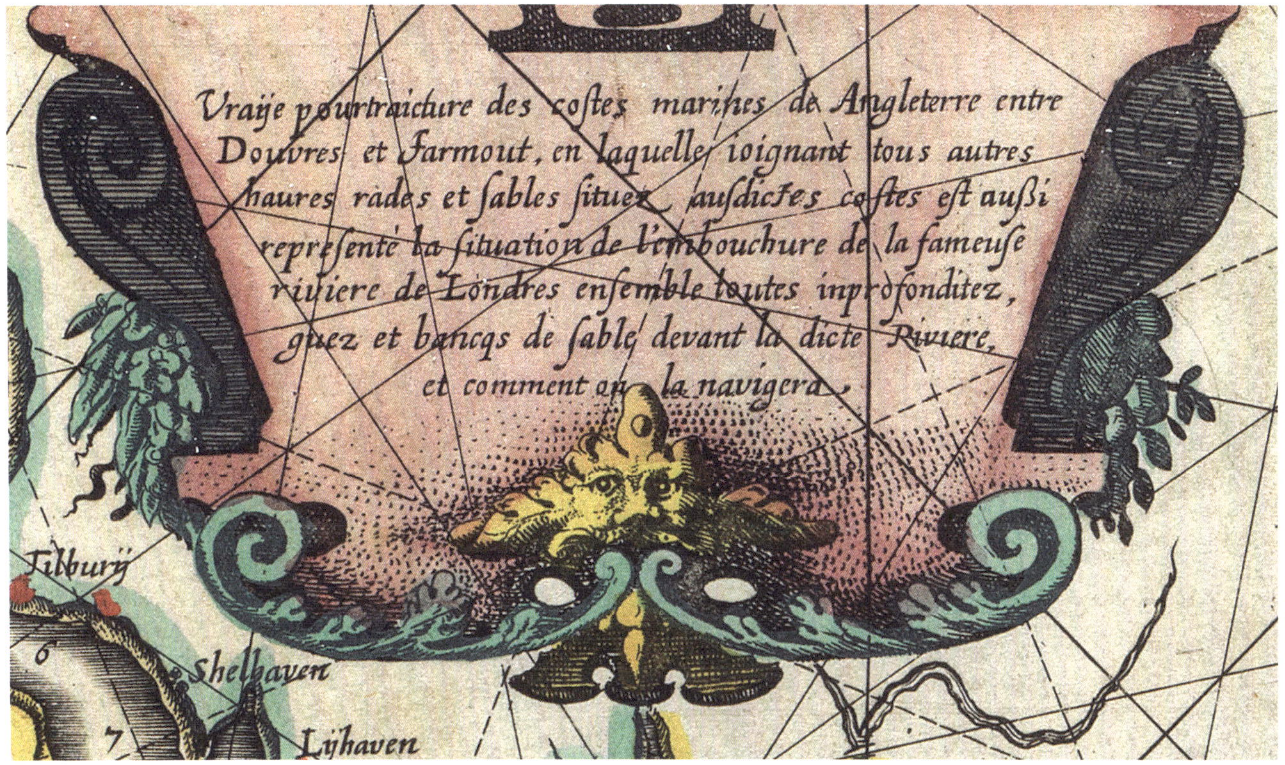
Vraije pourtraicture des costes marines de Angleterre entre Douvres et Jarmout, en laquelle ioignant tous autres haures rades et sables situez ausdictes costes est aussi represente la situation de l'embouchure de la fameuse riviere de Londres ensemble toutes inprofonditez, guez et bancqs de sable devant la dicte Riviere, et comment on la navigera.
Tilbury
Shelhaven
Lijhaven

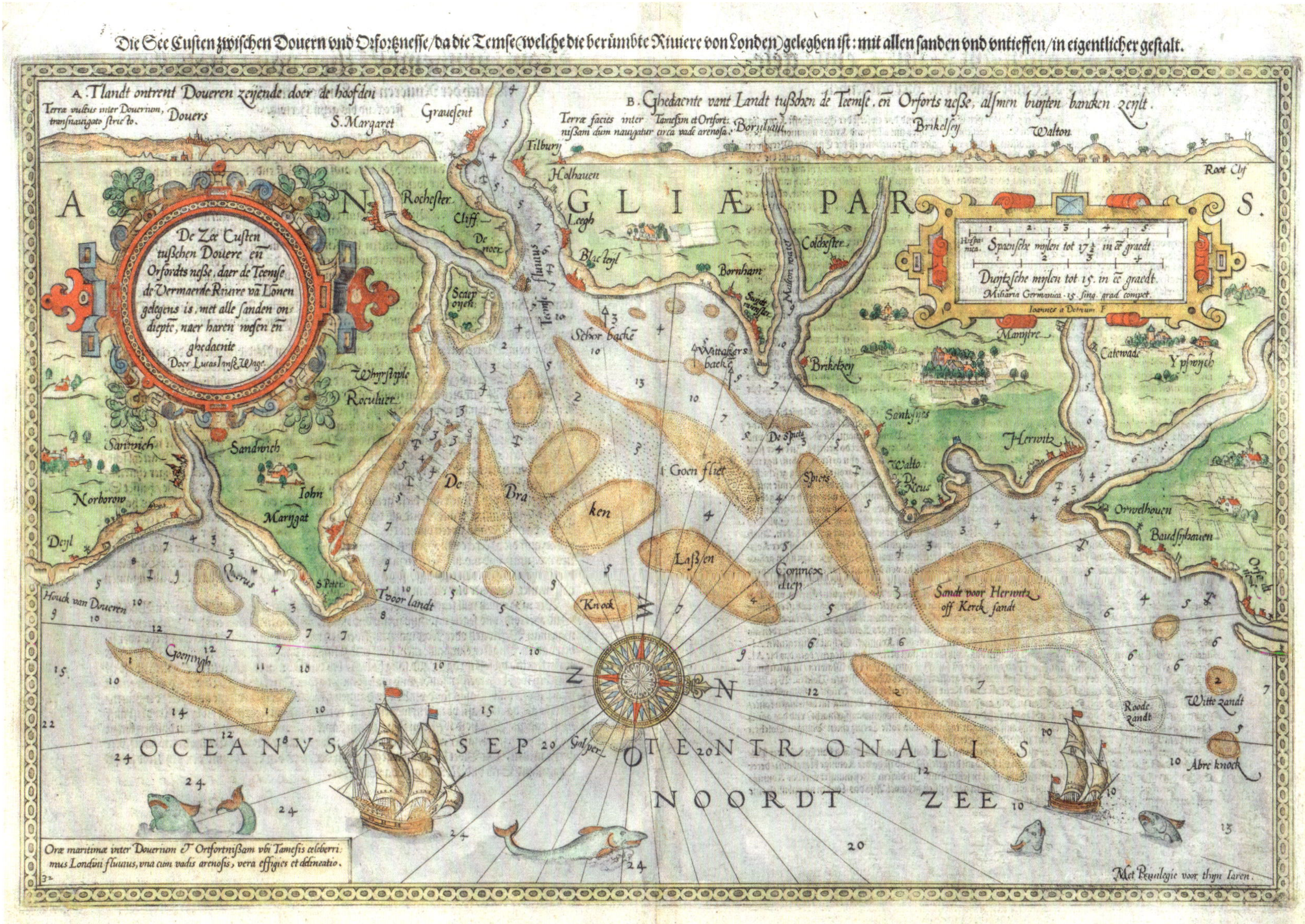

Lucas Jansz. Waghenaer and Joannes van Doetecum
De Zee Custen tußchen Dovere en[de] Orfordts neße, daer de Teemse de Vermaerde Rivire va[n] Lon[d]en gelegens is, met alle sanden ondiepte[n], naer haren wesen en[de] ghedaente, Amsterdam, 1588
Sea chart of the coast of eastern England, originally from the German edition of Lucas Jansz. Waghenaer's guide for sailors, Dess Spiegels der Seefart
Hand-coloured copper engraving and etching, 390 × 521 mm
ANTWERP, THE PHOEBUS FOUNDATION

Longitude and latitude

If you want to know where you are going, you first need to know where you are. Determining your position was crucial when sailing across the open sea. Nowadays, satellite-based GPS tells you precisely where you are at any moment of the day. Since ancient times, however, people resorted to various other techniques for this purpose. One of the many instruments invented over the centuries was the astrolabe, which was chiefly used to calculate latitude (⟶ p. 292). It was too refined, however, to do its job correctly in rough seas. A nautical astrolabe was, therefore, specially adapted for mariners, and it could also measure the angle of the sun. It was a simplified, heavier and more rugged version of the land astrolabe that was better able to cope with the motion of the sea.

Calculating longitude was considerably more complex. The clocks that were available were not yet sufficiently reliable to calculate the time differences caused by the earth's rotation. They went haywire aboard ship, so sailors had to rely on hourglasses instead, which was far from ideal. The problem grew so pressing that the States General – the Dutch Republic's parliament – offered a prize to whoever could come up with a solution. Willem Jansz. Blaeu – an acclaimed scientist and instrument maker as well as a mapmaker – sat on the jury. In the end, Blaeu himself compiled a booklet containing deviations and a method for calculating longitude: the *Stuurmans luywagen*, especially for mariners. Longitude could then be calculated using compass variations between the position of the sun in the morning and the evening, combined with accurate tables. The invention of the compass and that of longitude measurement are depicted in the *Nova Reperta* series of prints devoted to great innovations of the Renaissance. The Latin text accompanying the print depicting the invention of longitude measurement praises Petrus Plancius for making it possible to find a port everywhere with the help of a compass.

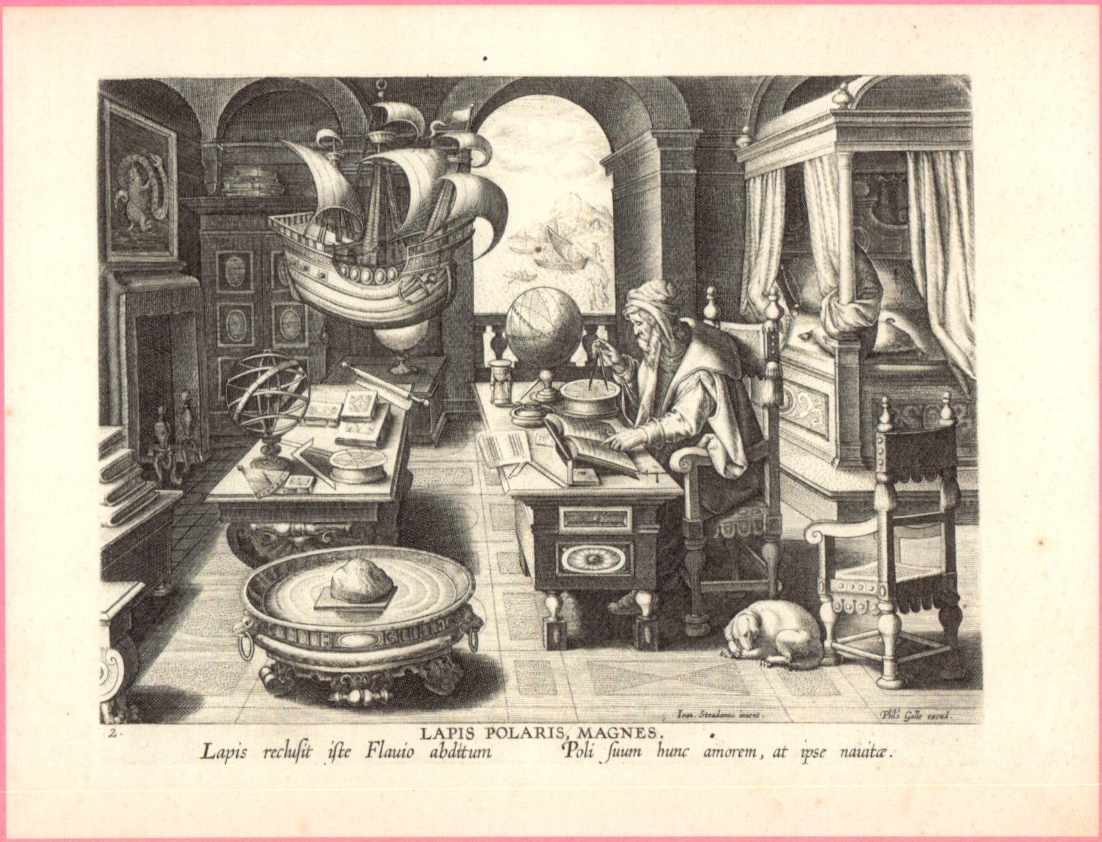

Hans Collaert and Theodoor Galle after Johannes Stradanus
Nova Reperta: Lapis Polaris Magnes, c.1590
The invention of the compass
Copper engraving, 263 × 340 mm
ANTWERP, THE PHOEBUS FOUNDATION

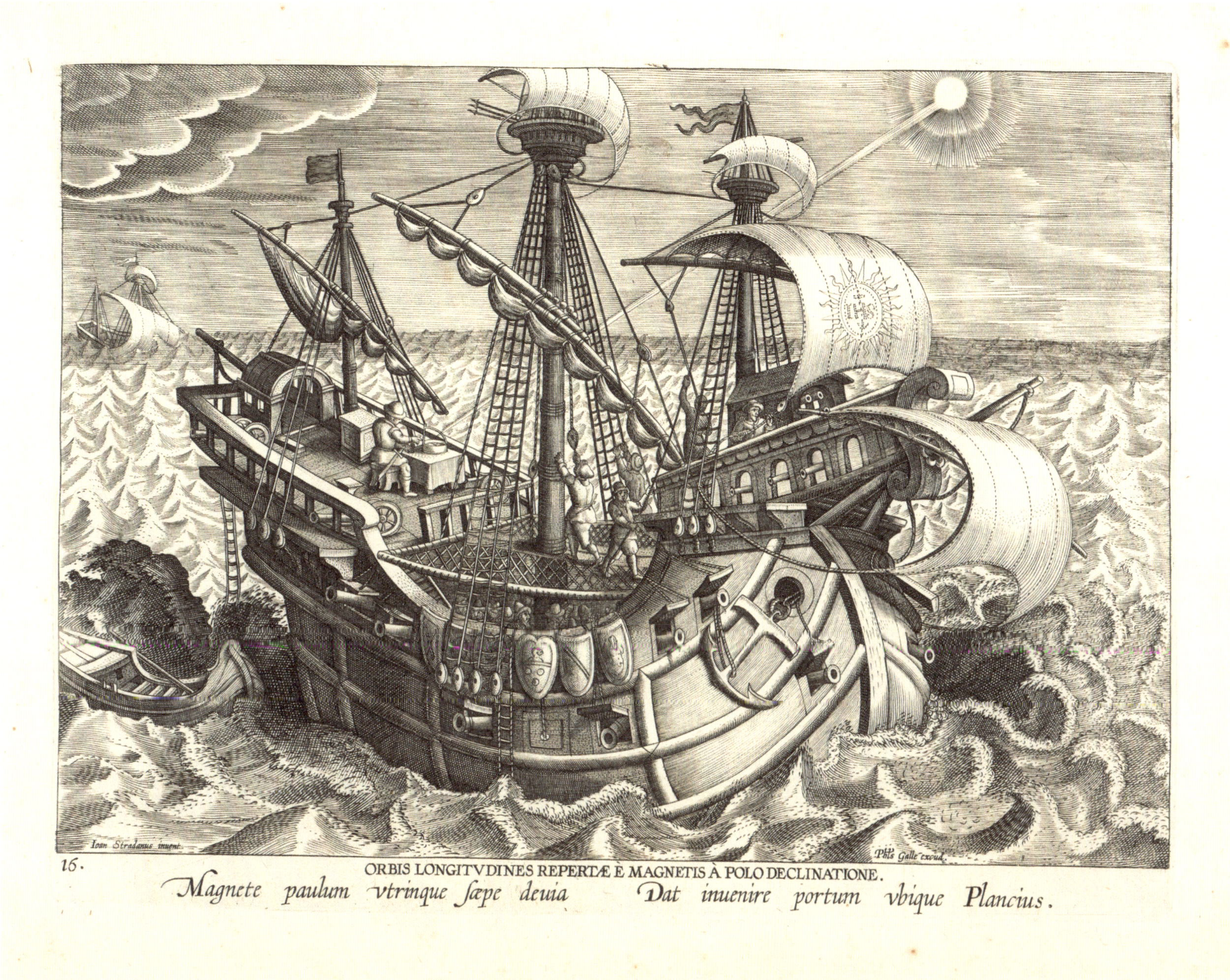

Hans Collaert and Theodoor Galle after Johannes Stradanus
Nova Reperta: Orbis Longitudines Repertae è Magnetis à Polo Declinatione, c.1590
The invention of longitude
Copper engraving, 263 × 340 mm
ANTWERP, THE PHOEBUS FOUNDATION

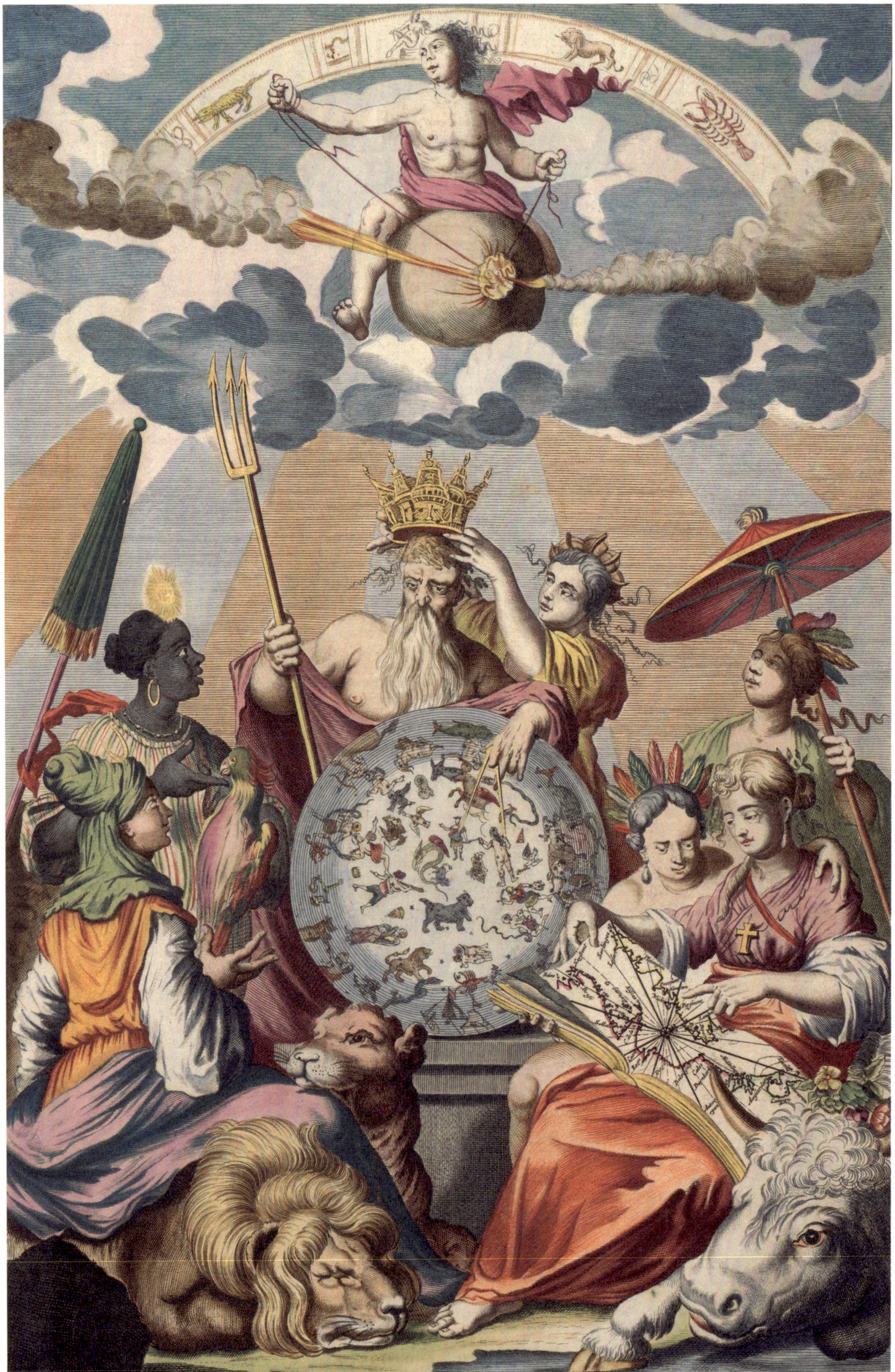
Luyken invenit Aernout Naghtegael

JOHANNES VAN KEULEN'S BURNING TORCH

Waghenaer, Barentsz. and Blaeu's pilot guides were useful aids to navigation. Practicality was key to books like this, but the richly decorated maps they contain suggest that they were also appreciated for their aesthetic value. The artistic quality of nautical charts reached fresh heights with the sea atlases of Johannes van Keulen (1654-1715), including *De Groote Nieuwe Vermeerderde Zee-atlas ofte Water-werelt* ('The big new expanded sea atlas or water world'; Amsterdam, 1680), in which maps took precedence over the texts and navigational instructions.

This was the first of his sea atlases. The Van Keulen family had been quick to master the art of maritime cartography. Johannes teamed up with the mathematician and surveyor Claes Jansz. Vooght (died 1696), who engraved original and innovative maps with immense precision and a great many decorative details, and who also penned the concise accompanying texts. The title cartouches, for instance, were illustrations in their own right, incorporating flora and fauna from the depicted area along with stereotypical images of the local population (⟶ p. 190).

The copy in The Phoebus Foundation's collection belongs to the later, expanded edition of 1688. Where the first 1680 edition ran to 36 maps, this edition contained no fewer than 135, covering parts of the world from the Mediterranean to Greenland and from Ceylon to the South Pacific. The atlas continued to evolve until 1734 with countless variations, such that no two editions are identical.

At the same time as he produced this atlas, Van Keulen was working on his more practical pilot guide – the *Zee-fakkel* ('Sea Torch') – which also allowed him to keep adding new maps to his *Zee-atlas*. In a manner of speaking, therefore, the two books were able to cross-fertilise one another. The *Zee-atlas* clearly targeted a different readership: it is a refined publication hand-painted in multiple colours, decorated with highly artistic cartouches and portraits, heightened with gold leaf and bound in a luxurious binding. Although sea charts contained technical information and landmarks for the mariner, this was an expensive book that will not have been taken aboard ship. Which is not to say that the atlases were any the less 'useful': they were also employed indirectly as reference works, to plan routes, or as a source of inspiration when planning future journeys.

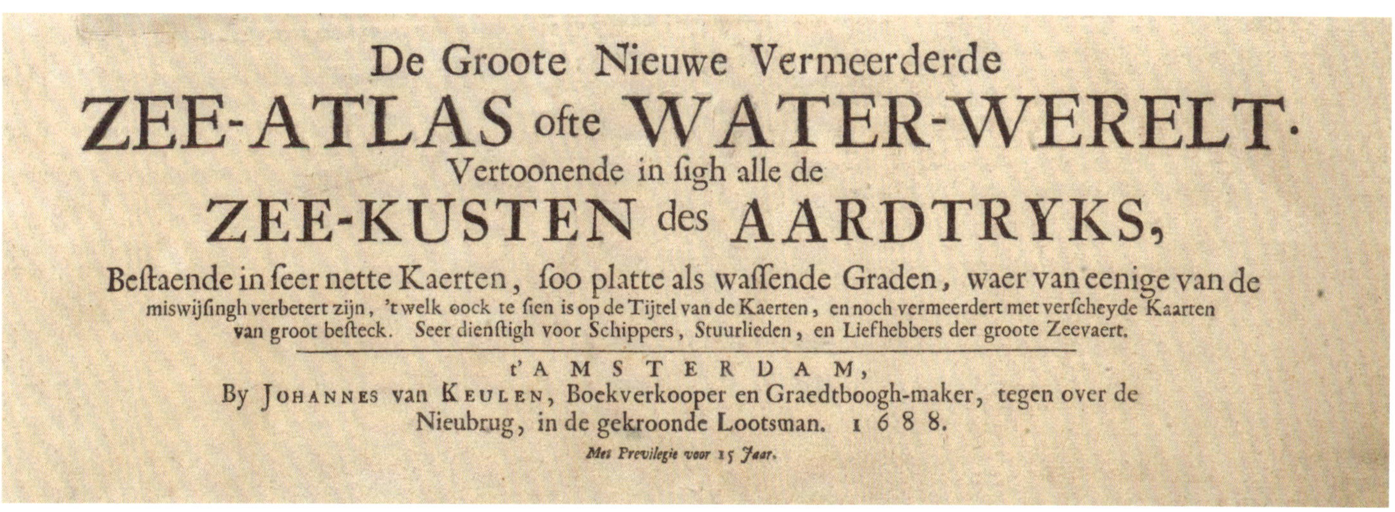

De Groote Nieuwe Vermeerderde
ZEE-ATLAS ofte WATER-WERELT.
Vertoonende in ſigh alle de
ZEE-KUSTEN des AARDTRYKS,
Beſtaende in ſeer nette Kaerten, ſoo platte als waſſende Graden, waer van eenige van de miswijſingh verbetert zijn, 't welk oock te ſien is op de Tijtel van de Kaerten, en noch vermeerdert met verſcheyde Kaarten van groot beſteck. Seer dienſtigh voor Schippers, Stuurlieden, en Liefhebbers der groote Zeevaert.
t'AMSTERDAM,
By JOHANNES van KEULEN, Boekverkooper en Graedtboogh-maker, tegen over de Nieubrug, in de gekroonde Lootsman. 1688.
Met Previlegie voor 15 Jaar.

Title page (details)
From: Johannes van Keulen, *De Groote Nieuwe Vermeerderde Zee-atlas ofte Water-werelt*, 1688
ANTWERP, THE PHOEBUS FOUNDATION

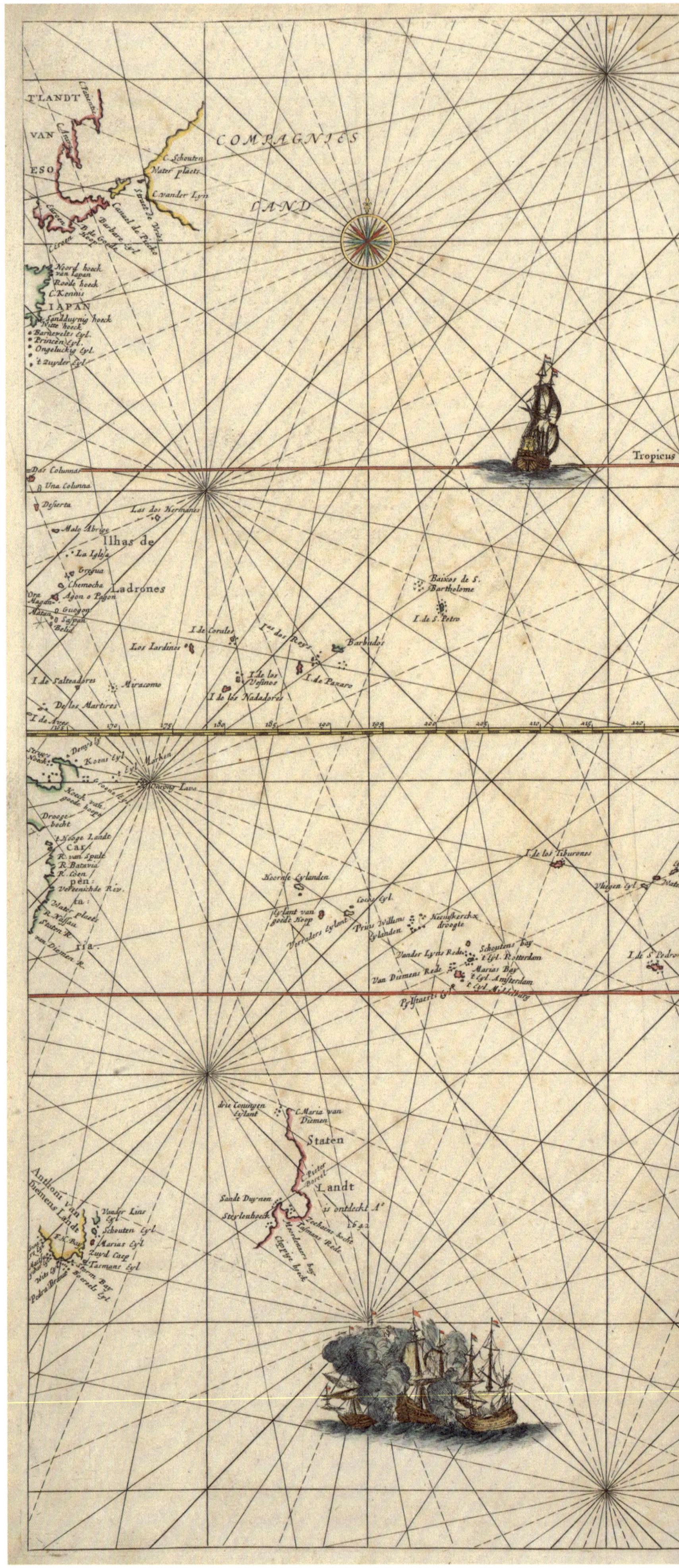

Pascaert vande Zuydzee en een gedeelte van Brasil Van Ilhas de Ladronos tot R. de la Plata
Sea chart of the Pacific Ocean and the coasts of North and South America in the marine atlas of Johannes van Keulen
From: Johannes van Keulen, *De Groote Nieuwe Vermeerderde Zee-atlas ofte Water-werelt*, 1688
ANTWERP, THE PHOEBUS FOUNDATION

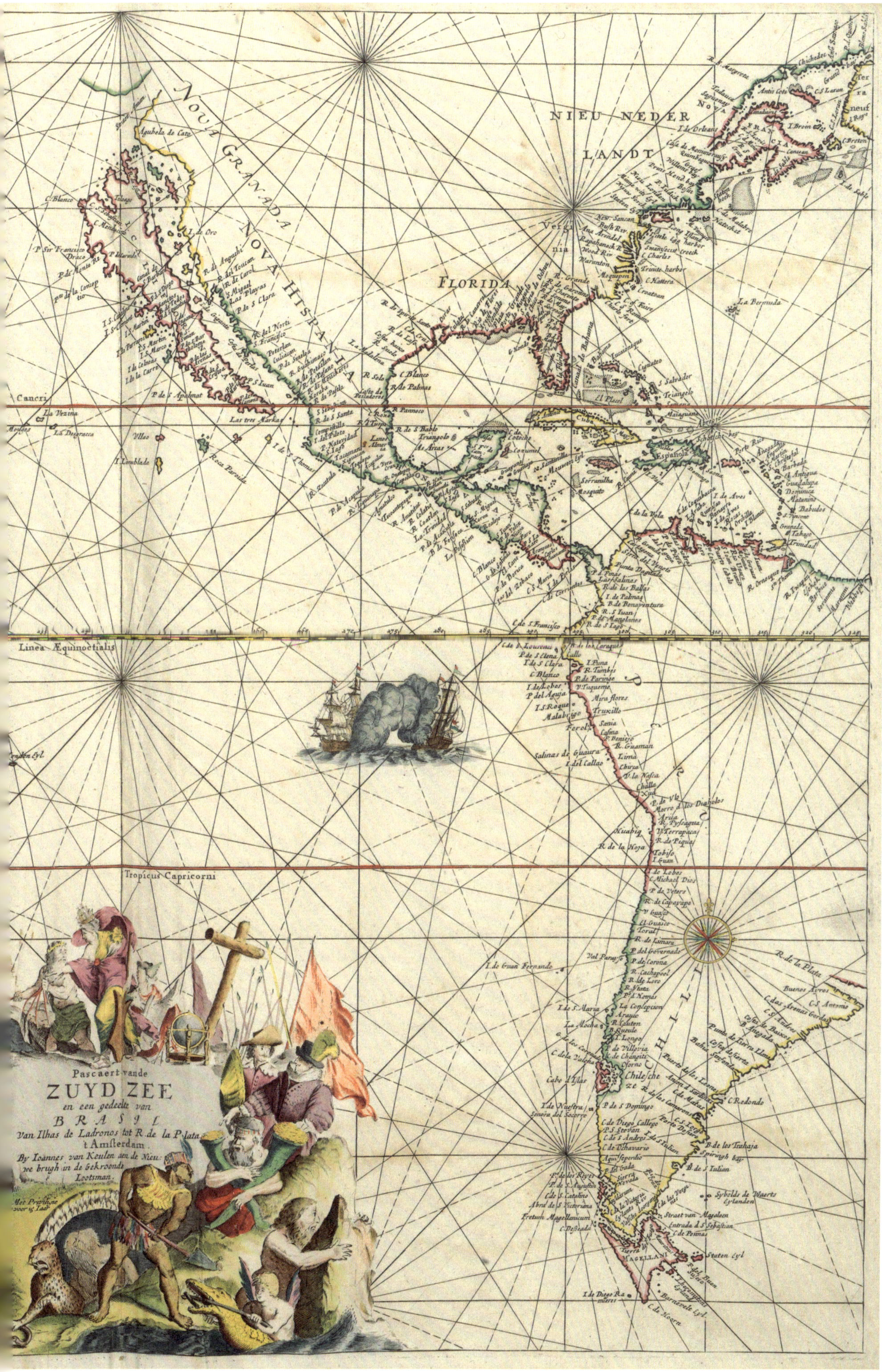
Pascaert vande
ZUYD ZEE
en een gedeelte van
B R A S I L
Van Ilhas de Ladronos tot R. de la Plata
t Amsterdam.
By Ioannes van Keulen aen de Nieuwe brugh in de Gekroonde Lootsman.
NOVA GRANADA
NOVA HISPANIA
NIEU NEDER LANDT
FLORIDA
Cancri
Linea Æquinoctialis
Tropicus Capricorni

'RICH IN SLAVES'

One map in particular that stands out in the collection of deluxe sea charts from Van Keulen's atlas is the one showing the 'Gold Coast' of Guinea, modern-day Ghana on the Bay of Benin. It is included among other maps of the African coast and is the only one in the book that also displays extensive information regarding the interior.

Inland areas were generally left blank in nautical charts, as these were maps intended for sea voyages. The space was then used for details of coastal areas, large title cartouches or decorations. The map of the Ghanaian coastline is an exception in this regard: the section of the interior shown on the map is not left blank but has been roughly divided into separate, demarcated zones that are either named or labelled as 'unknown territory' (*onbekent land*). A brief description has also been included of what can be traded in each region or, as the title explains, of the local customs: 'in which is shown all inland produce, what merchandise is traded there and the manners and customs of the inhabitants. All accurately studied among the blacks of the country.'

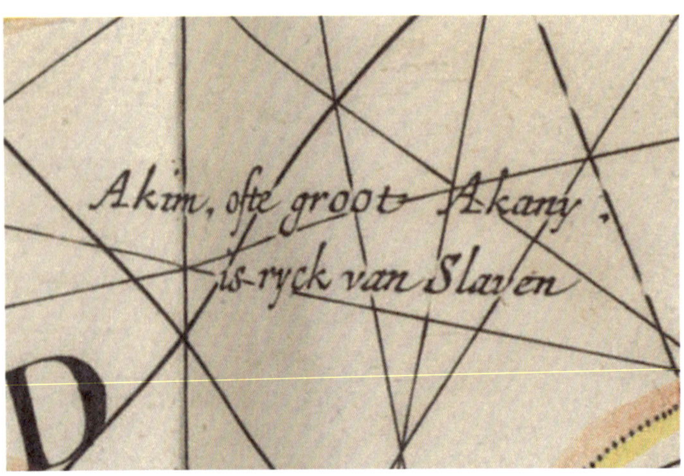

'Akim, ofte groot Akany, is ryck van Slaven'
Detail from: *Pas-caert vande goudcust in Guinea*, 1688 (p. 189)

The texts do not focus purely on the gold, clothing, carpets or horses that the Dutch evidently purchased there: the map also explicitly notes locations that were 'rich in slaves' (*ryck van slaven*), such as Akim or '*Cleyn Tropassa*'. It also presents degrading stereotypes of the population: which peoples were 'stupid' and which were 'thieving'. Another text refers to 'Agwana, or the land of Concomo, much inclined to war', offering a direct clue that the presence of the Dutch was not always welcome and that people actively sought to resist it. The map sheds a different light on, for instance, the one in this book showing Africa (⤚⟶ p. 267) and other overseas territories, which does not allude quite so explicitly to the triangular trade in enslaved people between Europe, Africa and the Americas. All the same, many of those maps too were created within a context of colonialism and slavery.

The riches evoked by the atlas are a painful reminder, therefore, of the motives behind the sea voyages and the 'achievements' they entailed: Dutch mariners did not simply navigate their way along the coast but also ventured inland to pursue their objectives through violence and slavery, as we will discuss further in the next chapter. It was through activities such as these that the merchants of Amsterdam were able to amass the wealth that paid for expensive sea atlases.

The maps in Johannes van Keulen's sea atlases appear entirely functional: full of compass points and lines of longitude and latitude, with the smallest ports minutely mapped along the coast. Yet the sumptuous colouring, gold highlights and decorative cartouches were of little use to a helmsman at sea. A practical pilot guide with detailed sailing instructions and woodcuts with small profiles of the coast would have been far more to his liking. The reference materials in the wheelhouse of a ship were practical tools, just like the navigation instruments. Deluxe sea atlases, by contrast, were kept safely in the libraries of merchants and naval officers, their quality often remaining pristine.

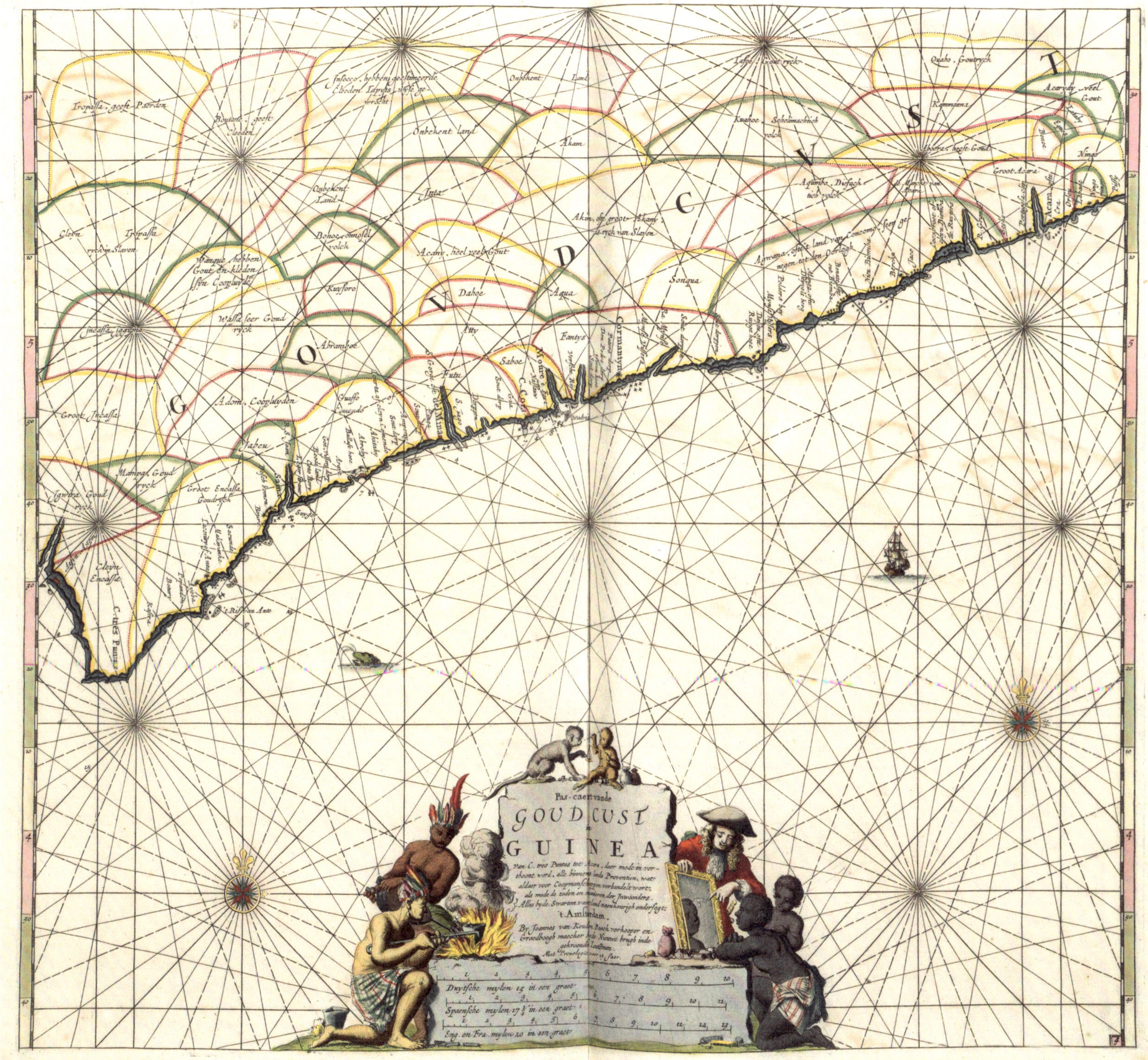

Pas-caert vande goudcust in Guinea
Sea chart of the coast of Ghana in the marine atlas of Johannes van Keulen
From: Johannes van Keulen, *De Groote Nieuwe Vermeerderde Zee-atlas ofte Water-werelt*, 1688
ANTWERP, THE PHOEBUS FOUNDATION

In the margins

Few explicit depictions of the practices of slavery have survived (⟼ p. 214). We can, however, look for references in the cartouches and other margin decorations outside Africa (see, for instance, the map of Africa on p. 267). The figures they depict are not done from life but are stereotypes. Africans were commonly presented as scantily clad and living among seemingly wild animals and native vegetation, a practice termed 'decorative exoticism'. People of colour were also framed as subservient, passive, or submissive to a white figure, handing over precious export products to Western merchants.

Take the men intended to represent the original population in the allegorical cartouche on the map of Ghana in Johannes van Keulen's *Zee-atlas*. The 'Horn of Plenty' is another common motif, overflowing with merchandise or gold and alluding to the 'fruits' of colonization. An example is found in the cartouche on the *Pascaert van de Zuydzee* ('Sea Chart of the South Sea'). The margins of the maps in the atlas overflow with stereotypes. These were supposed to represent the original inhabitants of the mapped areas but only did so with considerable imagination and projection – since very little reliable information had been gathered about these peoples – and certainly no illustrations. The lack of more firmly substantiated depictions and the recycling of these images in the arts and media perpetuated such stereotypes and stimulated racist thinking. So while deluxe atlases were viewed chiefly by members of the social elite, they also served for many years to define Western Europeans' perception of people of colour.

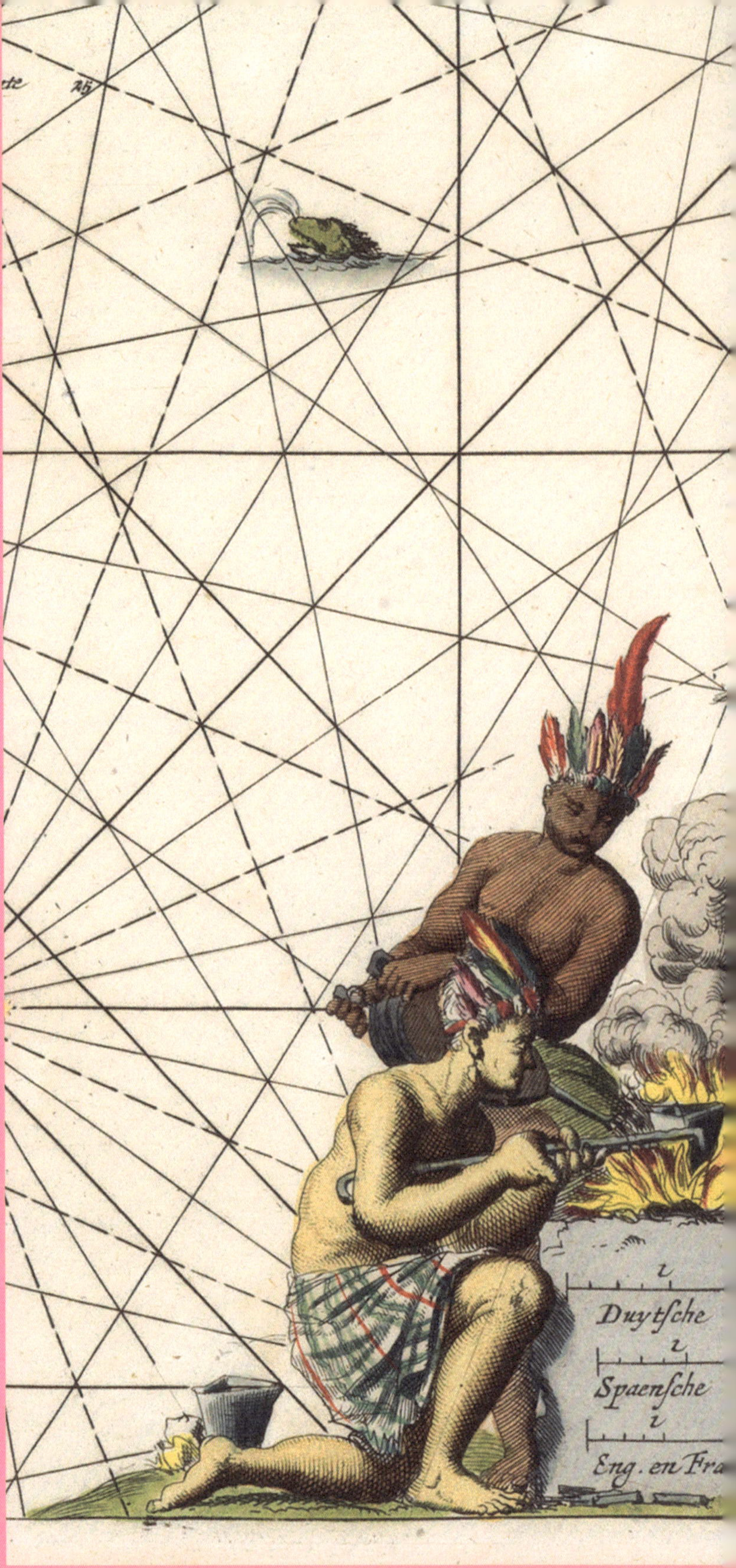

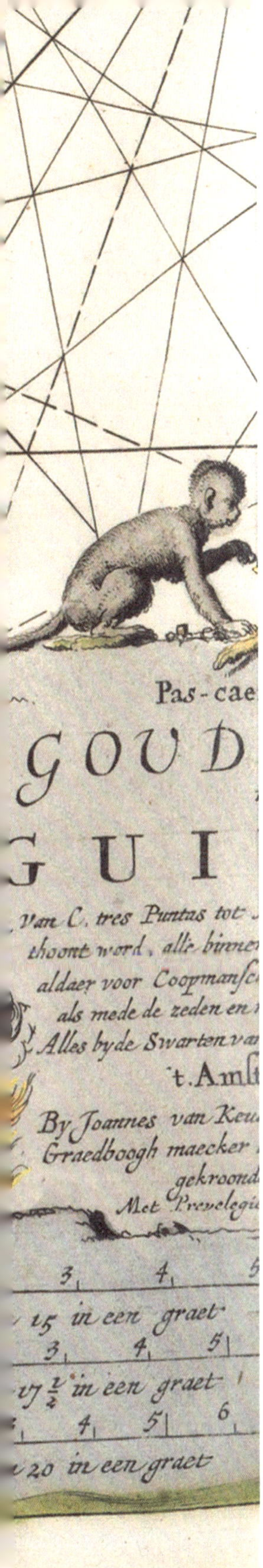

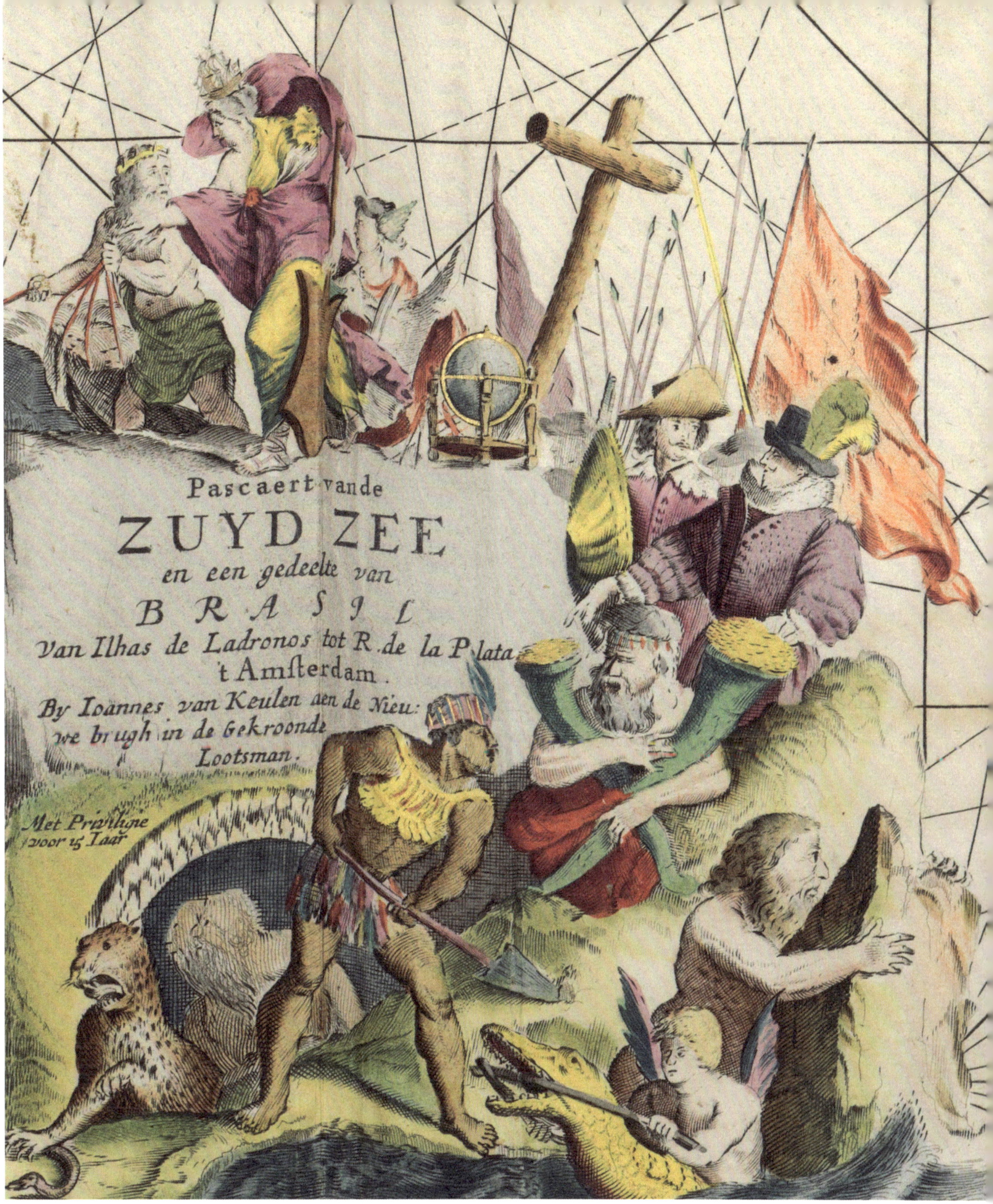

Allegorical cartouches depicting the riches that Western merchants hoped to acquire through the triangular trade
Details from: *Pascaert vande Zuydzee en een gedeelte van Brasil Van Ilhas de Ladronos tot R. de la Plata*, 1688 (pp. 186-187) and *Pas-caert vande goudcust in Guinea*, 1688 (p. 189)

LITERATURE

P. De Maeyer, 'Cartografie, de zee en de rol van de Vlamingen', *VLIZ, De Grote Rede*, 31, 2011: 11-16.

M. Donkersloot-de Vrij, *Drie generaties Blaeu: Amsterdamse cartografie en boekdrukkunst in de zeventiende eeuw*, Amsterdam, 1992.

E. Keulen & W.F.J. Mörzer Bruyns, *'In de Gekroonde Lootsman': het kaarten-, boekuitgevers en instrumentenmakershuis Van Keulen te Amsterdam 1680-1885*, Utrecht, 1989.

C. Koeman, *The Sea on Paper: The Story of the Van Keulens and Their "Sea-Torch"*, Amsterdam, 1972.

E. Kolfin, 'Zwart in de kunst van Rembrandts tijd', in: S. Archangel, E. Kolfin, M. Ponte et al., *Zwart in Rembrandts tijd*, Zwolle, 2020: 12-37.

W.F.J. Mörzer Bruyns & H. Hooijmaijers, *Tussen hemel en horizon: een korte geschiedenis van navigatie op Nederlandse schepen*, Haarlem, 2012.

G. Schilder, *Early Dutch Maritime Cartography. The North Holland School of Cartography (c. 1580-c. 1620)*, Leiden / Boston, 2017.

G. Schilder & M. van Egmond, 'Maritime Cartography in the Low Countries during the Renaissance', in: D. Woodward (ed.), *The History of Cartography, vol. 3. Cartography in the European Renaissance*, Chicago / London, 2007, vol. 2: 1384-1432.

G. Schilder & H. Kok, *Sailing Across the World's Oceans. History & Catalogue of Dutch Charts Printed on Vellum 1580-1725*, Leiden / Boston, 2019.

E. Sutton, *Capitalism and Cartography in the Dutch Golden Age*, Chicago, 2015.

K. Zandvliet, *De wereld van de familie Blaeu*, Zutphen, 2022.

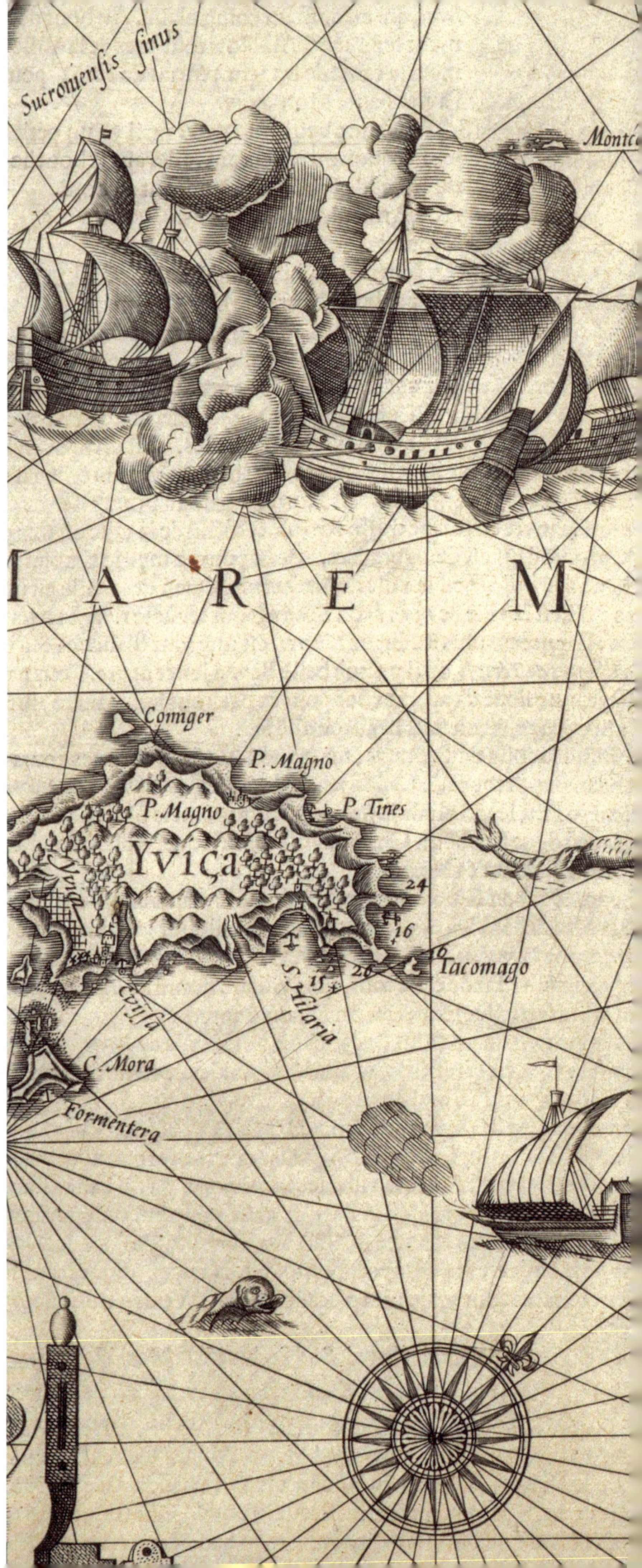

Detail from: Pieter Van den Keere, *Hydrographica descriptio*, 1609 (pp. 178-179)

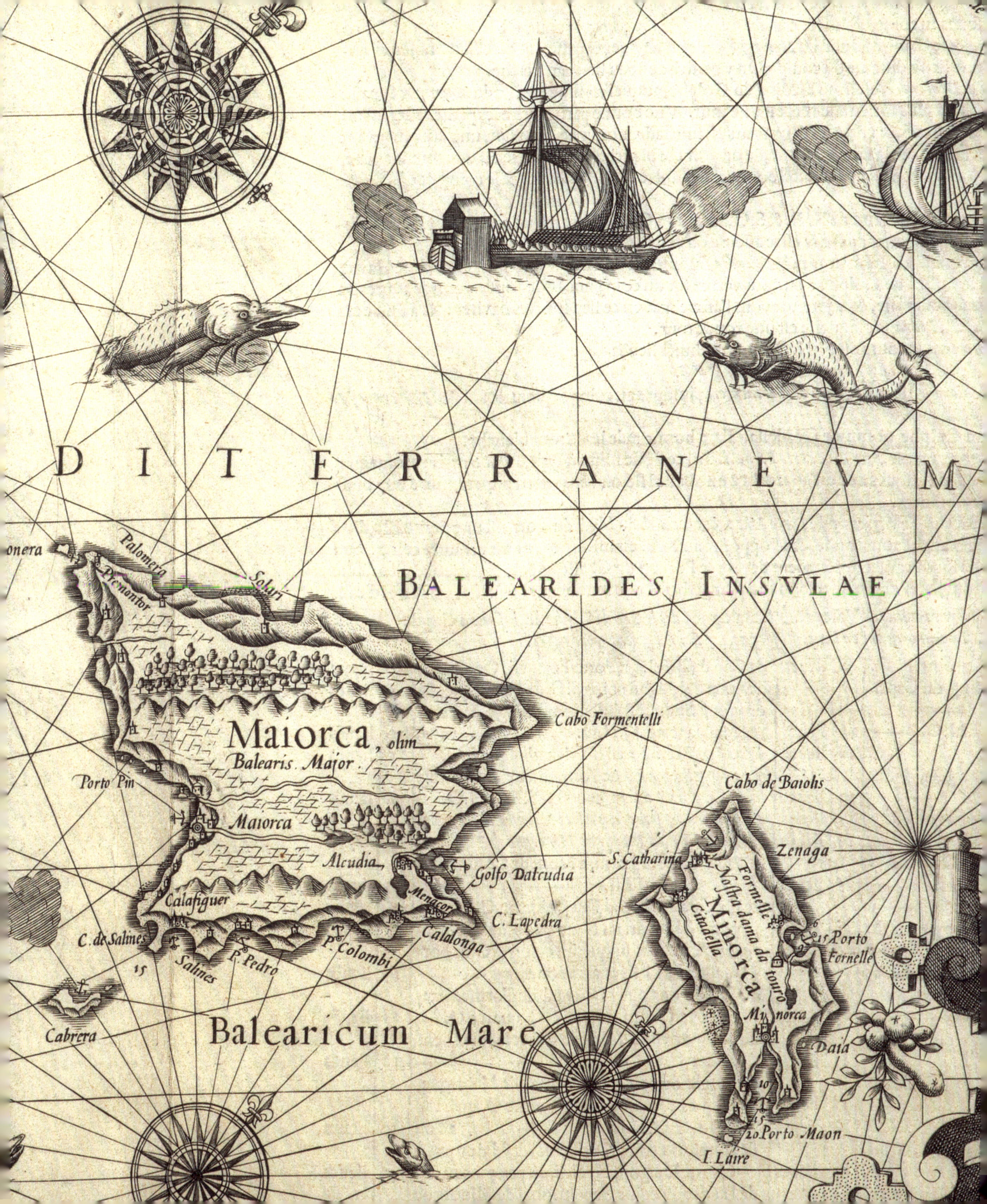
D I T E R R A N E V M
onera
Palomera
Premontor
Solari
BALEARIDES INSVLAE
Cabo Formentelli
Maiorca, olim
Balearis Major
Porto Pin
Maiorca
Alcudia
Golfo Dalcudia
Calafiguer
C. de Salines
Salines
P. Pedro
P. Colombi
Calalonga
C. Lapedra
Cabrera
Balearicum Mare
Cabo de Baiolis
S. Catharina
Zenaga
Fornelle
Nostra dama da touro
Minorca
Citadella
Porto
Fornelle
Minorca
Daia
Porto Maon
I. Laire

OCEAN VOYAGES, 'discoveries' on the far side of the world and the new improved maps being produced were not nearly as romantic as they were often presented over time, as the previous chapter already showed. First and foremost, they were commercial affairs, fuelled by a drive for money and power. Distant voyages were rarely initiated by the courageous mariners or curious scientists who actually made them. It was often administrators, merchants and companies that were behind such ventures, each with their own interest in a successful campaign.

The driving forces in Amsterdam behind colonial expansion and overseas cartography were the Dutch East India (founded in 1602) and West India Companies (founded in 1621). What is less well known, however, is that the roots of many important cartographers and seafarers lay in the Southern Netherlands. In the late sixteenth century, it was Antwerp that led the way in the fields of nautical science and map and book printing. It was fertile ground for several key figures from the South who would play an immensely important role in cartography and voyages to Asia and America, among them Petrus Plancius, Cornelis Claesz., Johannes De Laet (1581-1649) and Jacob le Maire (*c.*1585-1616), all of whom later emigrated to the North. How did these figures contribute to new trade links, overseas expansion and geographical knowledge of the Americas and Asia?

CHAPTER VIII

TERRA INCOGNITA ?

NEW OCEANS, NEW WORLDS

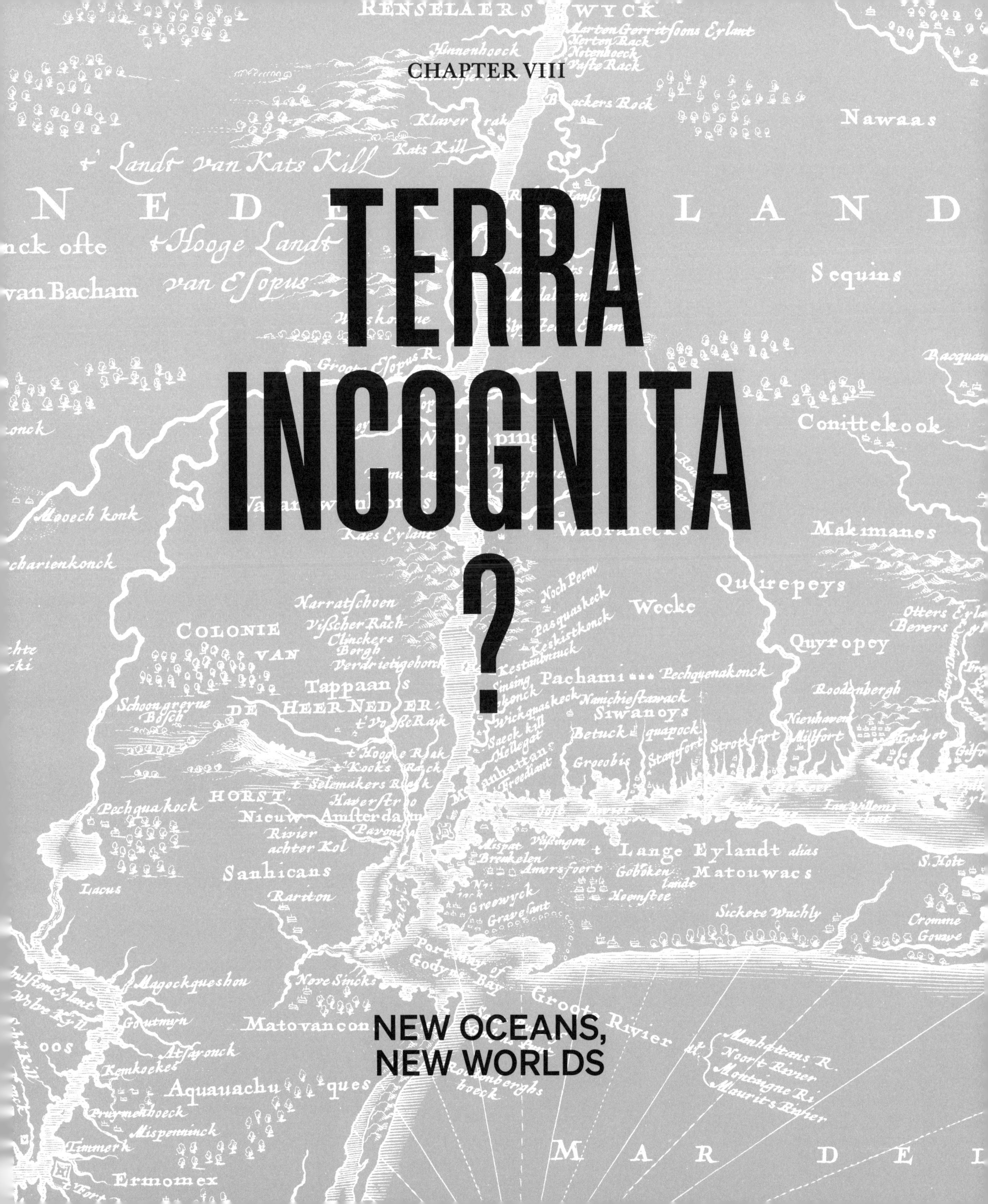

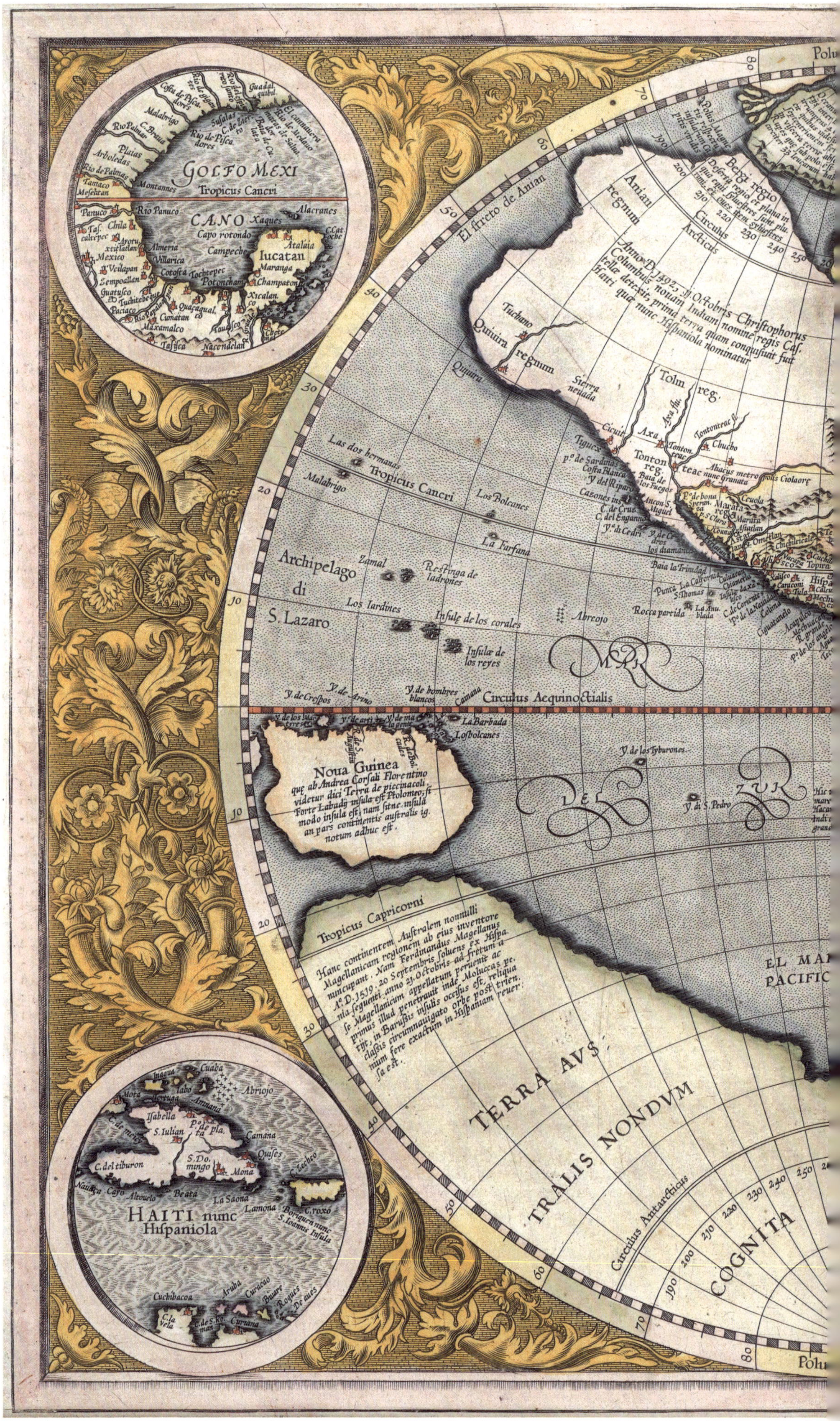

Michael Mercator
America sive India Nova, 1595 or later
Map of America
Hand-coloured copper engraving, 373 × 468 mm
ANTWERP, THE PHOEBUS FOUNDATION

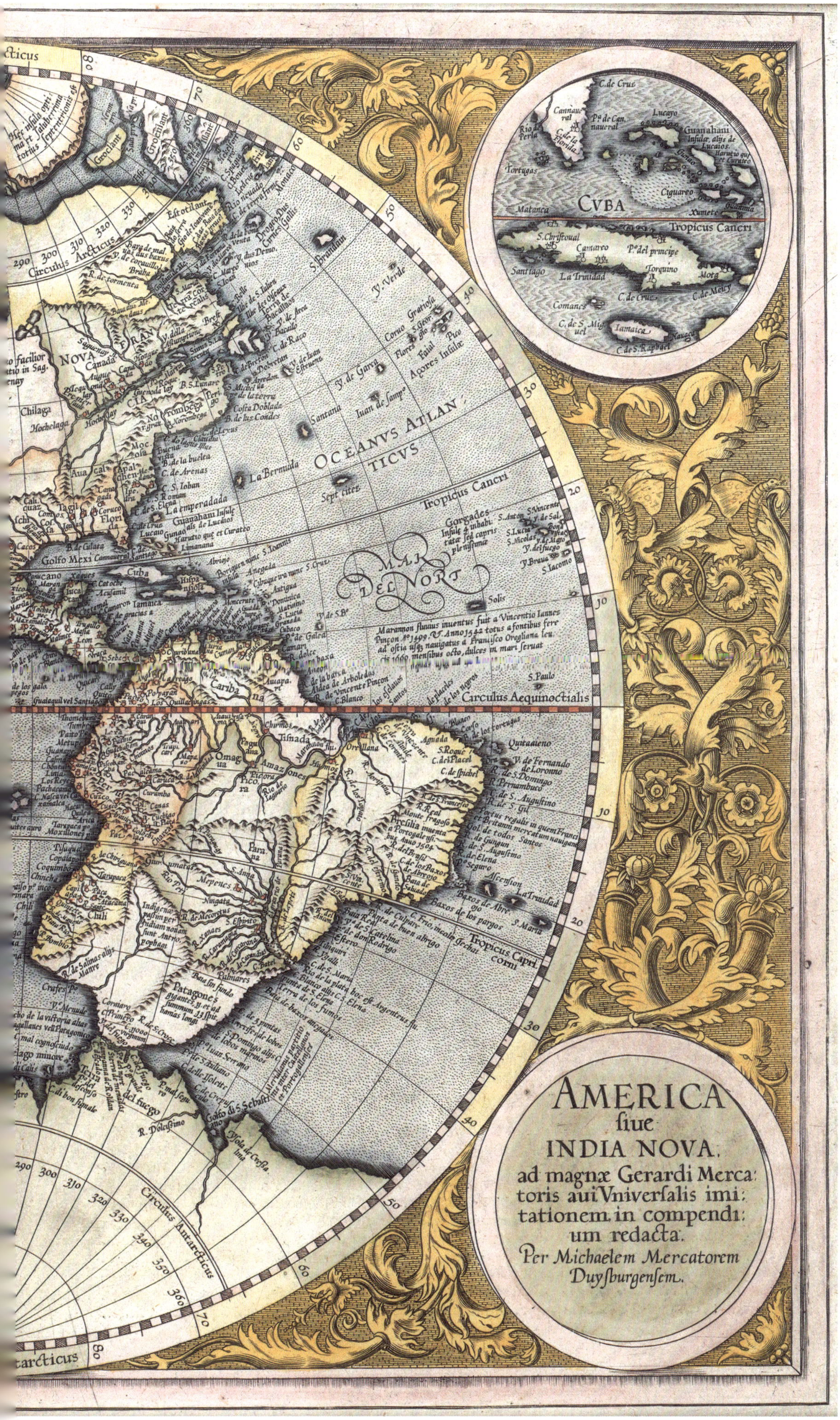
AMERICA
siue
INDIA NOVA,
ad magnæ Gerardi Merca:
toris aui Vniversalis imi:
tationem in compendi:
um redacta.
Per Michaelem Mercatorem
Duysburgensem.
CVBA
Tropicus Cancri
OCEANVS ATLANTICVS
MAR DEL NORT
Circulus Aequinoctialis
Tropicus Capricorni
Circulus Antarcticus

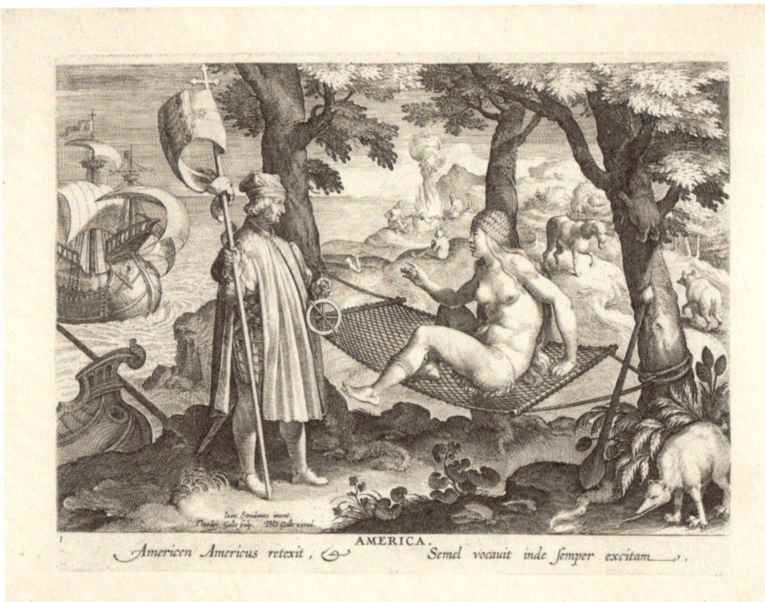

Hans Collaert and Theodoor Galle after Johannes Stradanus
Nova Reperta: America, c.1590
Amerigo Vespucci discovers America
Copper engraving, 263 × 340 mm
ANTWERP, THE PHOEBUS FOUNDATION

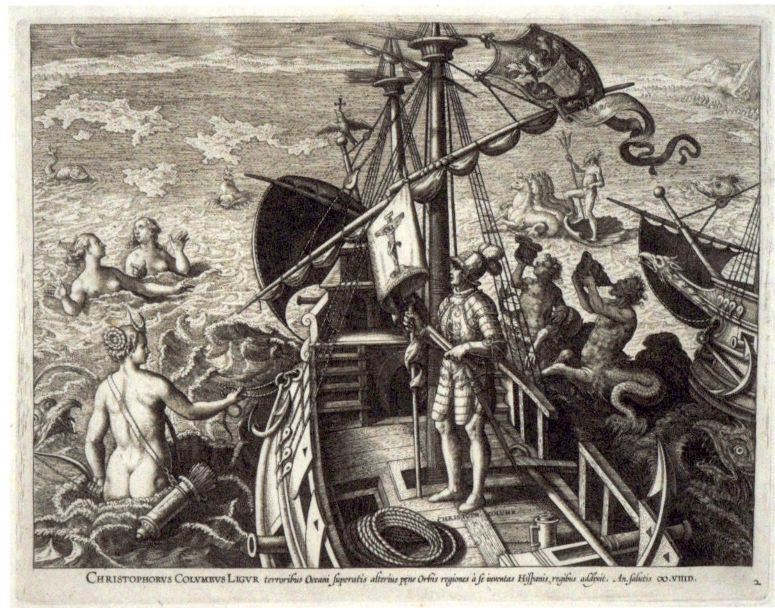

Adriaen Collaert after Johannes Stradanus
Americae Retectio: Christophorus Columbus Ligur, c.1589
The discovery of America by Christopher Columbus
Copper engraving, 237 × 304 mm
ANTWERP, THE PHOEBUS FOUNDATION

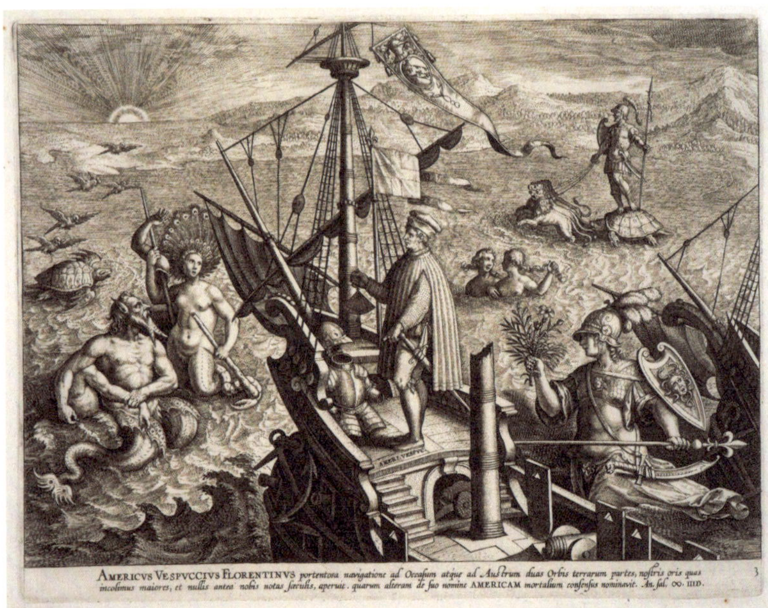

Adriaen Collaert after Johannes Stradanus
Americae Retectio: Americus Vespuccis Florentinus, c.1589
The discovery of America by Amerigo Vespucci
Copper engraving, 237 × 304 mm
ANTWERP, THE PHOEBUS FOUNDATION

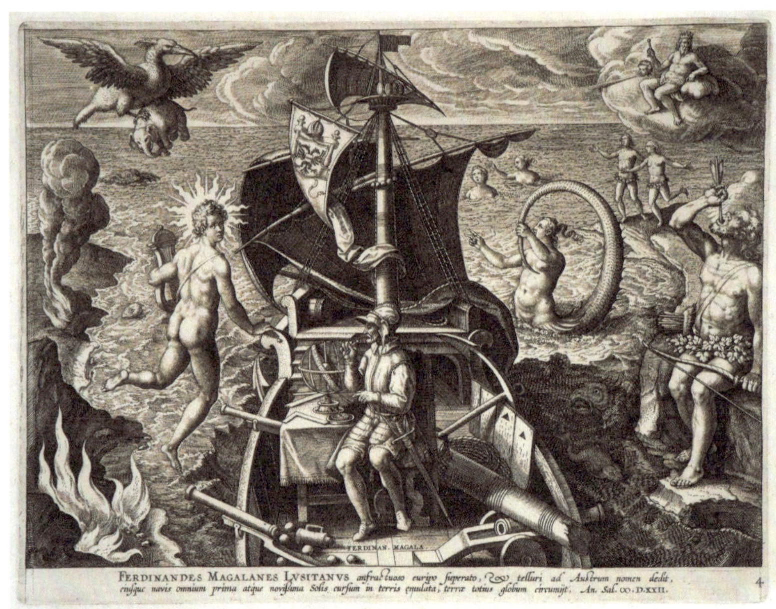

Adriaen Collaert after Johannes Stradanus
Americae Retectio: Ferdinandes Magalenes Lusitanus, c.1589
The discovery of America by Ferdinand Magellan
Copper engraving, 237 × 304 mm
ANTWERP, THE PHOEBUS FOUNDATION

SETTING FOOT IN THE AMERICAS

The year 1492 has gone down in history for the 'discovery' by Christopher Columbus (1451-1506) of the Americas during his voyage in search of a western route to the Indies. His venture was based on the geography of Claudius Ptolemy (*c*.100-*c*.170) and the idea of a spherical earth (⟼ Chapter IX). Heading west from Europe, it was believed, there was just a single ocean to cross in order to reach Asia from the other side. While a diverse range of seafarers contributed to the 'discovery' of the Americas over time, Columbus was swiftly crowned the pioneer, with 1492 adopted as the legendary starting point. Another key event was the 'discovery' in 1519 of the Strait of Magellan at the southernmost point of South America connecting the Atlantic to the Pacific, via which Asia could be reached from the west.

Both events are marked on the map that Michael Mercator (*c*.1567-1600), grandson of the famous Gerard, made of America in 1595, shortly after his grandfather's death. It was an attempt to complete the world atlas envisioned by the first Mercator. *America sive India Nova* is the only map by Michael that we know of today. The map places the voyage of Ferdinand Magellan (*c*.1480-1521) near Tierra del Fuego, below Patagonia, which the Portuguese mariner had come across. It was still believed at the time to be the upper part of the southern continent, *Terra Australis* (⟼ p. 256). Columbus's voyage is noted to the northwest of North America – not a location he had anything to do with, but probably because the map had space available as the area had yet to be explored.

The heroic status enjoyed by seafarers is also apparent from the *Nova Reperta* print series depicting Renaissance inventions, which the Antwerp artist Adriaen Collaert (*c*.1560-1618) engraved after a design by Johannes Stradanus and Philip Galle published *c*.1590. The series includes a print showing the Italian Amerigo Vespucci (1454-1512), who disputed the claim that Columbus had discovered Asia and confirmed that he had actually stumbled on a continent previously unknown to Europeans. The print shows him going ashore with banner and astrolabe (⟼ p. 292) in hand, encountering the scantily clad female personification of America while people roast a human body part over a fire in the background (⟼ p. 198 upper left).

There were frequent allusions to cannibalism in the visual image that was formed of the Americas. The oldest maps of the region in The Phoebus Foundation's collection include a woodcut from Sebastian Münster's *Cosmographia* (Basel, 1561) which features, more or less where Brasil lies, a questionable representation of cannibalism. It is one of the earliest known depictions of the continent showing it as a single entity. His other map of the world, for instance (⟼ Chapter IX), still shows a large ocean between North and South America.

The *Americae Retectio* series (*c*.1589-1590) follows on from the *Nova Reperta* and shows the successive explorers who 'discovered' the Americas – Columbus, Vespucci and Magellan – in action. Columbus is depicted on the deck of his Spanish ship, surrounded by the terrors of the ocean (*terroribus oceani*, as the caption puts it). Out of the waves rise the mythical figures of Diana, goddess of wild nature, birth and the moon, and Neptune, god of the sea. In the distance, we see the outlines of islands and a coastline intended to represent the 'New World'. Amerigo Vespucci holds a quadrant, a navigational instrument, as exotic birds fly overhead and turtles swim alongside him, alluding to the fauna of the Americas.

The print with Ferdinand Magellan contains a similar scene, this time placing him on a ship sailing towards the southernmost tip of South America. The Portuguese mariner sits at a table on deck with a compass, divider and astrolabe to show he is busy navigating. Apollo, god of the sun, is shown on the left guiding the ship, just as the actual sun was used to determine the ship's position and course. Mythical figures fly through the air and swim in the water. A male figure on the right, intended to represent an indigenous inhabitant of the Americas, is depicted in an exoticising manner. The flames blazing up from a pointed piece of land symbolise Tierra del Fuego (literally 'Land of Fire'), the southernmost tip of America, which Magellan had reached.

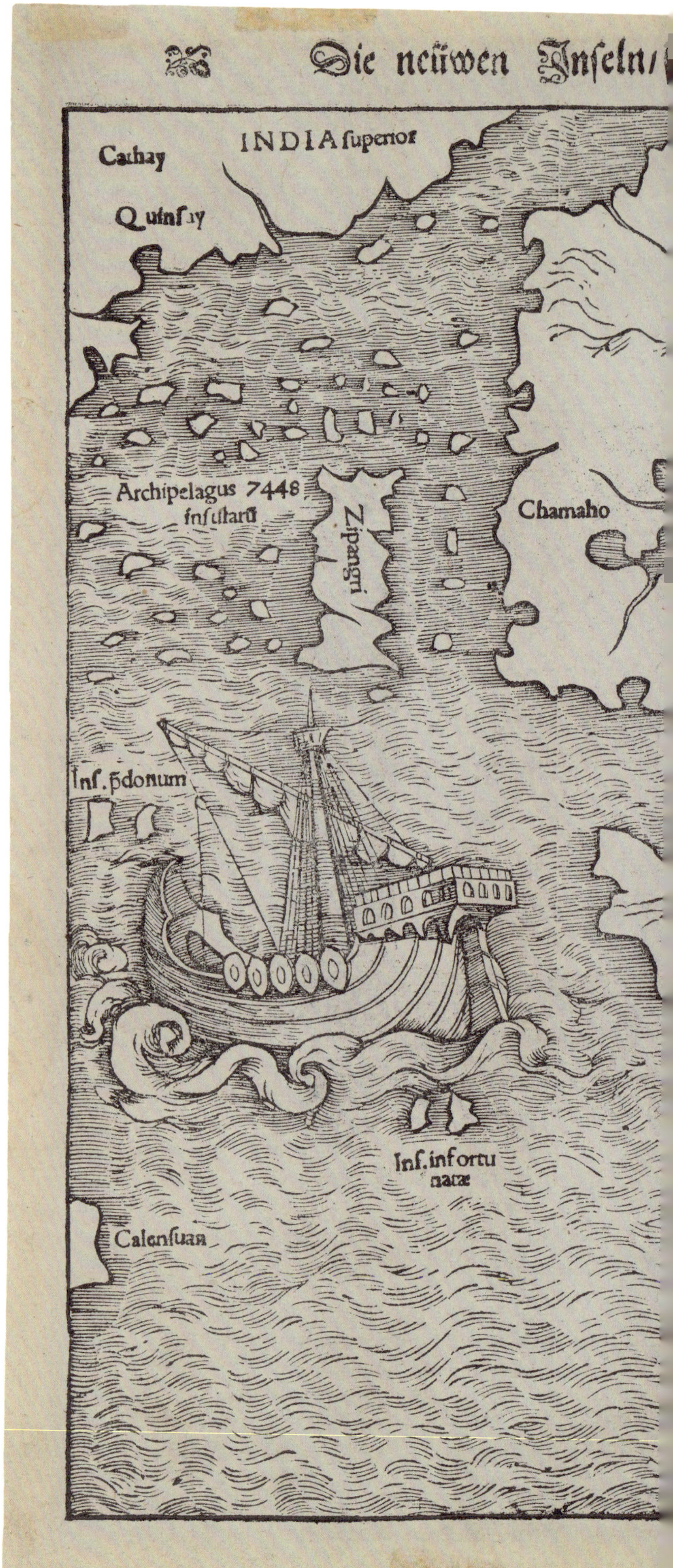

Sebastian Münster
Die neüwen Inseln, so hinder Hispanien gegen Orient bey dem land Indie ligen, 1561
Map of North and South America
Woodcut, 255 × 337 mm
ANTWERP, THE PHOEBUS FOUNDATION

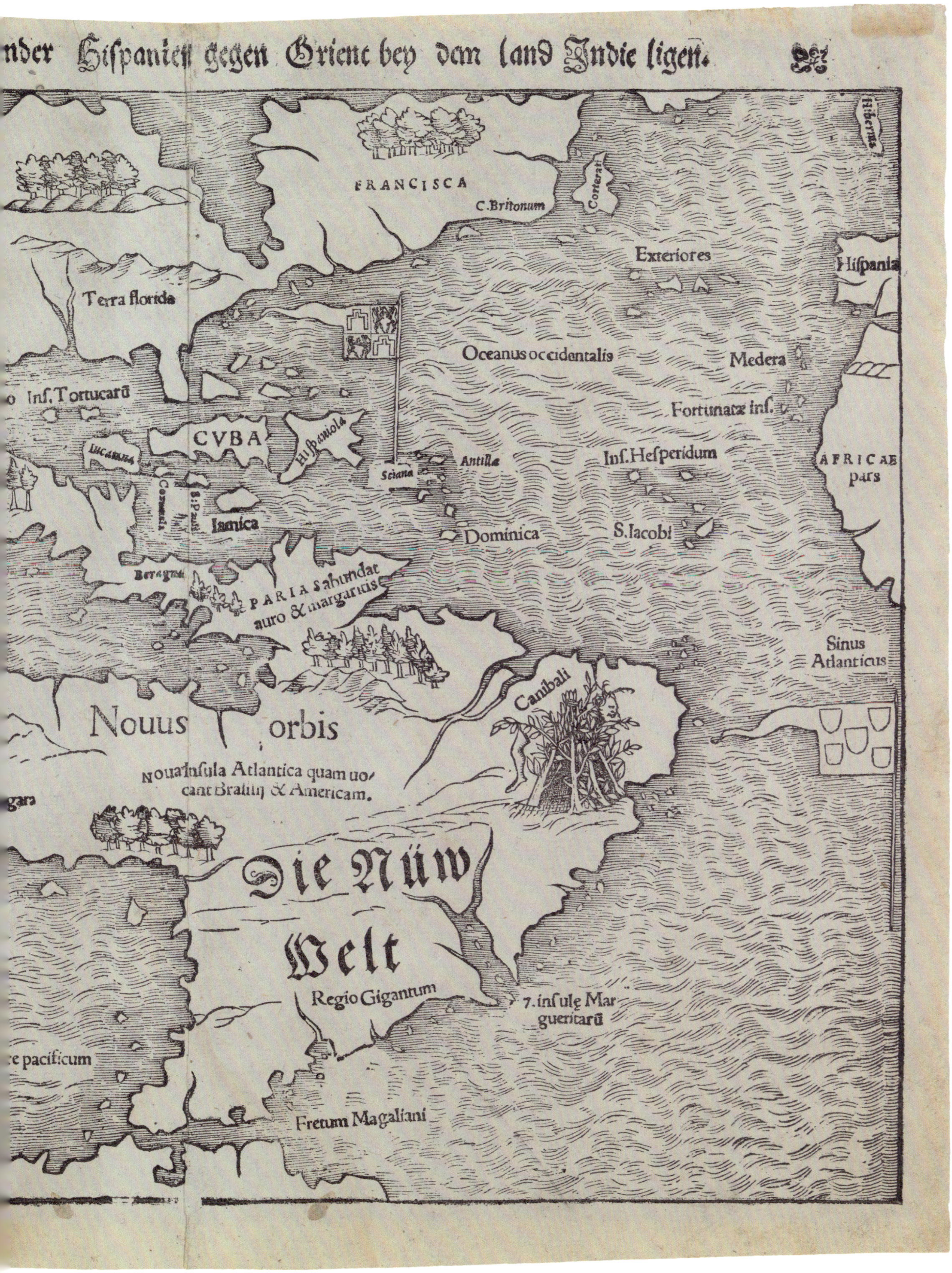
nder Hispanien gegen Orient bey dem land Indie ligen.
FRANCISCA
C.Britonum
Corterati
Hiberna
Exteriores
Hispania
Terra florida
Oceanus occidentalis
Medera
Inſ. Tortucarū
Fortunatæ inſ.
CVBA
Hiſpaniola
Iucarna
Sciana
Antillæ
Inſ. Heſperidum
AFRICAE pars
Iamica
Dominica
S. Iacobi
Beragna
PARIA abundat auro & margaritis
Sinus Atlanticus
Canibali
Nouus orbis
Noua inſula Atlantica quam uocant Braſilij & Americam.
Die Nüw Welt
Regio Gigantum
7. inſulę Marguertarū
e pacificum
Fretum Magaliani

Gerard Mercator and the Mercator projection

Gerard de Cremer (1512-1594) was born in Rupelmonde, in the Waasland area of the county of Flanders. He came from a good family and was able to attend the local grammar school, where he learned Latin and Greek, before going on to study at the university in Leuven. It was there that he discovered his love of science and mathematics and where he met other scholars, including the instrument maker Gaspard van der Heyden (c.1496-c.1549) and Gemma Frisius (⟶ p. 28), with whom he made globes and other instruments. Although he had been producing maps since 1537, he was first and foremost a scientist, mathematician and cosmographer, seeking to reconcile his religious, Catholic worldview with Ptolemy's ancient insights, science and the 'discoveries' flowing from the great voyages of exploration.

Mercator – the Latinised version of his surname – was of great and innovative importance to cartography in numerous respects but is best known for the projection he devised that allowed the three-dimensional, spherical shape of the earth to be transposed into two dimensions, making it extremely useful for navigating at sea. The system is also known as the 'cylindrical projection' as the meridians and parallels of the earth's surface are shown as a grid, with latitudinal intervals increasing towards the pole. In terms of land area, however, the projection creates significant distortions: the further away particular areas are from the equator, the larger they appear on the map than they really are. The North and South Poles, for example, are considerably stretched, while Russia and Greenland appear very large. The Mercator projection is familiar to this day from the world maps printed in school atlases, depicted in the media and hanging on walls.

It was Mercator's ambition to complete a larger project – a cosmographic work addressing the origins and history of the earth and the universe. The first volume of his 'modern geography' was published in Duisburg in 1585 and consisted of maps of Western Europe. It was followed in 1589 by a second, devoted to the countries of Southern and Eastern Europe. His death in 1594 meant that the atlas he had envisaged failed to materialise. One of his two sons, Rumold Mercator (c.1545-1599), published a more or less complete edition, supplemented by maps of the remaining parts of Europe and scaled-down versions of Gerard's world map. The book was given Mercator's intended title: *Atlas sive Cosmographicae Meditationes de Fabrica Mundi et Fabrica Figura* ('Atlas, or cosmographic reflections on the creation of the world and the depiction of that creation').

Although Mercator's atlas came later than that of his friend and colleague Abraham Ortelius (⟶ p. 126), it was of greater scientific quality, having been compiled using a different method. Mercator sought out the best maps available, critically edited them and used them to design a better, more homogeneous collection of maps of his own. His heirs later sold the copper plates of the *Atlas* to the publisher Jodocus Hondius I in Amsterdam. The Hondius family – in collaboration subsequently with Johannes Janssonius – expanded the ensemble of Mercator maps in the seventeenth century to create several commercial atlases with numerous translations. It was these publishers who passed the legacy on to the general public.

Gerard Mercator made several other contributions to cartography, including a method of writing in italics to allow very small texts to be inserted on a map. He also coined the term 'atlas' to describe a uniformly compiled selection of maps, explaining that he named his work after a cosmographic scholar and king of Mauritania, on the northwest coast of Africa. 'Atlas' later came to be associated with the mythical titan of that name, who had to carry the celestial globe on his back as punishment for a lost battle against Zeus and other gods. It is possible that the human and the divine Atlas ultimately derived from the same legendary figure.

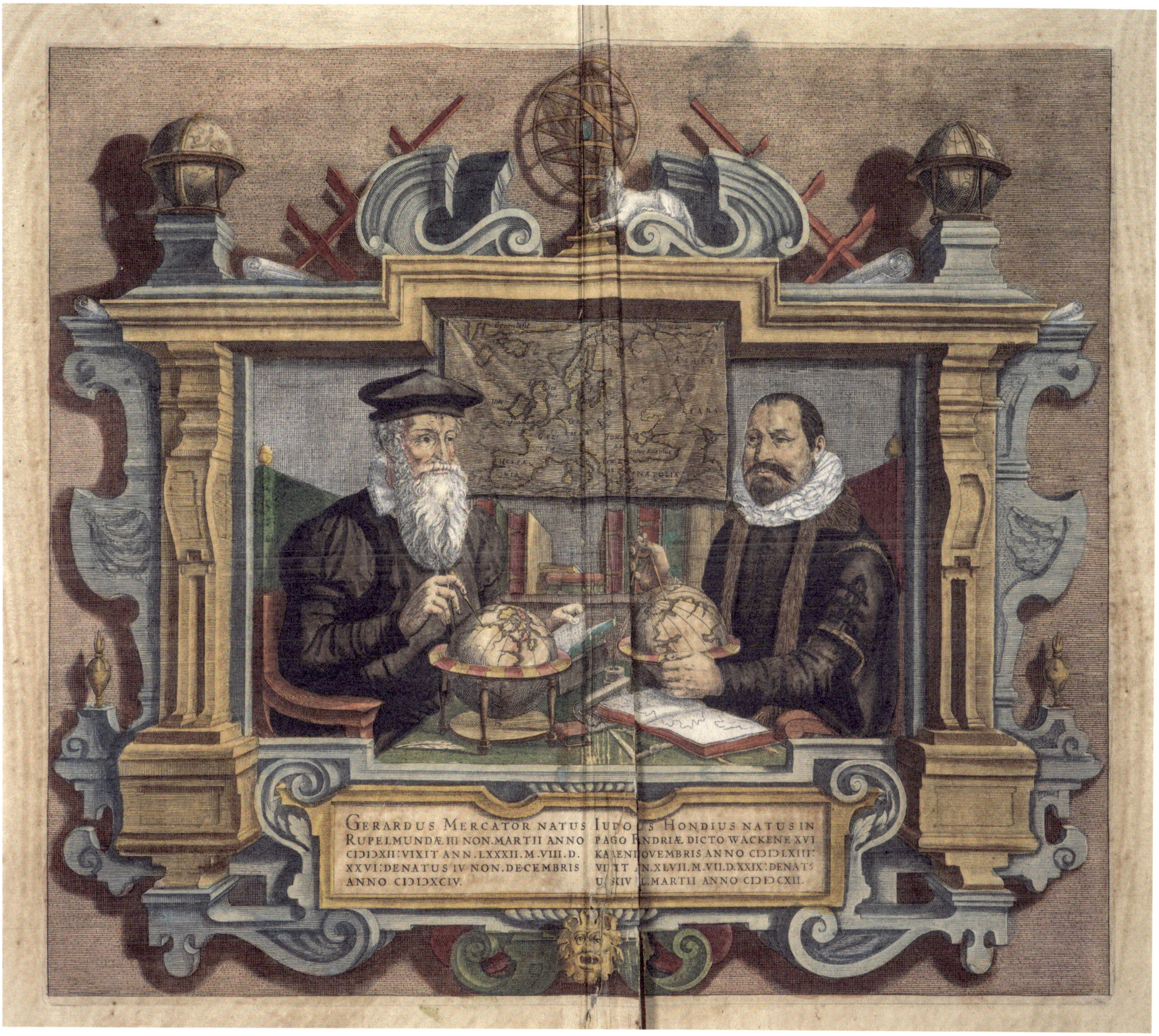

Portrait of Gerard Mercator and Jodocus Hondius I
From: Gerard Mercator and Jodocus Hondius I, *Atlas ou Representation du Monde Universel, et des Parties d'Icelui, Faicte en Tables et Descriptions Tres-Amples, et Exactes: Divise en Deux Tomes*, 1633
ANTWERP, THE PHOEBUS FOUNDATION

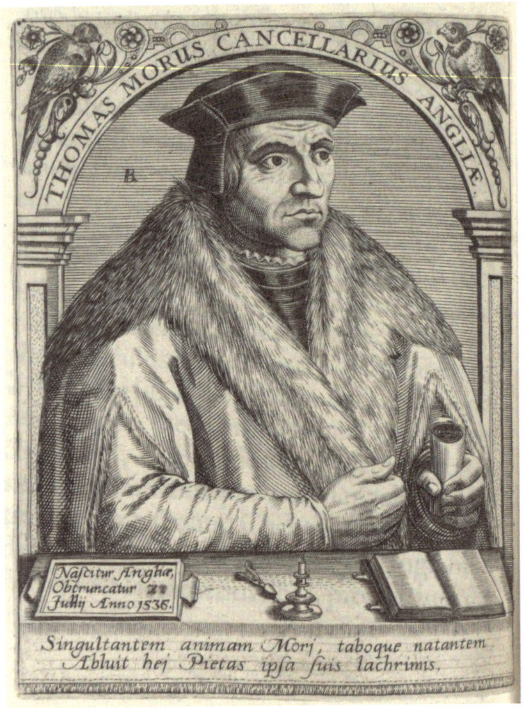

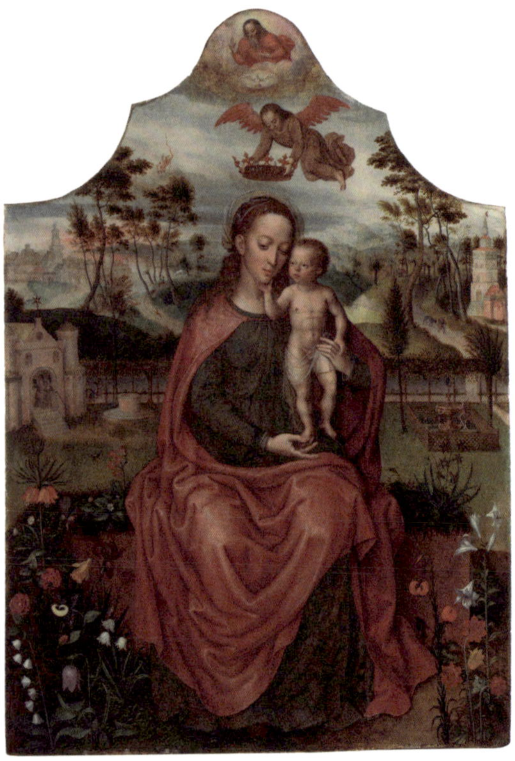

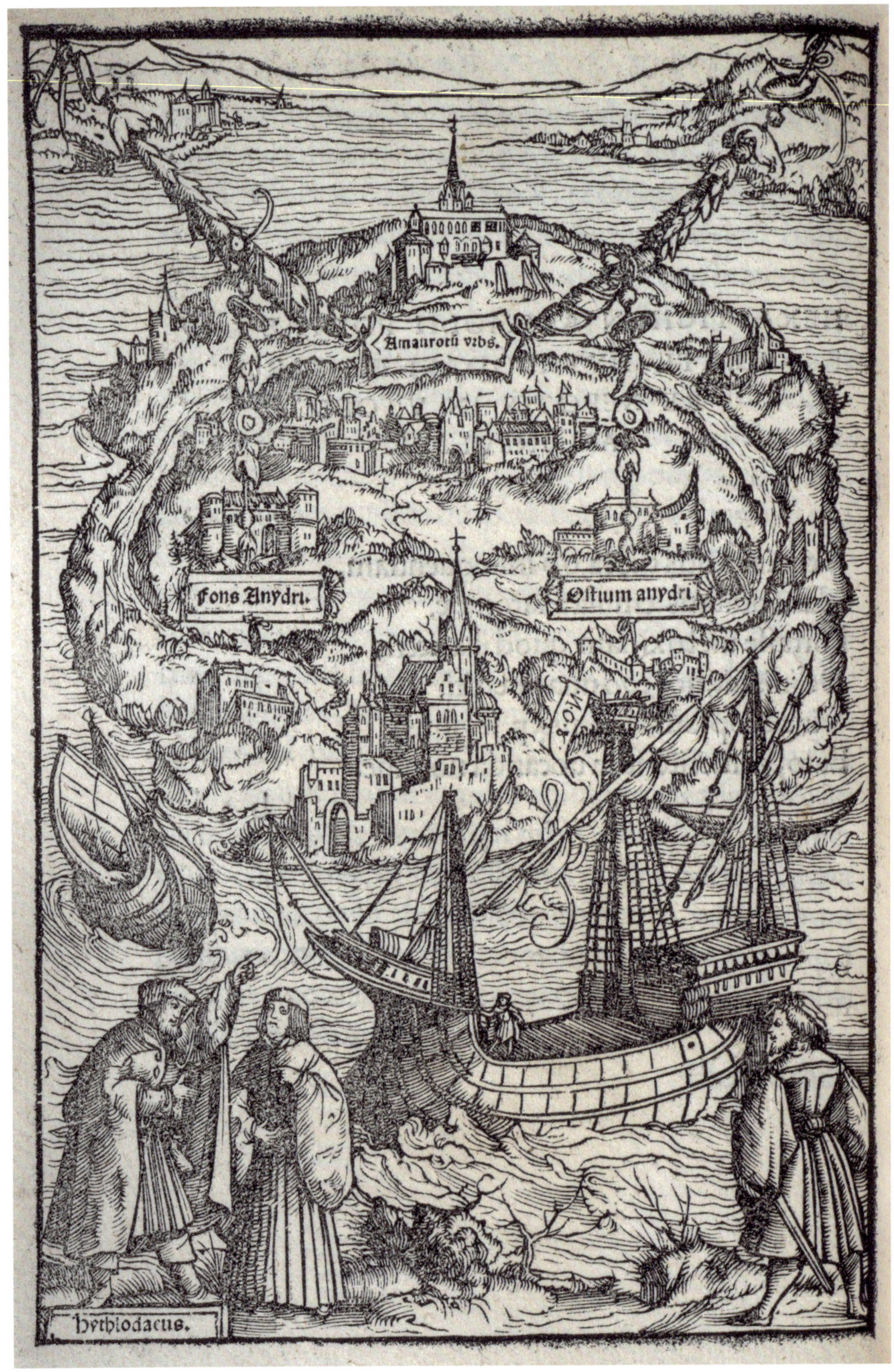

Portrait and description of Thomas More
From: Jean Jacques Boissard, *Icones quinquaginta virorum illustrium*, 1597-1599
ANTWERP, THE PHOEBUS FOUNDATION

Pieter Claeissens II
Virgin and Child in a Garden / Virgin with a Llama, 1614-1623
Oil on panel, 96.6 × 66.4 cm
ANTWERP, THE PHOEBUS FOUNDATION

Ambrose Holbein
Map of Utopia
From: Thomas More, *Utopia*, 1518
ANTWERP, THE PHOEBUS FOUNDATION

UTOPIA: AN IDEAL NEW WORLD

The news that Columbus had discovered islands on the other side of the Atlantic spread like wildfire. A letter he wrote to announce his achievement was swiftly translated into Latin and published in Rome a year later, in 1493. Before long, the text was also available in Antwerp, where the printer Dirk Martens (*c.*1446/1447-1534) became the first to publish it in the Southern Netherlands. Amerigo Vespucci's letters confirming the 'discovery of America', which he called the *Mundus Novus* (New World), were also on sale in Antwerp shortly afterwards. The Americas were not, of course, a virgin, unexplored territory: people had called the continent home for centuries. The term 'New World' is, therefore, obsolete. All the same, the thrilling tales of mysterious, unknown places captured the imagination of Europeans.

Where were these inviting fairy-tale places to be found? As Europeans embarked on ever-longer sea voyages, new, imaginary realms emerged alongside the paradises, gardens of delight and visions of hell of Hieronymus Bosch (*c.*1450-1516) and the Brueghel family, on which contemporary religious morality could be pegged. It was around this same period, in the early sixteenth century, that the English jurist and diplomat Thomas More (1478-1535) spent time in the Southern Netherlands, including Antwerp, where he came up with the idea for his famous book *Utopia* (Leuven, 1516). In it, he described a voyage to an imaginary island of the same name – surrounded by an ocean rather than protective walls or thorny hedges like the Garden of Eden – that was home to an ideal society.

It was no coincidence that More chose Antwerp, of all places, to set the framing narrative for his imaginary travelogue. He describes an encounter with the fictional Portuguese mariner Raphael Hythlodaeus, who has just returned to Europe after a series of long voyages in the company of Amerigo Vespucci. The sailor offers More a vivid account of his visit to a highly unusual island he has come across in the 'New World', somewhere in the Atlantic, south of the equator: Utopia, a name derived from Greek and meaning 'Nowhere' or 'No-Place'.

Hythlodaeus's tales gave More a vehicle to express his philosophical ideas about politics, economics and society. Everyone in Utopia was focused on the public interest rather than their own, and people lived together in harmonious, fraternal equality. Christian principles were successfully observed, and land was owned communally. More even created a fictional language for the inhabitants to speak. There was no hunger on the island: the authorities provided the people with unlimited healthy food and clean drinking water. The travel report recalls the medieval fantasy of the Land of Cockaigne, where people had all the food, drink, clothing, money, beauty and youth they could wish for. Stories and travelogues circulated about that place too, locating it 'three miles beyond Christmas' and describing it as only reachable by 'eat[ing] through a mountain of millet porridge which is about three miles thick'.

More had his friend Desiderius Erasmus (*c.*1467-1536) edit the text, and the first edition was published in Leuven in 1516 by Dirk Martens, who had relocated his printing business there from Antwerp. The third and fourth editions were published under Erasmus's supervision by Johann Froben (1460-1527), a Swiss publisher in Basel. The copy in The Phoebus Foundation's collection is the fourth edition, with Erasmus's epigrams appended. It includes a woodcut map of the imaginary island designed by Ambrose Holbein (*c.*1494-*c.*1519).

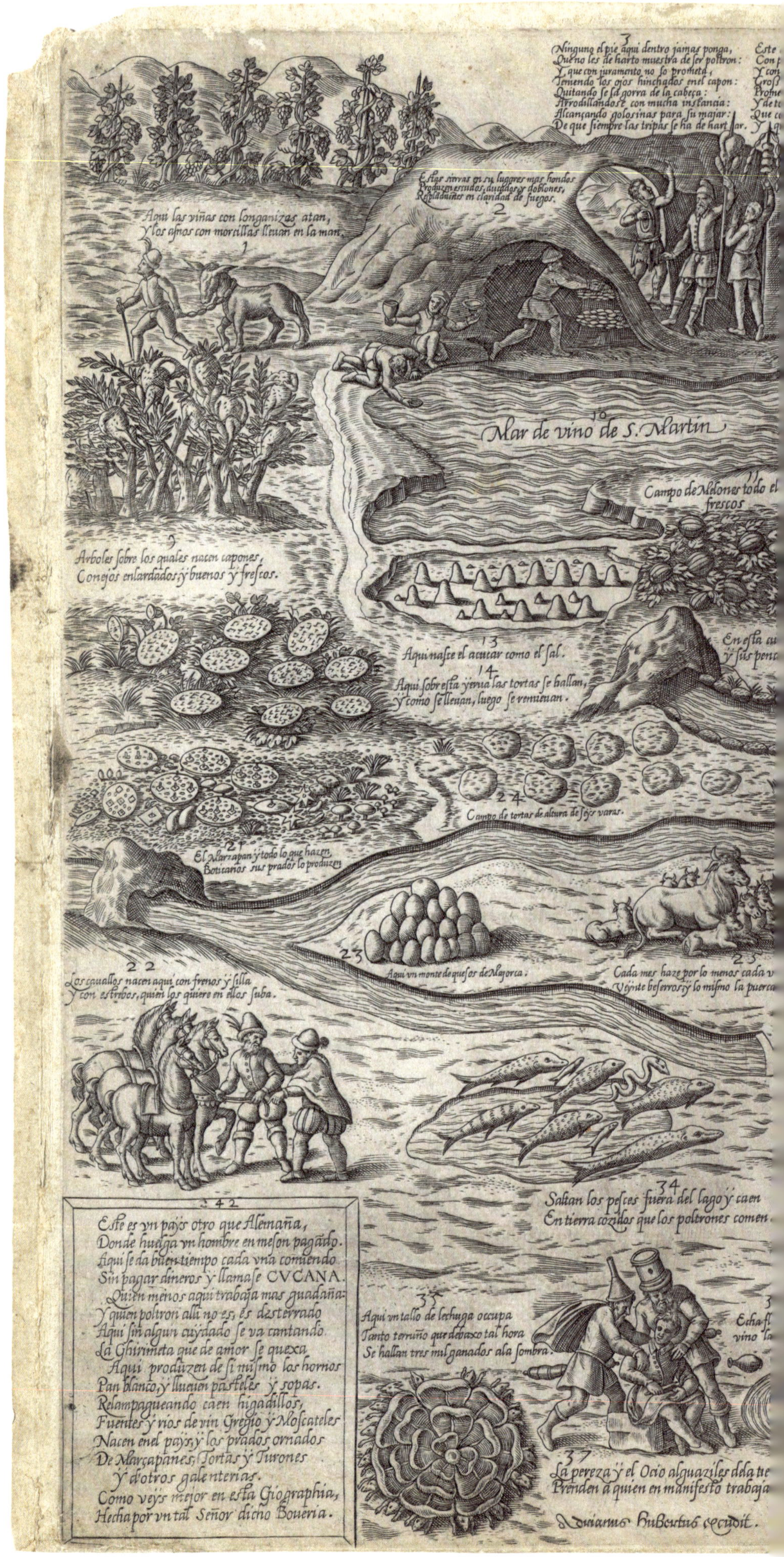

Adriaen Huybrechts
Land of Cockaigne, 1564
Copper engraving, 408 × 557 mm
ANTWERP, THE PHOEBUS FOUNDATION

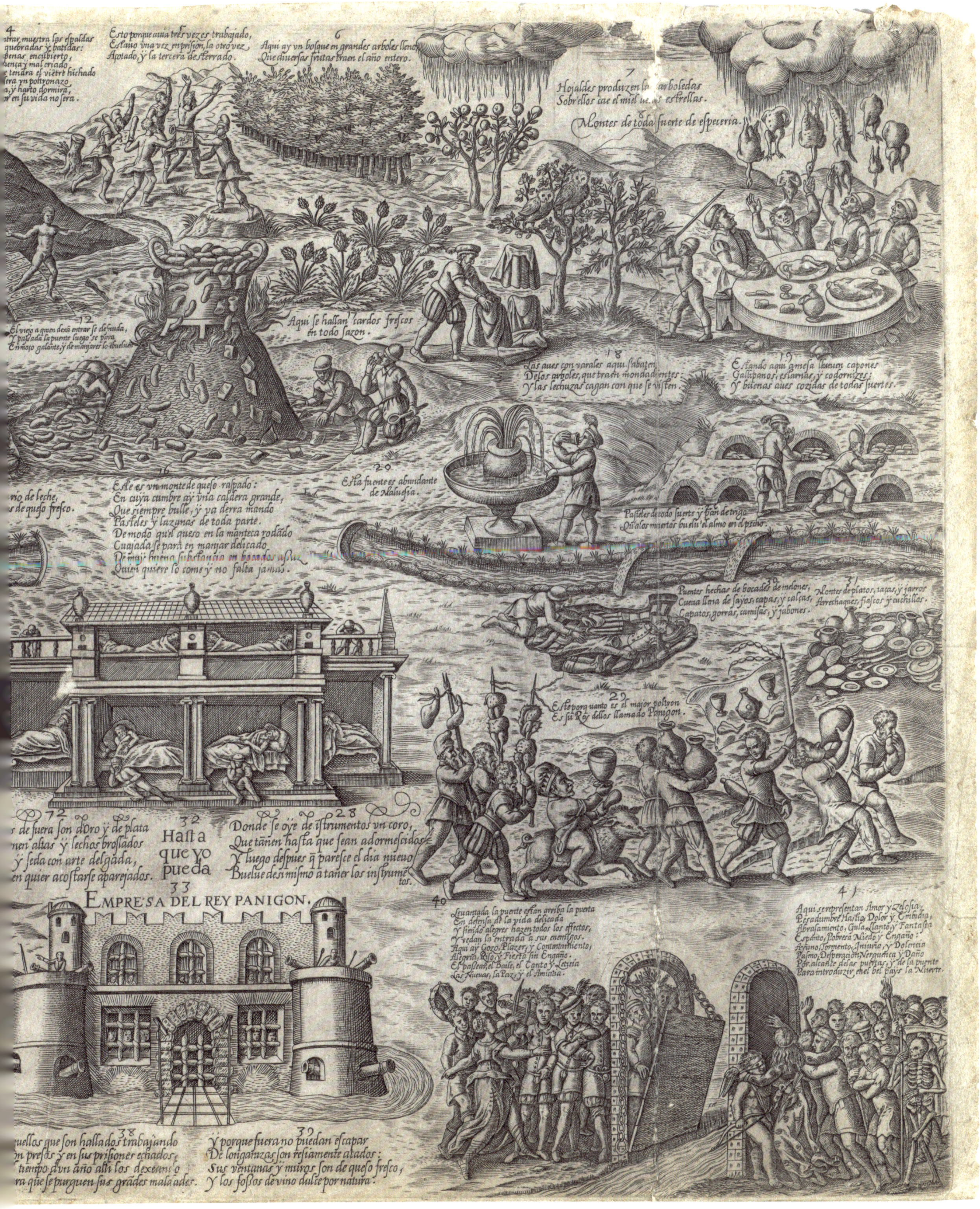
Esto porque una tres vezes trabajado, Estuuo vna vez en prision, la otra vez Açotado, y la tercera desterrado.
6 Aqui ay vn bosque en grandes arboles llenos Que diuersas frutas traen el año entero.
7 Hojaldes produzen las arboledas
8 Montes de toda suerte de especeria.
Aqui se hallan cardos frescos en todo sazon.
18 Las aues con varales aqui se baten Delos arboles, que traen mondadientes Y las lechuzas cagan con que se visten.
19 Estando aqui a mesa llueuen capones Gallipauos, estarnas, y codornizes; Y buenas aues cozidas de todas suertes.
20 Esta fuente es abundante de Maluasia.
16 Este es vn monte de queso raspado: En cuya cumbre ay vna caldera grande, Que siempre bulle, y va derramando Pasteles y lazañas de toda parte.
Puentes hechas de bocados de melones, Cuenta llena de sayos, capas, y calças, Çapatos, gorras, camisas, y jubones.
31 Montes de platos, tazas, y jarros
22 Esto porque esto es el mejor poltron Es su Rey dellos llamado Panigon.
28 Donde se oye de istrumentos vn coro, Que tañen hasta que sean adormescidos Y luego despues q̃ paresce el dia nueuo Buelue de si mismo a tañer los instrumentos.
32 Hasta que yo pueda 33
EMPRESA DEL REY PANIGON.
39 Y porque fuera no puedan escapar De longanizas son estrechamente atados; Sus ventanas y muros son de queso fresco, Y los fossos de vino dulce por natura.

Phantom islands

Maps played an important role not only in the growth and dissemination of geographical knowledge but also that of persistent geographical myths. One special category among these 'cartographic misunderstandings' consists of 'phantom islands'. Islands have a special appeal to the imagination, possibly because they are literally isolated (from the Latin *insula*, meaning 'island') – separated from the mainland and harder to find and to reach. Who knew what surprises awaited? Perhaps even a previously undiscovered paradise on earth? One of the most famous phantom islands is Baja California, which was long charted as a separate land mass, when it is actually a peninsula, now part of Mexico, to the south of the US state of California. The myth arose in the early sixteenth century and was associated with a Spanish folk legend that made it home to a kind of Garden of Eden. Gerard Mercator and Abraham Ortelius did show the whole of California as a peninsula, however, and the myth was also refuted by several sailors.

But people often preferred fiction to fact. The engraver, mapmaker and publisher Henricus Hondius (1597-1651), for instance, jumped on the bandwagon and turned California into an island in the Mercator-Hondius atlases he printed (⟶ Chapter X). California Island continued to crop up in the seventeenth century in Frederik de Wit's richly decorated and coloured map, first published around 1675. The same map contains several other phantoms, including the fictitious island of Bermeja in the Gulf of Mexico and the non-existent Lake Parime in the northern part of South America, linked to the legend of Eldorado, the mythical land of gold. The cartouche contains equally fictitious and stereotypical representations of the indigenous people of the Americas (⟶ p. 190, for more on this kind of stereotyping). It is not easy to determine whether mapmakers and publishers in the Low Countries actually believed that these locations were real or whether they preferred not to know. Whatever the case, De Wit's map does incorporate a number of improvements compared to earlier maps of the Americas, notably the five Great Lakes and the western coastline of Hudson Bay, indicating that fantasy and newly gathered 'true' knowledge need not be mutually exclusive.

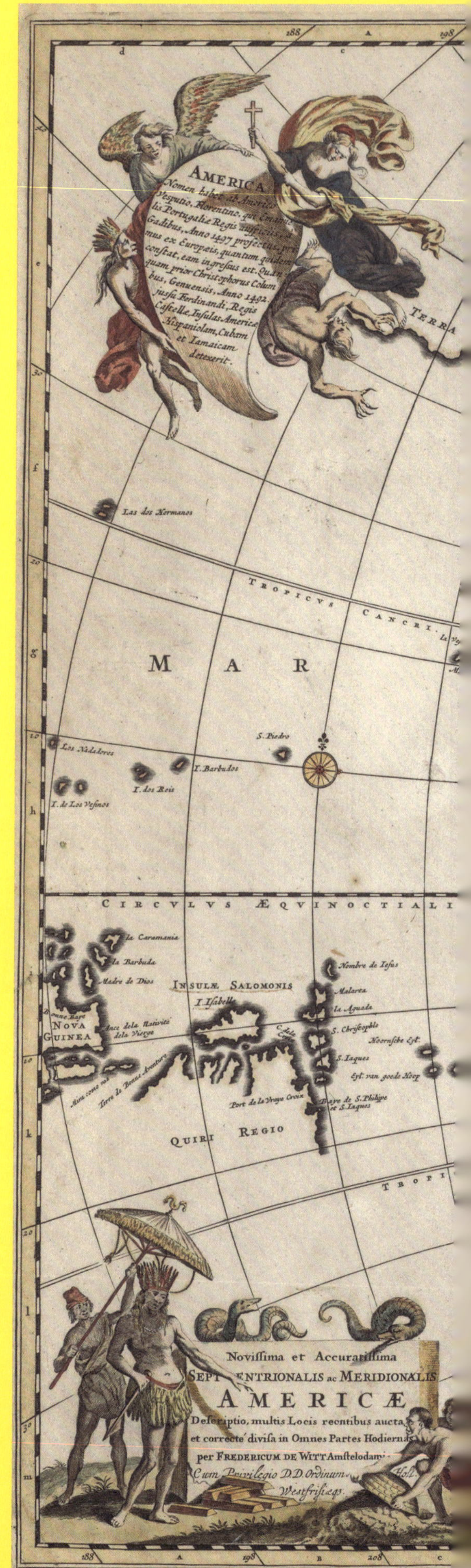

Frederik de Wit
Novissima et Accuratissima Septentrionalis ac Meridionalis Americae Descriptio, c.1690
Map of North and South America and the Pacific
Hand-coloured copper engraving, 490 × 590 mm
ANTWERP, THE PHOEBUS FOUNDATION

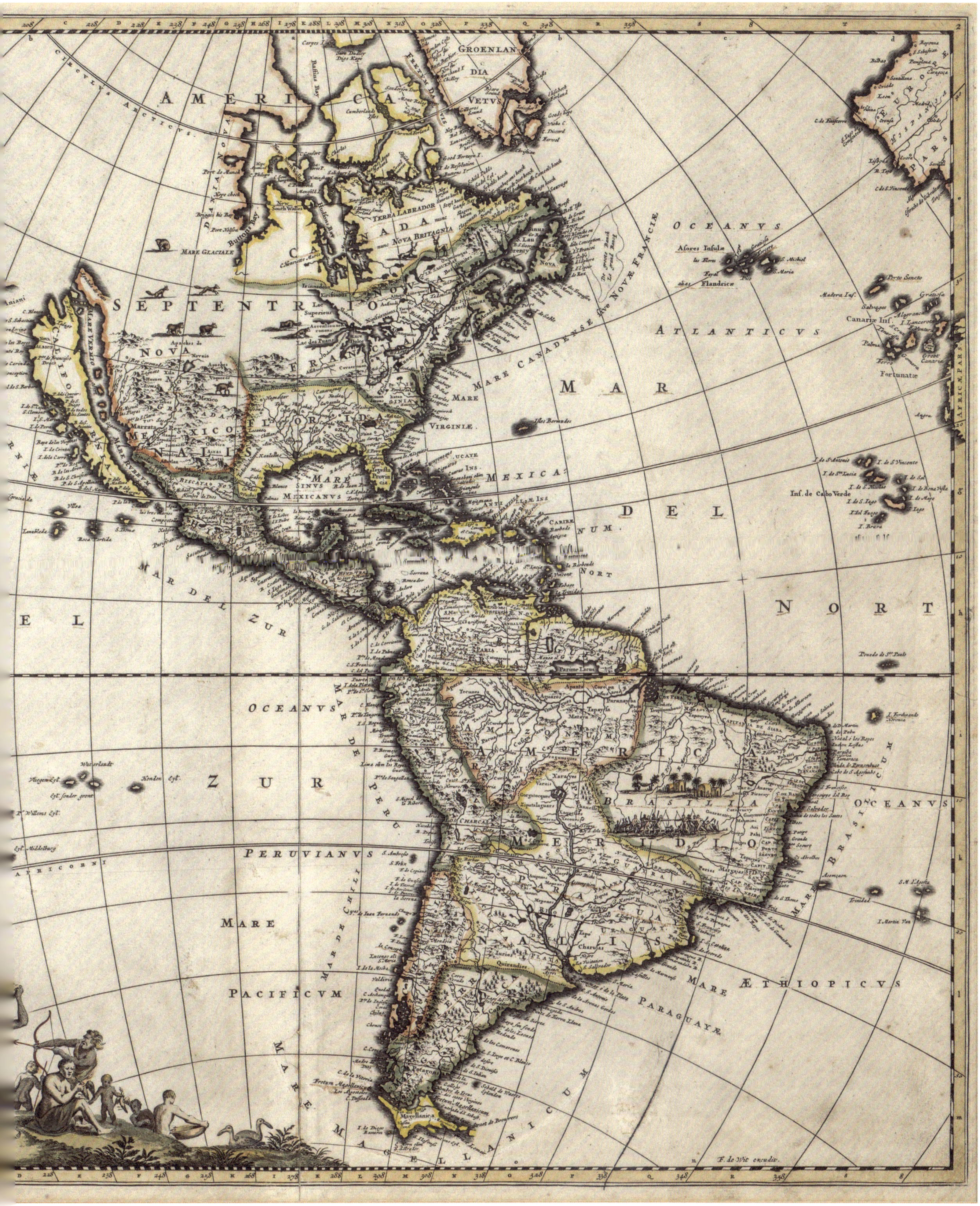
AMERICA
SEPTENTRIONALIS
NOVA FRANCIA
NOVA BRITANNIA
TERRA LABRADOR
GROENLANDIA
MARE GLACIALE
OCEANUS
ATLANTICUS
MAR DEL NORT
MARE CANADENSE sive NOVÆ FRANCIÆ
MARE VIRGINIÆ
MARE MEXICANUM
SINUS MEXICANUS
FLORIDA
NOVA MEXICO
MAR DEL ZUR
OCEANUS ZUR
PERUVIANUS
MARE PACIFICUM
AMERICA MERIDIONALIS
BRASILIA
OCEANUS BRASILICUS
MARE ÆTHIOPICUS
MARE PARAGUAYÆ
MARE MAGELLANICUM
MARE DE CHILI
MAR DE PERU
Inf. de Cabo Verde
Fortunatæ
F. de Wit excudit.

PETRUS PLANCIUS, CORNELIS CLAESZ. AND THE *ITINERARIO*

In the Southern Netherlands, then, when it came to 'voyages of discovery', facts mixed with fiction. Tales were spun by the sailors who gathered in Antwerp and immortalised in the maps and books now being published en masse. In the sixteenth century, however, the centre of international trade in the region shifted to Amsterdam, as did the Flemish geographer and preacher Petrus Plancius, a convinced Protestant. Plancius – Pieter Platevoet as he was born – became part there of a stimulating environment of fellow scientists, globemakers and cartographers. His linguistic skills and international contacts made him a central figure in the Amsterdam network of cartographers, printers, publishers and sailors. He gathered knowledge, produced maps, encouraged fresh voyages and when they returned, assimilated that knowledge too to make newer and better maps.

Plancius contributed to the planning, for instance, of the three expeditions in search of the Northwest Passage, to reach Asia via the North Pole. The route via Nova Zembla and Svalbard turned out not to be the solution, however. It ended in fiasco – a tragic expedition that cost its leader, Willem Barentsz., his life (⟶ Chapter VII). In the meantime, Plancius was also thinking about other routes to the Asian continent, namely the solution found by the Portuguese, which involved sailing around the southernmost tip of Africa. Plancius gained access to Portuguese sea charts drawn up by the Portuguese cartographer Bartholomeo Lasso, cosmographer and master of navigation to King Philip II of Spain. Portugal and Spain were among the first European countries to make the great sea voyages to the Americas and Asia and to establish overseas colonies there. They joined forces in 1580, after which the Spanish king also ruled over the Portuguese overseas possessions. The Portuguese charts contained crucial information about the voyages to Asia that the Iberians had been making for some time. To this day, it remains a mystery how Lasso's maps came into Petrus Plancius's hands.

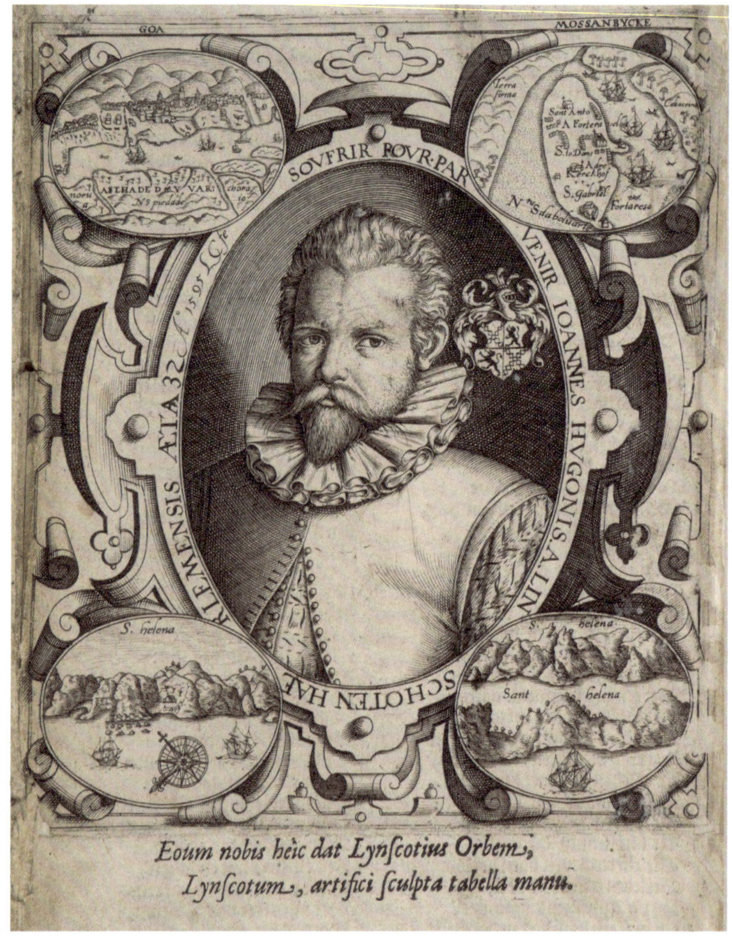

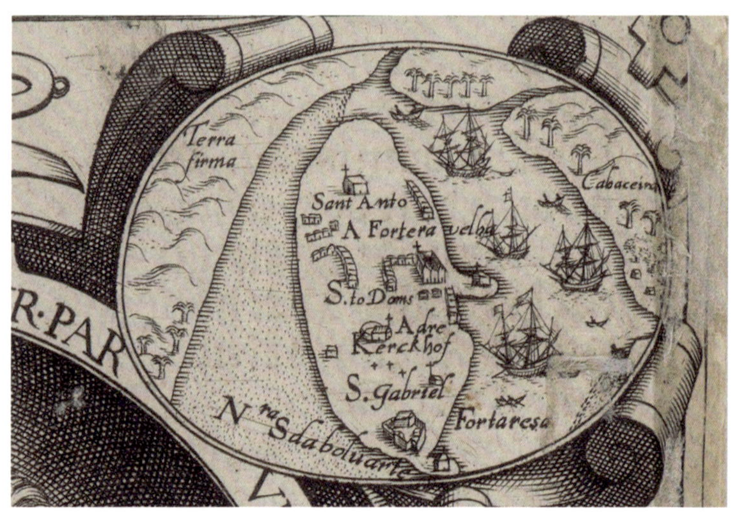

Portrait of Jan Huygen van Linschoten and detail
From: Jan Huygen van Linschoten, *Itinerario, Voyage ofte Schipvaert*, 1595-1596
ANTWERP, THE PHOEBUS FOUNDATION

Plancius collaborated with the printer Cornelis Claesz. and sailor Jan Huygen van Linschoten (*c.*1563-1611) to publish the important navigational reference *Reys-gheschrift vande navigatien der Portugaloysers in Orienten* ('Account of the voyages of the Portuguese in the East'; Amsterdam, 1595). Claesz. – thought to have been born in Leuven – had emigrated from the Southern Netherlands to Amsterdam, just as Plancius did. *Reys-gheschrift* is a pilot guide with practical navigation and sailing instructions for the Cape route along the coast of Africa and then via the Indian Ocean and the South China Sea to Japan.

Van Linschoten did not make all these voyages in person, although he was employed for some years as secretary to the Portuguese archbishop in Goa (the former Portuguese India), where he picked up relevant information on seafaring, trading and colonial administration. Having returned home to the Dutch port of Enkhuizen, he set out the knowledge he had gained in a book. Merchants in Amsterdam were already hatching plans to follow the Portuguese example, which made Van Linschoten's *Reys-gheschrift* the go-to book for the *Eerste Schipvaart* ('First Voyage'), organised to the East Indies in 1595-1597 by a forerunner of the Dutch East India Company (founded in 1602).

While Dutch ships were sailing off all over the world, Van Linschoten continued to work on a comprehensive edition of his book in three parts, with *Reys-gheschrift* forming the second part. The first was *Itinerario, Voyage ofte Schipvaert* ('Itinerary, voyage or navigation'; Amsterdam, 1596). It consisted of a report on the voyage from Lisbon to Goa (and back again), with details of the geography, customs, culture and trade of the Portuguese colonialists, together with 'inside information' on the weak points of their regime, of which the Dutch could take advantage.

The scenes from the colony were drawn by Van Linschoten himself but were engraved by father and son Joannes and Baptista van Doetecum (who died in 1605 and 1611 respectively). They address themes such as local culture, customs, clothing and shipping. Claesz. – the printer and publisher of the book – recognised that readers would need information about other areas too – the world did not finish at Goa, after all. Dirck Gerritsz. Pomp (1544-1608), also known as 'Dirck China', contributed to the Chinese section, while the Enkhuizen physician and collector Bernardus Paludanus (1550-1633) more or less wrote the third part: *Beschryvinghe van de gantsche custe van Guinea, Manicongo, Angola, Monomatapa* ('Description of the entire coast of Guinea, Manicongo, Angola, Monomatapa'; Amsterdam, 1596), which dealt with routes to the Indies but also to Africa, the Americas (Brazil) and the Caribbean.

This trilogy in a single volume, known simply as the *Itinerario*, became a standard work containing all the information a sailor could need about voyages to the then-known corners of the earth. The book itself also found its way all over the world in numerous reprints and translations. The Phoebus Foundation has a copy of the first Dutch edition. The six maps it contains, together with details of indigenous populations and nature, provided the information needed not only to sail around the north-west coast of Africa to China and Japan but also to find one's way around on arrival. Arnold (died *c.*1644) and Hendrik van Langren (*c.*1574-1648) engraved the maps in the book.

The world map in the copy of *Itinerario* features the name of Jan Baptist Vrients in Antwerp as publisher and the Van Langrens as engravers. However, the map had been copied from Plancius's smaller 1594 world map, which was produced separately but was also included in the first edition of the *Itinerario*, published by Claesz. (⟶ pp. 212-213). The fact that Vrients's imprint appears on the map as well as in several copies of the *Itinerario* suggests that Vrients and Claesz. worked together as co-publishers and marketed their products in both Amsterdam and Antwerp.

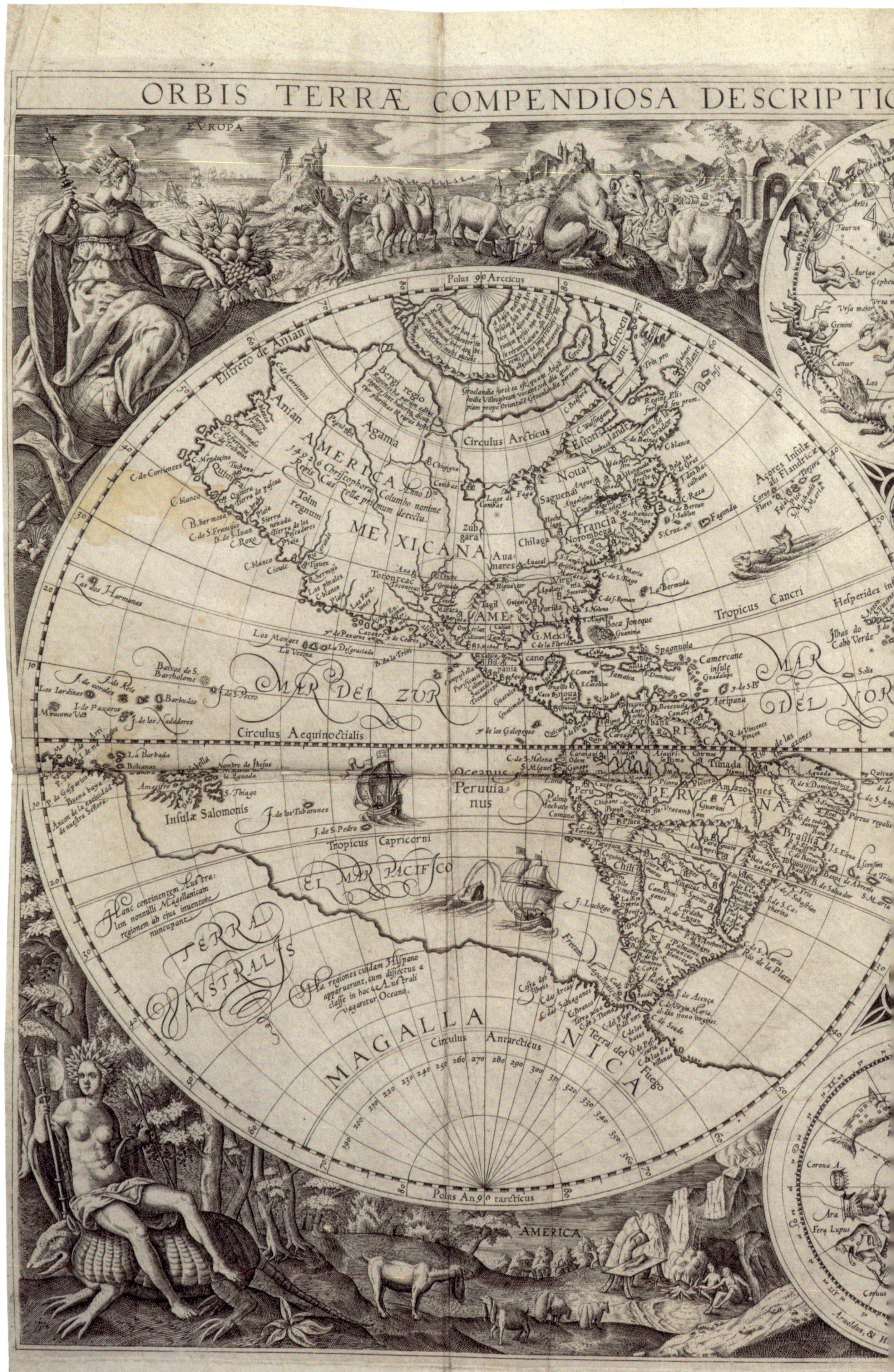

Jan Baptist Vrients after Petrus Plancius
Orbis terrae compendiosa descriptio
World map
From: Jan Huygen van Linschoten,
Itinerario, Voyage ofte Schipvaert, 1595-1596
ANTWERP, THE PHOEBUS FOUNDATION

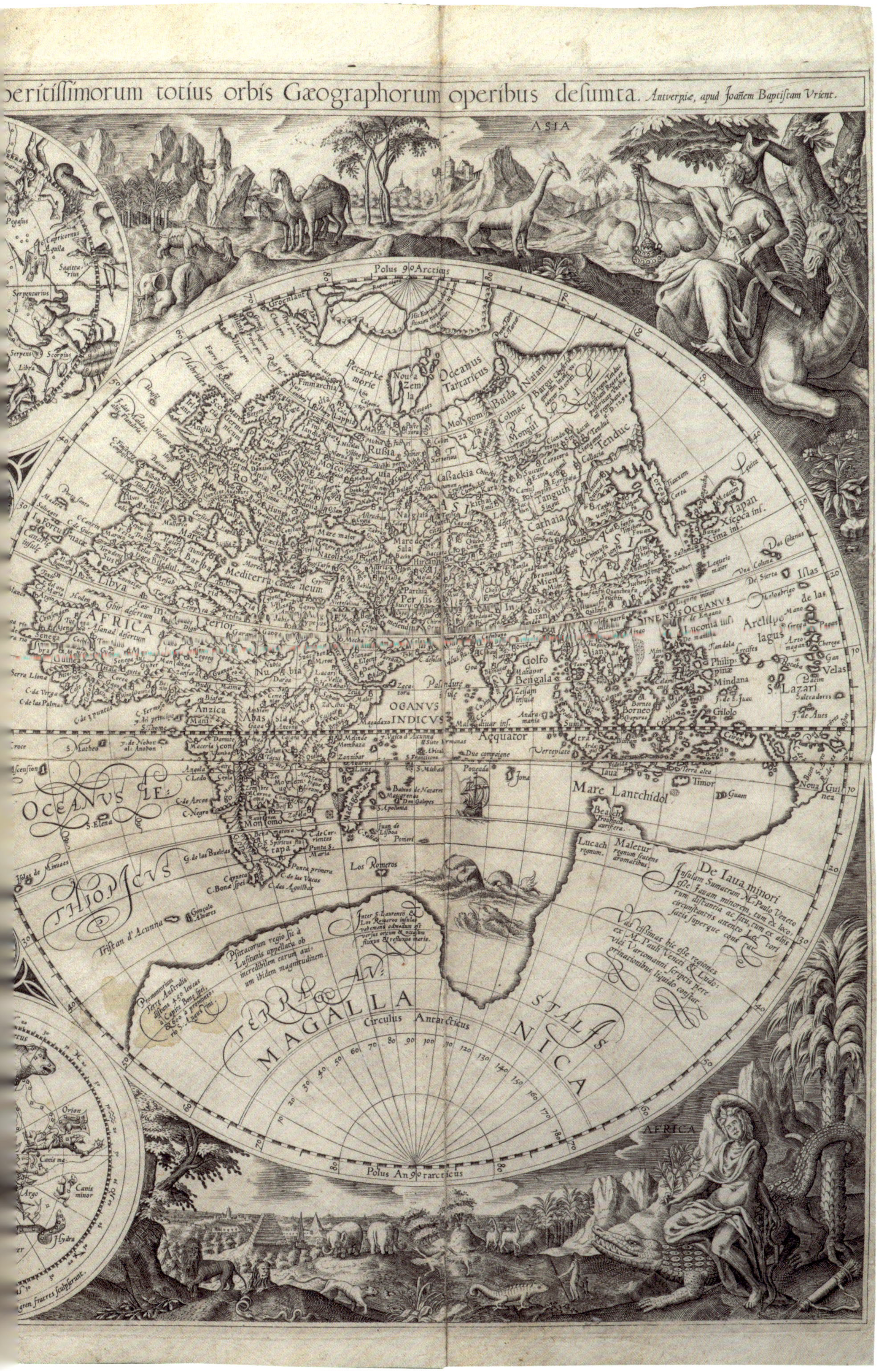
peritissimorum totius orbis Gæographorum operibus desumta. Antverpiæ, apud Ioañem Baptistam Vrient.
ASIA
Polus Arcticus
Oceanus Tartaricus
Petzorke morie
Noua Zembla
Rusia
ASIA
Tanguth
Cathaia
Iapan
Xicoca ins.
Libya
AFRICA
Mediterr
SINENSIS OCEANUS
Archipelagus
Philippinæ
Mindana
S. Lazari
Golfo Bengala
OCEANUS INDICUS
Æquator
Mare Lantchidol
Timor
Noua Guinea
OCEANUS ÆTHIOPICUS
S. Elena
Tristan d'Acunna
Los Romeros
De Iaua minori
TERRA AUSTRALIS MAGALLANICA
Circulus Antarcticus
Polus Antarcticus
AFRICA

Images of slavery

Jan Huygen van Linschoten's *Itinerario* includes a series of prints showing the everyday lives of Portuguese colonists in Goa, the former Portuguese India. The travel guide also contains an illustration of an affluent couple making a nocturnal visit to church – one of the few moments when Portuguese women in Goa left their homes. The enslaved people walking with them, swathed in fabric and armed with spears, are an intriguing element. The Portuguese are known to have enslaved and traded Africans long before the Dutch did. Van Linschoten was aware of this too, writing in his *Itinerario* that Africans were 'strongest in all the East' and were thus transported to the Indies and other countries to perform dirty and heavy work. There has, however, been little emphasis historically on the fact that besides the Portuguese, the Dutch too played a role in the international slave trade even then. The forerunners of the Dutch East India Company (founded in 1602) already traded in African and Asian people during the first voyages to the East Indies, and the practice grew in scale a few years later with the establishment of the West India Company's triangular trade between Africa, the Americas and Europe.

The Amsterdam theologian and preacher Arnold Montanus (1625-1683) wrote a voluminous work on the Americas and the Dutch colonies: *De Nieuwe en Onbekende Weereld, of Beschryving van America en 't Zuid-Land* ('The new and unknown world: or description of America and the Southland'; Amsterdam, 1671). He never set foot on American soil himself but relied on the accounts of travellers. It is also one of the first books to focus on slavery, in both words and pictures. His book includes a view of New Amsterdam which shows a black man suspended by his middle on some kind of tall crane from which he is being swung around. This abuse was a method of punishment for enslaved people who rebelled or escaped.

Dutch Brazil was the first major Dutch colony in which slavery was widely and systematically practised on sugar plantations, during the regime of Johan Maurits of Nassau-Siegen (1604-1679). Sugarcane processing is also described in the book. Blaeu's *Atlas Maior* (⟶ Chapter X) features scenes of slavery in the decorations of the maps showing Pernambuco and Rio Grande. These decorations were based on drawings by Frans Post (1612-1680), who had visited Dutch Brazil in the company of John Maurice between 1636 and 1644. They show the work in a sugar mill, with a cheerful scene of enslaved people dancing in the background – a whitewashed version of how onerous and inhuman life was in reality.

The Dutch banned transatlantic human trafficking in 1814, but only in 1863 did slavery become illegal in the Netherlands and its colonies, one of the last countries in Europe where this occurred. Even then, it was a long time before slavery ceased in practice in the Dutch colonies. In total, the Dutch traded around 600,000 enslaved Africans, many of whom did not survive the sea voyage.

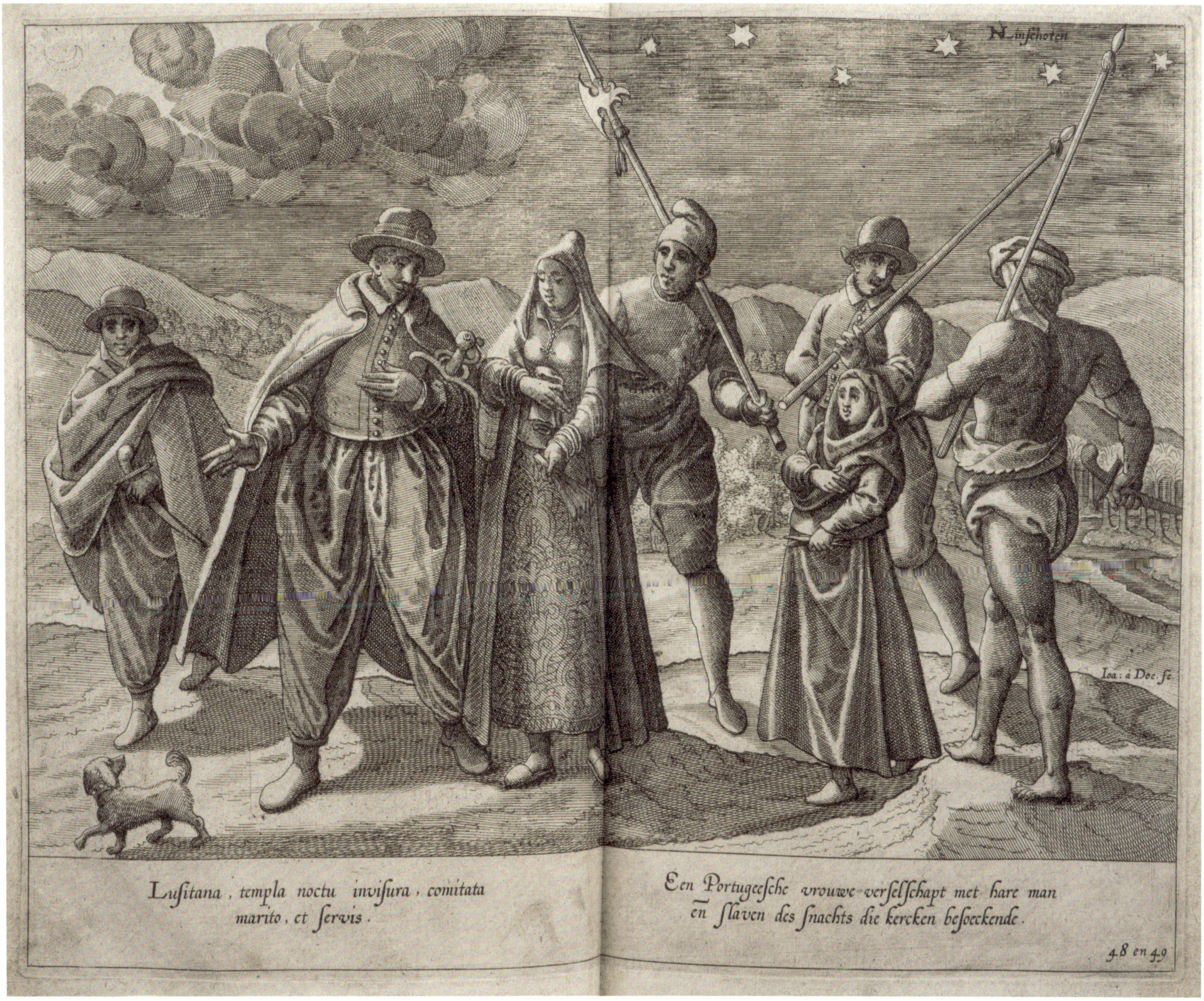

Jan Huygen van Linschoten and Joannes van Doetecum
Portuguese couple visiting the church in Goa, Portuguese-Indies, at night
From: Jan Huygen van Linschoten, *Itinerario, Voyage ofte Schipvaert*, 1595-1596
ANTWERP, THE PHOEBUS FOUNDATION

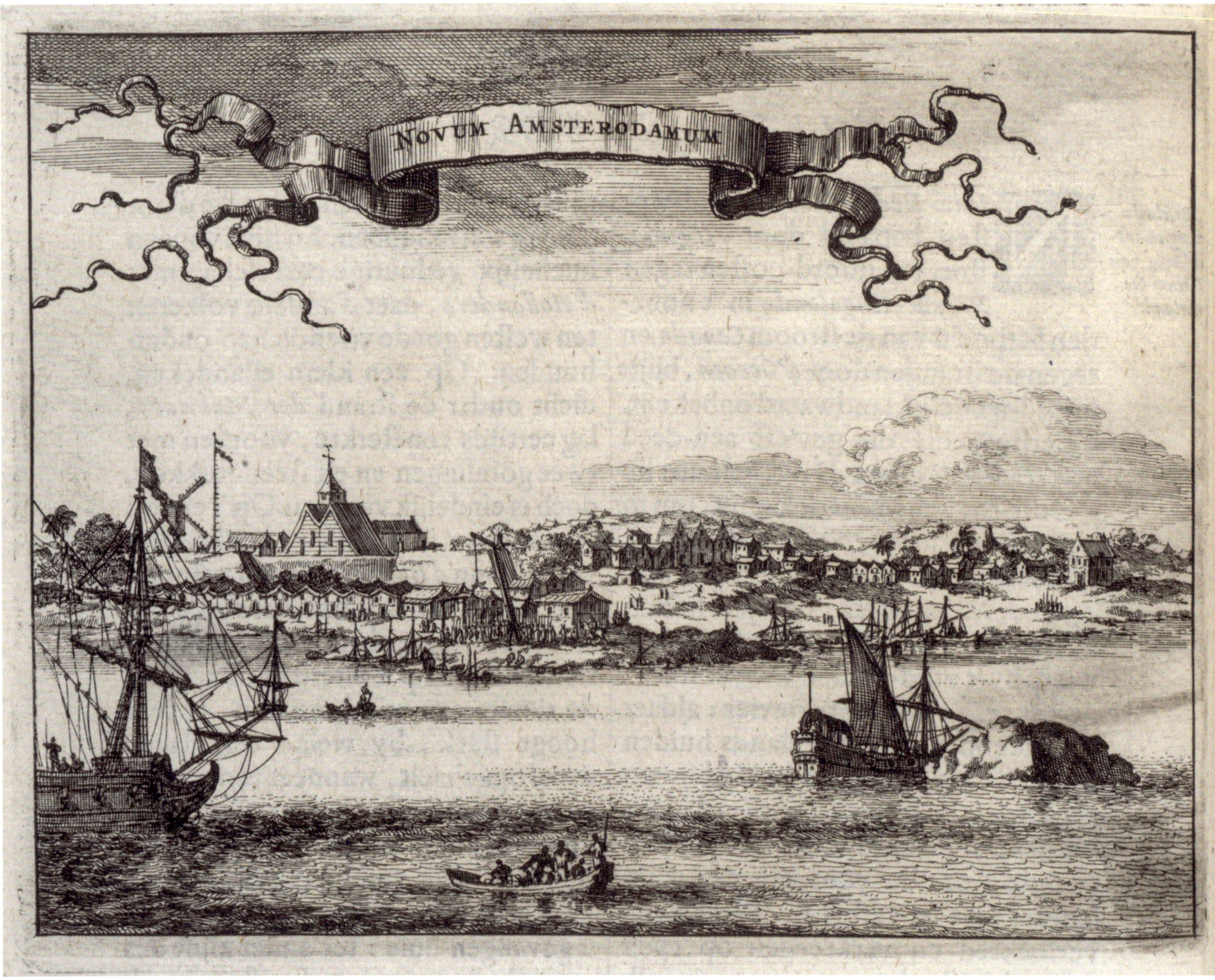

Novum Amsterodamum
View of New Amsterdam
From: Arnold Montanus, *De Nieuwe en Onbekende Weereld, of Beschryving van America en 't Zuid-Land*, 1671
ANTWERP, THE PHOEBUS FOUNDATION

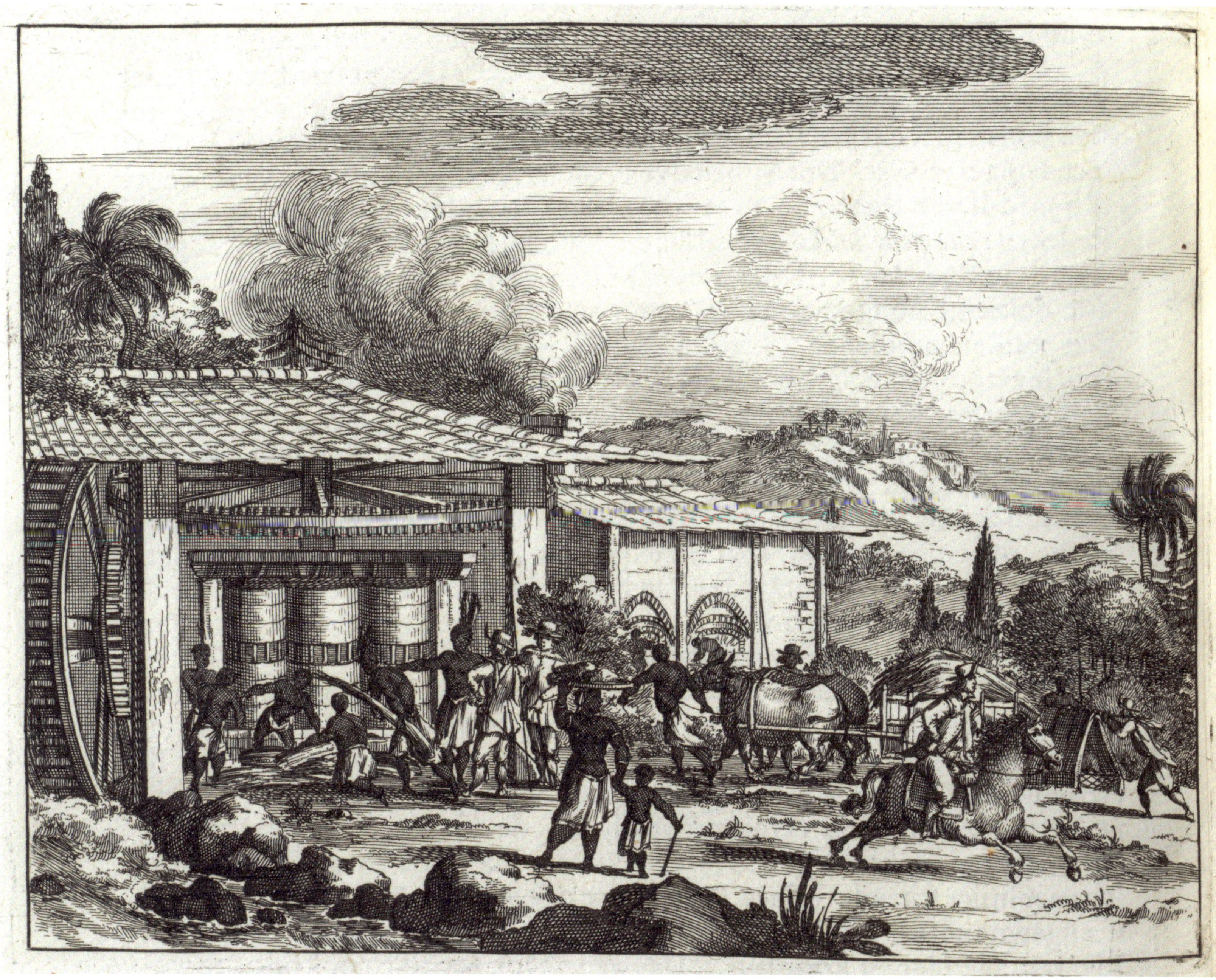

Print of a sugar mill
From: Arnold Montanus, *De Nieuwe en Onbekende Weereld, of Beschryving van America en 't Zuid-Land*, 1671
ANTWERP, THE PHOEBUS FOUNDATION

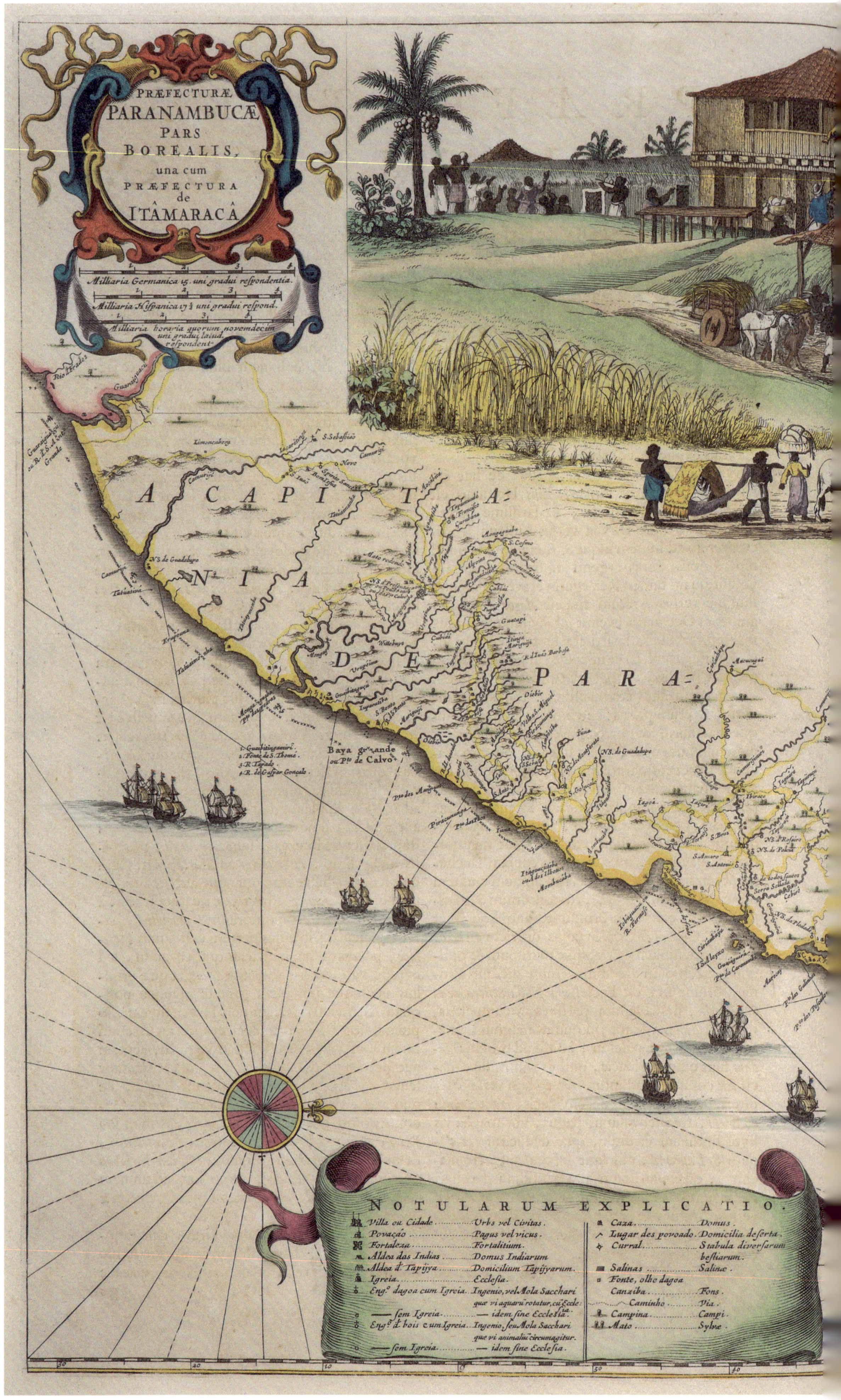

Praefecturae Paranambucae Pars Borealis una cum Praefectura de Itamaraca
Map of Pernambuco, Dutch Brazil
From: Joan Blaeu, *Atlas Maior* [volume 11], 1662
ANTWERP, THE PHOEBUS FOUNDATION

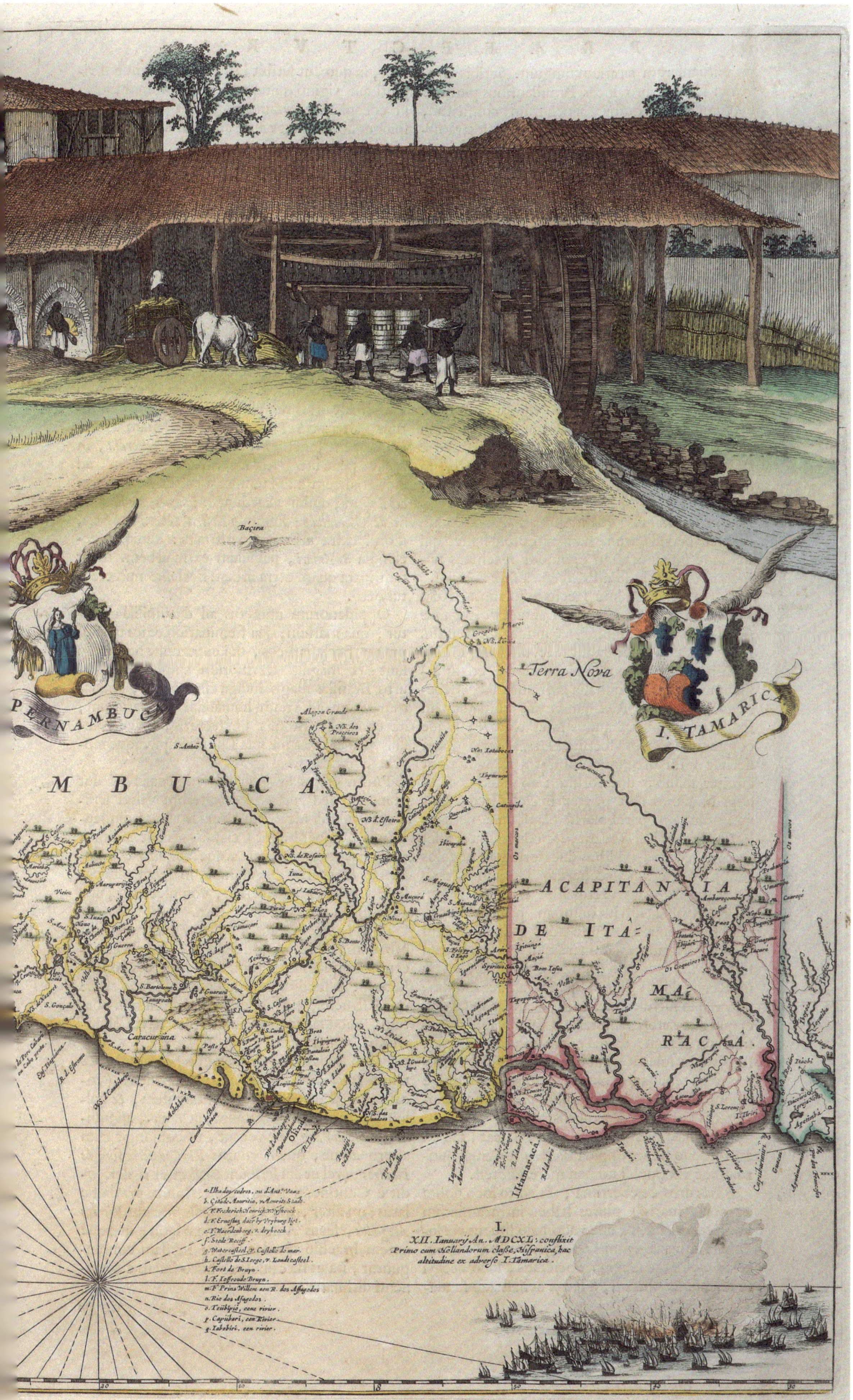
PERNAMBUC
M B U C A.
Terra Nova
I. TAMARICA
ACAPITANIA DE ITAMARACA
I.
XII. Ianuarij An. MDCXL: conflixit
Primo cum Hollandorum classe, Hispanica, hac
altitudine ex adverso I. Tamarica.

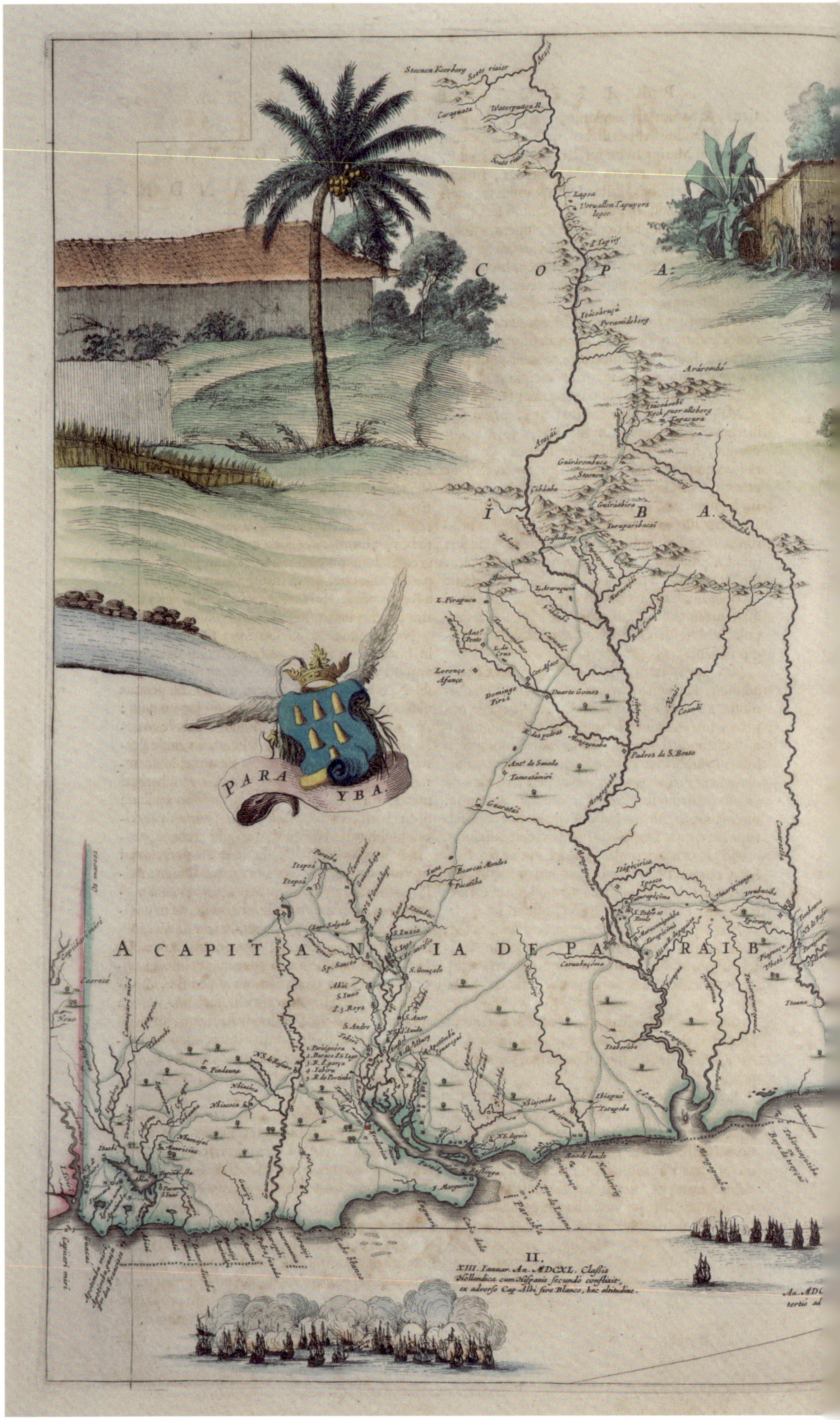

Praefecturae de Paraiba, et Rio Grande
Map of the Rio Grande, Dutch Brazil
From: Joan Blaeu, *Atlas Maior* [volume 11], 1662
ANTWERP, THE PHOEBUS FOUNDATION

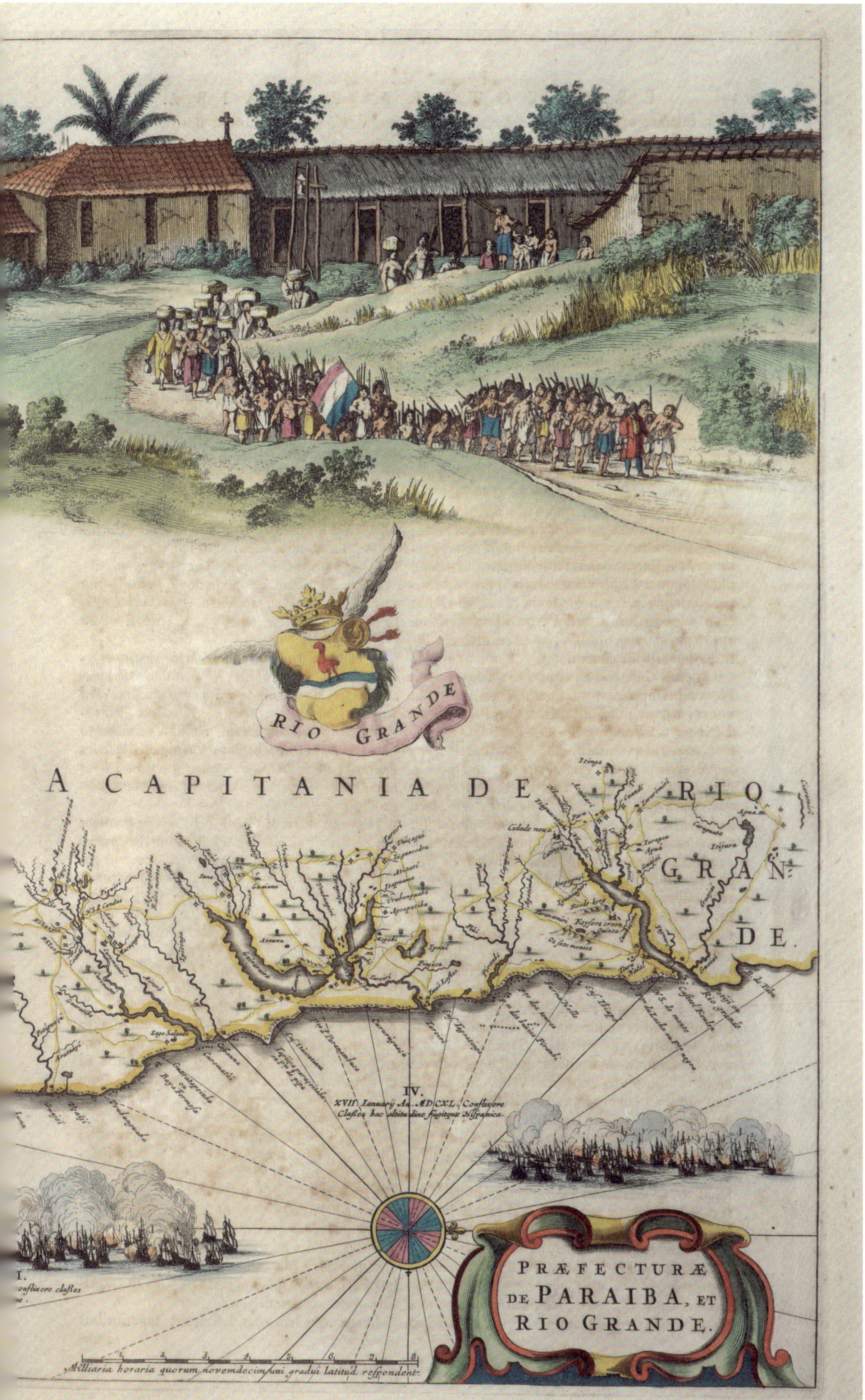
RIO GRANDE
A CAPITANIA DE RIO GRANDE.
IV.
XVII. Januarij An. MDCXL. Conflixere Classes hac altitudine fugitque Hispanica.
PRÆFECTURÆ DE PARAIBA, ET RIO GRANDE.
Milliaria horaria quorum novemdecim uni gradui latitud. respondent.

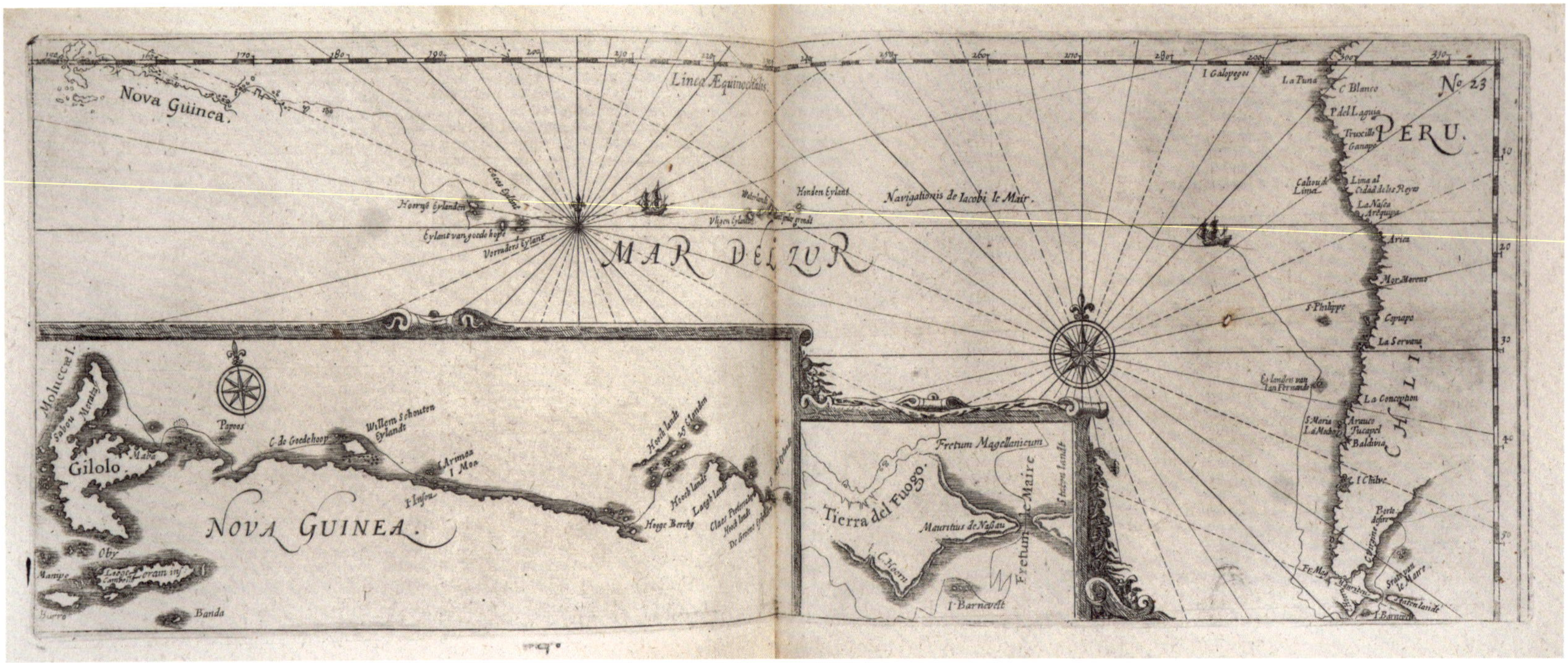

Map of Jacob le Maire's voyage through the Le Maire Strait, 1615-1617
From: Isaac Commelin, *Begin ende Voortgangh van de Vereenighde Nederlantsche Geoctroyeerde Oost-Indische Compagnie*, 1646
ANTWERP, THE PHOEBUS FOUNDATION

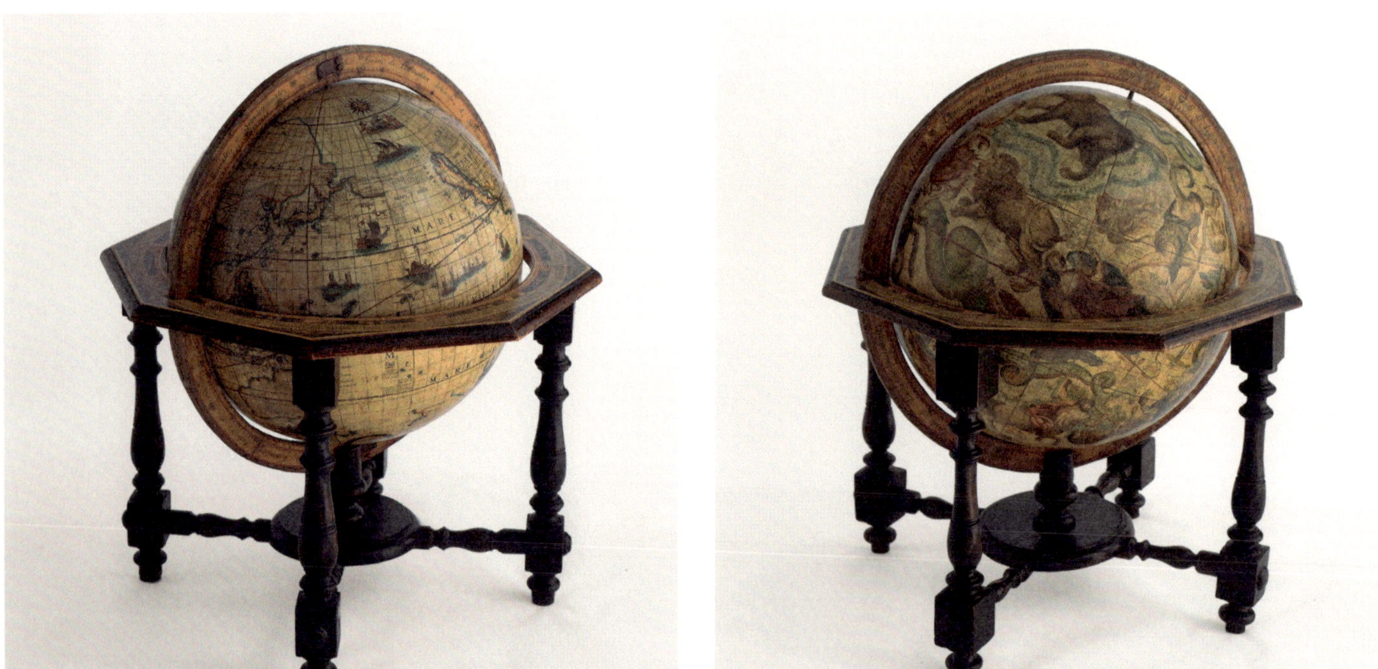

Vincenzo Coronelli
Terrestrial globe and celestial globe, 1696
Hand-coloured copper engraving and wood, 75 × 70 × 70 cm (diameter: 47 cm, each)
ANTWERP, THE PHOEBUS FOUNDATION

AROUND THE WORLD

Jacob le Maire was another Antwerp migrant who sought refuge in Amsterdam. Together with several fellow merchants from the Southern Netherlands and Amsterdam, he founded the New Brabant Company, which later became part of the Dutch East India Company. His father Isaac le Maire (1558/1559-1624), who was also a merchant, achieved great success in the family's new home. He was one of the founders of the East India Company, but following a dispute, he was pushed out and went on to start a company of his own. In 1615, he dispatched his son Jacob with two ships to the southernmost tip of South America to see whether it offered a viable alternative to the South African route to Asia, over which the East India Company enjoyed a monopoly. Willem Cornelisz. Schouten (*c*.1567-1625) from Hoorn also went with him, together with his younger brother Jan.

They reached what became known as Cape Horn in January 1616, naming it for Schouten's birthplace and one of their two ships, the *Hoorn*. They also found a new passage to the Pacific, later called the Le Maire Strait. Having eventually reached the Dutch East Indies, they were detained in Batavia by the East India Company for allegedly infringing its trade monopoly, and their ship, *De Eendracht*, was seized.

The collection of travel reports compiled by Isaac Commelin (1598-1676) – *Begin ende Voortgangh van de Vereenighde Nederlantsche Geoctroyeerde Oost-Indische Compagnie* ('The origin and progress of the Chartered United Dutch East India Company'; Amsterdam, 1646) – contains a fold-out map in the section devoted to Jacob le Maire and Willem Schouten's voyages to Australia in which the Le Maire Strait can be seen. The accompanying text includes the following explanation: 'No. 23: The map of the voyage of Jacob le Maire demonstrates in what manner the aforesaid le Maire passed the South Sea via a new strait as far as the Indies.' A small inset map shows 'the proof', as the explanatory text puts it, of this 'discovery' of the passage in black and white. The chapter in Commelin's volume is a reprint of the previously published travel report. The map shown here was, however, an invention of the publisher Janssonius. He combined original maps to create a new one, thus saving space, effort and expense.

This legendary voyage would live on in a great many maps for decades and even centuries. The Italian cartographer and globemaker Vincenzo Coronelli (1650-1718) designed a pair of globes in London in 1696, the terrestrial one of which shows Le Maire's journey in dotted lines. The Venetian-born monk travelled all over Europe during his lifetime, creating a furore primarily as a globemaker, although he also produced maps and books. In 1696, he visited England, the Low Countries and the Holy Roman Empire, where he purchased materials, gained inspiration, made contacts and learned the latest tricks of the cartographic trade. His globes not only reflected the most recent geographical discoveries and voyages around the world, they were objects of beauty: Coronelli made them colourful works of art, abundantly decorated with mythological figures, magnificent ships and sea monsters.

The map of South America published in Augsburg by Tobias Conrad Lotter (1717-1777) also shows several routes based on celebrated voyages and the corresponding mariners and dates. The green line marks the route taken by '*Le Maire et de Schouten An. 1616*'. Le Maire's name also features at the very bottom as 'discoverer' of the new southern ocean below South America: '*Novum mare australe detectum per le Maire*'.

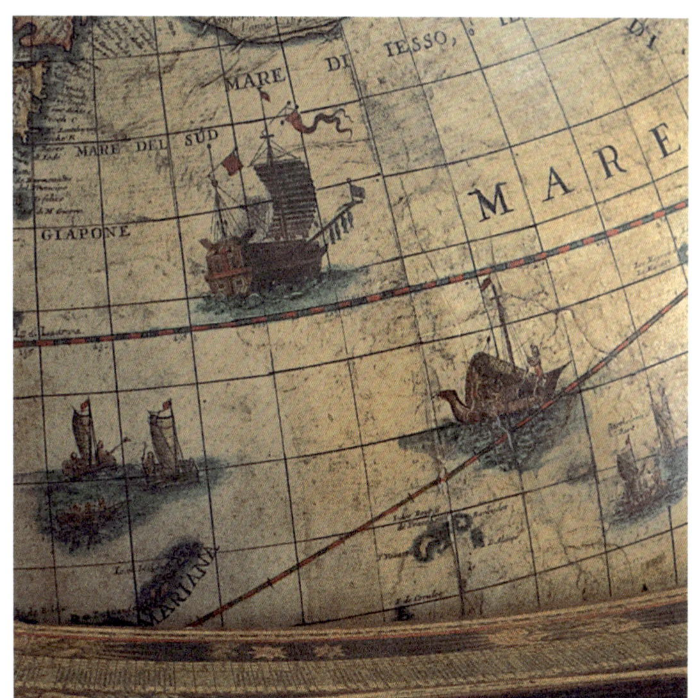

Detail from: Vincenzo Coronelli, terrestrial globe, 1696 (p. 222)

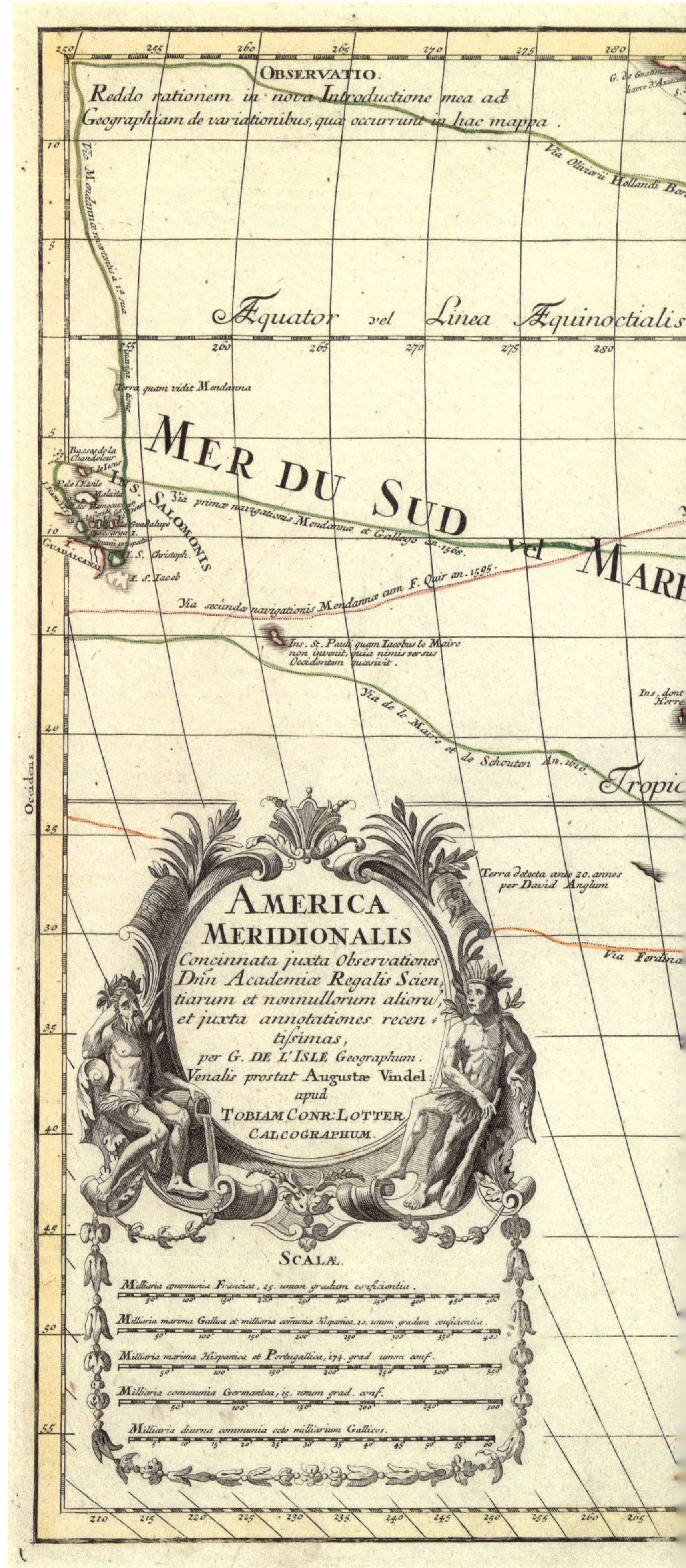

Georg Friedrich Lotter after Guillaume Delisle
America Meridionalis, c.1760-1770
Map of South America
Hand-coloured copper engraving, 470 × 615 mm
ANTWERP, THE PHOEBUS FOUNDATION

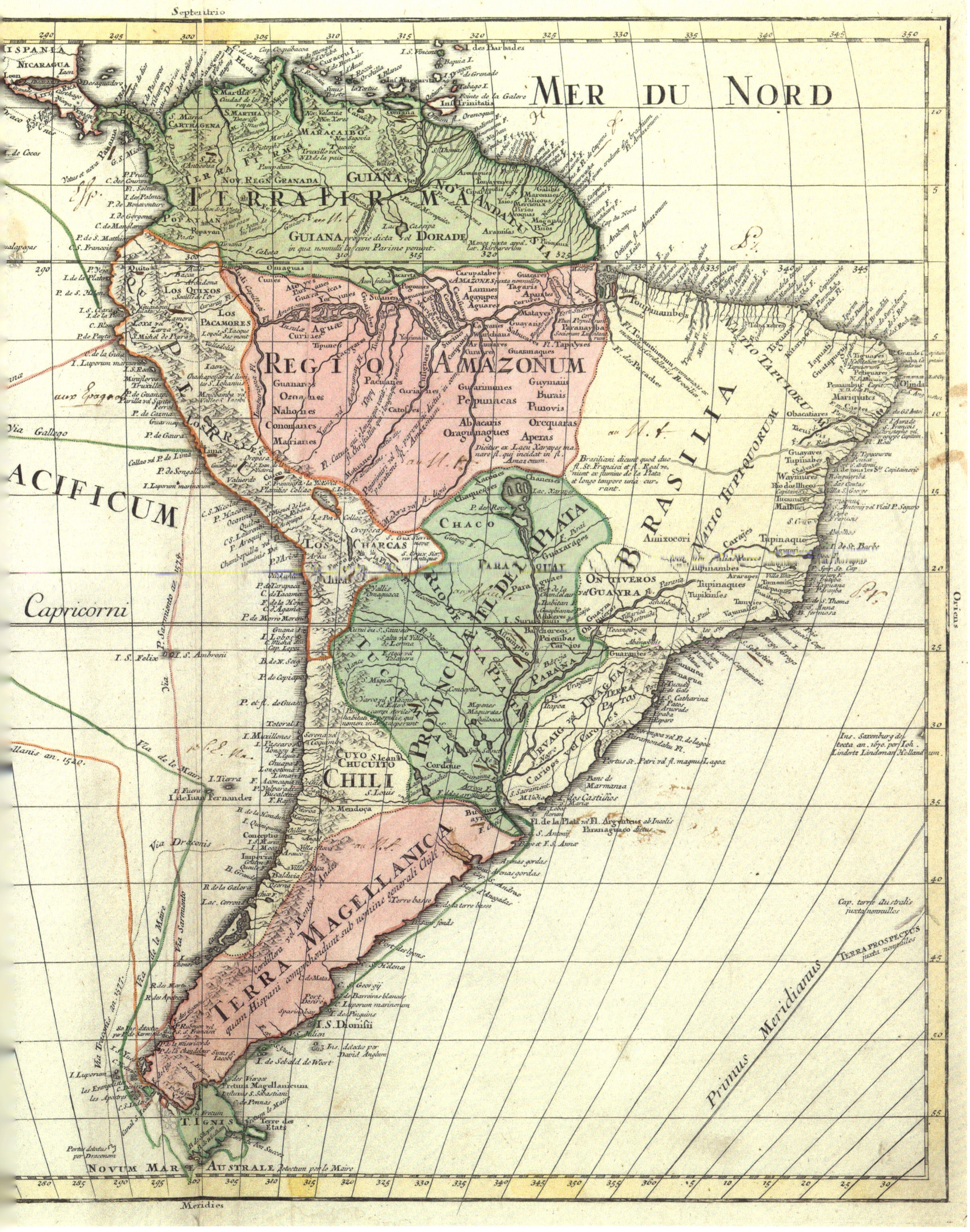
Septentrio
Mer du Nord
Terra Firma
Nov. Regn. Granada
Guiana
Guiana propriè dicta vel Dorado
Regio Amazonum
Brasilia
Peru
Chaco
Charcas
Paraguay
Provincia Fl. de la Plata
Chili
Cuyo
Terra Magellanica
Acificum
Capricorni
Primus Meridianus
Terra Prospectus
Novum Mare Australe
T. Ignis
Meridies
Oriens

Pleasure for the armchair traveller

The territory of the Dutch East India Company stretched from the Cape of Good Hope in South Africa to the Pacific Ocean. The company embarked on numerous voyages from 1600 onwards, aided by developments in international trade and navigation. An endless flow of travelogues were published for those back home, often richly illustrated and provided with maps.

Commelin's *Begin ende Voortgangh* is a two-part collection of over thirty voyages (with twenty-one travel reports) to the East Indies made by various travellers under the company's auspices between 1594 and 1636. The Amsterdam historian Isaac Commelin compiled and edited the book, and it was published by Johannes Janssonius, but the original authors were a varied lot. Commelin did not only draw on previously published travel reports: as many as twelve out of the twenty-one were published here for the first time. He did not have direct access, however, to the East India Company's archives and is likely to have relied on an informant to supply the necessary material.

Some of the prints in the work are new impressions of copper plates by the Amsterdam printer and publisher Michiel Colijn (1584-1637), who previously printed a collection of travel reports. Janssonius supplemented these, however, with a number of new copper engravings in more or less the same style, bringing him to an impressive total of 230 engraved maps and plates. A trial run in 1644 was followed by the first edition in 1645. Judging by the unamended second edition published as early as 1646 – of which The Phoebus Foundation has a copy – it must have sold very fast indeed. It was also translated into English and French. Its many illustrations made it an expensive publication, out of reach of most wallets, as reflected in the gold-bound copy owned by The Phoebus Foundation. Other publishers, therefore, saw a market for cheaper reprints, which meant that accounts of travel to the Dutch East Indies did manage to reach a wide readership from different walks of life in the second half of the seventeenth century.

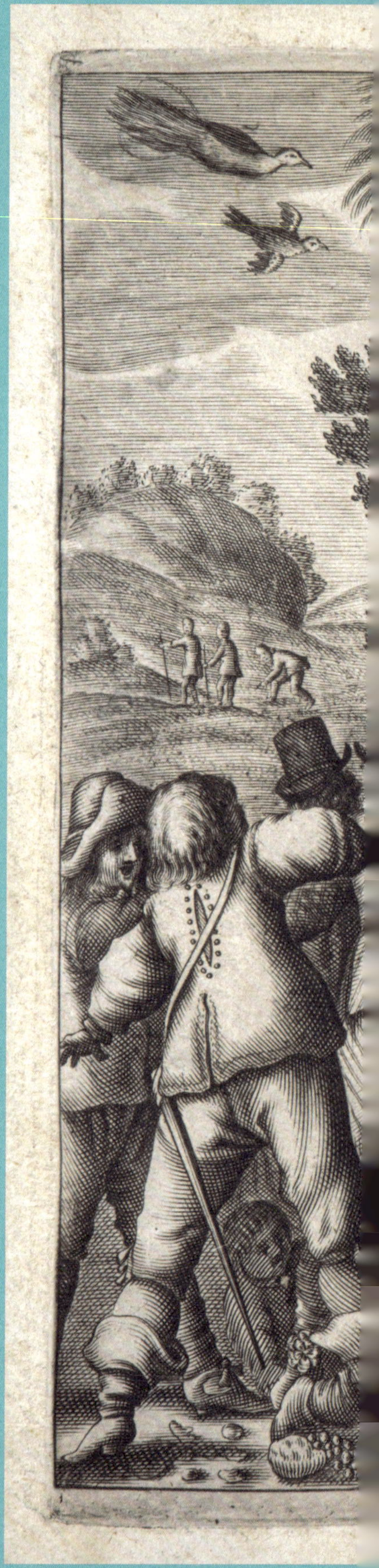

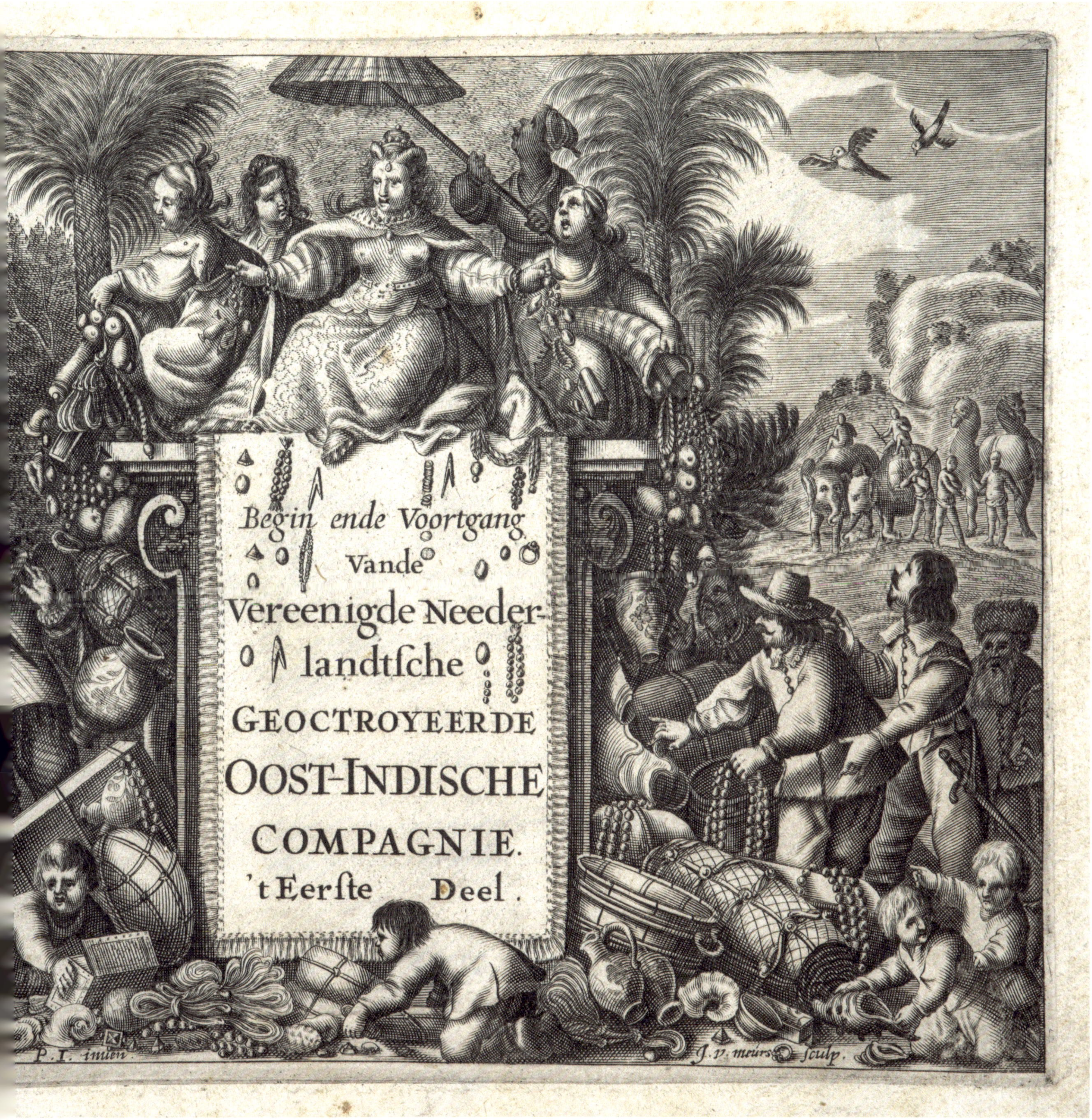

Title page
From: Isaac Commelin, *Begin ende Voortgangh van de Vereenighde Nederlantsche Geoctroyeerde Oost-Indische Compagnie*, 1646
ANTWERP, THE PHOEBUS FOUNDATION

ſorum, caſtorum & ſimilium ferarum tegunt. Arma ipſis ſunt arcus & ſagittæ ligneæ clavæ & parmæ corio tectæ. Icones à Champlanio expreſſas hic adjungere viſum, in quibus habitus marium & arma ipſorum videre eſt.

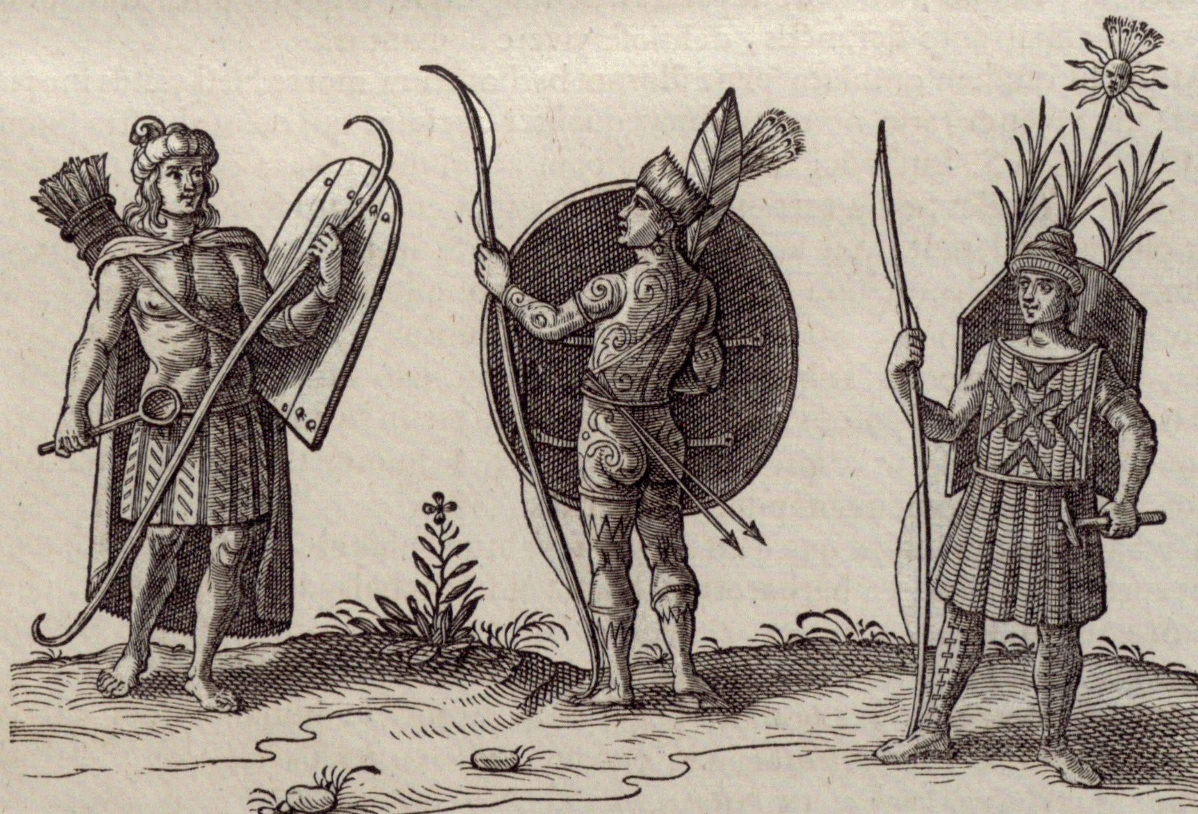

Adoleſcentulæ poſtquam decimum & quartum aut quintum ætatis annum attigerunt, omnibus promiſcue ſui copiam faciunt, dein poſt quinquennium aut ſexennium, tam vagæ libidinis pertæſæ, unum ſibi virum deligunt, cum quo reliquam vi-

Wood-cut illustration of the indigenous people of North America
From: Johannes De Laet, *Novus orbis seu Descriptionis Indiae occidentalis*, 1625
ANTWERP, THE PHOEBUS FOUNDATION

COLONISE AND DOMINATE

Reading More's *Utopia*, we can readily immerse ourselves in the stories of Raphael Hythlodaeus – stories that are fanciful yet still inspired by reality. We risk losing sight, however, of the fact that those 'voyages of discovery' were anything but exciting adventures into the unknown. 'Discovery' was merely the beginning; in reality it was all about monopolising trade with and power over the colonies. This was true not only of the Portuguese and Spanish explorers who led the drive to establish overseas possessions but of the Dutch too. Dutch colonialists used force to conquer territories in the 'New World' on the opposite side of the Atlantic, and they had few qualms about engaging in human trafficking and slavery. The Dutch East India and West India Companies might have been headquartered in Amsterdam, but immigrants from the Southern Netherlands were also numbered among the inhabitants of the newly established colonies. A group of Protestant Walloons, for instance, arrived in New Netherland in 1624. They were dispatched there by the West India Company to begin a new life, and it was these same 'pioneers' who founded New Amsterdam a few years later on the island of Manhattan.

The book *Novus orbis seu Descriptionis Indiae occidentalis* ('New world or description of West India'; Leiden, 1633) was an extensive description of 'West India' and played a part in the campaign to tempt people to emigrate from the Low Countries to the 'New World'. This geographical and historical description of New Netherland was penned by the Antwerp-born theologian and historian Johannes De Laet, who never visited the Americas in person but drew on his extensive personal library of books from the Low Countries and abroad. As a member of the Dutch West India Company's board, he was also able to draw on its archives, including the journal of Henry Hudson (*c*.1565-*c*.1611) and its description of the 'discovery' of Manhattan.

The maps in *Novus orbis* were made by Hessel Gerritsz. (*c*.1581-1632), the official cartographer of both the Dutch East India and West India Companies, who was responsible for maintaining, correcting and compiling all their cartographic material. Gerritsz. did not create all the maps in person but also drew on the work of other mapmakers, including the Englishman John Smith (1580-1631), who had mapped his country's North American colonies. Where necessary, he adjusted Smith's maps. De Laet's book served to justify and promote Dutch domination of the area. He described the original inhabitants and their language and culture, but also referred to them condescendingly: they were, he claimed, devil worshippers who lived in a disorderly society.

Beschryvinge van Nieuw-Nederlandt ('Description of New Netherland'; Amsterdam, 1656) was a similar book by Adriaen van der Donck (*c*.1618-1655), to which the publisher Evert Nieuwenhof added a map of New Netherland based on a drawing made around 1650 of New Amsterdam, which would later become New York. The map of New Netherland shows the newly established Dutch colonies amid original Native American settlements and also contains one of the earliest views of New Amsterdam. The city view would continue to feature in different maps of New Netherland for many years: The Phoebus Foundation has a copy from the latter part of the nineteenth century, which was used to illustrate a bibliography on New Netherland. Similar to the many portraits, profiles and panoramas made of cities in the Low Countries (⟶ Chapter II), the picture fed into the narrative marketing New Amsterdam as a kind of 'Amsterdam in America'. The books of Van der Donck and De Laet, along with the city view of New Amsterdam, were intended to attract new settlers.

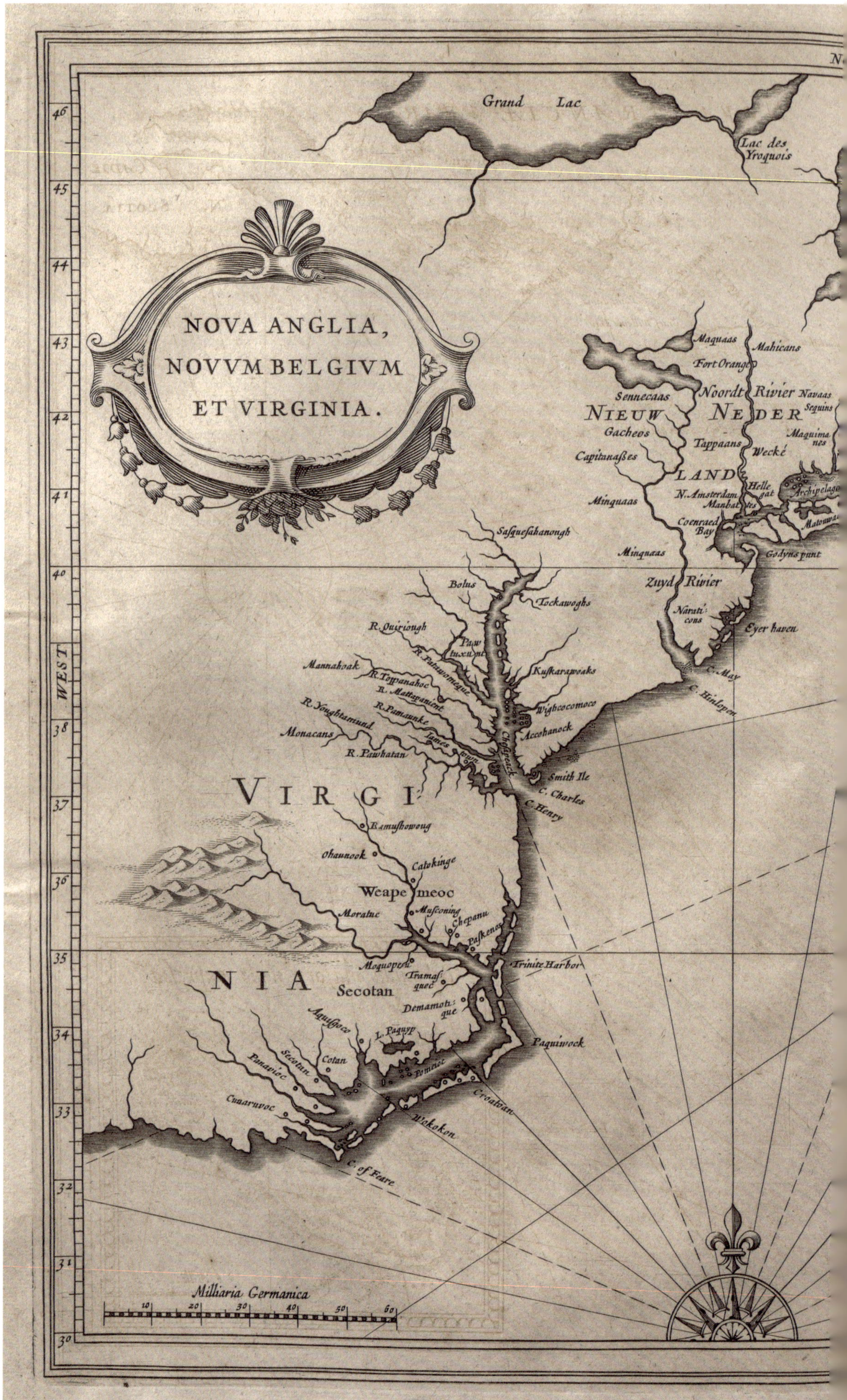

Nova Anglia, Novum Belgium et Virginia
Map of the east coast of North America
From: Johannes De Laet, *Novus orbis seu Descriptionis Indiae occidentalis*, 1633
ANTWERP, THE PHOEBUS FOUNDATION

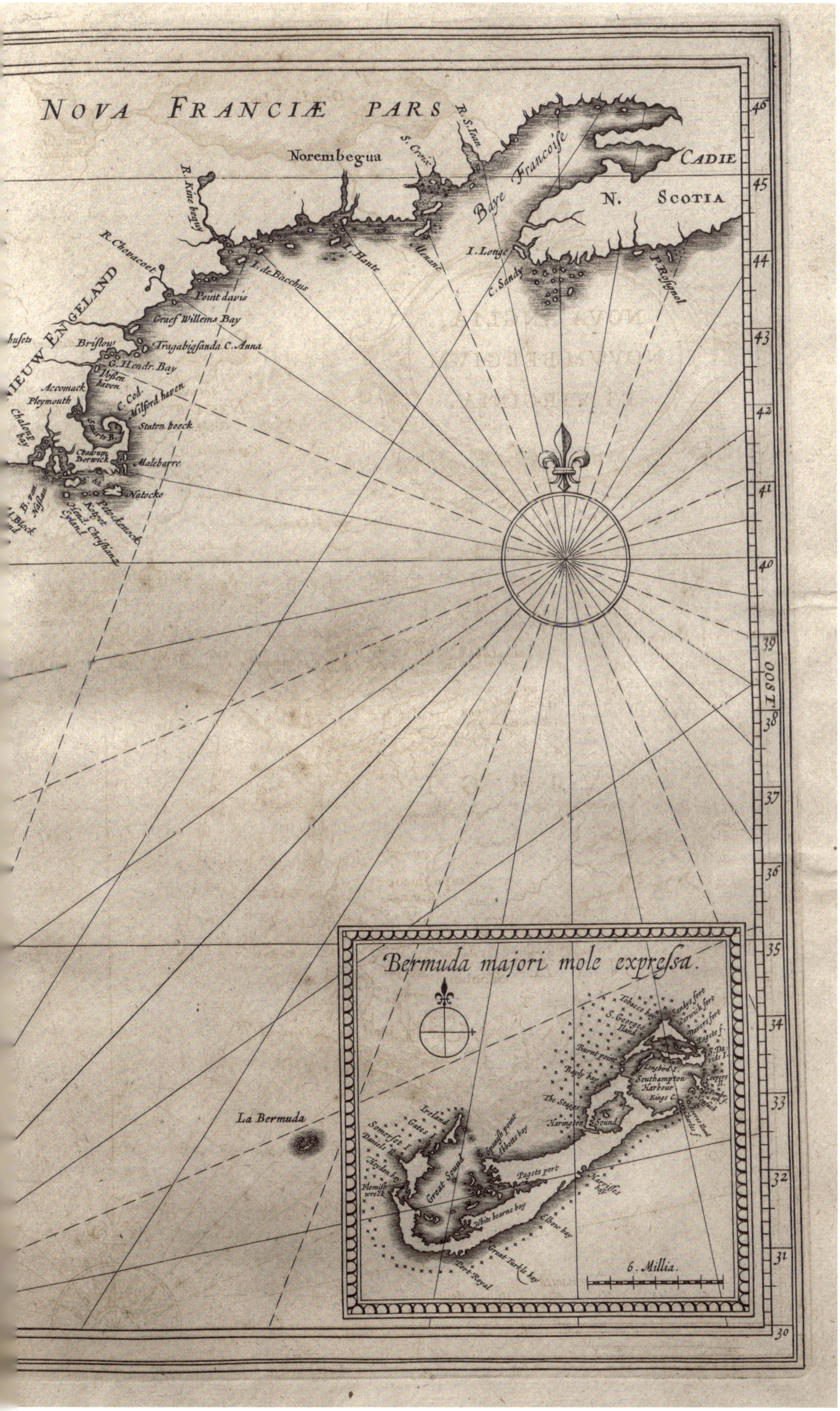
NOVA FRANCIÆ PARS
Norembegua
CADIE
N. SCOTIA
Baye Francoise
I. Longe
C. Sandy
I. Hante
I. de Bacchus
Pount davis
Graef Willems Bay
NIEUW ENGELAND
Bristow
G. Hendr. Bay
Trugabigsanda C. Anna
Accomack
Pleymouth
C. Cod.
Milford haven
Staten hoeck
Malebarre
Natocke
La Bermuda
Bermuda majori mole expressa.
Great Sound
Southampton Harbour
Pagets fort
Port Royal
6. Millia.

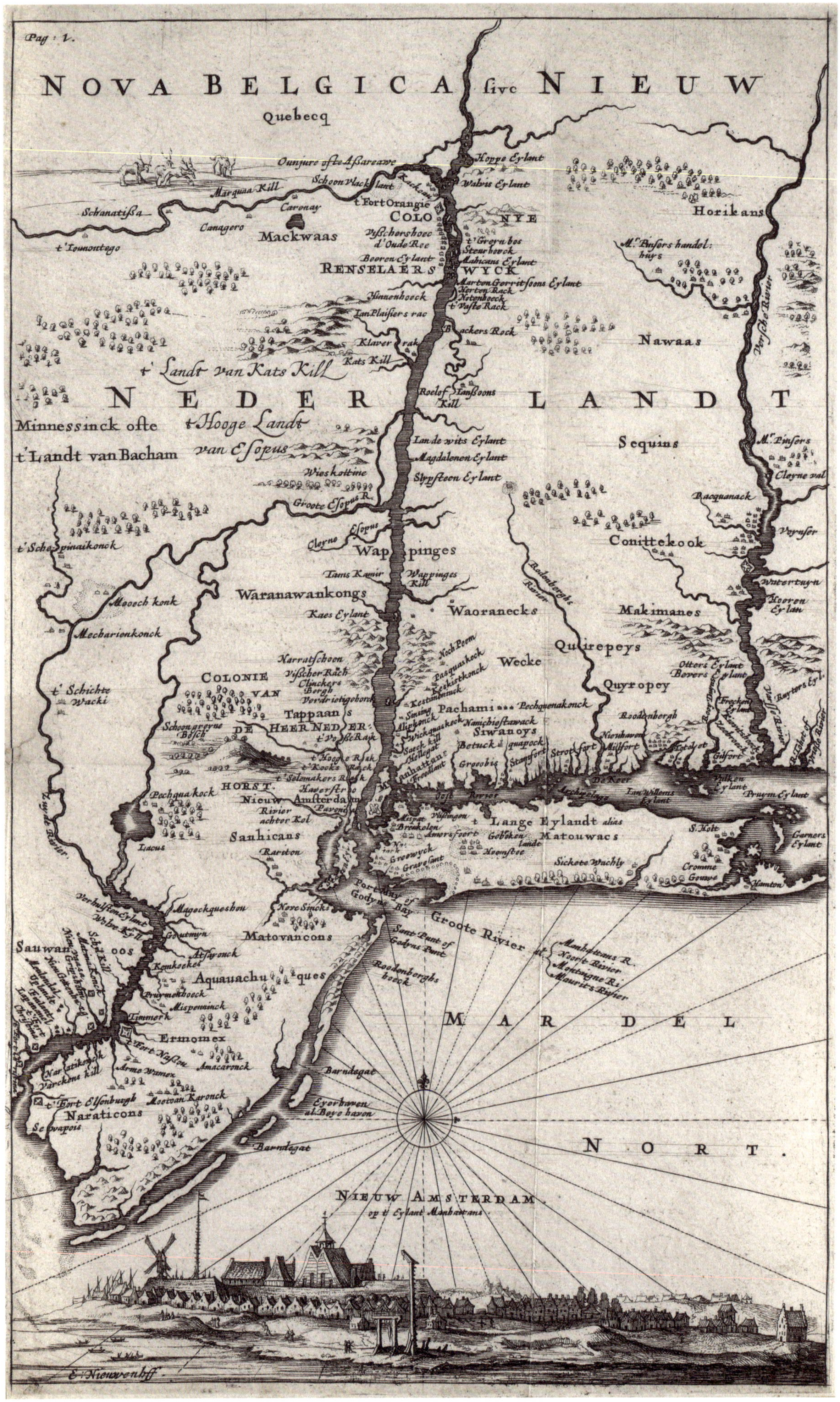

Evert Nieuwenhof
Nova Belgica sive Nieuw Nederlandt, 1656
Map of New Amsterdam
Copper engraving and etching, 330 × 380 mm
ANTWERP, THE PHOEBUS FOUNDATION

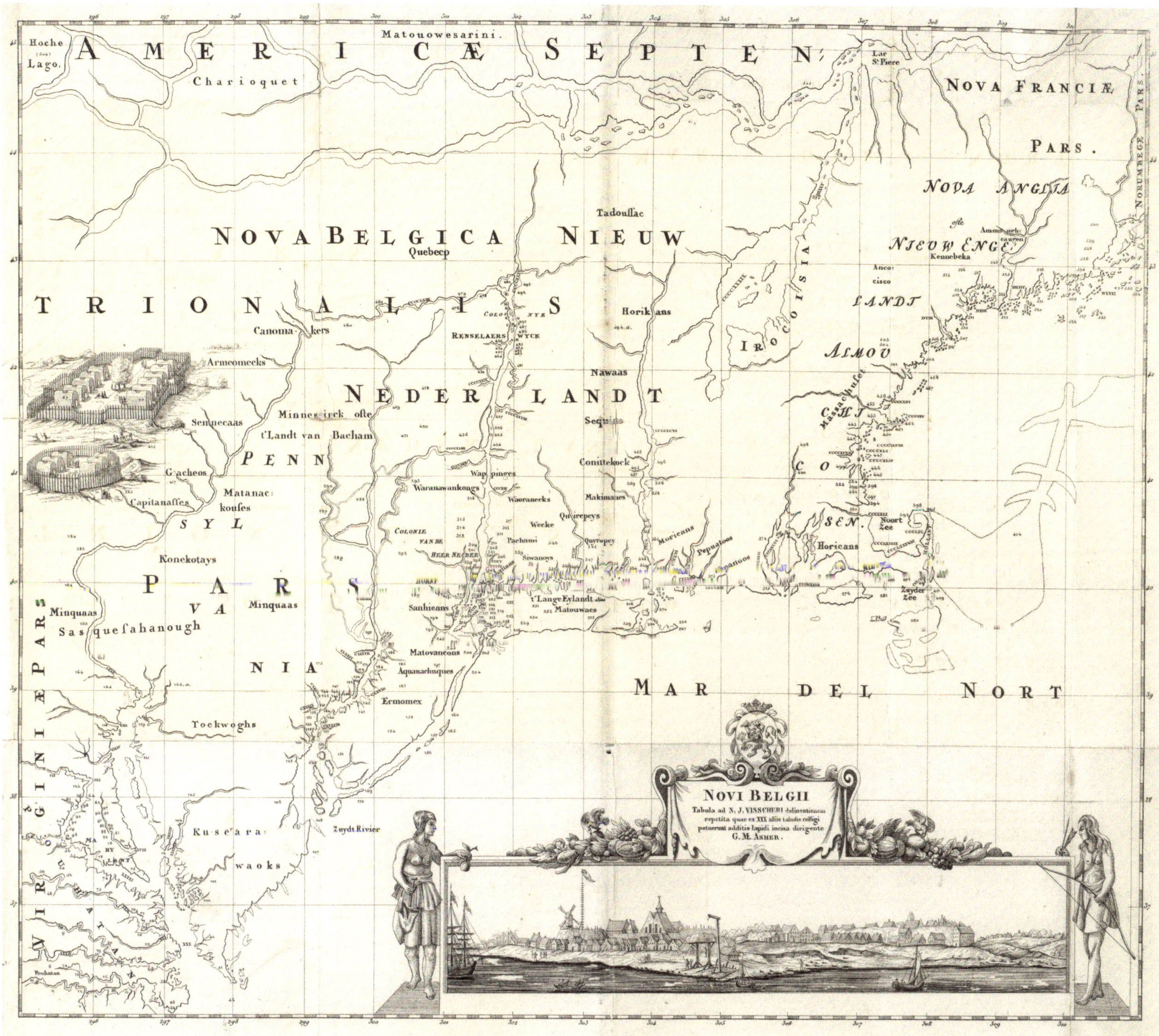

G. M. Asher after Nicolaas Visscher II
Novi Belgii, c.1854
Map of the east coast of North America, with an inset view of New Amsterdam
Copper engraving and etching, 500 × 595 mm
ANTWERP, THE PHOEBUS FOUNDATION

Disaster at sea

It goes without saying that by no means every maritime voyage turned out successfully. Not infrequently, ships ran into storms; were attacked by rivals, enemies or pirates; or were wrecked. While many of the accounts promoted by the Dutch East India Company preferred to focus on the success and glory their voyages achieved, there was an avid readership too for reports of less fortunate incidents, which offered readers back home a combination of thrills and illicit pleasure in others' misfortune.

Such adventures were offered, among others, by the Amsterdam publishers Jacob van Meurs (*c.*1618-1679) and Johannes van Someren (1632-1678). In 1675, for instance, they published an account of the disastrous voyage by the vessel *Ter Schelling*, which ran aground off the coast of Bengal (present-day India and Bangladesh). A raft carrying survivors washed up on a desert island, where they suffered from starvation and faced cannibalism. Frans Jansz. van der Heiden, one of the group members, wrote an account of their experiences.

Shortly afterwards, Van Meurs and Van Someren published the journal kept by Jan Jansz. Struys (1630-1694) during a voyage to Asia. Struys was also known as an adventurer who continued to set off on new campaigns despite all the setbacks and danger he had suffered. His personal account of his journey revels in violent storms, shipwrecks, plunder, capture, slavery, starvation, threats and all the other tribulations he experienced along the way. It was not only success stories, therefore, that sold well: misadventure, failure and catastrophe – often implausibly exaggerated – were clearly lapped up by readers too. Struys's book includes several views and maps of the places in the Middle East that he visited.

In 1751, the general and sea captain Hendrik Cornelis Steenis was dispatched by the Dutch government on a military mission to the Mediterranean and the coast of North Africa. His warship *Het Huys in 't Bosch* was caught in a violent storm and ran aground off the Moroccan coast, upon which the crew was imprisoned and enslaved for a year. Their evocative story was published in the form of a richly illustrated travel report.

Detail from: The mariners are stranded on an uninhabited island
From: Frans Jansz. van der Heiden, *Vervarelyke schip-breuk van't Oost-Indisch jacht Ter Schelling, onder het landt van Bengale*, 1675
ANTWERP, THE PHOEBUS FOUNDATION

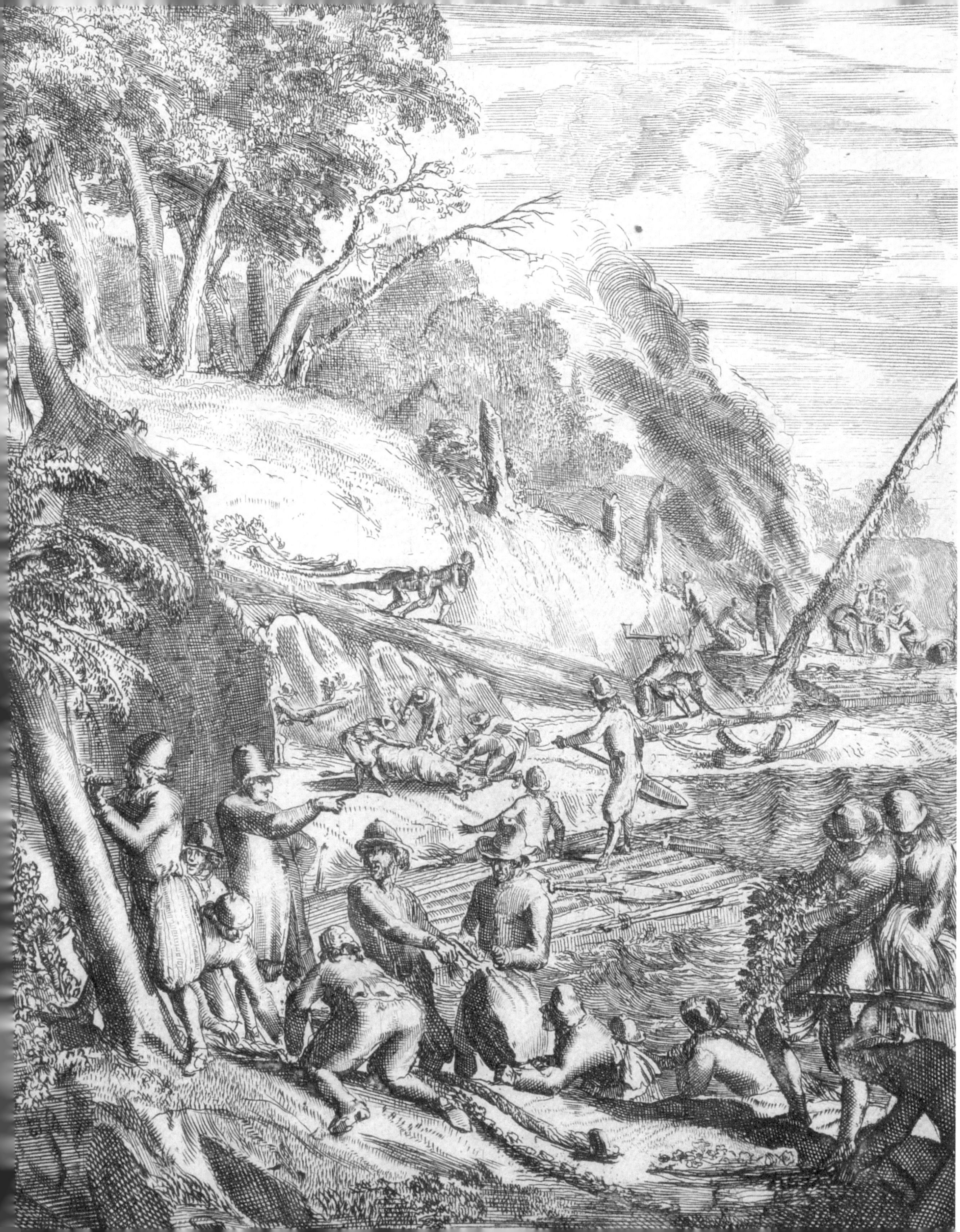

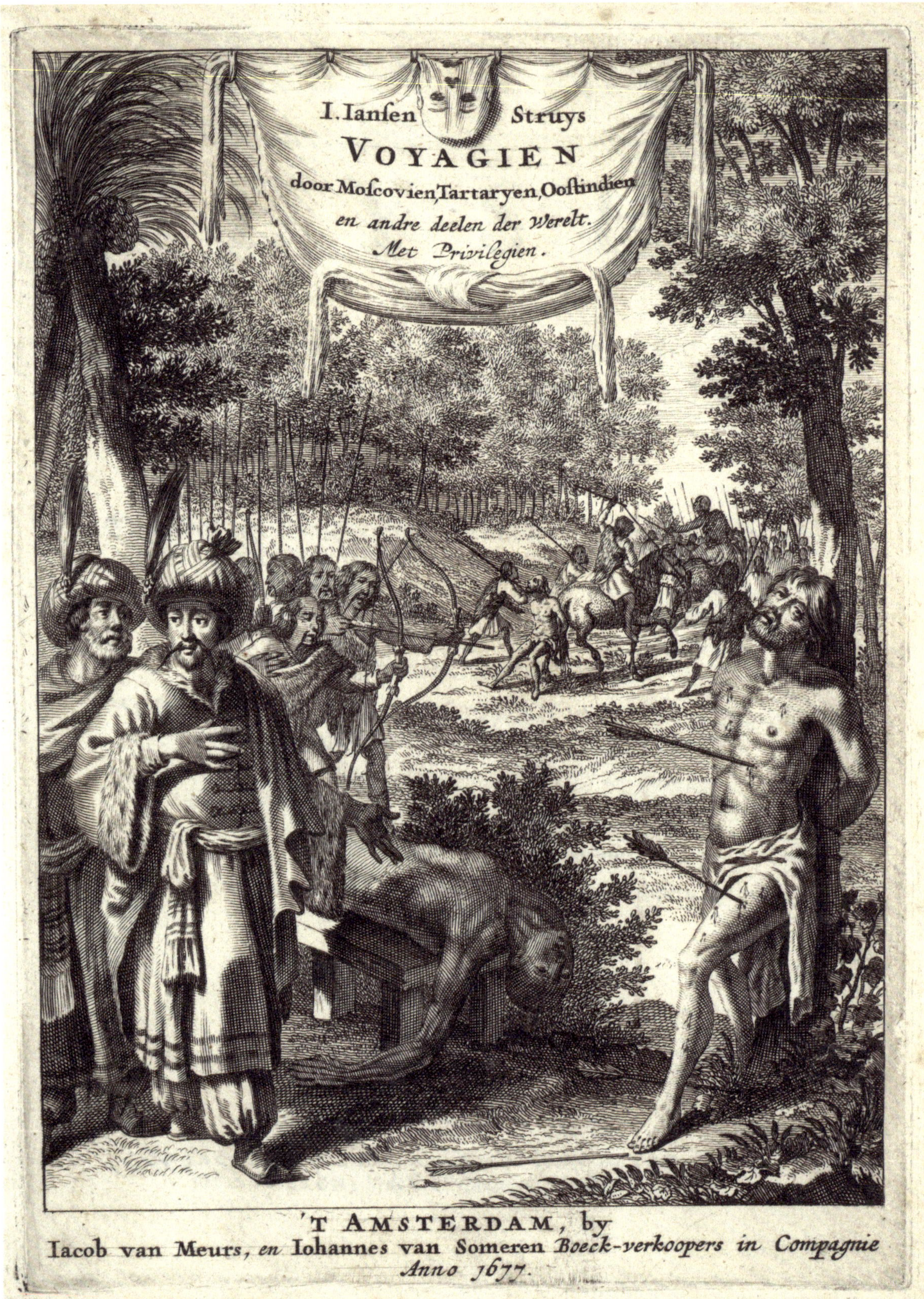

Title page
From: Jan Jansz. Struys, *Drie aanmerkelijke en seer rampspoedige reysen, door Italien, Griekenlandt, Lijflandt, Moscovien, Tartarijen, Meden, Persien, Oost-Indien, Japan, en verscheyden andere Gewesten*, 1676
ANTWERP, THE PHOEBUS FOUNDATION

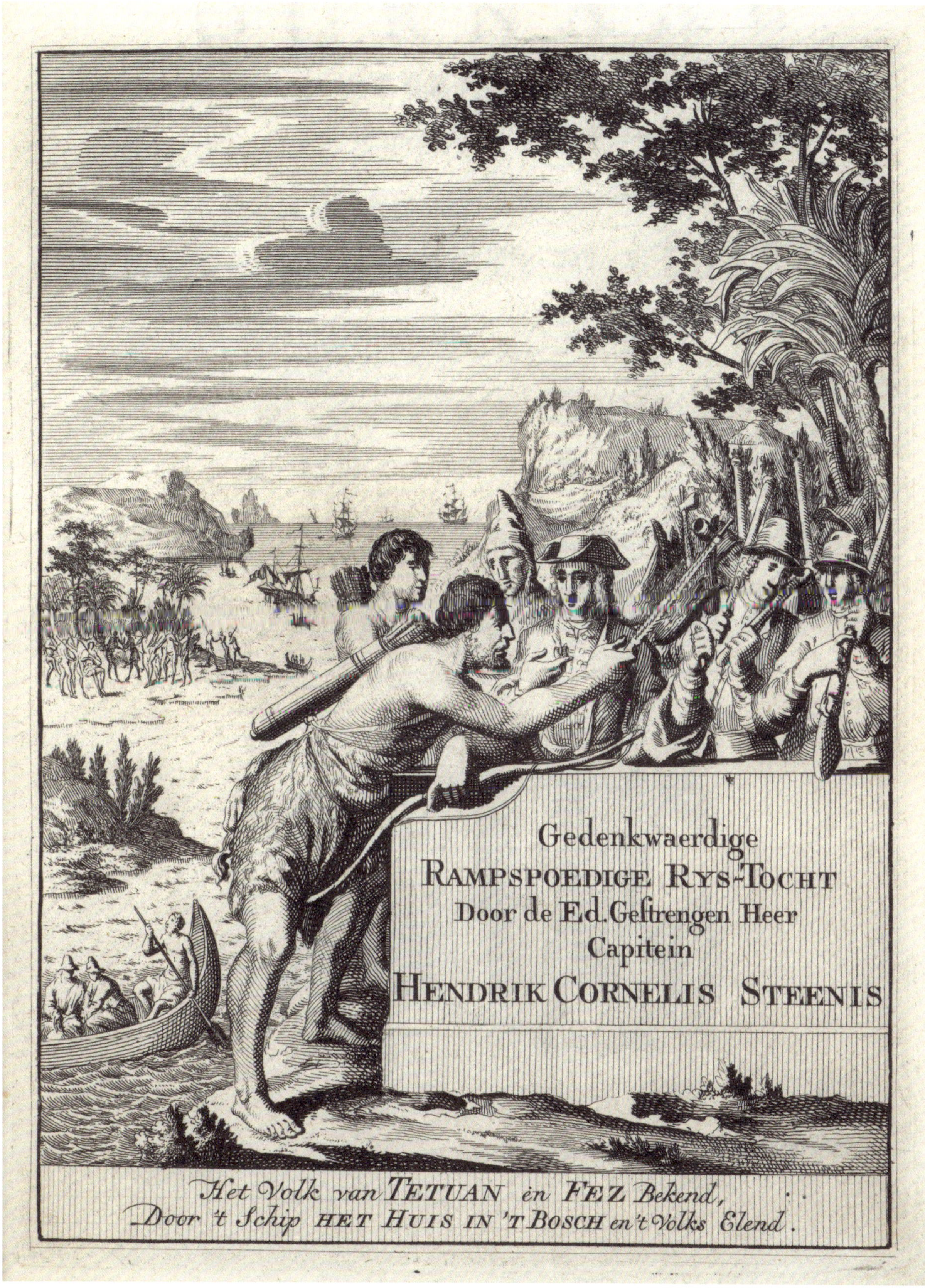

Title page
From: Hendrik Cornelis Steenis, *Journaal wegens de rampspoedige reys-tocht [...] met het oorlogschip genaamt het Huys in 't Bosch, gestrand op de Moorsche Kunst in Afrika, tusschen Ceuta en Kaap Porkes*, 1753
ANTWERP, THE PHOEBUS FOUNDATION

LITERATURE

P. Burden, *The Mapping of North America: a List of Printed Maps*, Rickmansworth, 1996.

J. De Koning, 'From Van Der Donck to Visscher', *Mercator's world*, 5, 4, 2000, 28-33.

H. Devisscher & W. Wauters (ed.), *In Search of Utopia: Art and Science in the Era of Thomas More*, Leuven, 2016.

J. Easterbrook, 'Cosmopolitanism and Adriaen van Der Donck's A Description of New Netherland', *Early American Literature*, 49, 1, 2014, 3-36.

J. Jacobs, *New Netherland: A Dutch Colony in Seventeenth-Century America*, Leiden, 2005.

A. Johnson, *America Explored: A Cartographical History of the Exploration of North America*, New York, 1974.

J. Parmentier, *De wereld gespiegeld: wereldkaarten van de middeleeuwen tot nu*, Kontich, 2015.

V. Roeper & D. Wildeman, *Ontdekkingsreizen van Nederlanders (1590-1650)*, Utrecht, 1993.

V. Roeper, D. Wildeman & K. Cramwinckel, *Reizen op papier: journalen en reisverslagen van Nederlandse ontdekkingsreizigers, kooplieden en avonturiers*, Amsterdam, 1996.

G. Schilder, *Early Dutch Maritime Cartography. The North Holland School of Cartography (c. 1580-c. 1620)*, Leiden / Boston, 2017.

G. Schilder & H. Kok, *Sailing Across the World's Oceans. History & Catalogue of Dutch Charts Printed on Vellum 1580-1725*, Leiden / Boston, 2019.

G. Schilder & M. van Egmond, 'Maritime Cartography in the Low Countries during the Renaissance', in: D. Woodward (ed.), *The History of Cartography, vol. 3. Cartography in the European Renaissance*, Chicago / London, 2007, vol. 2: 1384-1432.

R. Shirley, *The Mapping of the World: Early Printed World Maps, 1472-1700*, Riverside, CT, 2001.

R. Shorto, *The Island at the Center of the World: The Epic Story of Dutch Manhattan and the Forgotten Colony That Shaped America*, New York, 2004.

C. Van Zandt, *Brothers among Nations: The Pursuit of Intercultural Alliances in Early America 1580-1660*, Oxford, 2008.

P. Whitehead & M. Boeseman, *A portrait of Dutch Brazil. Animals, plants and people by the artists of Johan Maurits of Nassau*, Amsterdam / Oxford / New York, 1989.

K. Zandvliet, 'Mapping the Dutch World Overseas in the Seventeenth Century', in: D. Woodward (ed.), *The History of Cartography, vol. 3. Cartography in the European Renaissance*, Chicago / London, 2007, vol. 2: 1433-1462.

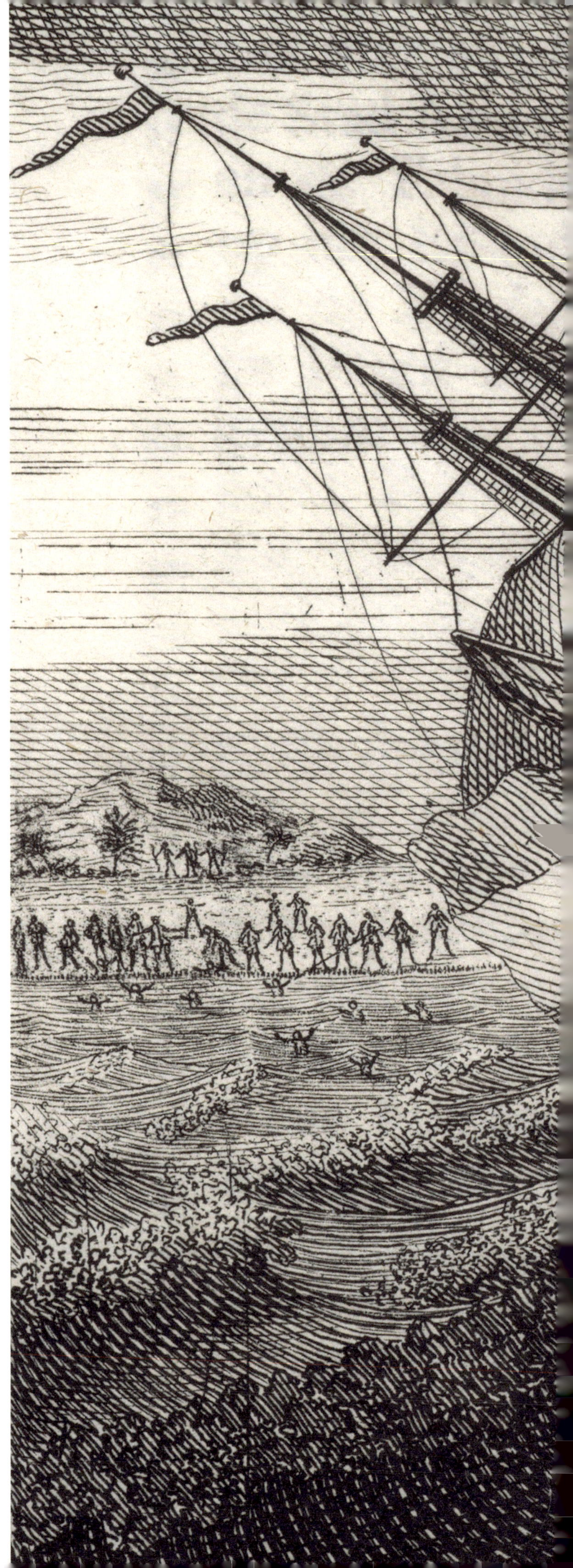

Detail from: *'t Geweld der winden werpt 's lands oorlog schip op strand, en 't vry gevochten volk in hunne vyands hand*
The battleship Huys in 't Bosch is shipwrecked
From: Hendrik Cornelis Steenis, *Journaal wegens de rampspoedige reys-tocht [...] met het oorlogschip genaamt het Huys in 't Bosch, gestrand op de Moorsche Kunst in Afrika, tusschen Ceuta en Kaap Porkes*, 1753
ANTWERP, THE PHOEBUS FOUNDATION

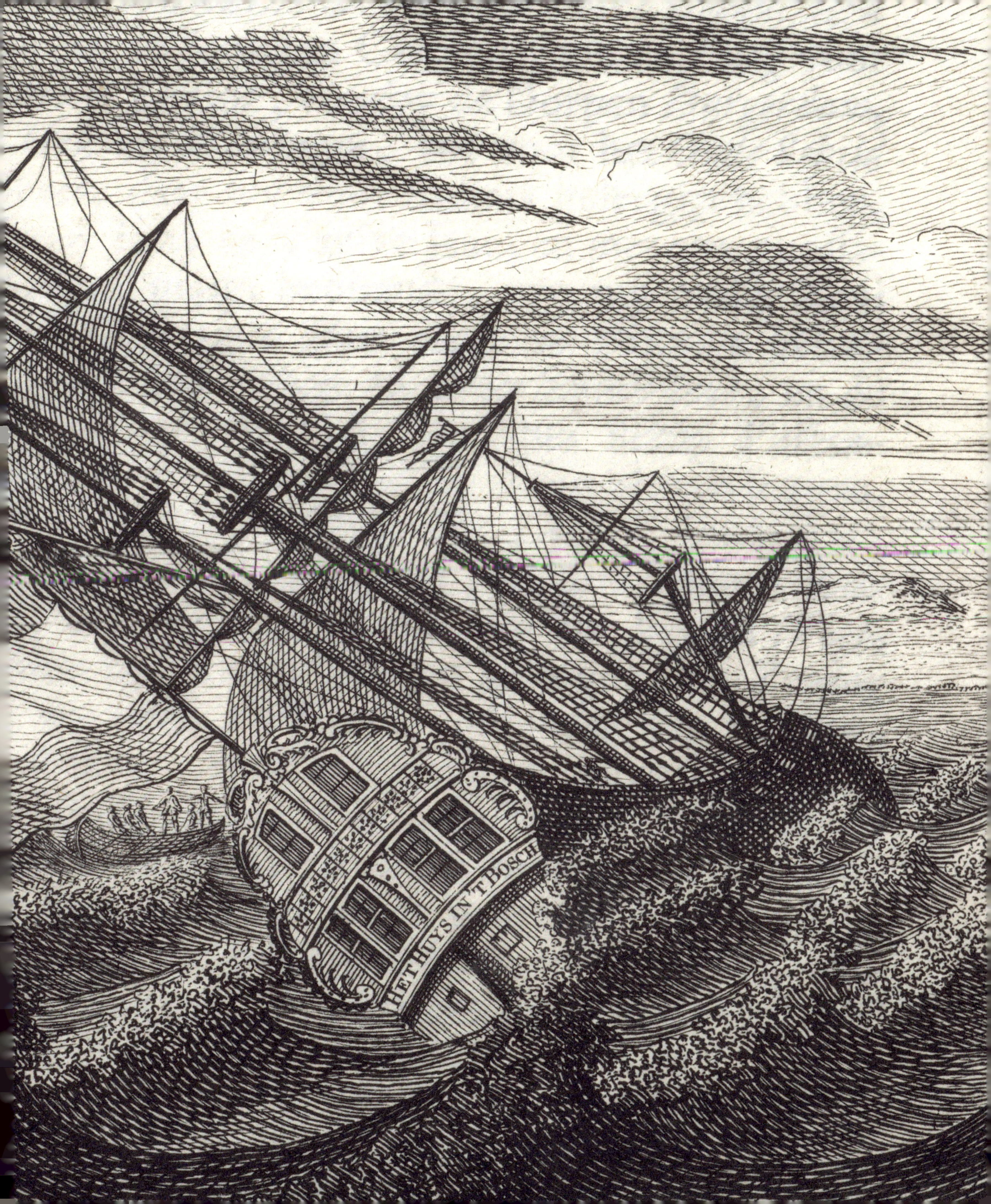
HET HUYS IN 'T BOSCH

You don't need to travel the globe to create a world map: images of the earth were made even when little of its actual geography was known. These were entirely conceptual and schematic, or else consisted simply of terrain that *had* been identified. Tens of thousands of years ago, cave dwellers were already drawing parts of the landscape and stars on rock walls. A Babylonian surveyor later incised the world in his clay tablet in the shape of a star, with the part he inhabited shown as a disc in the middle, surrounded by water. The ancient Greeks, meanwhile, believed that the earth was shaped like a drum.

Every world map has a logic of its own. Nowadays, we have Google Earth, in which every inch of the planet has seemingly been captured and turned into a pixel. We can wander endlessly among the serene plains of the most extensive deserts before immersing ourselves moments later in the crowds thronging the streets of the biggest metropolises. The earliest maps in The Phoebus Foundation's collection were intended to show the 'whole' world. They date from the late fifteenth century and were strongly influenced by the Ptolemaic view of the world and by medieval *mappae mundi*.

The invention of printing – or, more accurately, the rediscovery of moveable type – by Johannes Gutenberg (*c.*1400-1468) in the German town of Mainz in the mid fifteenth century played an important part in disseminating this view of the earth. The first cities in Europe in which books were printed were also German, among them Ulm, Nuremberg and Münster, which we discuss in this chapter. They were joined in the following century by the Low Countries. As empirical science continued to develop in the early modern era and sea voyages ranged further and further afield, the prevailing image of the world had to be constantly updated. While cartographers and geographers improved the map of the world, old myths about distant lands and peoples proved stubbornly persistent. For many years, therefore, new discoveries and insights continued to exist alongside traditional and religious ideas.

CHAPTER IX

WORLDS OF DIFFERENCE

MAPS OF THE ANCIENT, MEDIEVAL AND MODERN WORLD

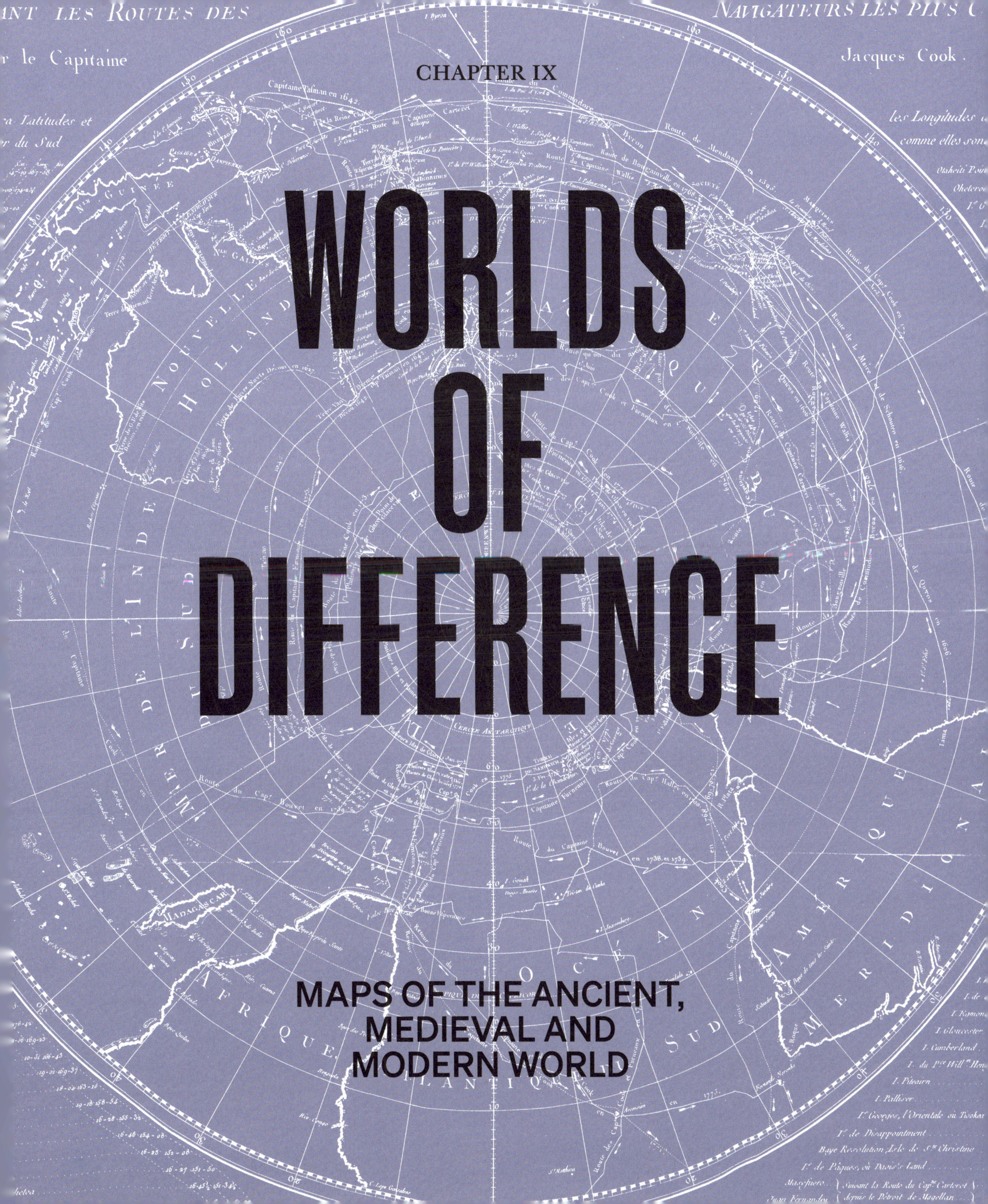

All roads lead to Rome

The ‘Peutinger map’, created by the ancient Romans, is the only known world map from that era. It shows the Roman Empire around the reign of Caesar Augustus (until 14 CE) and does so on a remarkable scale: the entire map is said to have run to twelve sheets with a total length of about seven metres and a width of thirty centimetres, showing the network of roads across the then known world, from Britain in the West to the Ganges in the East. It was thus more of a road map than anything else. The original, which continued to be updated until the fourth century, is believed lost, but medieval copies have survived. The oldest copy still extant, made by a thirteenth-century French monk, found its way into the possession of the sixteenth-century German collector and humanist Konrad Peutinger (1465-1547), whose name it now bears. Johannes Janssonius made his own version of the Peutinger map for his atlas of the ancient world, the *Accuratissima orbis antiqui delineation*, published from 1652 onwards.

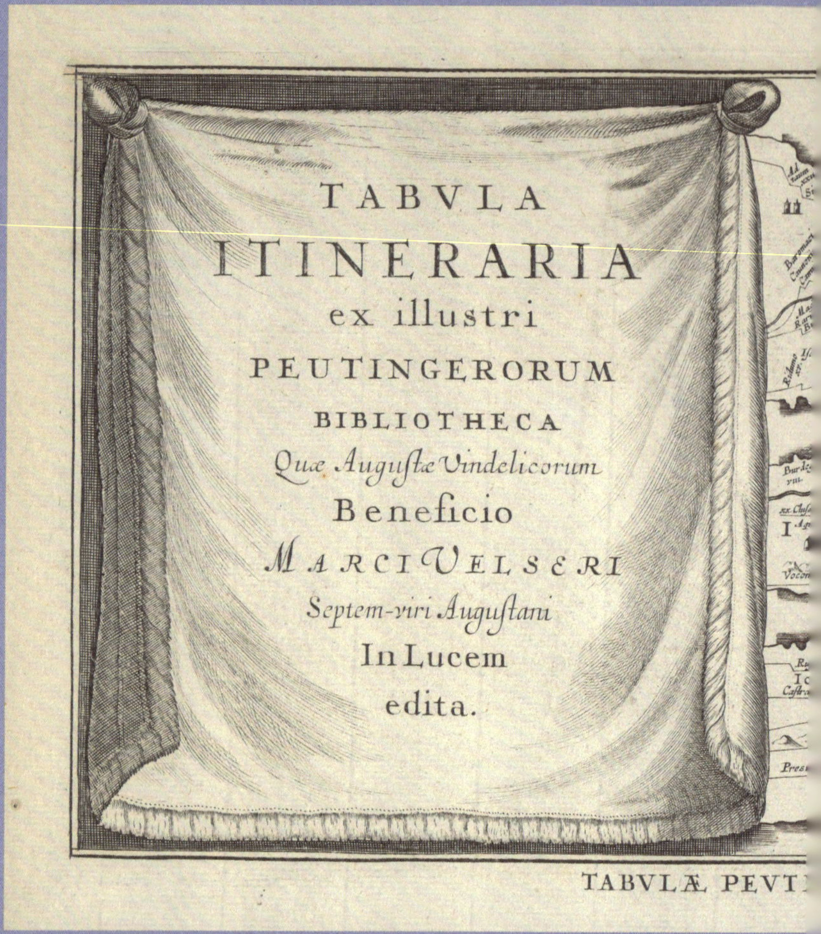

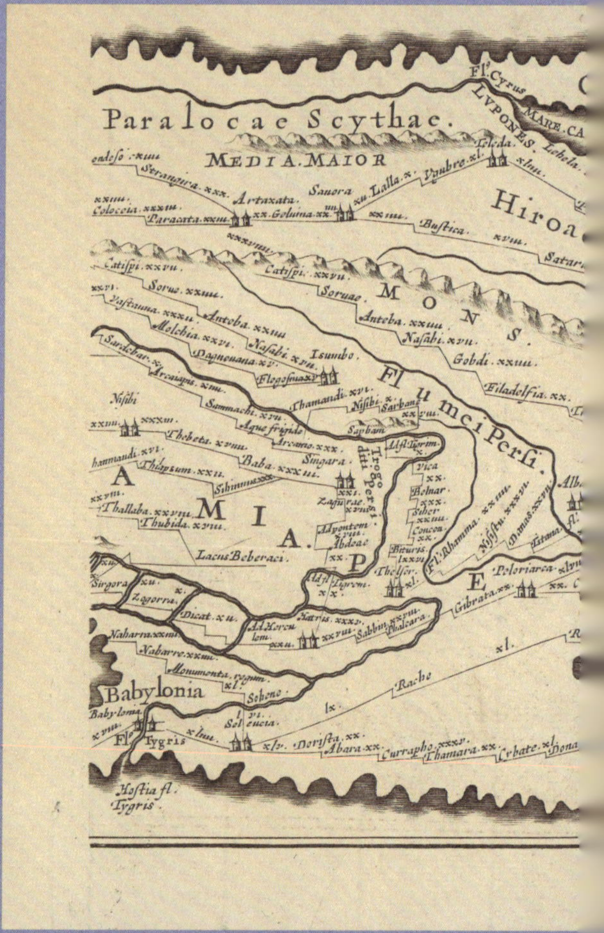

Johannes Janssonius after Marcus Welser, after Konrad Peutinger
Tabula itineraria ex illustri Peutingerorum, 1652
The first segment and last segment of the Peutinger map (total of eight segments)
Copper engraving, 400 × 510 mm (segment) / 470 × 580 mm
ANTWERP, THE PHOEBUS FOUNDATION

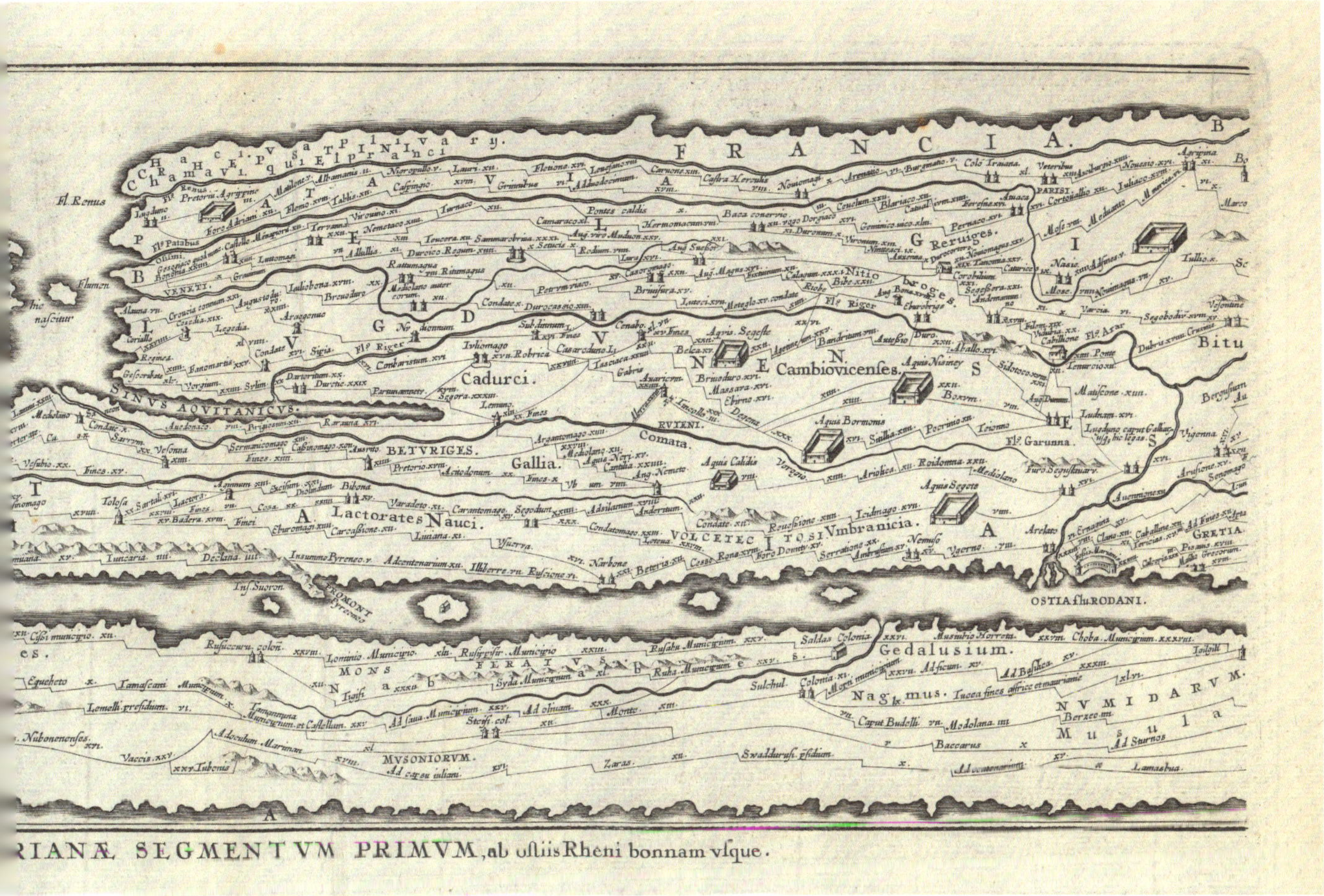
FRANCIA.
Cadurci.
Cambiovicenses.
Lactorates Nauci.
Gallia.
Comata.
Vimbrancia.
SINVS AQVITANICVS.
BETVRIGES.
Gedalusium.
NVMIDARVM.
MVSONIORVM.
OSTIA flu. RODANI.
...RIANÆ SEGMENTVM PRIMVM, ab ostiis Rheni bonnam vsque.

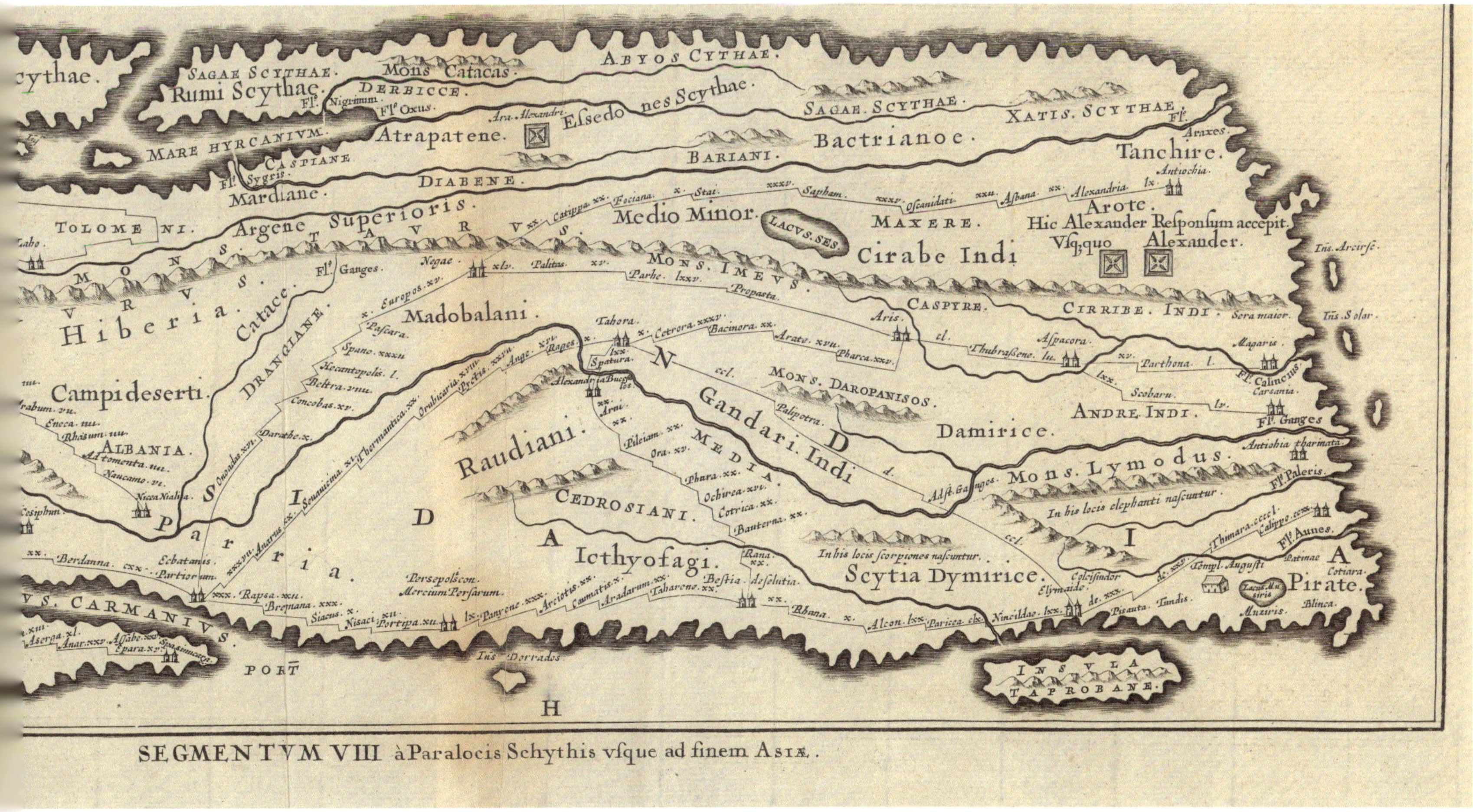
ABYOS CYTHAE.
Rumi Scythae
Mons Catacas
DERBICCE.
MARE HYRCANIVM.
CASPIANE
Atrapatene.
Essedo nes Scythae
SAGAE SCYTHAE.
XATIS SCYTHAE
Bactrianoe.
BARIANI.
Tanchire.
Mardiane
DIABENE.
TOLOME NI.
Argene Superioris.
Medio Minor
LACVS SES
MAXERE.
Cirabe Indi
Arote.
Hic Alexander Responsum accepit.
Vsq quo Alexander.
Hiberia.
Catace.
Madobalani
MONS IMEVS.
CASPYRE.
CIRRIBE. INDI.
Campideserti.
DRANGIANE
MONS DAROPANISOS.
ALBANIA.
Raudiani.
Gandari Indi
ANDRE INDI.
Damirice.
Mons Lymodus.
CEDROSIANI.
In bis locis elephanti nascuntur.
Icthyofagi.
Paria.
In his locis scorpiones nascuntur.
Scytia Dymirice.
Pirate.
CARMANIVS.
INSVLA TAPROBANE.
SEGMENTVM VIII à Paralocis Schythis vsque ad finem Asiæ.

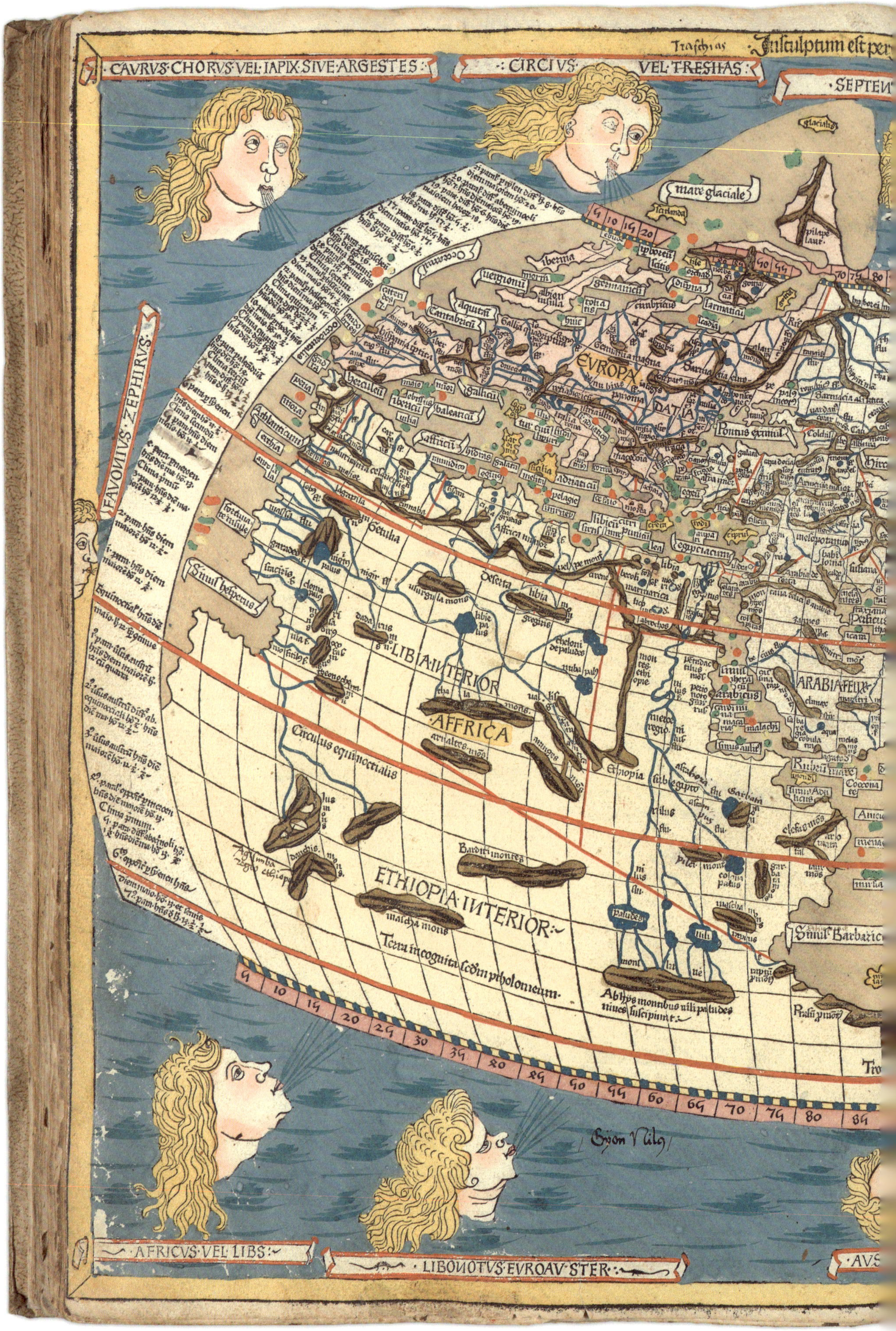

World map
From: Nicolaus Germanus, *Cosmographia*, 1486
ANTWERP, THE PHOEBUS FOUNDATION

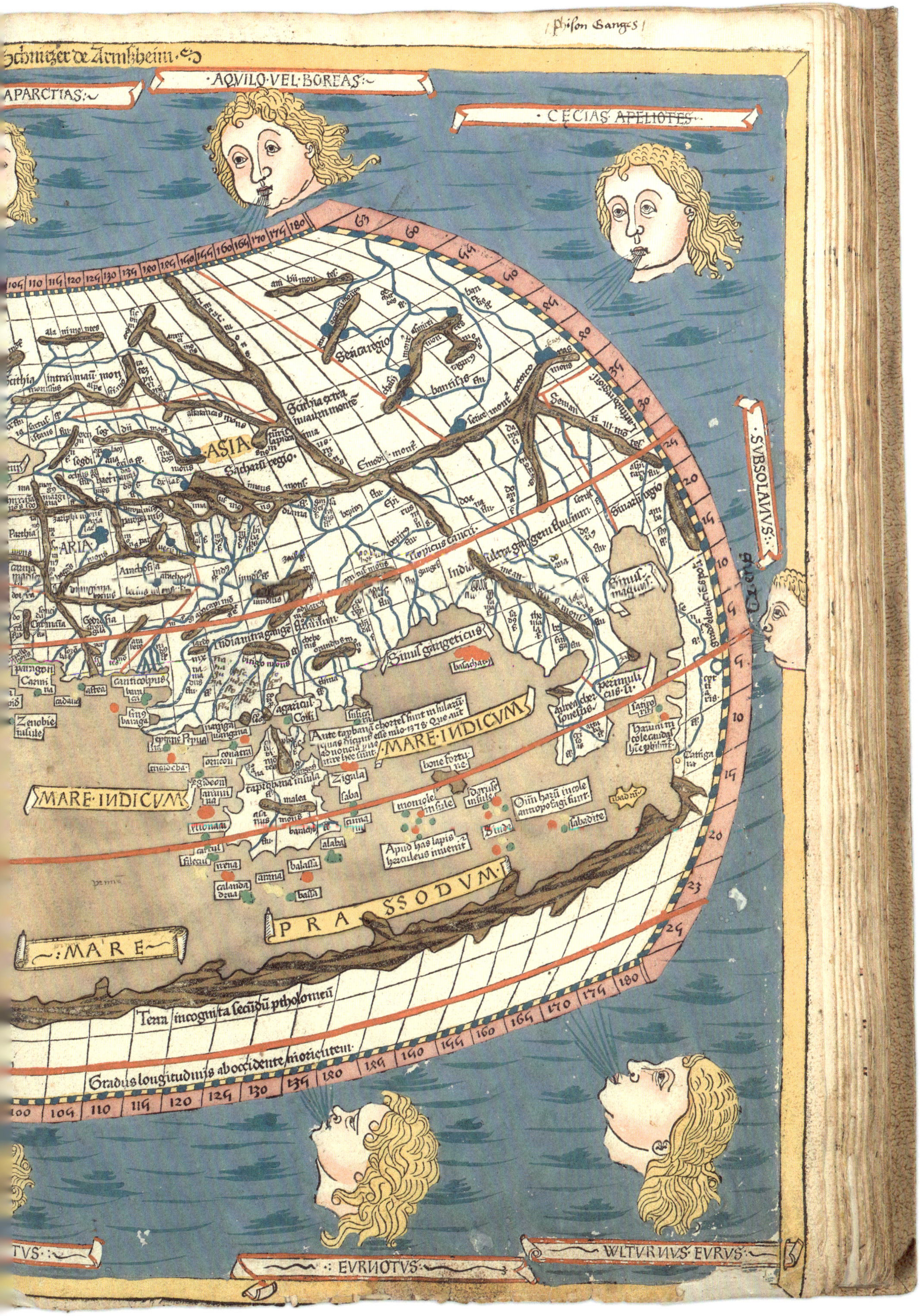
AQVILO·VEL·BOREAS
APARCTIAS
CECIAS APELIOTES
SVBSOLANVS
ASIA
MARE·INDICVM
MARE·INDICVM
PRA SSODVM
MARE
EVRNOTVS
VLTVRNVS·EVRVS

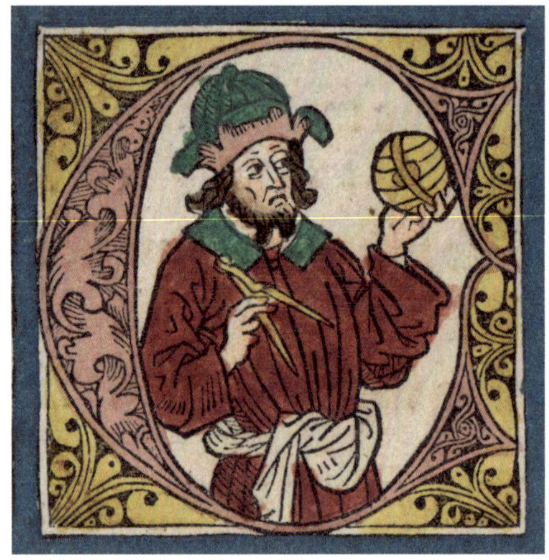

Portrait of Ptolemy
From: Nicolaus Germanus, *Cosmographia*, 1486
ANTWERP, THE PHOEBUS FOUNDATION

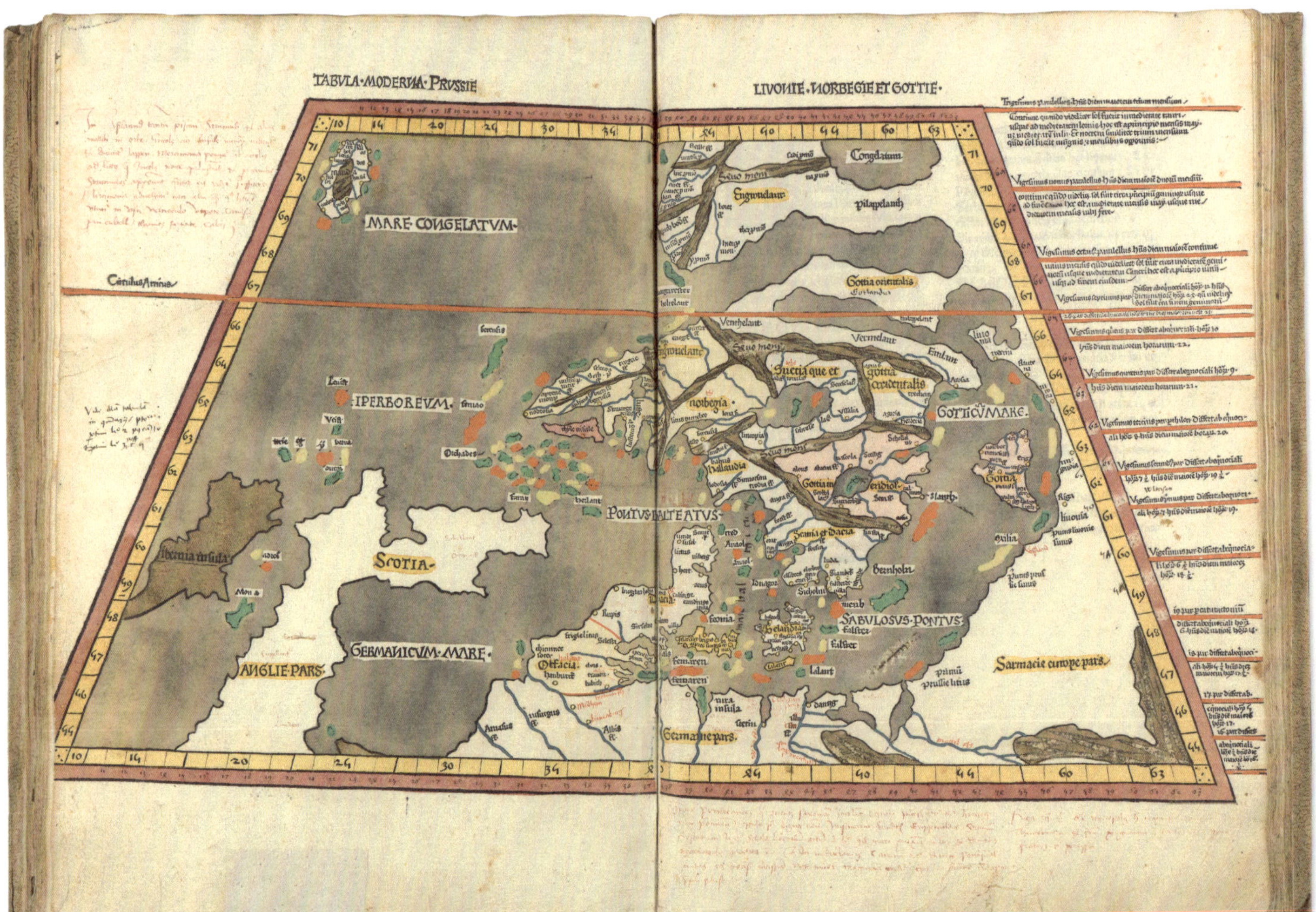

Tabula Moderna Prussie, Livonie, Norvegie et Gottie
Map of Northern Europe, Scandinavia and the British Isles
From: Nicolaus Germanus, *Cosmographia*, 1486
ANTWERP, THE PHOEBUS FOUNDATION

PTOLEMY'S WORLD

Claudius Ptolemy (*c*.100-*c*.170) was a Greek scholar, mathematician and astronomer who lived in Alexandria. Sometime around 150, he wrote his famous *Geographia*, which later editors also titled *Cosmographia*. A comprehensive, geographical description of the classical world, it is also considered one of the first ever atlases. Ptolemy did not devise the content as such but built instead on the work of ancient predecessors like Marinus of Tyre (who came up with the idea of meridians and parallels), Strabo and Hipparchus. Ptolemy's work was more a summary of centuries of geographical thinking in ancient Greece. An important aspect was the world view he proclaimed: to his mind, the world did not consist of a concept or idea of how it ought to be but was determined by what was known. In this way, he allowed scope for that view to be expanded and supplemented with fresh information, which is precisely what his many followers went on to do.

Ptolemy's original texts no longer exist, nor do we know whether he personally drew up maps to accompany his treatise. Later copies of his work (the earliest surviving examples of which date from the twelfth to the thirteenth centuries) contain instructions for making maps, a list of around eight thousand locations and their coordinates, and a foundational map of the world. The text experienced a major revival in the Renaissance, when it was translated into Latin and then retranslated and disseminated in other languages too. A variety of scholars produced modified versions illustrated with woodcuts.

The Phoebus Foundation has a copy of the second Ulm edition (first edition: Ulm, 1482). The German scholar Nicolaus Germanus (*c*.1420-1490) modernised Ptolemy's text to include several new projections, additional maps, and other innovations. The book is an incunabulum – a 'cradle book', as early printed books were known – in that it was published shortly after the invention of the printing press in 1450. In some respects, it still resembles a manuscript: an owner at the time annotated it in Latin in red and black ink in the margins and sometimes in the map itself: the word 'London', for instance, has been written inside the map of England. Traces of use like this make the copy unique.

The book contains thirty-two woodcut maps, the same as the 1482 first edition, but the texts on the reverse have been reset. The world map expresses the fifteenth-century view of the world: America had yet to be 'discovered' by Europeans and so does not feature. On the other hand, this is the first Ptolemaic map in which the rudimentary outlines of Scandinavia can be seen in a protruding lobe in the north, above Europe. The Phoebus Foundation's copy was richly coloured at the time it was produced.

The world map by the German scholar Martin Waldseemüller (*c*.1470-1520) comes from his edition of the *Geographia*, published in Strasbourg. The 1513 woodcut is one of the earliest maps to show the Americas. The outlines of the 'New World' were drawn following Columbus's voyages in 1492-1504 (⟶ Chapter VIII). Waldseemüller had already published a large twelve-sheet world map in 1507 which is thought to be the first to feature the name 'America', a reference to Amerigo Vespucci, who had 'discovered' the Atlantic coastline of South America in 1499-1501. Interestingly, his later map shows America in a more sketchy, 'unfinished' manner, as if he had begun to doubt its shape.

His fellow countryman Sebastian Münster (1488-1552) also published adapted editions of Ptolemy. The Phoebus Foundation has a copy of the first edition, *Geographia universalis, vetus et nova* (Basel, 1540). Münster revised the translations and the Ptolemy maps already in circulation and added several new ones: roughly half of the maps are said to be by Ptolemy, the other half by himself. The book contains a map of the Ptolemaic world and one of the modern world. The latter does now show America, albeit in an odd form (North America in particular). It also marks a northwest passage to the Moluccas that was still believed to exist at that time. It is the first printed map to name the 'Pacific Ocean' as such. Münster himself was working on his own *Cosmographia* at the time (first printing: Basel, 1544).

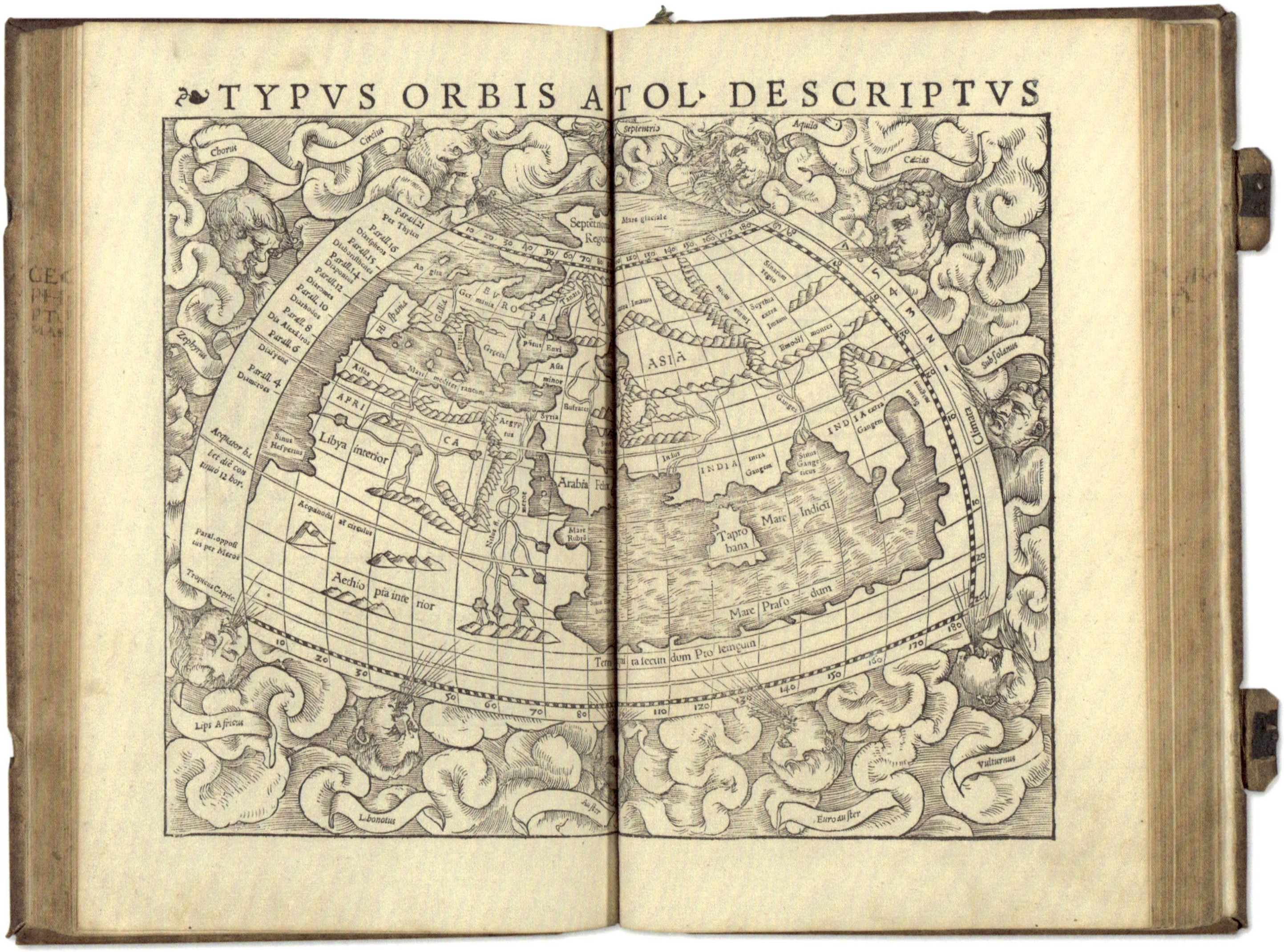

Sebastian Münster after Claudius Ptolemy
Typus Orbis a Ptol. Descriptus
World map after Ptolemy
From: Sebastian Münster, *Geographia universalis, vetus et nova*, 1540
ANTWERP, THE PHOEBUS FOUNDATION

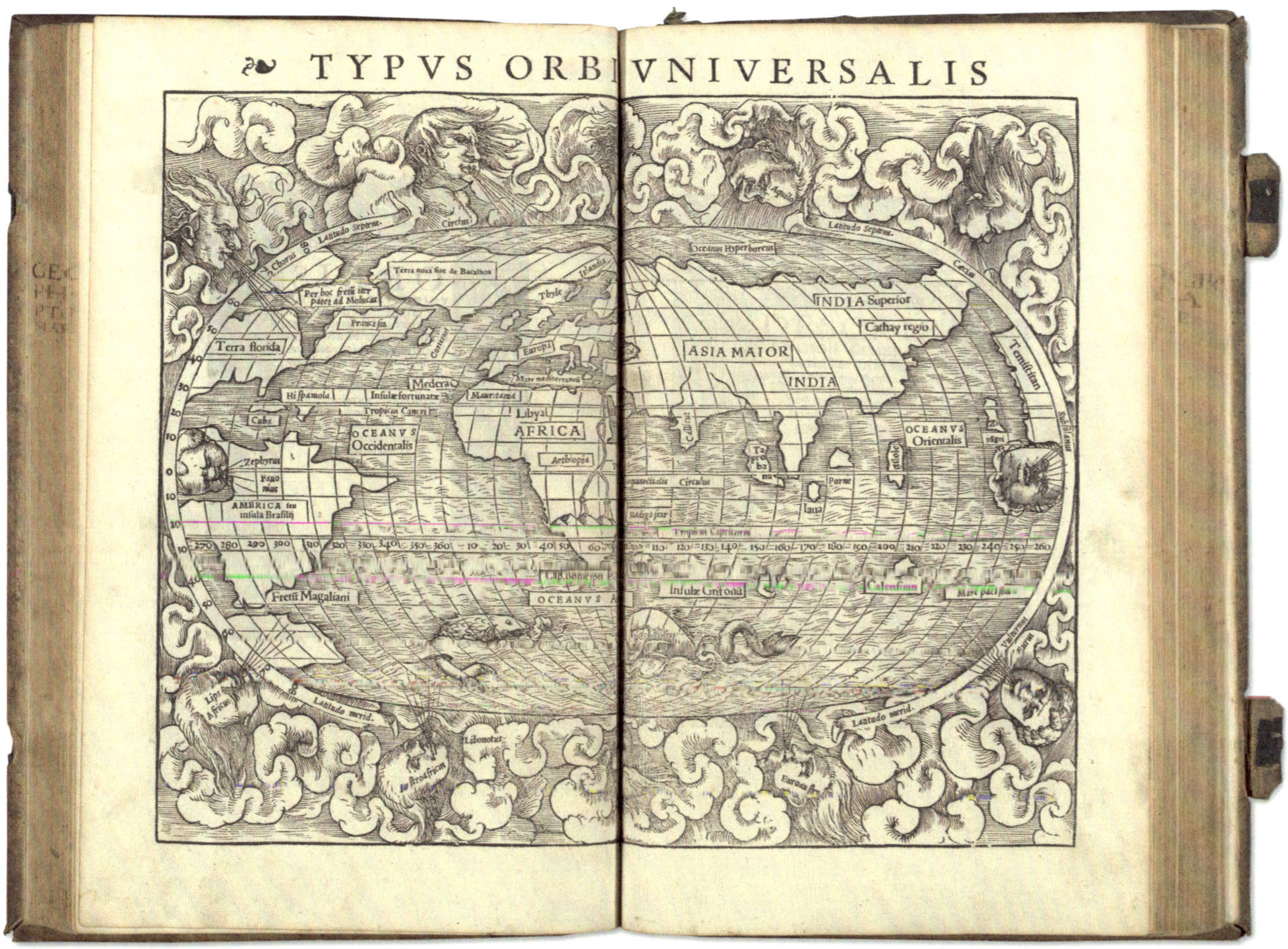

Sebastian Münster after Claudius Ptolemy
Typus orbis universalis
World map
From: Sebastian Münster, *Geographia universalis, vetus et nova*, 1540
ANTWERP, THE PHOEBUS FOUNDATION

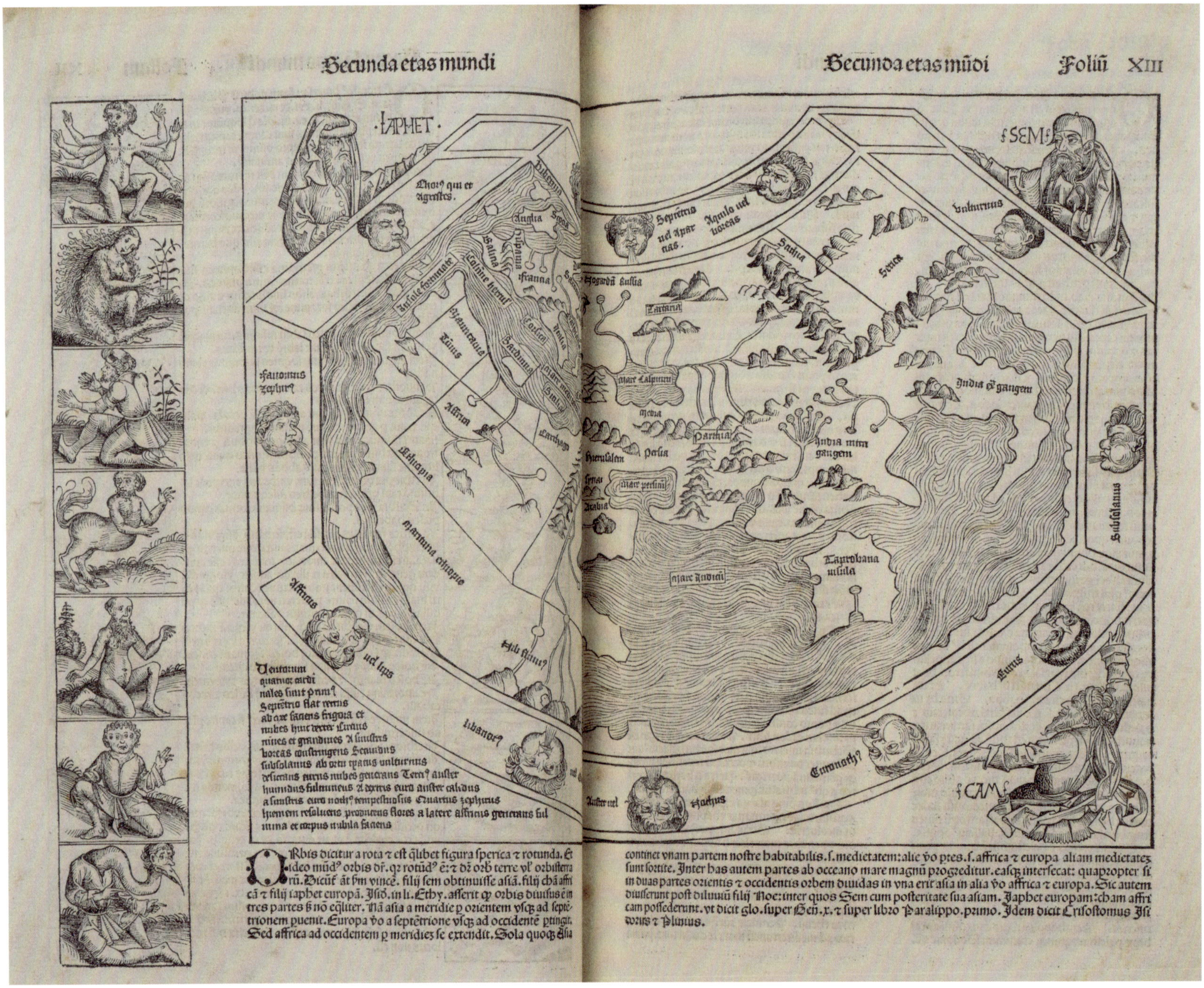

World map
From: Hartmann Schedel, *Liber Chronicarum*, 1493
ANTWERP, THE PHOEBUS FOUNDATION

TIME TRAVELLING ROUND THE WORLD

The Nuremberg physician and humanist Hartmann Schedel (1440-1514) was another who built on Ptolemy's work. Inspired by other authors, he went further, putting a spin on it entirely of his own. He completed his masterpiece, the *Liber Chronicarum* ('Book of Chronicles'), in 1493 with the financial backing of his Nuremberg patrons Sebald Schreyer and Sebastian Kammermeister. The scholar created a narrative picture of the world from the distant past to the present and even the future. Schedel discussed the history of the world in six epochs, by analogy with the Six Days of Creation, to which he added a further two: his own era as the seventh and the end of the world as the eighth – the future. The Bible account served as a guide but is interwoven with secular, classical history and myths from the Arab world.

The late-fifteenth-century incunabulum is a huge tome – immense in its influence and fame – but also in its format. It runs to as many as 1,800 woodcuts of widely varying places, mostly city views (albeit many of which did not match reality). There is also a map of central and northern Europe, and a world map for which Ptolemy's map served as the model, resulting in the Indian Ocean being enclosed in the same way and there being no America nor any ocean to the east of Asia. Noah's sons Shem, Ham and Japheth – the three Old Testament progenitors, representing the continents of Asia, Africa and Europe respectively – look on from the corners. They belong to Hartmann Schedel's second global epoch. Especially striking are the images of strange figures in the border on the left: fabulous creatures that were believed, based on travellers' tales and the fables of classical authors like Pliny and Herodotus, to live in the furthest corners of the earth. We make out a centaur, for instance, while the six-armed man recalls the Hindu gods; the four-eyed man is thought in turn to come from the stories of Ethiopian tribes.

The world map reflects the book's ambivalent, synthesising content. The 'Nuremberg Chronicle' belongs to the *imago mundi* genre, which blends a historical and geographical description of the world with other forms of knowledge, including the mystical, astrological and magical. No clear distinction is drawn between theology and science: the work is an amalgamation of religion and empiricism, classical and modern, origin and downfall, past and future, the natural and the supernatural. The genre of world history was a humanist monument in which a contemporary, critical spin was given to Christian values and history, but this time with human beings at the centre. Schedel's work offers a glimpse of the modern view that was taking shape of the world and of humanity.

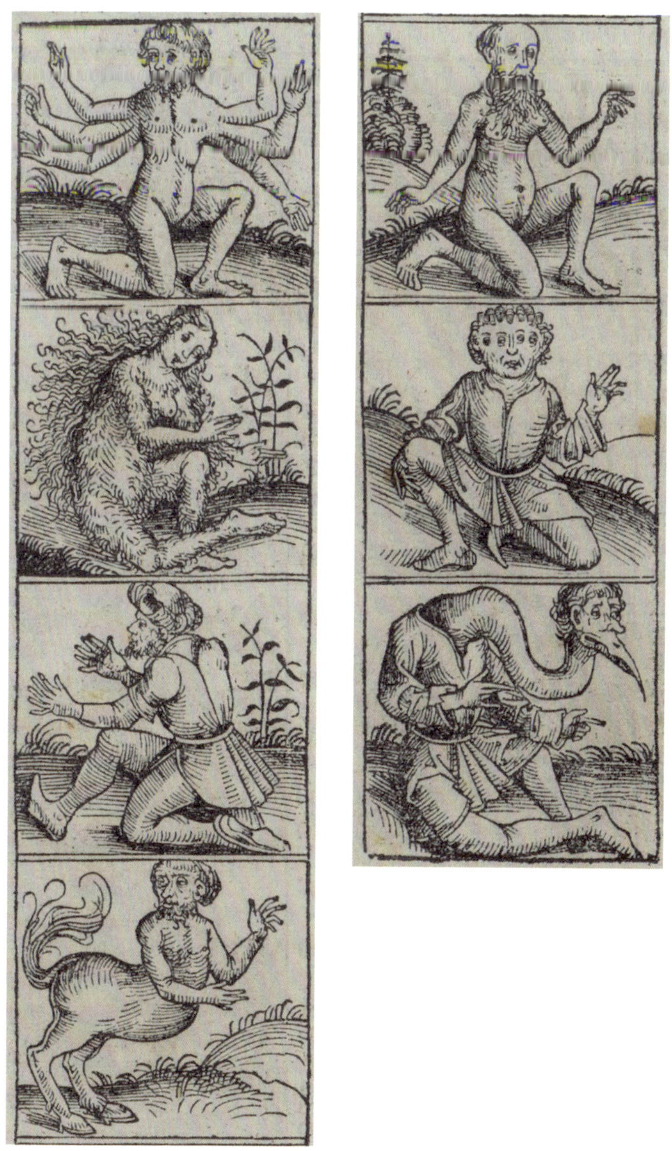

Details from world map
From: Hartmann Schedel, *Liber Chronicarum*, 1493 (p. 250)

MEDIEVAL *MAPPAE MUNDI*

A *mappa mundi* is a medieval world map. Many variants are known, their form and content differing widely. What they do often have in common is that they consist of a complex combination of time and space in which Christian and secular histories are placed within a matrix of topographical connections. One well-known form of *mappa mundi* is the 'T-O map', which splits the world into three zones separated by a T-shaped Mediterranean, with Asia as the upper half. Europe and Africa form the two lower sections. Together, and with the city of Jerusalem at the centre, they are surrounded by a world ocean. The earliest printed world map is a schematic diagram of this kind, first printed at Augsburg in 1472 as a woodcut in the *Etymologiae* of Isidore of Seville (Isidorus Hispansis).

Isidore's conception resonates through other medieval world maps too. It appears in more elaborate form, for instance, in the woodcut T-O map in *La Mer des Histoires* – a French translation of the anonymous *Rudimentum Novitiorum* ('Handbook for Novices'; Lübeck, 1475), which offers a survey of events from world history. The woodcut maps it contains are seen as the first printed world maps to attempt to represent countries and landforms in topographical relationship to each other, albeit underpinned by a powerful Christian medieval tradition. East is placed at the top of the world map, with countries marked as individual hills. Several pictorial elements are incorporated in the map, including the pope in Rome (centre), a number of kings, four rivers flowing out of a walled Garden of Eden (top) and the devil in northern Asia (left).

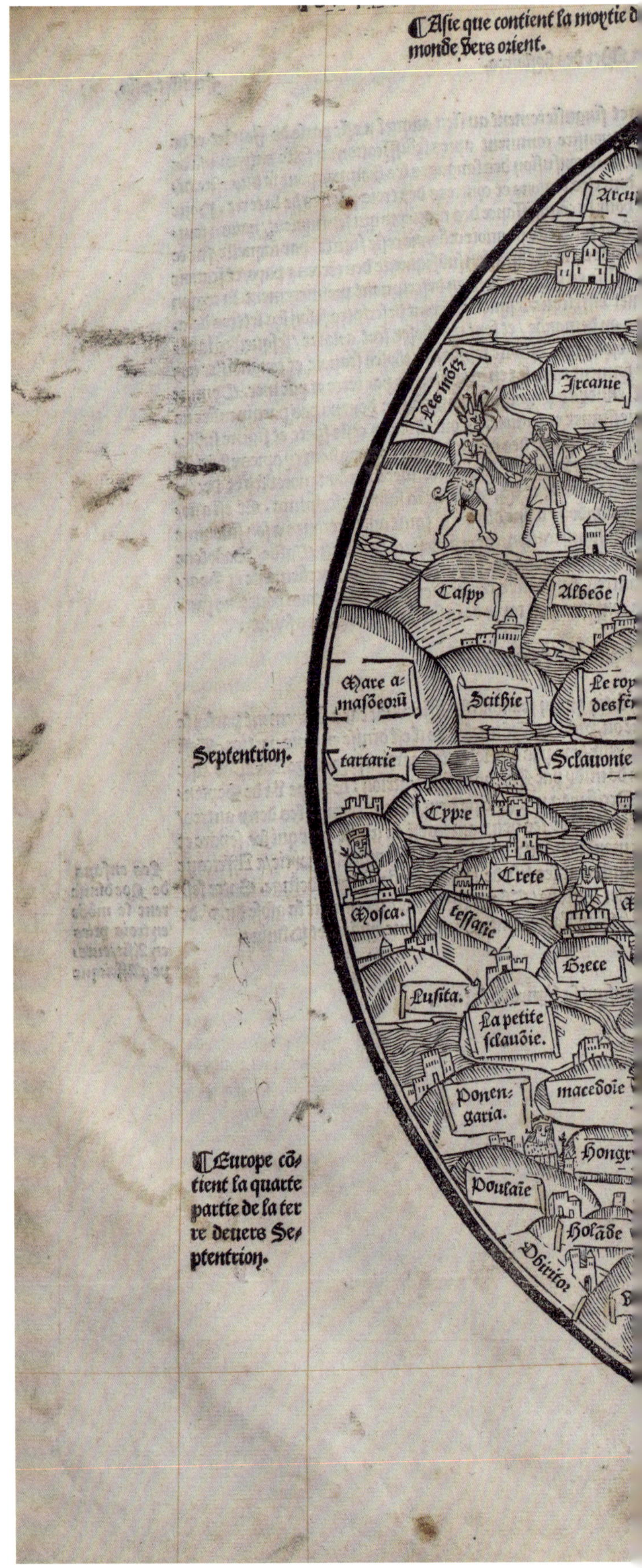

T-O map
From: *La Mer des Histoires*, 1536
ANTWERP, THE PHOEBUS FOUNDATION

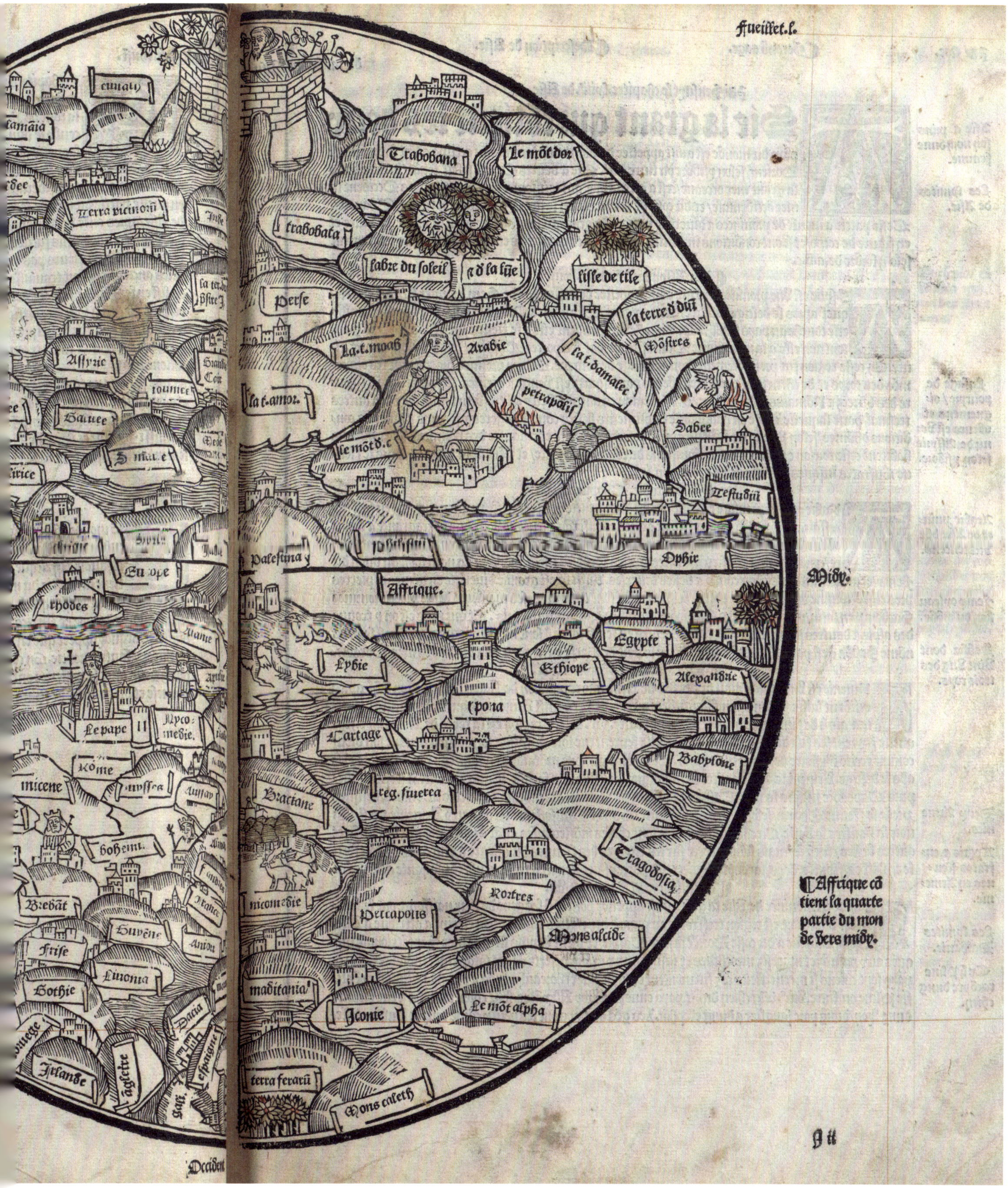
Fueillet.l.
Trabobana
Le mōt dor
trabobata
lisle de tile
Perse
Arabie
Sabee
Palestina
Ophir
Midy.
Affrique.
Egypte
Lybie
Ethiope
Alexandrie
Cartage
Babylone
Bractane
reg. finerea
Tragodosia
nicomedie
Mons alcide
maditania
Iconie
Le mōt alpha
terra ferarū
Assyrie
Europe
rhodes
Le pape
Rome
micene
Frise
Gothie
Irlande
Occident
Affrique cōtient la quarte partie du mon de vers midy.

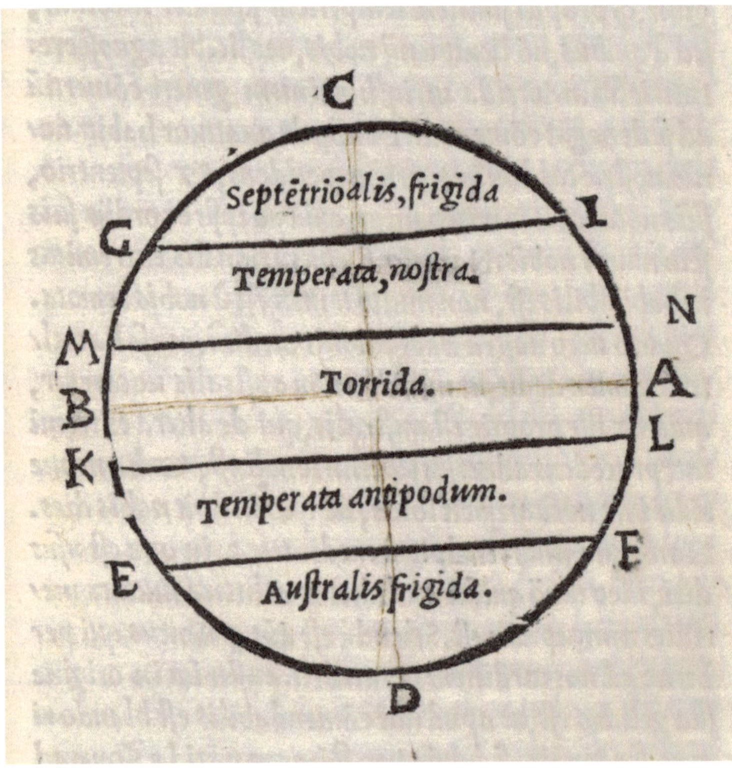

CLIMATE ZONES

Macrobius's zonal world map (fourth to fifth century CE) is a very different example of a *mappa mundi*. The Roman author and philosopher penned a commentary on Cicero's *Dream of Scipio*, which became popular in the Middle Ages. It has survived in an estimated 150 copies, roughly a hundred of which also contain a small map. Humanist authors in the Renaissance adapted the text in turn, as did Niccolo Angeli, who compiled the Latin version of Macrobius's text which was published in Florence in 1515 by Filippo Giunta.

The book includes twee images of the world that are somewhat different again from the T-O type. They are two examples of a zonal map, a diagram of the world in five climate zones, separated in the middle by an ocean at the equator. The central, hot and uninhabitable zone (*perusta* or *torrida*) is divided into uneven parts by the impassable ocean. In the north lies the temperate and inhabited zone (*temperata nostra*), with India to the left and the Mediterranean to the right; the south contains the inaccessible unknown temperate zone (*temperata antipodum nobis incognita*). Due to the cold, the polar regions (*septentrionalis frigida*) are uninhabitable, as is the hot centre. Strikingly, the southern hemisphere consists of a large land mass. Europe is situated on the right in Angeli's version, whereas the original map in the first edition from Brescia in 1483 places it on the left. Another addition to this variant is the island of Great Britain, shown in the top-right corner.

Zonal map
From: Niccolo Angeli after Ambrosius Macrobius, *Hoc volumine continentur Macrobii interpretatio in somnium Scipionis a Cicerone confectum*, 1515
ANTWERP, THE PHOEBUS FOUNDATION

DE SOMNIO SCIPIONIS.

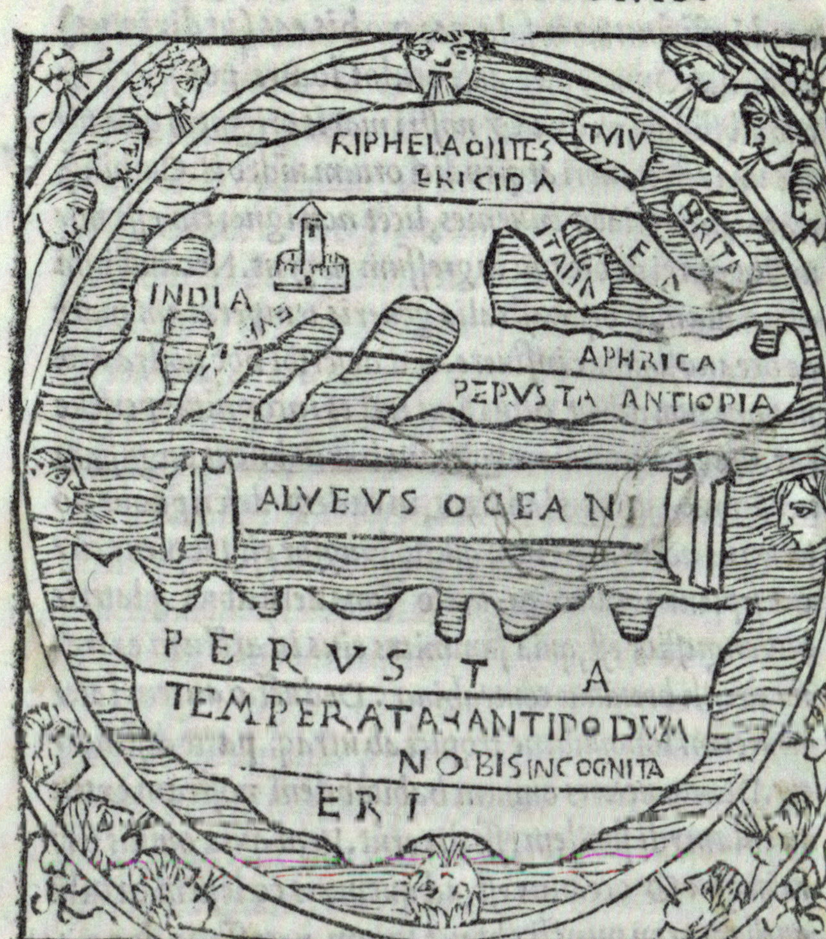

uir fortis intelligat, quæ in tam paruo magna esse nō poterit, quod doctrinæ propositum non minus in sequentibus apparebit, quinetiam si cupiat proles futurorum hominum deinceps laudes uniuscuiusq; nostrum acceptas à patribus posteris prodere, tñ pp eluuiones exustionesq; terrarum, quas accidere tpe certo necesse est, non modo non æternam sed ne diuturnā quidem gloriā assequi possumus. Virtutis fructum sapiens in ꝯscientia ponit, minus perfectus in gloria, unde Scipio perfectionem cupiens infundere nepoti auctor est, ut ꝯtentus cō-

Virtutis fructus

Zonal map
From: Niccolo Angeli after Ambrosius Macrobius, *Hoc volumine continentur Macrobii interpretatio in somnium Scipionis a Cicerone confectum*, 1515
ANTWERP, THE PHOEBUS FOUNDATION

Terra Australis

Ptolemy claimed in his *Geographia* that there was a large unknown land mass in the southern hemisphere, arguing that the weight of such a continent was needed to balance the land masses of Europe and Asia in the northern hemisphere. Without it, a top-heavy world would tip over. Hence the reason that until the seventeenth century mapmakers continued to insert an immense continent at the bottom of the world map, without knowing for sure whether it existed.

In the Ulm edition of the *Geographia*, the world maps contain the text '*terra incognita secundum Ptolomeum*' ('unknown land according to Ptolemy') (⟶ pp. 244-245). On Abraham Ortelius's 1570 world map, this actually became '*terra australis nondum cognita*' ('as yet unknown southern land') (⟶ pp. 262-263). Ortelius expressed his doubts concerning the shape of this southern continent in the map itself, writing above New Guinea: 'recently discovered, but it is not known whether this is an island or part of the southern continent' ('*Nova Guinea nuper inventa que an sit insula an pars continentis australis incertum est*').

Due in part to the discoveries of Abel Tasman (1603-1659) in the seventeenth century, it was realised slowly but surely that New Guinea, Australia and New Zealand were all islands and not part of a larger continent. The Englishman James Cook (1728-1779) made what became known as his 'second voyage' around the southern hemisphere between 1772 and 1775 in search of the mysterious continent. The Phoebus Foundation has the first edition of Captain Cook's complete travel journals in French, printed and published in London. The second part (running to five volumes) covers his travels to Terra Australis. He crossed the Antarctic Circle with a large team of scientists and a crew of almost 200 aboard two ships in the years in question but failed to locate the continent, which he concluded did not exist, and struck it permanently from the map. Although Cook did make a close approach to Antarctica, he did not consider the continent to be a 'discovery'. It was not until 1820 that Russian, American and British explorers reached the coldest continent in quick succession, allowing it finally to be included in Western maps of the world.

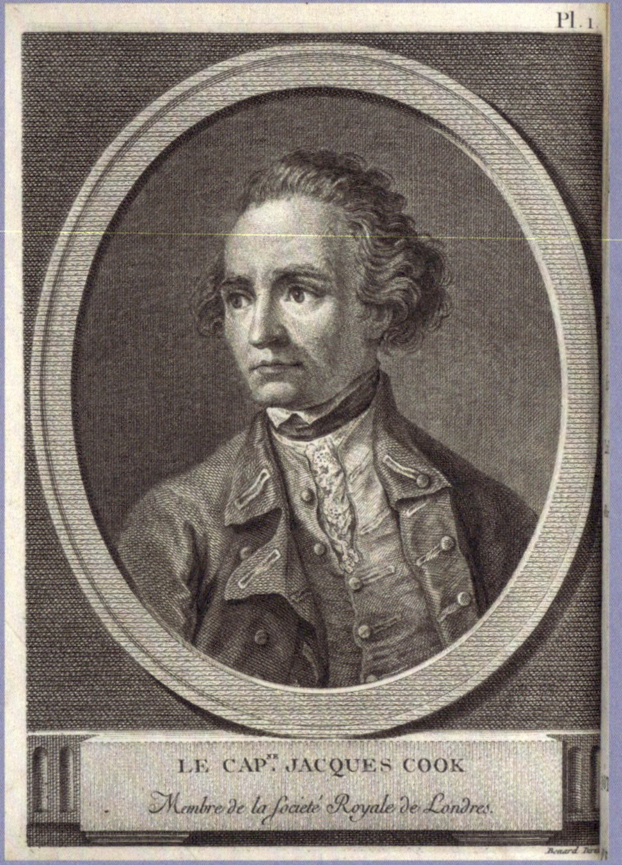

Portrait of James Cook
From: James Cook, *Voyage dans l'hémisphère austral, et autour du monde, fait sur les vaisseaux de [sic] roi, l'Aventure, & la Résolution, en 1772, 1773, 1774 & 1775*, 1778
ANTWERP, THE PHOEBUS FOUNDATION

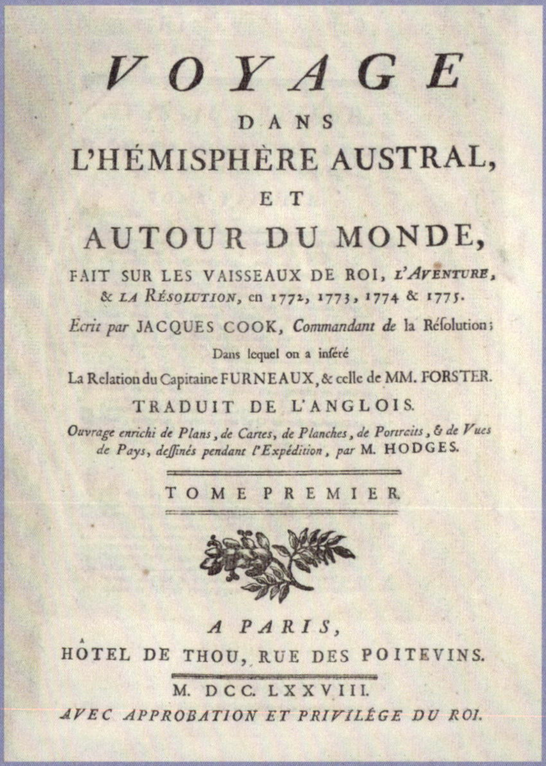

VOYAGE
DANS
L'HÉMISPHÈRE AUSTRAL,
ET
AUTOUR DU MONDE,
FAIT SUR LES VAISSEAUX DE ROI, L'AVENTURE, & LA RÉSOLUTION, en 1772, 1773, 1774 & 1775.
Écrit par JACQUES COOK, Commandant de la Résolution;
Dans lequel on a inséré
La Relation du Capitaine FURNEAUX, & celle de MM. FORSTER.
TRADUIT DE L'ANGLOIS.
Ouvrage enrichi de Plans, de Cartes, de Planches, de Portraits, & de Vues de Pays, dessinés pendant l'Expédition, par M. HODGES.
TOME PREMIER.
A PARIS,
HÔTEL DE THOU, RUE DES POITEVINS.
M. DCC. LXXVIII.
AVEC APPROBATION ET PRIVILÈGE DU ROI.

Title page
From: James Cook, *Voyage dans l'hémisphère austral, et autour du monde, fait sur les vaisseaux de [sic] roi, l'Aventure, & la Résolution, en 1772, 1773, 1774 & 1775*, 1778
ANTWERP, THE PHOEBUS FOUNDATION

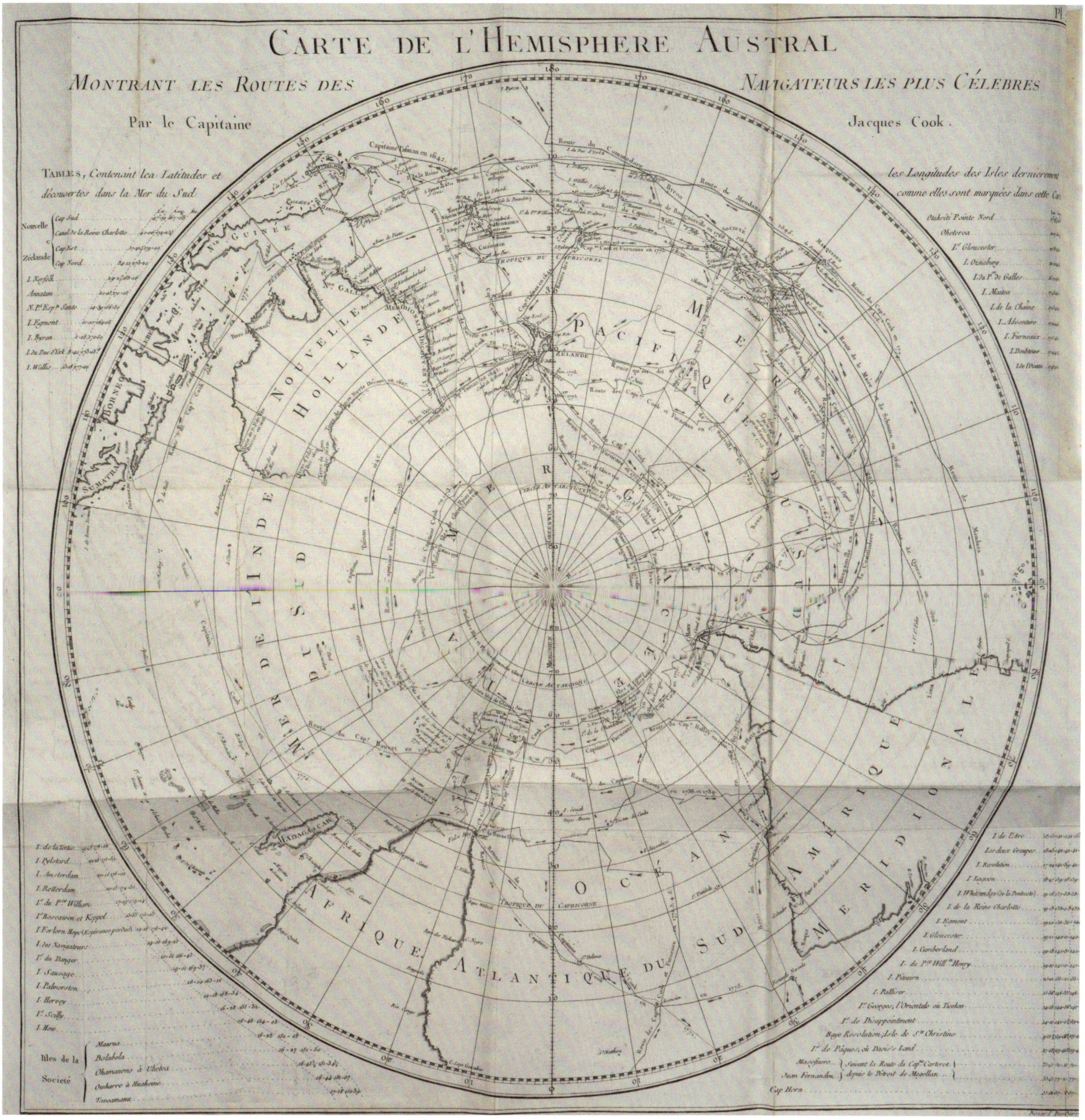

James Cook and Jacques-François Bénard
Carte de l'Hemisphere Austral montrant les routes de navigateurs les plus célebres
Map of the southern hemisphere showing the voyages of famous 'explorers'
From: James Cook, *Voyage dans l'hémisphère austral, et autour du monde, fait sur les vaisseaux de [sic] roi, l'Aventure, & la Résolution, en 1772, 1773, 1774 & 1775*, 1778
ANTWERP, THE PHOEBUS FOUNDATION

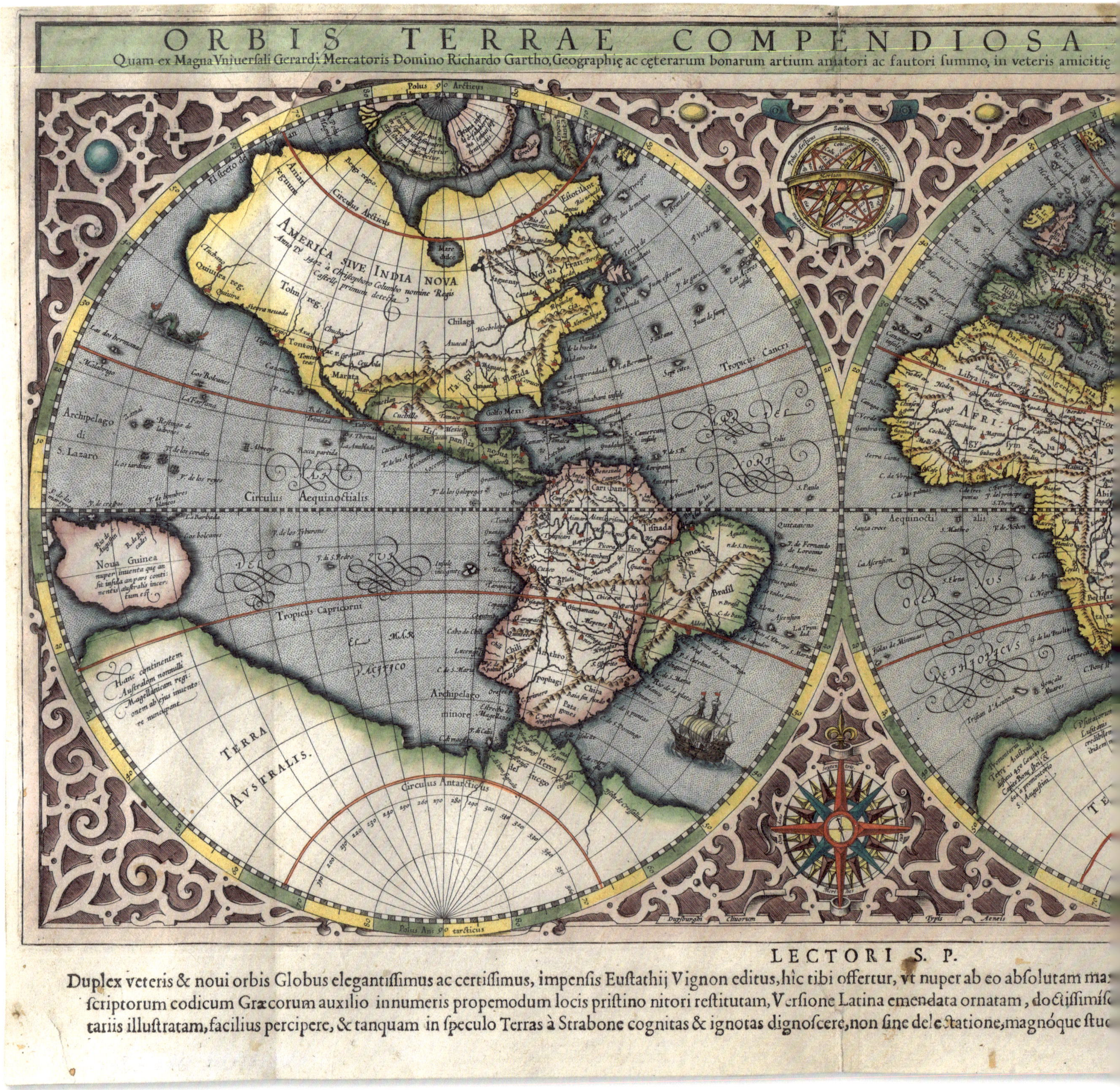

Rumold Mercator after Gerard Mercator
Orbis terrae compendiosa descriptio, 1587
World map
Hand-coloured copper engraving, 630 × 828 mm
ANTWERP, THE PHOEBUS FOUNDATION

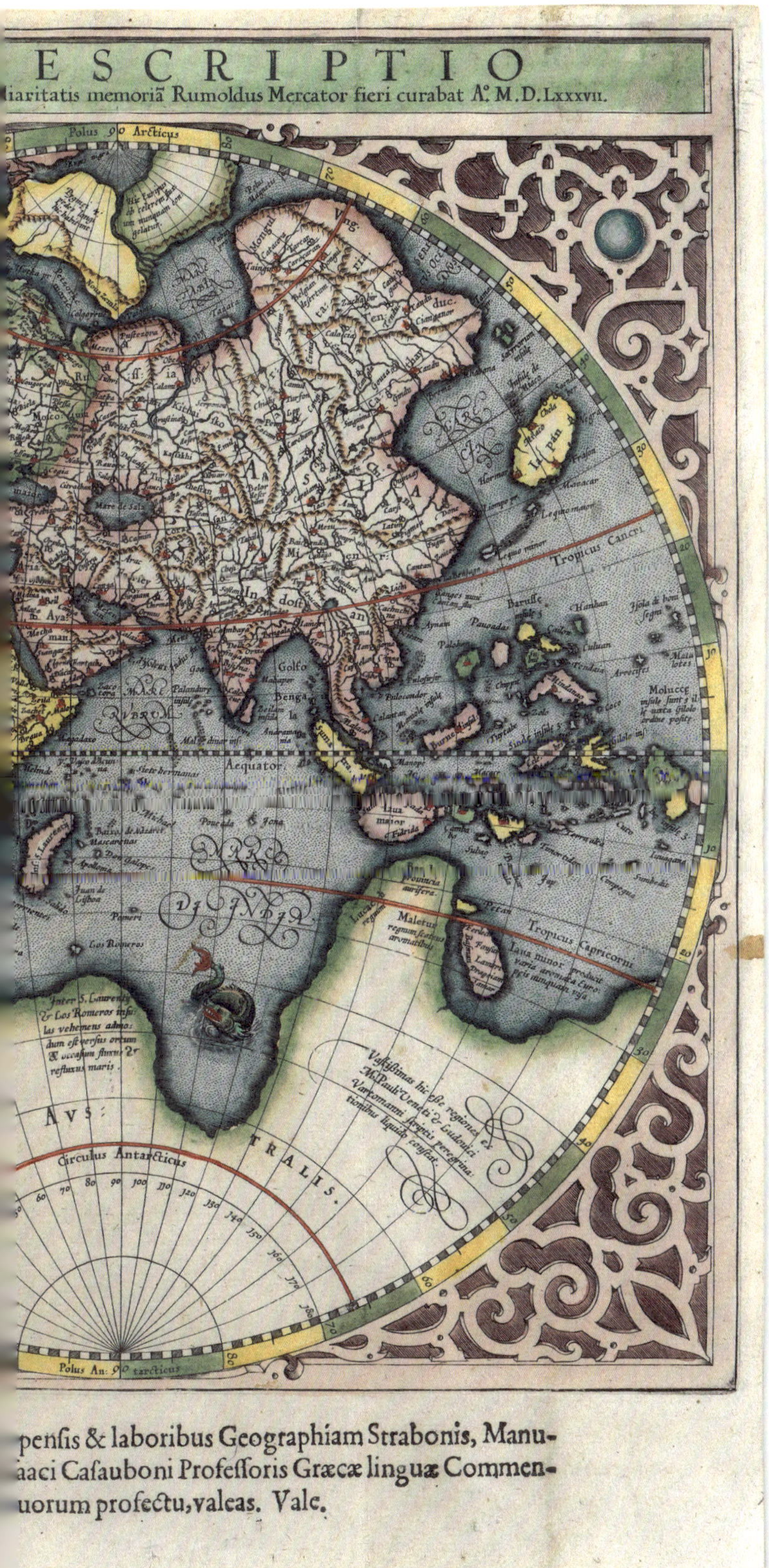

MERCATOR'S LEGACY

Gerard Mercator owed his fame, among other things, to the cartographic projection that bears his name (⟼ p. 202). He presented his projection via a monumental world map running to no fewer than eighteen sheets, which he engraved in his workshop and published in 1569: *Nova et aucta orbis terrae descriptio ad usum navigantium emendate accomodata* ('A new and improved representation of the world for accurate navigation'). Only three original copies of the map still exist.

Gerard's son Rumold Mercator produced a world map in 1587, commissioned by Isaac Casaubon (1559-1614) for his critical edition of *Geographika* by the Greek geographer and historian Strabo, which was published in Geneva in 1587. Rumold based his map on his father's huge and famous one of 1569 but reduced it in size considerably and applied stereographic rather than cylindrical projection. This method utilises lines drawn from a point on the sphere diametrically opposite the plane of projection. While the convex shape of the earth is flattened as a result, it retains its round shape, so that the poles are not as distorted as they are in the Mercator projection. World maps of this type would be adopted in the Mercator and Mercator-Hondius atlases published between 1595 and the 1630s (⟼ Chapter X).

Abraham Ortelius's first map in his world atlas *Theatrum Orbis Terrarum* was a fairly traditional representation of the world, without the application of the Mercator projection and in oval form. The grid lines are based on *Cosmographicus Liber* (Landshut, 1524) by Peter Apian (1495-1552) (⟼ Chapter XI). The coordinates of locations in America are indebted to Mercator's maps. The landmass to the south was still uncharted territory. Through the North Pole run two rivers which were believed (in the Middle Ages) to offer passage to ships en route for the East.

BLAEU'S (IN)FINITE WORLD

Joan Blaeu's *Atlas Maior* (Amsterdam, 1662) – the largest world atlas for many years – obviously needed to have a map of the world too. In 1648, Blaeu had published a large wall map of this kind (in several sheets) which formed the basis for the one in the atlas. He continued to use the same 1606 map for his atlases, so when it came to his magnum opus, it was time for a new one. This world map shows the earth and more: there are personifications of the four seasons in the foreground at the bottom, the scientist Galileo Galilei (1564-1642) from Italy and Tycho Brahe (1546-1601) from Denmark peer over the edge with their instruments, and a variety of classical Greek gods represent the planets and the realm between heaven and earth. On the one hand, his map shows the finite nature of the world, while on the other suggesting the sprawling, infinite expanse of the heavens and the universe beyond.

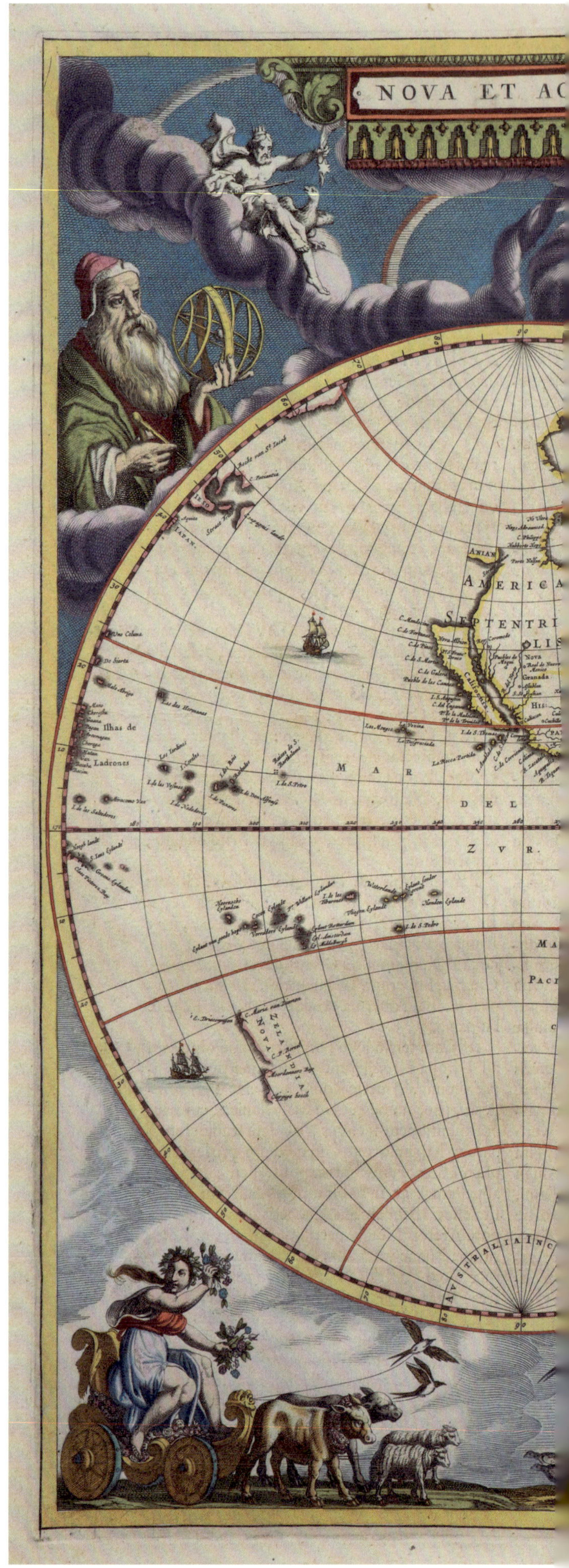

Nova et accuratissima totius terrarum orbis tabula
World map
From: Joan Blaeu, *Atlas Maior* [volume 1], 1662
ANTWERP, THE PHOEBUS FOUNDATION

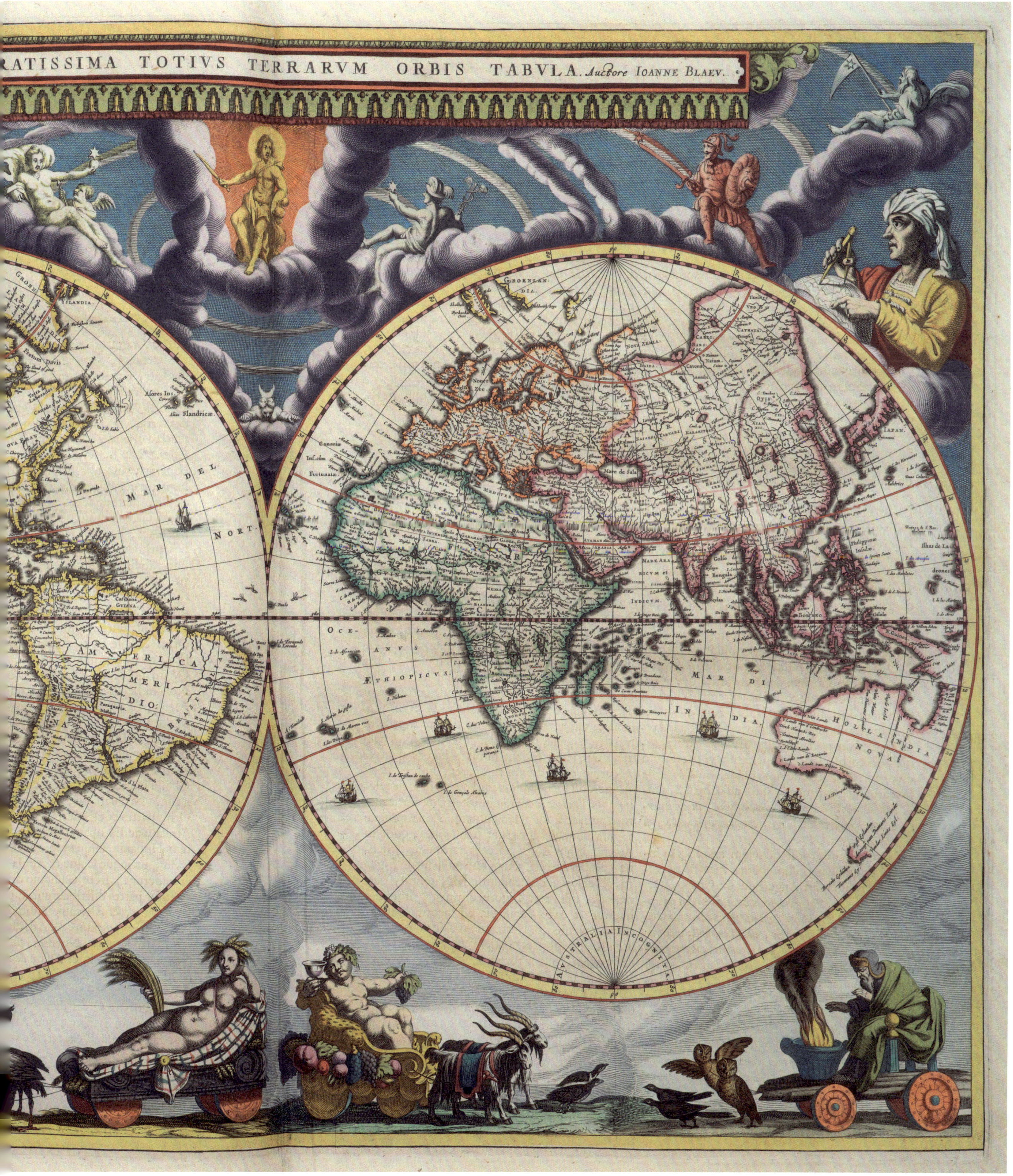
RATISSIMA TOTIVS TERRARVM ORBIS TABVLA. Auctore IOANNE BLAEV.

LITERATURE

D. Cosgrove, 'Mapping the World', in: J. Akerman & R. Karrow, *Maps. Finding your place in the world*, Chicago / London, 2007: 65-115.

E. Edson, *The World Map, 1300-1492. The Persistence of Tradition and Transformation*, Baltimore, 2007.

S. Füssel, *Die Welt im Buch: Buchkünstlerischer und humanistischer Kontext der Schedelschen Weltchronik von 1493*, Mainz, 1996.

A. Hiatt, 'The Map of Macrobius before 1100', *Imago Mundi*, 59, 2, 2007: 149-176.

C. Koeman, *The History of Abraham Ortelius and His Theatrum Orbis Terrarum*, Lausanne, 1964.

J. Parmentier (ed.), *De wereld gespiegeld: wereldkaarten van de middeleeuwen tot nu*, Kontich, 2015.

R. Shirley, *The Mapping of the World: Early Printed World Maps, 1472-1700*, Riverside, CT, 2001.

M. van den Broecke, P. van der Krogt & L. Voet, *Abraham Ortelius and the First Atlas: Essays Commemorating the Quadricentennial of His Death, 1598-1998*, Houten-'t Goy, 1998.

B. Wagner, *Worlds of Learning: The Library and World Chronicle of the Nuremberg Physician Hartmann Schedel (1440-1514)*, Munich, 2015.

P. Whitfield, *The Image of the World: 20 Centuries of World Maps*, London, 1994.

A. Wilson, J. Lancaster & Peter Zahn, *The Making of the Nuremberg Chronicle*, Amsterdam, 1976.

D. Woodward, 'Medieval Mappaemundi', in: J.B. Harley & D. Woodward (ed.), *Cartography in Prehistoric, Ancient, and Medieval Europe and the Mediterranean*, vol. 1, dl. 3, Chicago / London, 1987: 286-370.

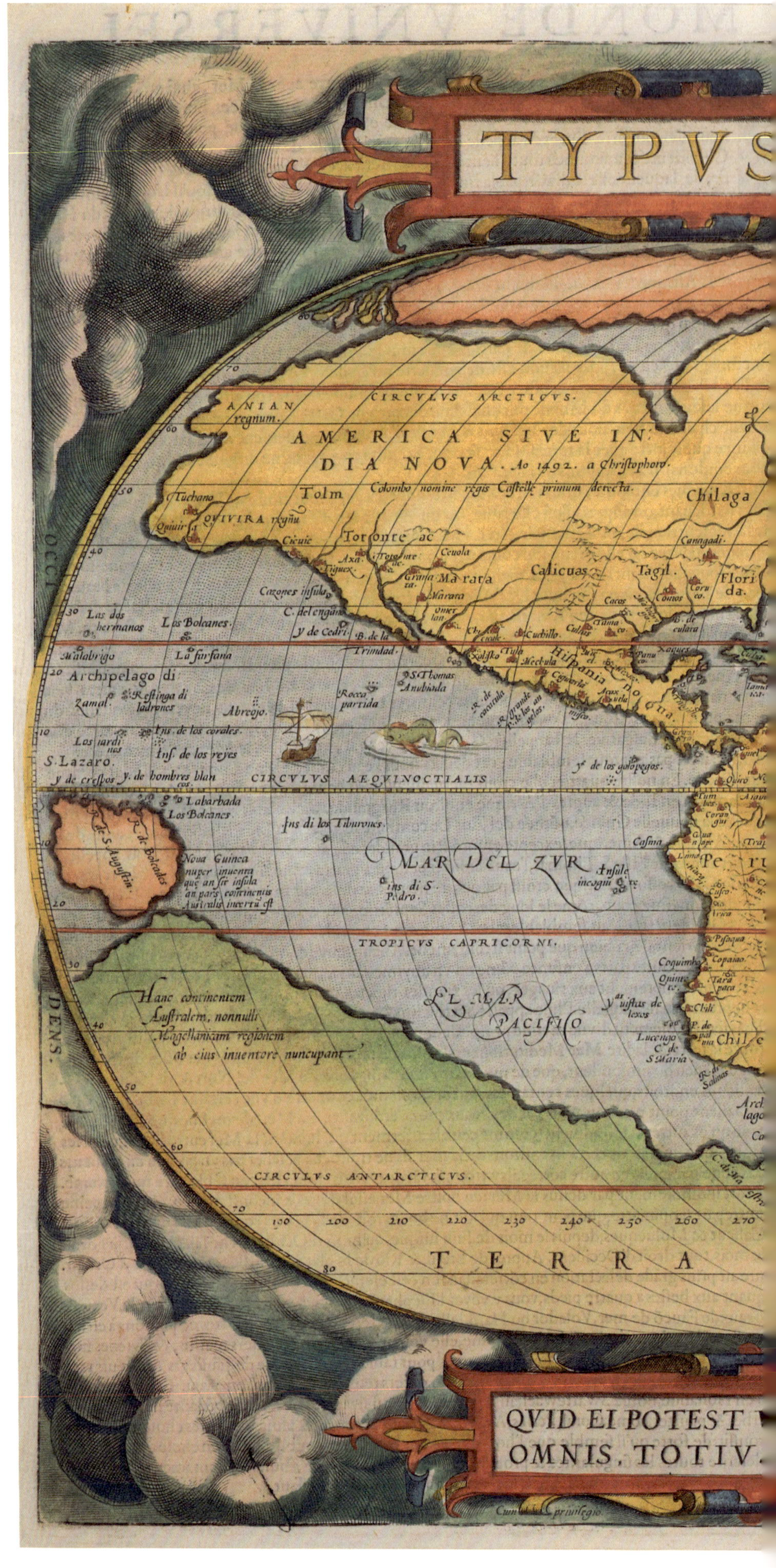

Abraham Ortelius and Frans Hogenberg
Typus orbis terrarum, 1570
World map
Hand-coloured copper engraving, 340 × 495 mm
ANTWERP, THE PHOEBUS FOUNDATION

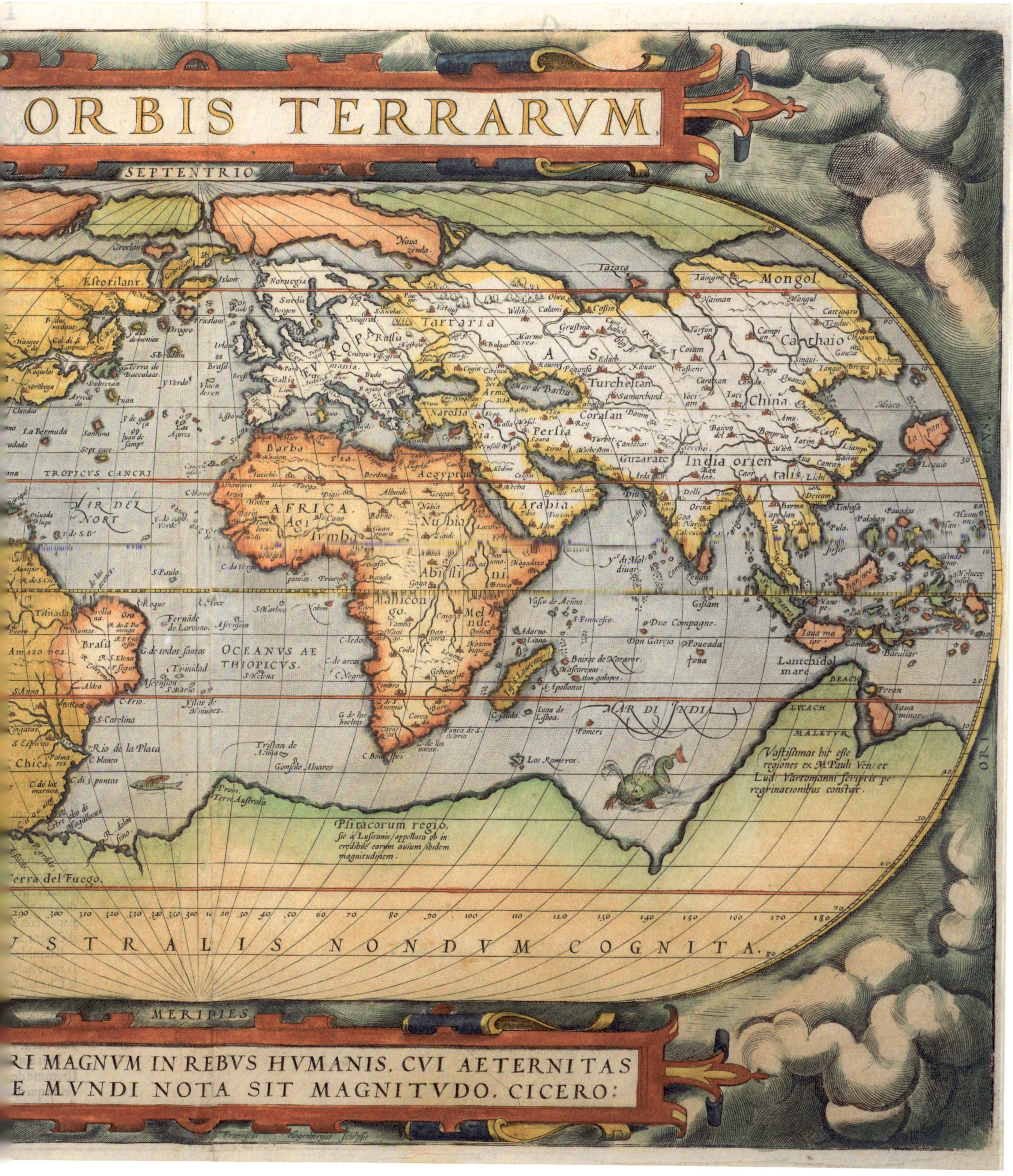
ORBIS TERRARVM
SEPTENTRIO
TROPICVS CANCRI
AFRICA
ASIA
Tartaria
Mongol
Cathaio
China
India orientalis
Persia
Arabia
OCEANVS AETHIOPICVS
MAR DI INDI
Psitacorum regio
Terra del Fuego
...STRALIS NONDVM COGNITA.
MERIDIES
...RI MAGNVM IN REBVS HVMANIS, CVI AETERNITAS
...E MVNDI NOTA SIT MAGNITVDO. CICERO:

COMMERCIAL ATLAS EDITIONS meant that scientific world maps like those of Gerard Mercator could also reach a wider readership. These printed collections of maps of the world, continents, countries, cities and the heavens sold like hot cakes, especially in the seventeenth century, which witnessed the heyday of commercial cartography in the Low Countries.

Joan Blaeu's celebrated eleven-volume *Atlas Maior* (Amsterdam, 1662) has been described as 'the greatest and most beautiful atlas ever published'. An initiative of this kind did not appear out of thin air: it was the culmination of the development of map production in the Low Countries, where mapmakers and publishers deftly imitated and vied with one another. Without this combination of market forces and intense competition, a cartographic pinnacle of this kind could never have been attained. The Amsterdam publisher stood on the shoulders of giants like Abraham Ortelius and Gerard Mercator, his work driven on by a neck-and-neck race with rival producers. While Amsterdam was the capital of cartography in the Low Countries around 1662, the Southern Netherlands accounted for a substantial share of the output. So how did Blaeu's masterpiece come about, and what Southern Netherlandish fingerprints can we detect in the acclaimed maps it contains?

CHAPTER X

BIG, BIGGER, BIGGEST

WORLD ATLASES AND THE CREATION OF THE *ATLAS MAIOR*

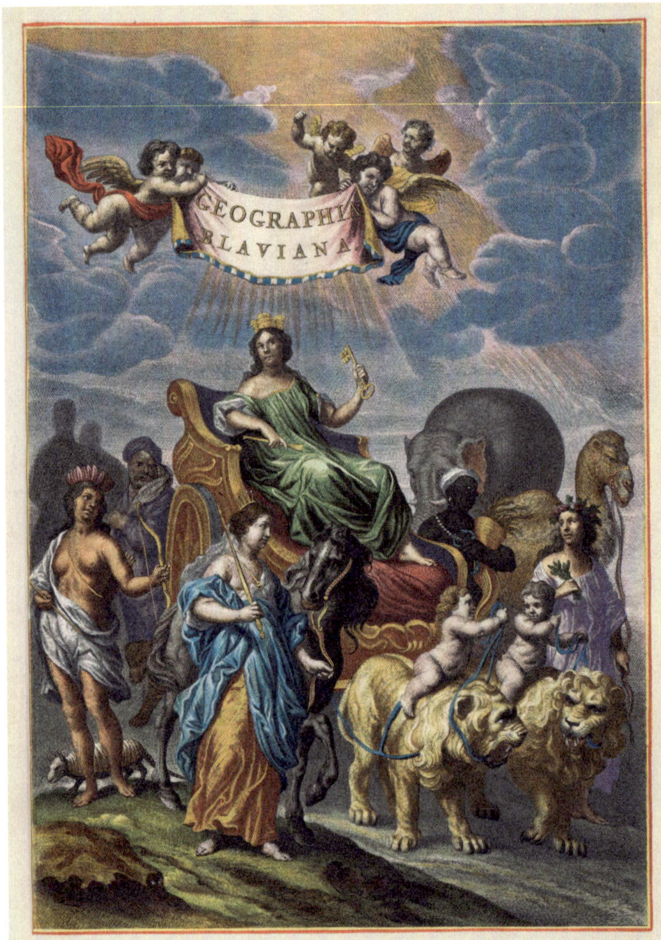

Title page
From: Joan Blaeu, *Atlas Maior* [volume 1], 1662
ANTWERP, THE PHOEBUS FOUNDATION

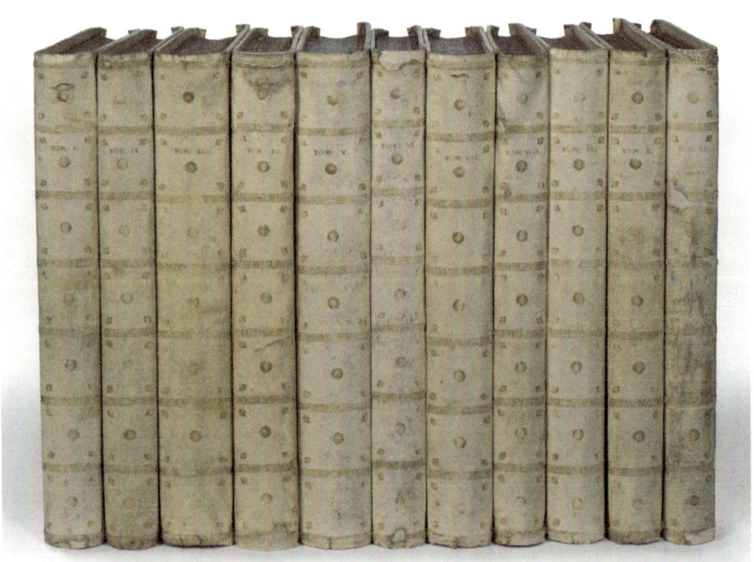

Joan Blaeu
Atlas Maior [11 volumes], 1662
ANTWERP, THE PHOEBUS FOUNDATION

THE LARGEST ATLAS

The Phoebus Foundation has a copy of Joan Blaeu's *Atlas Maior*. A single such copy runs to as many as 600 maps in eleven volumes, around 3,000 pages of geographical descriptions, each a generously sized 52.7 by 32.1 centimetres. In 1662, it was a product of unprecedented size in more ways than one. The atlas was admired for its beauty: the harmoniously balanced typography, the artistic and consistent design of the maps, the rich decorations and sumptuous colouring had all been taken to a new level. Special cabinets were constructed to house the precious parchment volumes.

Volumes one to eight cover Europe (plus the Arctic and Russia), volume nine Spain and Africa, volume ten Asia, and volume eleven the Americas. The set belonging to The Phoebus Foundation is an early variant of the first Latin edition. It had yet to be corrected, as many of the maps still have a title in German. Later variants have all the titles in Latin.

It goes without saying that Joan Blaeu did not draw up all the maps in this extensive collection in person. He built on decades of hard work by his father, Willem Jansz. Blaeu, who had a strong interest in science and geography, and on the achievements of his predecessors and contemporaries. Joan Blaeu was a successful businessman who boasted the necessary contacts and resources to procure new copper plates and maps for his atlas. He also owned enough presses and employees to sustain his output. The atlas brought together the very best of cartography that the Blaeu firm could find at the time, all presented in a highly attractive form. It is the culmination of years of geographical knowledge-gathering and cartographic invention.

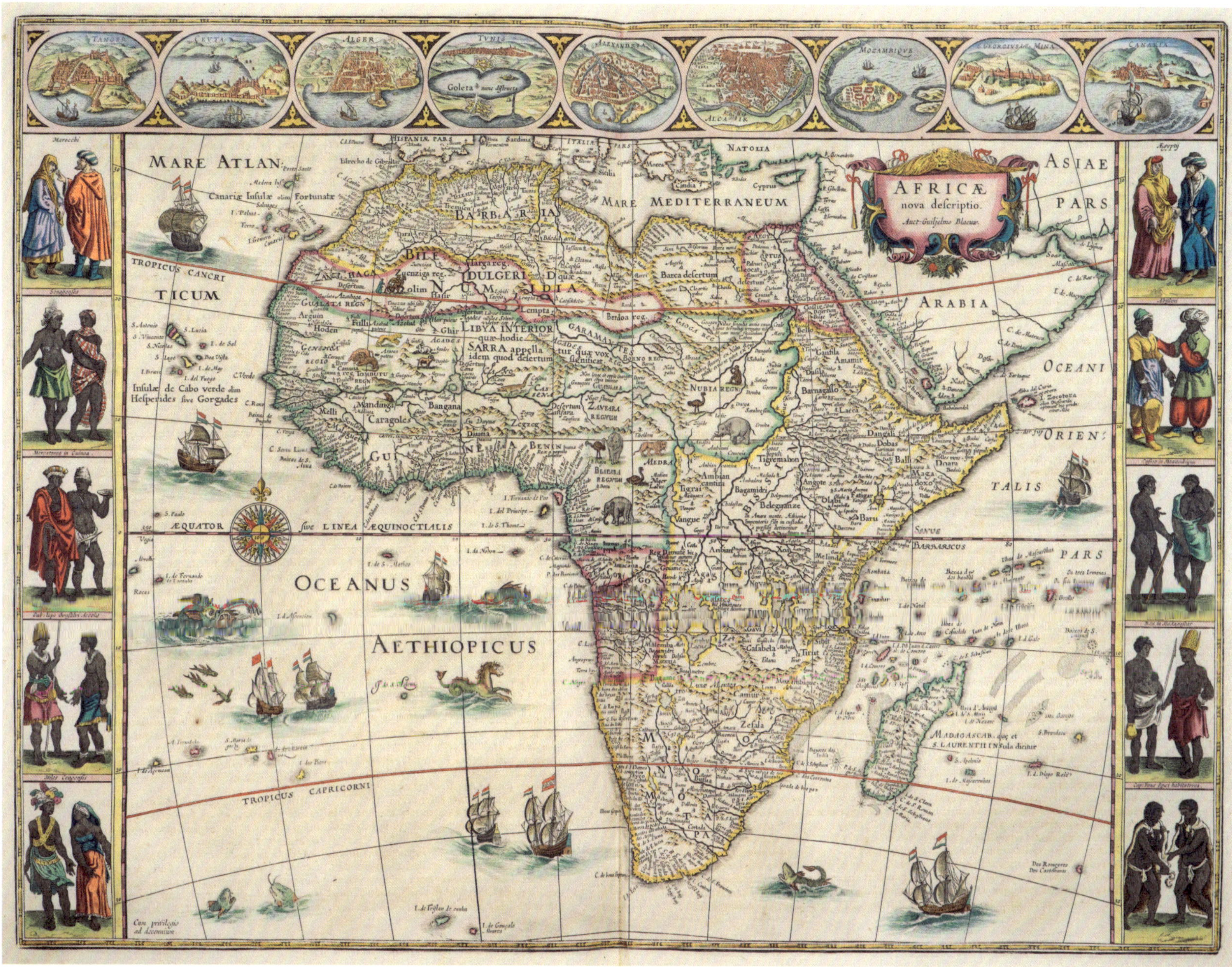

Africae nova descriptio
Map of Africa
From: Joan Blaeu, *Atlas Maior* [volume 9], 1662
ANTWERP, THE PHOEBUS FOUNDATION

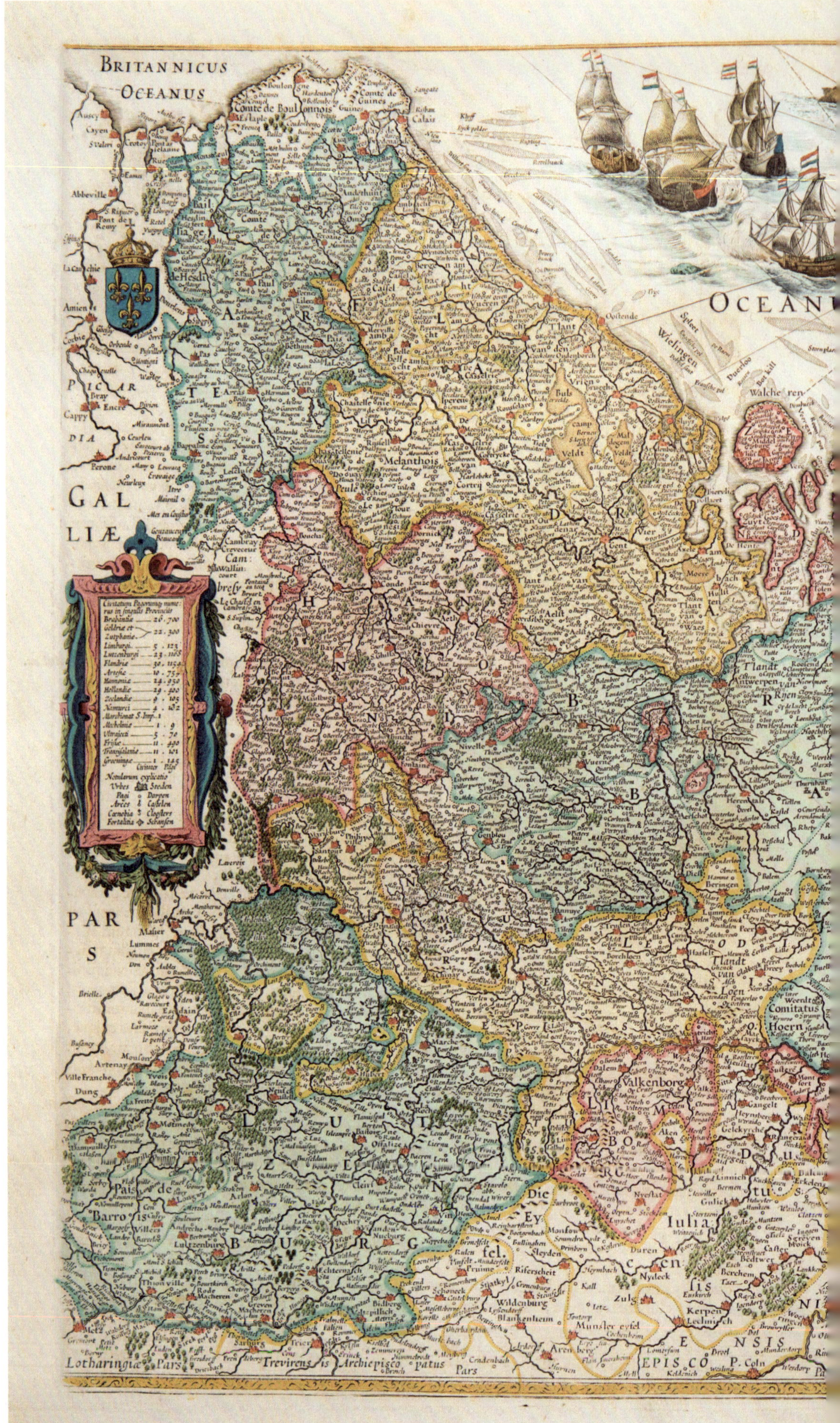

Novus XVII Inferioris Germaniae Provinciarum Typus
Map of the Low Countries
From: Joan Blaeu, *Atlas Maior* [volume 2], 1662
ANTWERP, THE PHOEBUS FOUNDATION

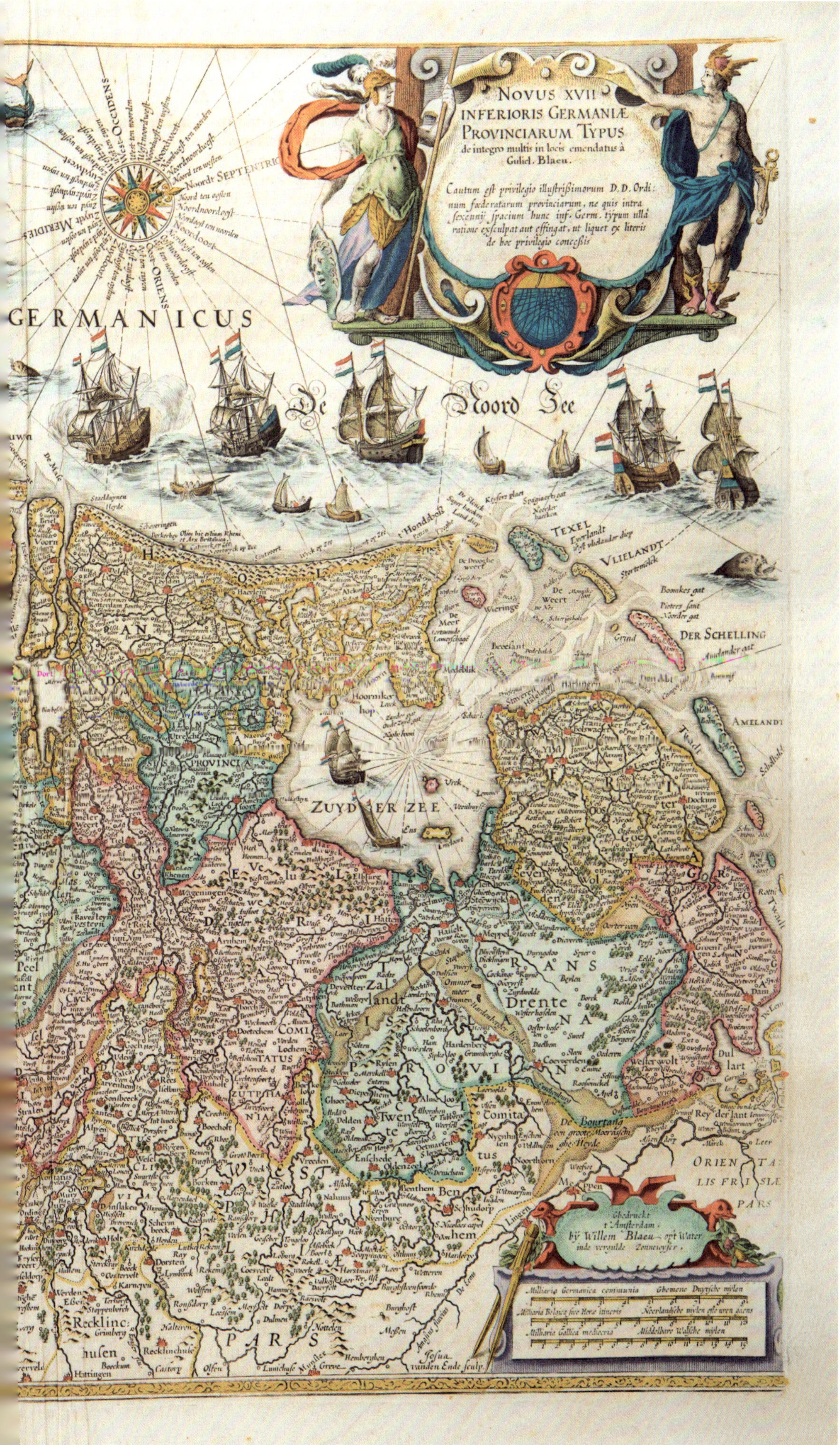
NOVUS XVII
INFERIORIS GERMANIÆ
PROVINCIARUM TYPUS
de integro multis in locis emendatus à
Guliel. Blaeu.
Cautum est privilegio illustrißimorum D.D. Ordinum fœderatarum provinciarum, ne quis intra sexennij spacium hunc inf. Germ. typum ullâ ratione exsculpat aut effingat, ut liquet ex literis de hoc privilegio concessis.
GERMANICUS
De Noord Zee
TEXEL
VLIELANDT
DER SCHELLING
AMELANDT
ZUYDERZEE
Drente
Gedruckt t'Amsterdam by Willem Blaeu op't Water inde vergulde Sonnewyser.
Josua vanden Ende sculp.

THE SHOULDERS OF GIANTS

To trace the principles of the *Atlas Maior*, we have to go back roughly a century, to the time of Gerard Mercator and Abraham Ortelius. Prior to then, the commercial publisher's atlas – a standardised collection of printed maps intended for open sale – did not exist. What there was, however, was the *atlas factice*, a trend from Italy, a custom atlas produced to order. The Phoebus Foundation has one of those too, a five-volume collector's atlas that was most likely compiled in Nuremberg in the late eighteenth century, with maps dating from 1695-1787. This atlas mostly contains maps of Europe, many of them made by the German mapmaker Johann Baptist Homann (1664-1724) and lifted from his *Atlas Germania* (Nuremberg, 1753).

The driving force behind the first-ever true atlas was Abraham Ortelius from Antwerp, who brought together the best maps and cartographic sources, had them re-engraved, wrote the accompanying texts and collated them in a book. Ortelius was the brain behind the venture, therefore, rather than the designer of the maps and, aware of this, was magnanimous in crediting the cartographic sources – Mercator among them – in a list of authors.

The first edition of his Latin *Theatrum Orbis Terrarum* ('Theatre of the World') was printed in Antwerp in 1570 by Gillis Coppens van Diest (*c.*1496-1572). The plates were engraved by Frans Hogenberg. It contained 53 maps at this stage and was an instant success, as evidenced by the reprints that followed that same year. The earliest Ortelius atlas in The Phoebus Foundation's collection is the fourth Latin edition of the *Theatrum*, printed by Gillis's son Anthonis Coppens van Diest in 1574. It is a complete copy, with all seventy of the maps the atlas contained by that time.

Ortelius continued to produce new editions: there were a total of forty all together (up to 1612), including Latin, Dutch, German, French, Spanish, English and Italian editions. The final edition of the world atlas ran to some 150 maps. The Phoebus Foundation owns several other copies too in addition to the copy from 1575, all of them printed in Antwerp: a Latin edition by Gillis van de Rade (died 1615) from 1575, a Latin edition by Christophe Plantin from 1579, a German edition by Jan Baptist Vrients from 1602 and another Latin edition by Vrients from 1603. While the first of these might not be the oldest, it is the most spectacular, containing all seventy coloured maps heightened in gold.

Gerard Mercator was likewise planning a world atlas around this time. For his part, he was more of a scientist than a strategic map dealer like Ortelius, and he devoted more time and attention to the pursuit of technical and geographical precision. He aspired to a comprehensive cosmography, describing the entire universe and its history. All the same, Mercator too built on the work of others. His map of Flanders, for instance, possibly drew on the manuscript maps of Jacob van Deventer, a cartographer from Mechelen who had criss-crossed the Low Countries earlier in the sixteenth century to map them accurately on the order of the king of Spain. The project was initially military in motivation but later provided a rewarding model for the further mapping of the Low Countries.

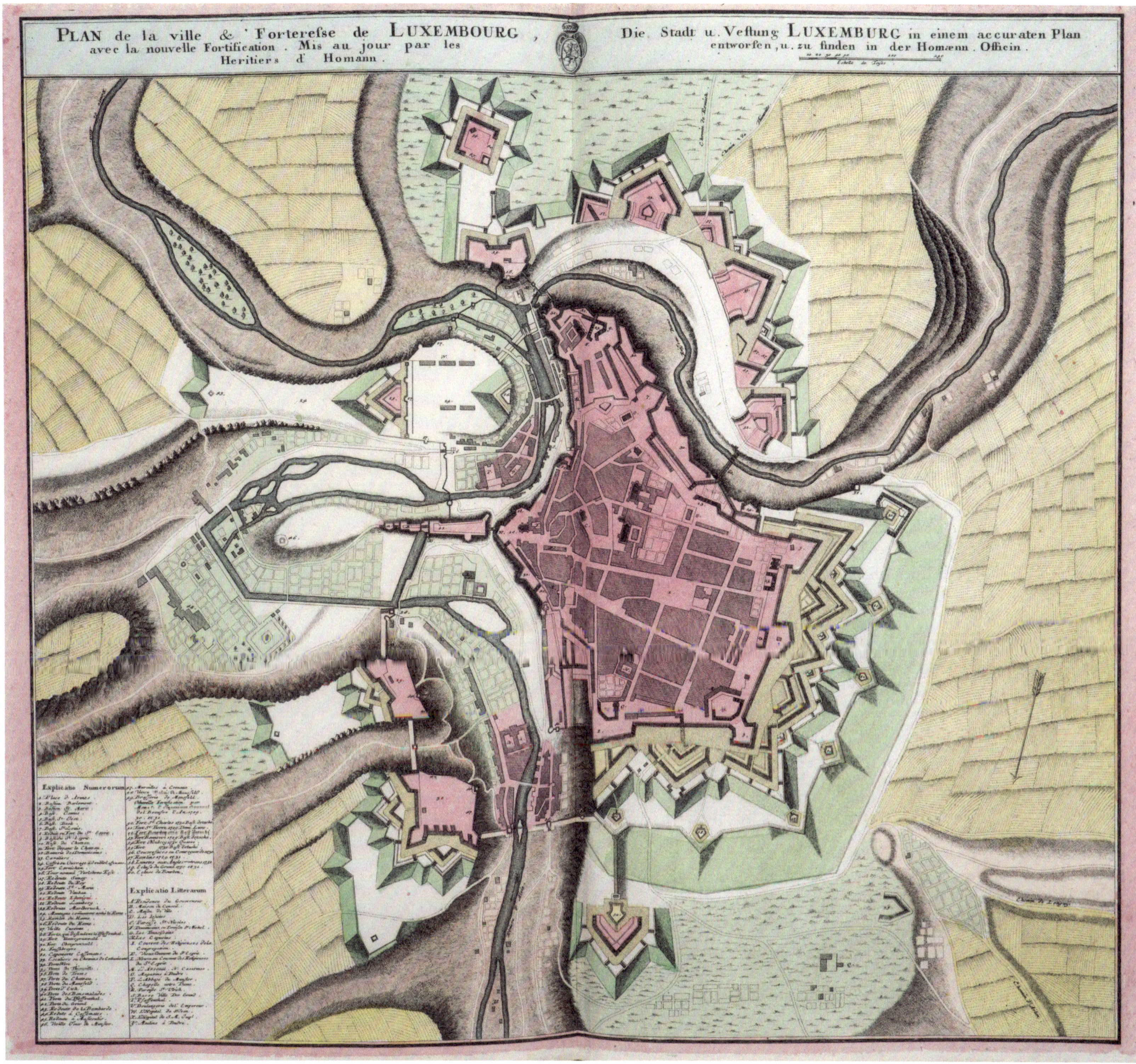

Johann Baptist Homann
Plan de la ville et Forteresse de Luxembourg
Map of Luxembourg City
From: *Atlas factice*, c.1695-1788
ANTWERP, THE PHOEBUS FOUNDATION

In the end, however, Mercator only published 51 of his planned 120 maps during his lifetime. His son Rumold Mercator issued as many maps as he could after his father's death (in 1594) in the *Atlas sive Cosmographicae Meditationes de Fabrica Mundi et Fabrica Figura* (1595), a posthumous work that also included the map of America by Gerard's grandson Michael Mercator (⟼ pp. 196-197). The cosmography that the original Mercator had envisaged never materialised, and the baton was passed on to the next generation.

It was picked up around 1604 by Jodocus Hondius I, born in the West Flemish town of Wakken, and Cornelis Claesz. from Leuven, who had the good fortune to secure Mercator's copper plates by way of his grandchildren. These Southern Netherlanders continued the tradition by compiling a new world atlas. Hondius engraved 37 new maps, Petrus Montanus (Pieter van den Berg; 1560-1625) wrote the accompanying texts and Claesz. printed the plates for a new atlas first offered for sale in 1606. The Phoebus Foundation owns a copy of the first French edition of 1609, which consists of 147 maps (of which two are missing in this copy). The translation highlights the publishers' ambition to distribute the atlases internationally, and many more translations and reprints of this atlas would follow, a series now known as the 'Mercator-Hondius Atlas'. Its makers retained the title 'Atlas' in tribute to their predecessor Mercator, who had coined the term that he used for his posthumously published book of maps.

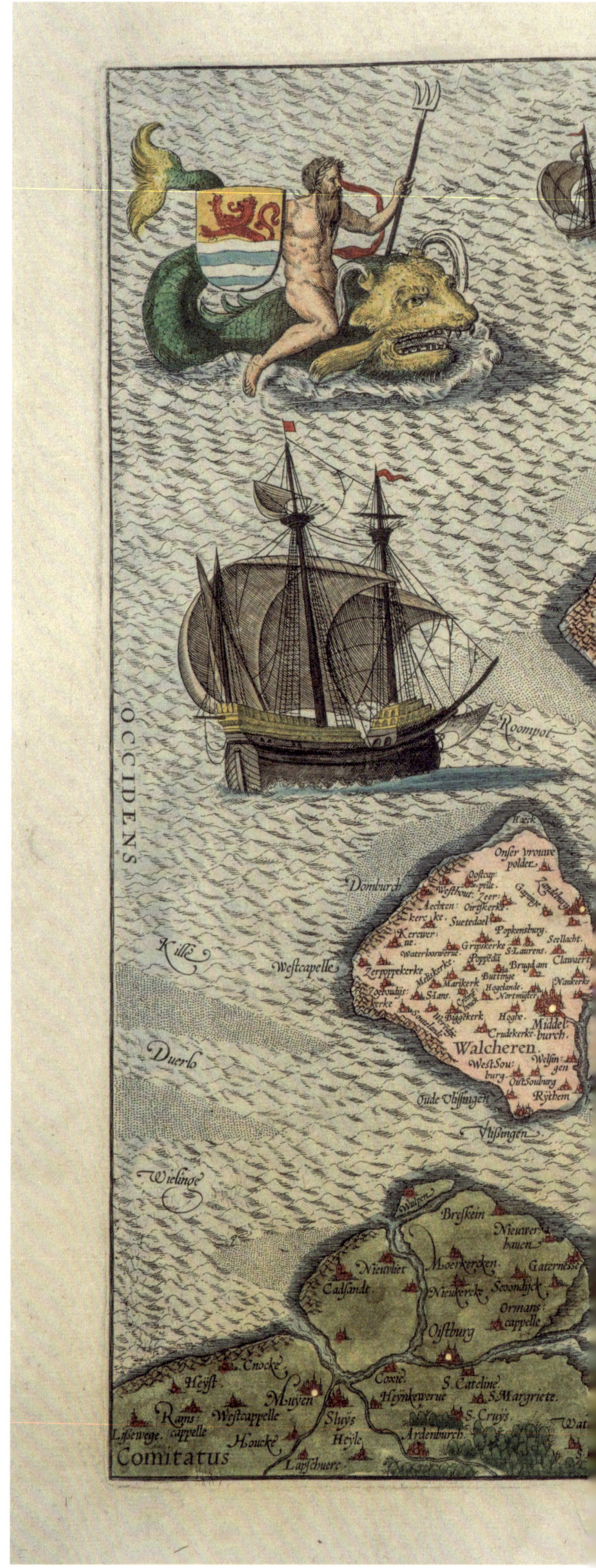

Abraham Ortelius after Jacob van Deventer
Zelandicarum Insularum Exactissima et Nova Descriptio
Map of Zealand
From: Abraham Ortelius, *Theatrum Orbis Terrarum*, 1575
ANTWERP, THE PHOEBUS FOUNDATION

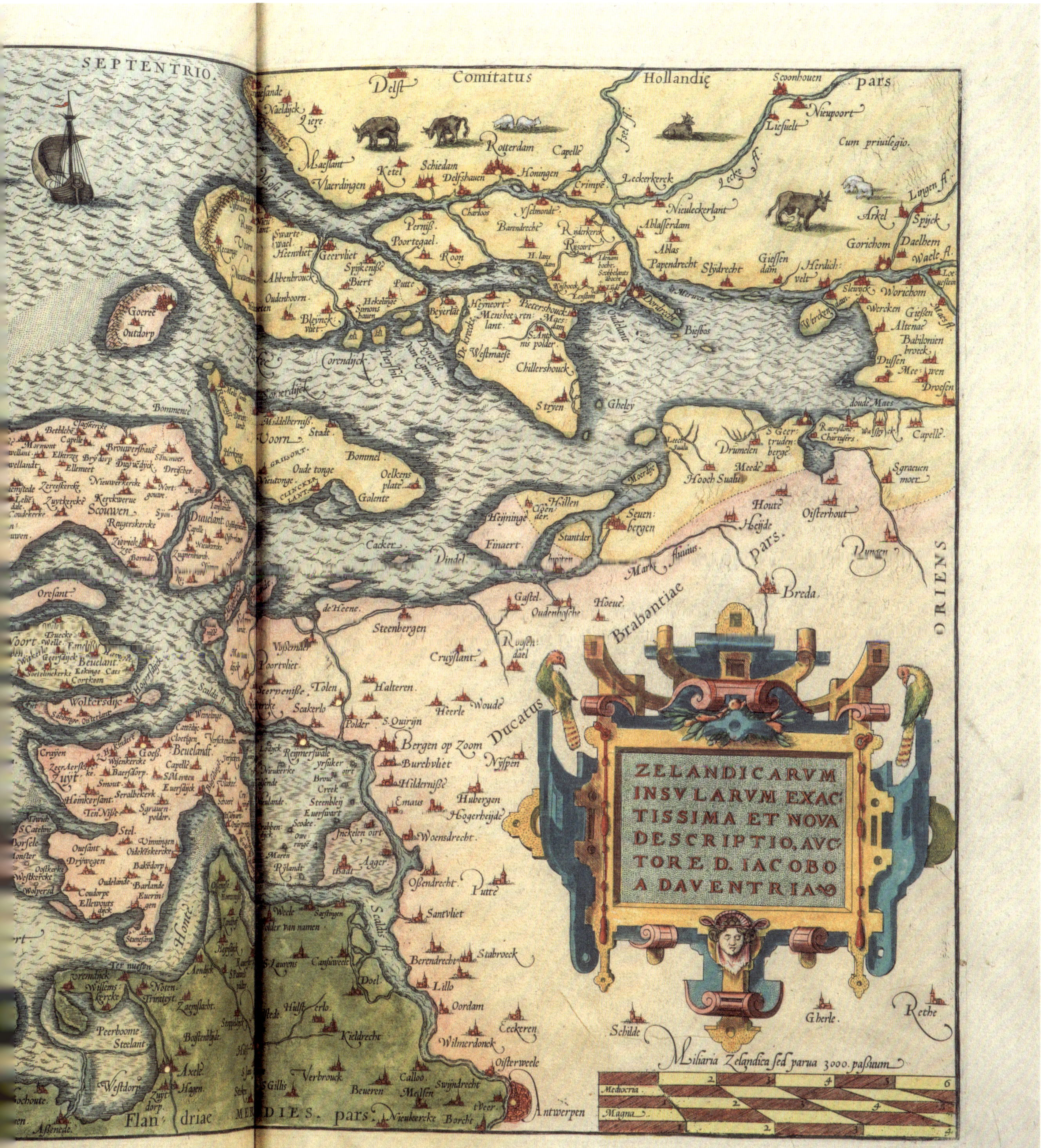

ZELANDICARVM INSVLARVM EXACTISSIMA ET NOVA DESCRIPTIO, AVCTORE D. IACOBO A DAVENTRIA
SEPTENTRIO.
Comitatus Hollandię pars
Cum priuilegio.
ORIENS
Ducatus Brabantiae pars.
Flandriae
MERIDIES. pars
Miliaria Zelandica sed parua 3000. passuum
Mediocria
Magna
Delft
Rotterdam
Schiedam
Vlaerdingen
Dordrecht
Gorichom
Nieupoort
Breda
Bergen op Zoom
Steenbergen
Antwerpen
Oisterhout
Woensdrecht
Santvliet
Lillo
Schilde
Goeree
Outdorp
Scouwen
Beuelant
Tolen
Axele

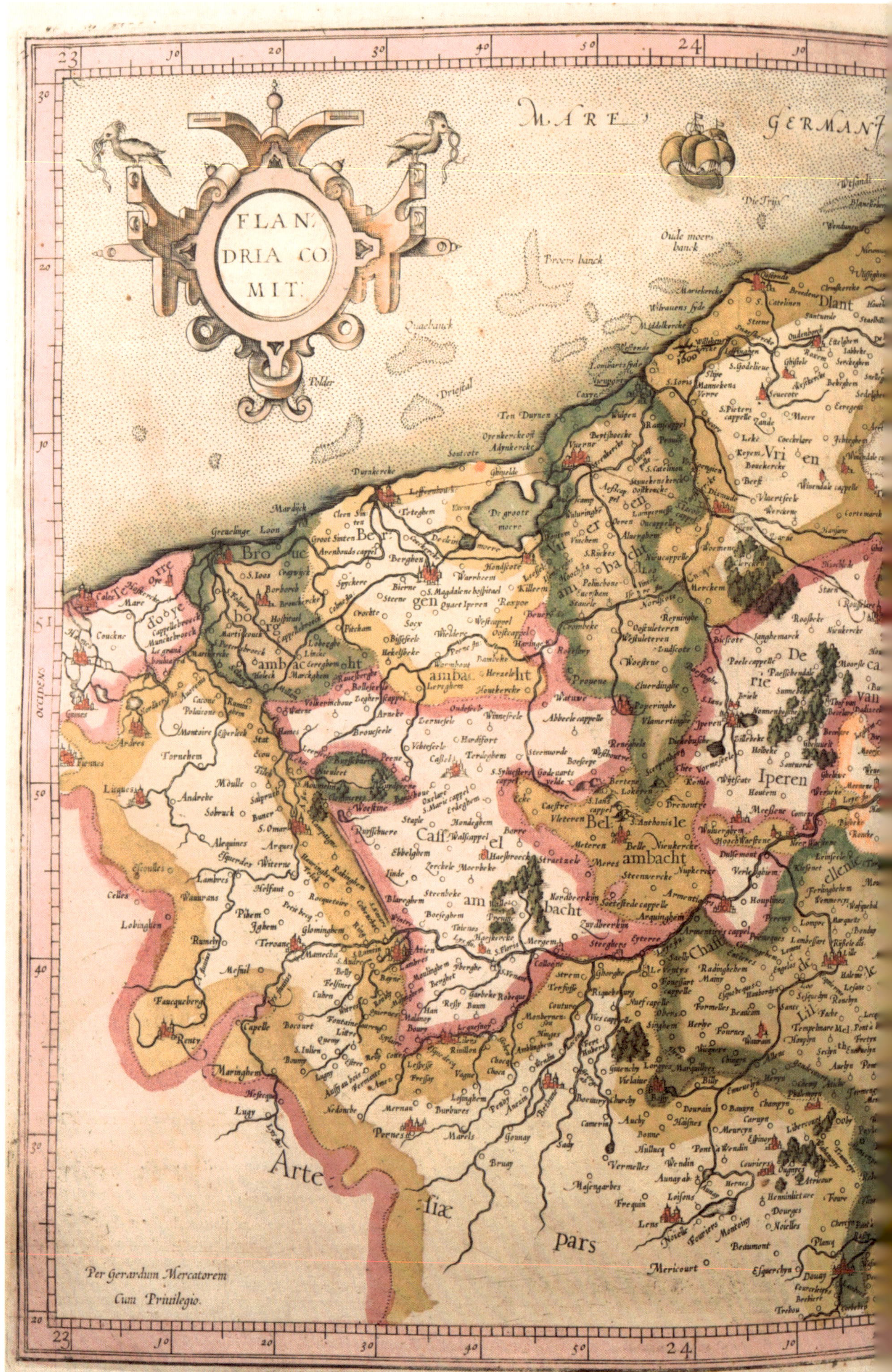

Gerard Mercator
Flandria comit
Map of Flanders
From: Jodocus Hondius I after Gerard Mercator, *Atlas ou Méditations Cosmographiques de la Fabrique du Monde et Figure d'Iceluy*, 1609
ANTWERP, THE PHOEBUS FOUNDATION

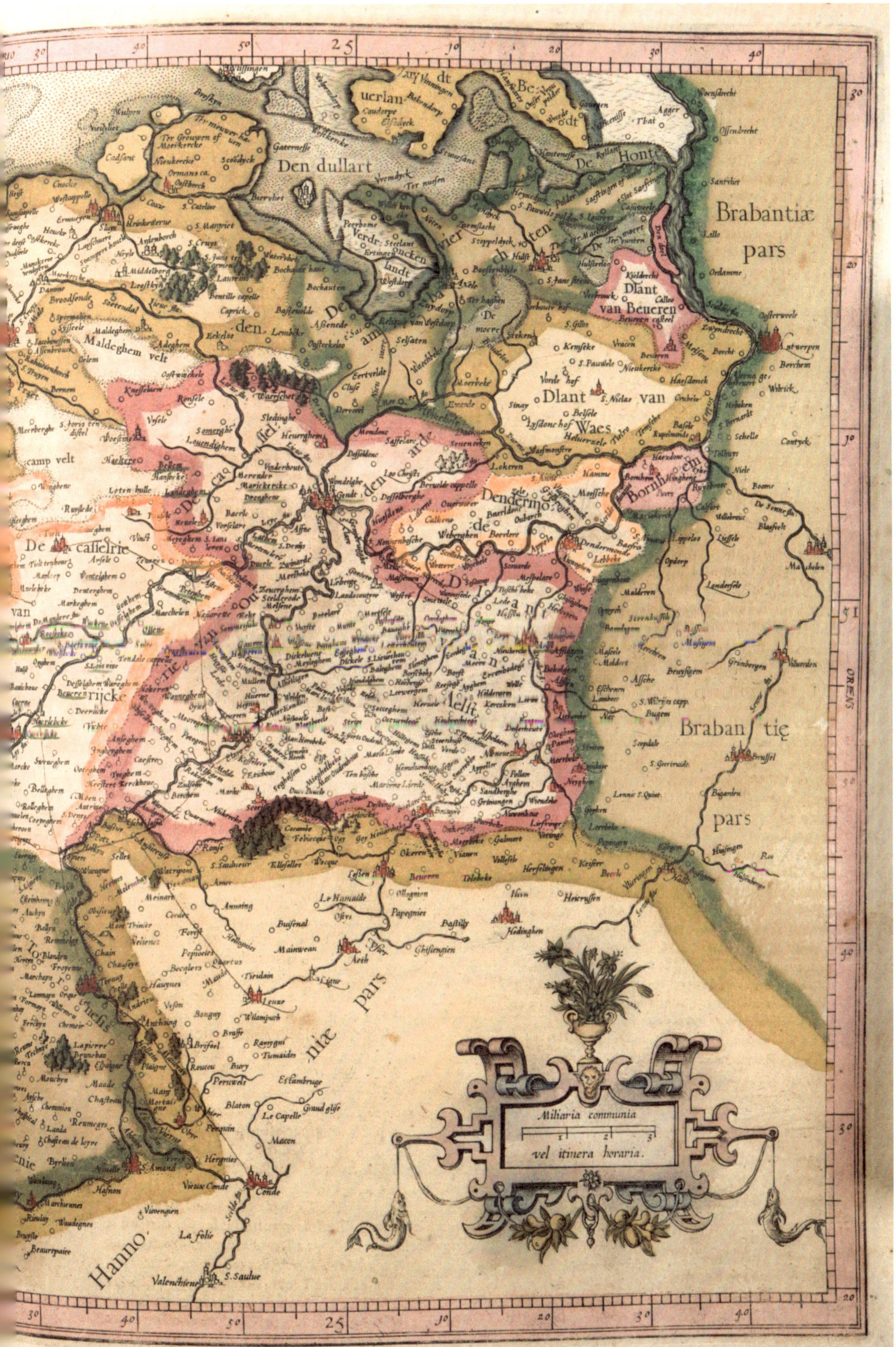
Den dullart
Brabantiæ pars
Dlant van Beueren
Dlant van Waes
Braban tię pars
Hanno niæ pars
Miliaria communia
vel itinera horaria.

OUTDOING THE COMPETITION

Having procured the rights to the Mercator maps, Jodocus Hondius I was able to dominate atlas production in the Northern Netherlands for a considerable period, and following his death in 1612, his sons Jodocus II (1593-1629) and Henricus continued the business. Jodocus II died in 1629, however, and rival mapmaker Willem Jansz. Blaeu was able to secure 34 copper plates from the Mercator atlas from his estate. Johannes Janssonius from Arnhem (1588-1664) had married Jodocus II and Henricus's sister, no doubt motivated in part by strategic business considerations, and a year after the death of Jodocus II, Henricus teamed up with his publisher brother-in-law Janssonius, changing tack now that a significant number of Mercator plates were no longer in his family's possession.

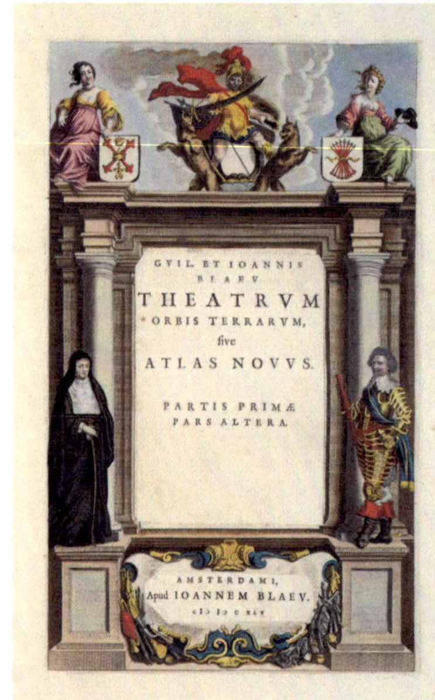

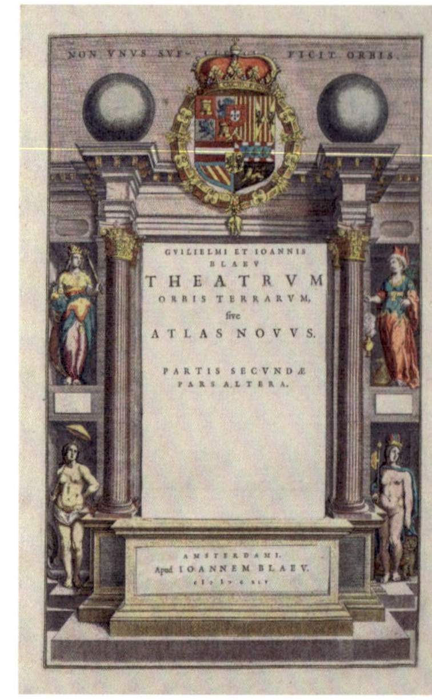

Blaeu's appearance on the world-atlas scene sparked fierce competition. The maritime expert had already been active in Amsterdam since around 1599 but had mainly dealt with nautical cartography, instruments, pilot guides and maps for the Dutch East India Company. His critical scientific eye saw considerable scope, however, for improvement in his peers' Mercator-Hondius atlases. Blaeu began by publishing the *Atlantis Appendix, sive pars Altera* in 1630, 37 of the sixty maps in which came from Hondius. It triggered an 'atlas race' lasting many years between Hondius-Janssonius and Blaeu, who continually outdid each other with atlases that were ever larger, with even more volumes, more maps and more text. They were not shy about pirating each other's material either.

In this way, the Blaeu and (Hondius-)Janssonius companies constantly expanded their multi-volume atlases as much as possible with each fresh edition all the way through to 1658. Willem Jansz. Blaeu died in 1638, and Henricus Hondius ceased publishing in the early 1640s, but Joan Blaeu and Johannes Janssonius carried on their respective work. The Phoebus Foundation has the first and second parts of Blaeu's *Nieuwe Atlas* of 1644-1645. Its title, *Theatrum Orbis Terrarum, sive Novus Atlas*, is a tribute to Ortelius, the founding father of the world atlas. This edition comprised two volumes with a combined total of 212 maps. In the years that followed, Blaeu continued to add new sections: a third one on Italy, a fourth on England, a fifth on Scotland and a sixth on China.

In 1646, Janssonius issued a fourth volume with 56 maps covering the British Isles. He just about outdid Blaeu by also including Scotland and Ireland in his atlas: Blaeu's own fifth volume on Scotland was not published until 1654.

One variant and reprint followed another in a variety of languages including Latin, French, Dutch and German. The publications were not especially well balanced: the focus was disproportionate, with many more maps of European nations, for instance, than those on other continents. Quantity took precedence, culminating in the largest world atlas in 1658 – Janssonius's *Atlas Novus*, which ran to 450 maps compared to his rival Joan Blaeu's 403.

Title page
From: Joan Blaeu and Willem Jansz. Blaeu,
Theatrum Orbis Terrarum, sive Atlas Novus [volume 1], 1644-1645
ANTWERP, THE PHOEBUS FOUNDATION

Title page
From: Joan Blaeu and Willem Jansz. Blaeu,
Theatrum Orbis Terrarum, sive Atlas Novus [volume 2], 1644-1645
ANTWERP, THE PHOEBUS FOUNDATION

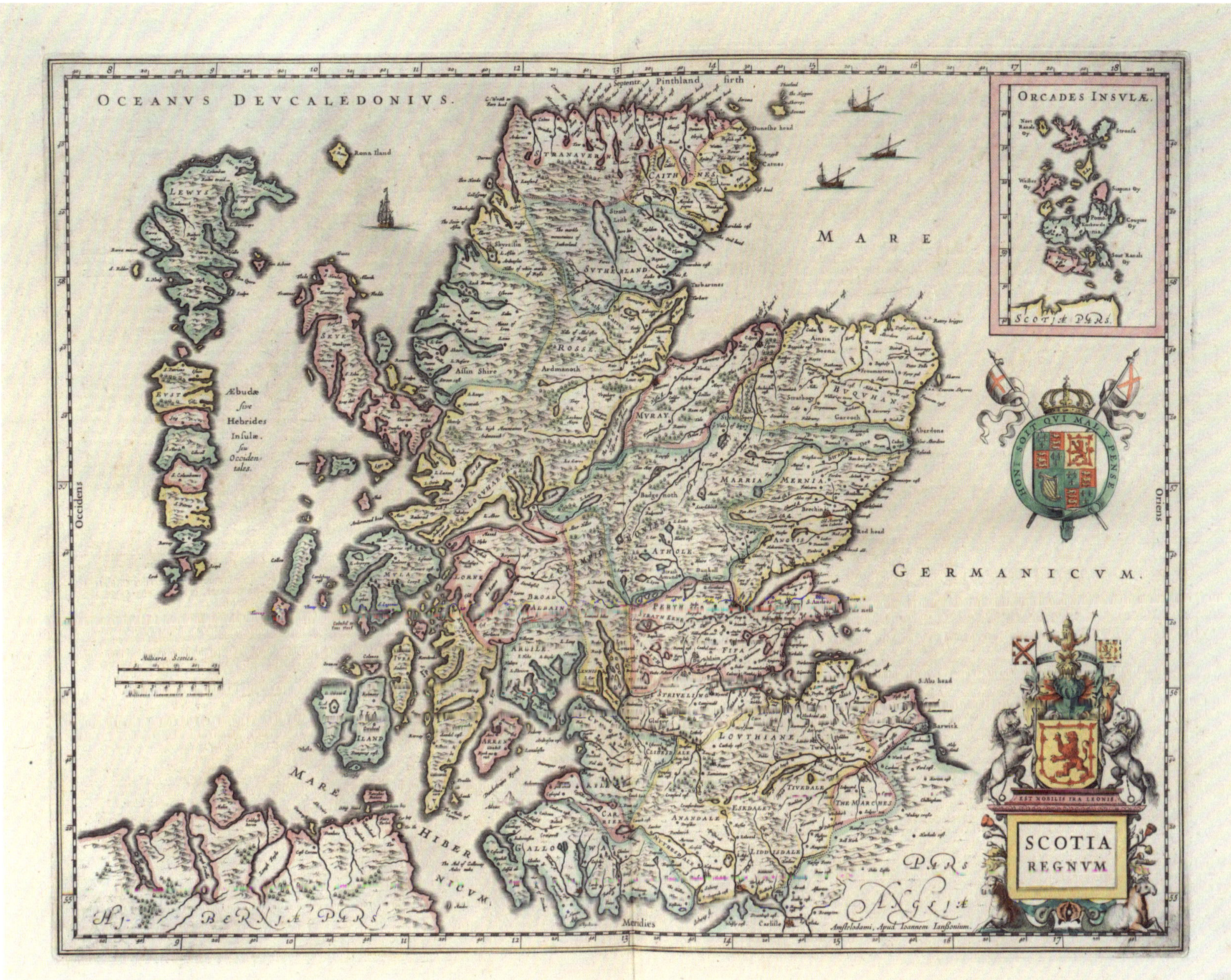

Scotia regnum
Map of Scotland
From: Johannes Janssonius, *Atlas Novus* [volume 4, Atlas of the British Isles], 1646
ANTWERP, THE PHOEBUS FOUNDATION

Globe race

Just as each successive new atlas was outdoing the last, a 'globe race' was also raging between the Blaeu family and the duo Hondius and Janssonius. Jodocus Hondius I kept up good relations with maritime cartographers like Petrus Plancius, who passed on to him astronomical knowledge and information gathered on distant voyages of discovery. Willem Jansz. Blaeu was a maritime expert and oversaw the scientific quality of the Blaeu globes. His was the last word when it came to size, surpassing Hondius's globes in 1616 with a celestial and earth globe measuring 68 centimetres in diameter.

Johannes Janssonius and Jodocus Hondius II likewise joined forces to produce globes. The younger Jodocus continued to build on his father's work for many years. The 1623 globe is an example of this collaboration: engraved by Abraham Goos (born 1589), it had a diameter of 44 centimetres and can be seen as a response to Willem Blaeu's pair, a new edition of which had been launched in 1622. Hondius and Janssonius set out to persuade users what exactly it was that made *their* globe worthwhile. The text on the globe states that it elegantly reflects the most recent changes, discoveries and corrections, and that it, thus, is the very latest and most accurate model at that point in time. The duo's world globe in The Phoebus Foundation's collection was not made at the same time as their celestial globe. The terrestrial version belongs to the first edition of 1623, while the celestial one is a later production – the third edition from 1648, engraved by Adriaan Metius (1571-1635) and marketed by Janssonius. In stylistic terms, it closely resembles Blaeu's great celestial globe of 1616, from which it probably also copied a lot of the contents.

The pair is dedicated to the Dutch West India Company. By 1623, the Dutch were actively seeking to secure colonies in America, and Henry Hudson's 'voyage of discovery' is commemorated by a text placed on the globe near Hudson Bay. Jodocus Hondius I had been involved in the search for a passage to Asia in 1609. Hudson was acquainted with Jodocus personally and visited his home with Petrus Plancius so that Jodocus could incorporate details of Hudson's voyages in his cartography and they could discuss planned voyages to the East Indies. The globe also features the Le Maire Strait, named after Jacob le Maire (⟶ Chapter VIII).

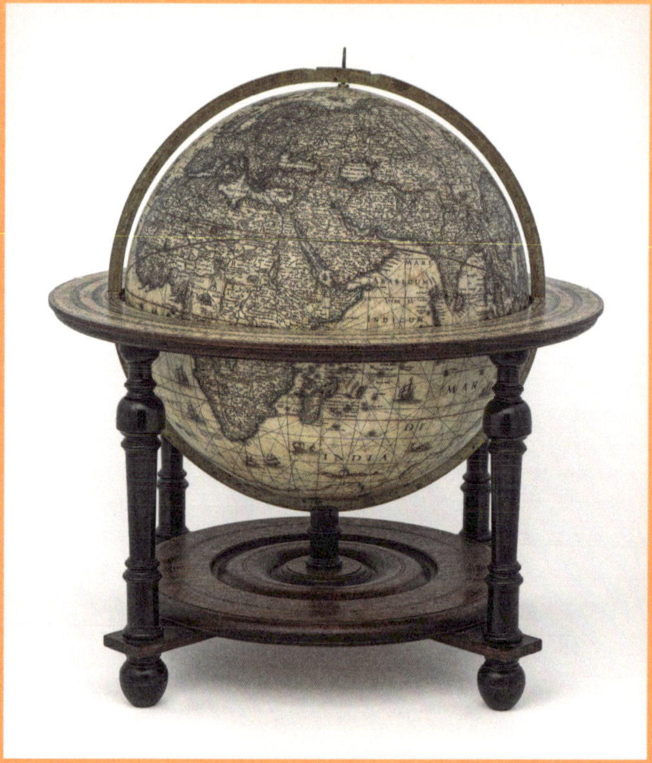

Abraham Goos, Jodocus Hondius and Johannes Janssonius
Terrestrial globe, 1623
Plaster, brass, wood and copper engraving, 65.5 × 58.5 × 58.5 cm
(diameter: 44 cm)
ANTWERP, THE PHOEBUS FOUNDATION

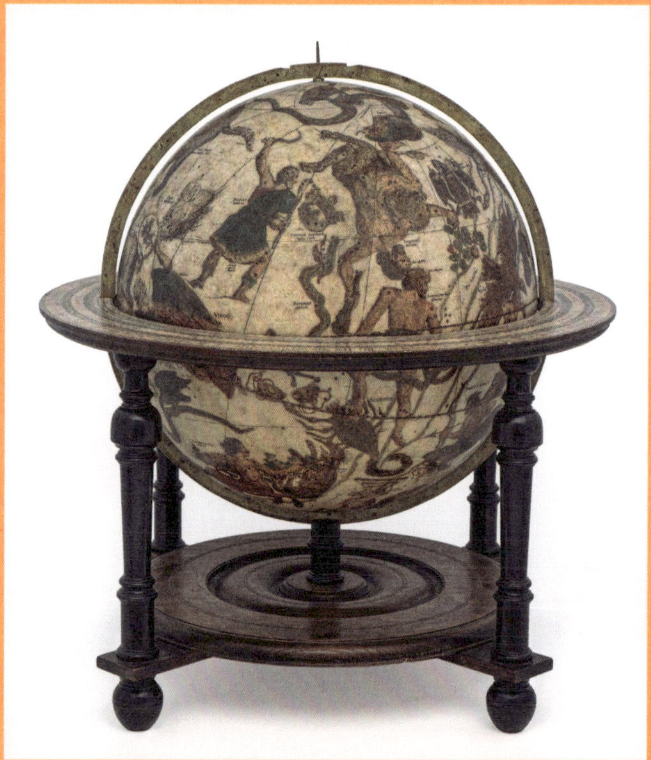

Adriaan Metius after Johannes Janssonius
Celestial globe, 1648
Plaster, brass, wood and copper engraving, 65.5 × 59.5 × 59.5 cm
(diameter: 44 cm)
ANTWERP, THE PHOEBUS FOUNDATION

BIGGER AND BIGGER

The greatest achievement in the atlas race was yet to come, however: a comprehensive cosmography in the form of maps, text and pictures. Janssonius made the first move with his biggest atlas of all; the *Novus Atlas Absolutissimus* (Amsterdam, 1658) was unprecedented in size and ran to eleven volumes and up to 547 maps, describing every city and country, the heavens and the earth. Only three copies of the work are known today. Yet it amounted to little more than an expanded version of the *Atlas Novus*, Janssonius having split each of its five volumes in two and added some new maps. It was lacking in both typographical consistency and flair, the 'X factor' that Blaeu's great atlas possessed in spades.

Joan Blaeu would top it a few years later with his *Atlas Maior*, consisting of eleven volumes and some six hundred maps. Its high price only made it a more desirable luxury product: an uncoloured copy of the series cost 350 guilders and a coloured copy 450 – amounts in the region of what a Reformed minister earned in a year. Despite the fact that it was unfinished (the atlas was actually intended to show the 'earthly realm, the sea and the heavens') and not every country was represented equally, the project proved an immense success. Nor did it seem to matter that half of the maps were old, pre-1638 productions. The *Grooten Atlas* would settle the rivalry between the Janssonius and Blaeu firms in Amsterdam once and for all. Joan Blaeu had taken Dutch cartography as far as it could go for the time being and died a few years after the atlas was published, shortly after a fire caused irreparable damage to his publishing house in 1672 – the Disaster Year for the Dutch Republic more widely, when it found itself at war with England, France and several German territories all at once. This was the beginning of the end for the Blaeu firm: while Joan's sons soldiered on for a while, they were unable to emulate their father's success.

The decline of the Blaeu company was matched by the stagnation of cartography in the Low Countries after the Disaster Year, and the centre of map production shifted in the late seventeenth and early eighteenth centuries to England, France and Germany. It was the end of an era, which possibly lent even greater symbolic importance to the *Atlas Maior*, a legendary atlas that represented a peak of Netherlandish cartography that would never be attained again.

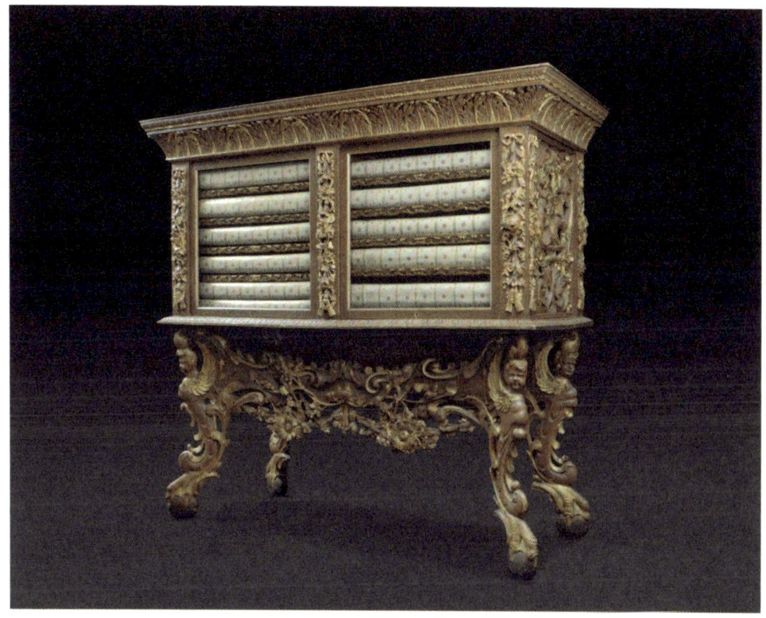

Ornamental cabinet for the atlases of Joan Blaeu, seventeenth century
170 × 160 × 70 cm
AMSTERDAM, COLLECTION ALLARD PIERSON, UVA

Pocket atlas

Not every atlas was as large as possible. The Amsterdam-based mapmaker and engraver Jan van Jagen (1709-1800) compiled a conveniently sized atlas that could be carried around in a pocket. The vignette on the title page shows two putti playfully consulting their maps, illustrating the purpose of this mini-atlas as something you could refer to yourself when travelling. As the subtitle points out, however, it was, besides adventurers, also aimed at homebodies following the news: 'Pocket atlas of the world for the traveller, merchant, soldier and mariner, and for convenience and utility when reading newspapers.' Henceforth, curious citizens reading reports in the press could immediately turn to a map of the region or country in question. The relevant map could be found by simply looking it up in the list at the front of the book.

The back of the pocket atlas, meanwhile, stated the title in French, and the list of all the maps was also accompanied by a version in that language, enabling an international readership to use the marvellous little publication too. The order in which the maps are presented in the atlas suggests that the intended users were expected to stay relatively close to home: a map of the earth and one of the heavens are followed by maps of Europe, including one showing all seventeen Netherlandish provinces, one of the Southern Netherlands and one of the Republic of the Seven United Provinces (⟶ Chapter V), before maps of all the individual regions. They were mostly designed and engraved by Van Jagen, with the earliest dating from 1759 and the last from 1793.

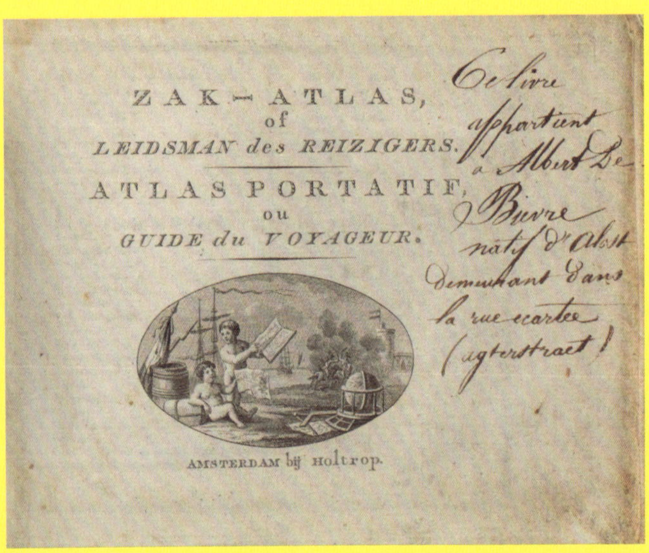
ZAK-ATLAS,
of
LEIDSMAN des REIZIGERS.
ATLAS PORTATIF,
ou
GUIDE du VOYAGEUR.
AMSTERDAM bij Holtrop.

Ce livre appartient a Albert De Bievre natif d'Alst demeurant dans la rue écartee (agterstraet)

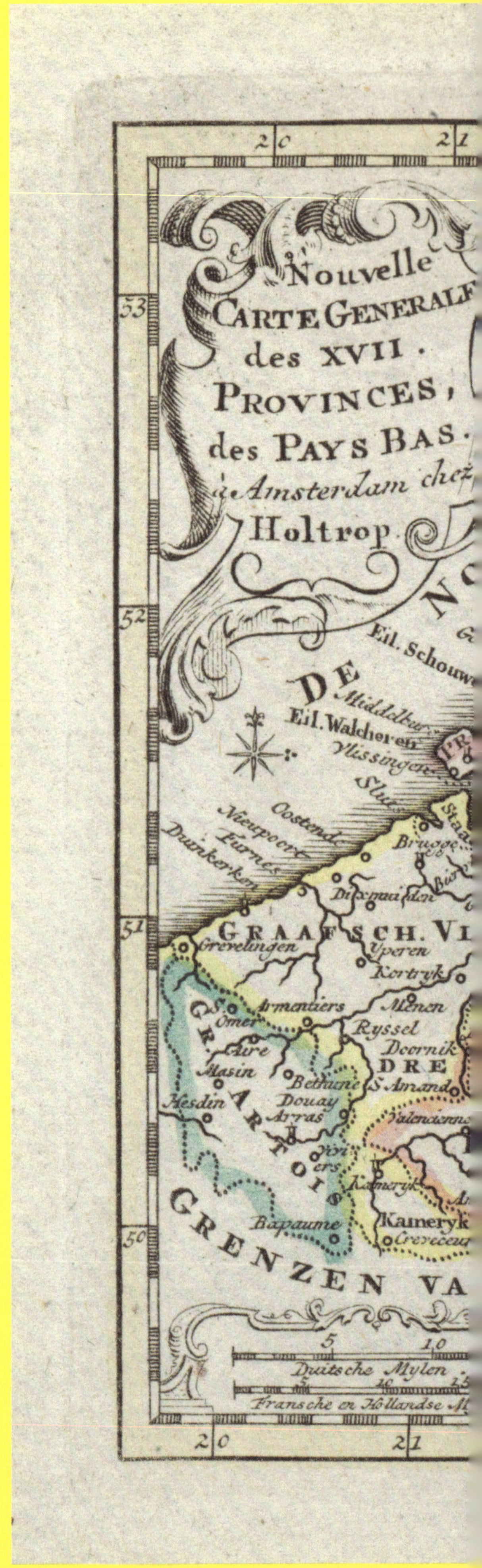

Title page
From: Jan van Jagen, *Zak-atlas, of Leidsman des reizigers*, after 1793
ANTWERP, THE PHOEBUS FOUNDATION

Nouvelle Carte Generale des XVII Provinces, des Pays Bas
Map of the Seventeen Provinces
From: Jan van Jagen, *Zak-atlas, of Leidsman des reizigers*, after 1793
ANTWERP, THE PHOEBUS FOUNDATION

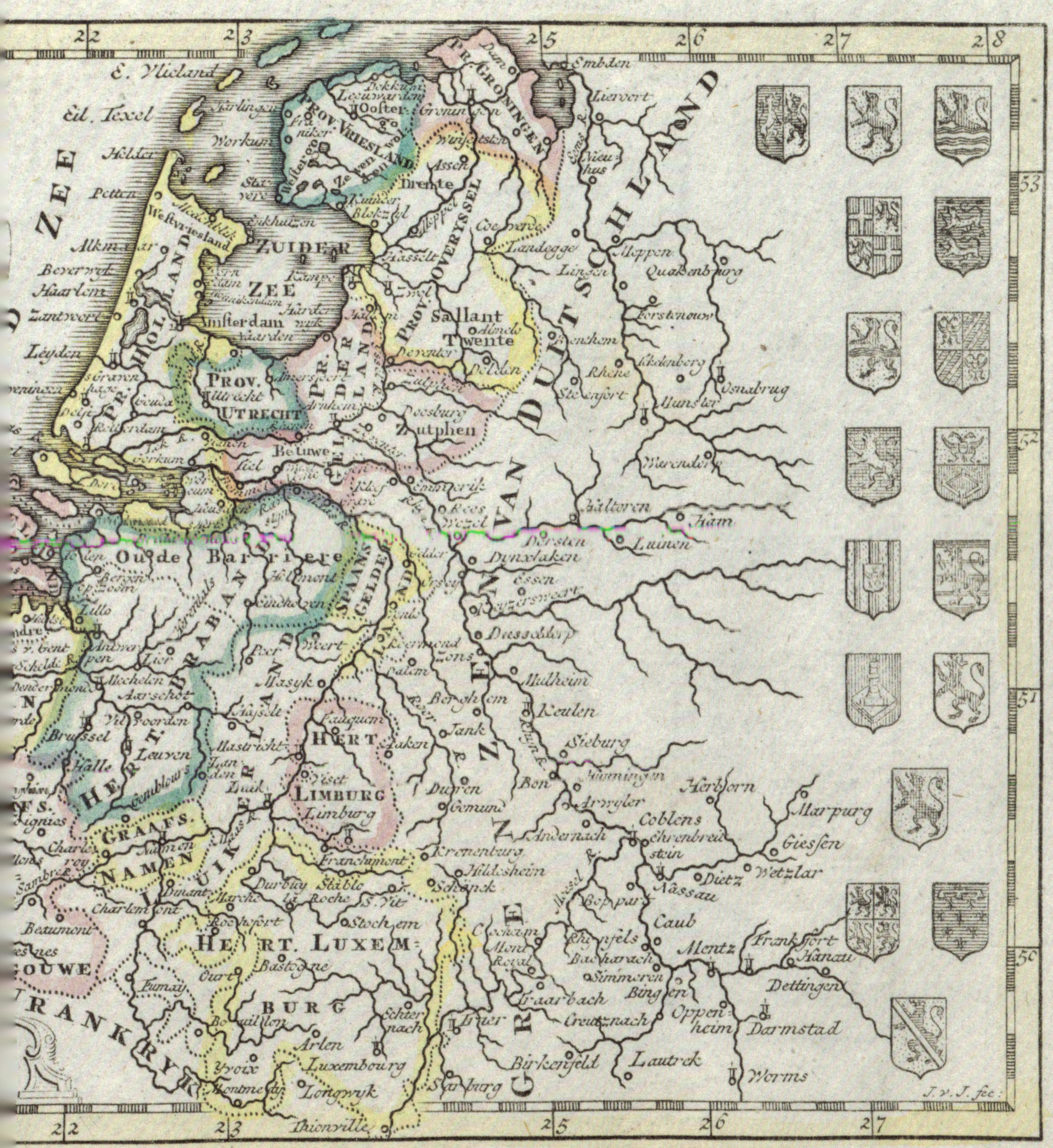

E. Vlieland
Eil. Texel
ZEE
Helder
Petten
Alkmaar
Beverwyk
Haarlem
Zantvoort
Leyden
Workum
Harlingen
Leeuwarden
PROV. VRIESLAND
Dam
GRONINGEN
Groningen
Embden
Assen
Drente
OVERYSSEL
ZUIDER ZEE
Amsterdam
Westvriesland
HOLLAND
PROV. UTRECHT
UTRECHT
Deventer
Sallant
Twente
Zutphen
Doesburg
Betuwe
Emmerik
Rees
Wezel
Lingen
Meppen
Quakenburg
Munster
Osnabrug
Steenfort
Hamm
Dorsten
Dinslaken
Essen
Dusseldorp
Mulheim
Keulen
Sieburg
Bon
Andernach
Coblens
Ehrenbreitstein
Herborn
Marpurg
Giessen
Wetzlar
Dietz
Nassau
Caub
Mentz
Frankfort
Hanau
Dettingen
Bacharach
Bingen
Oppenheim
Darmstad
Worms
Creutznach
Lautrek
Birkenfeld
Trier
Saarburg
DUITSCHLAND
VAN
Oude Barriere
BRABAND
SPAANS GELDERLAND
Eindhoven
Weert
Maastricht
HERT. LIMBURG
Limburg
Brussel
Leuven
Halle
Aarschot
Mechelen
Lier
Zoom
NAMEN
GRAAFS.
Charleroy
Dinant
Marche
Rochefort
HERT. LUXEMBURG
Bastogne
Luxembourg
Arlen
Longwyk
Thionville
Montmedy
Yvoix
Bouillon
FRANKRYK
Beaumont
J. v. J. fec.

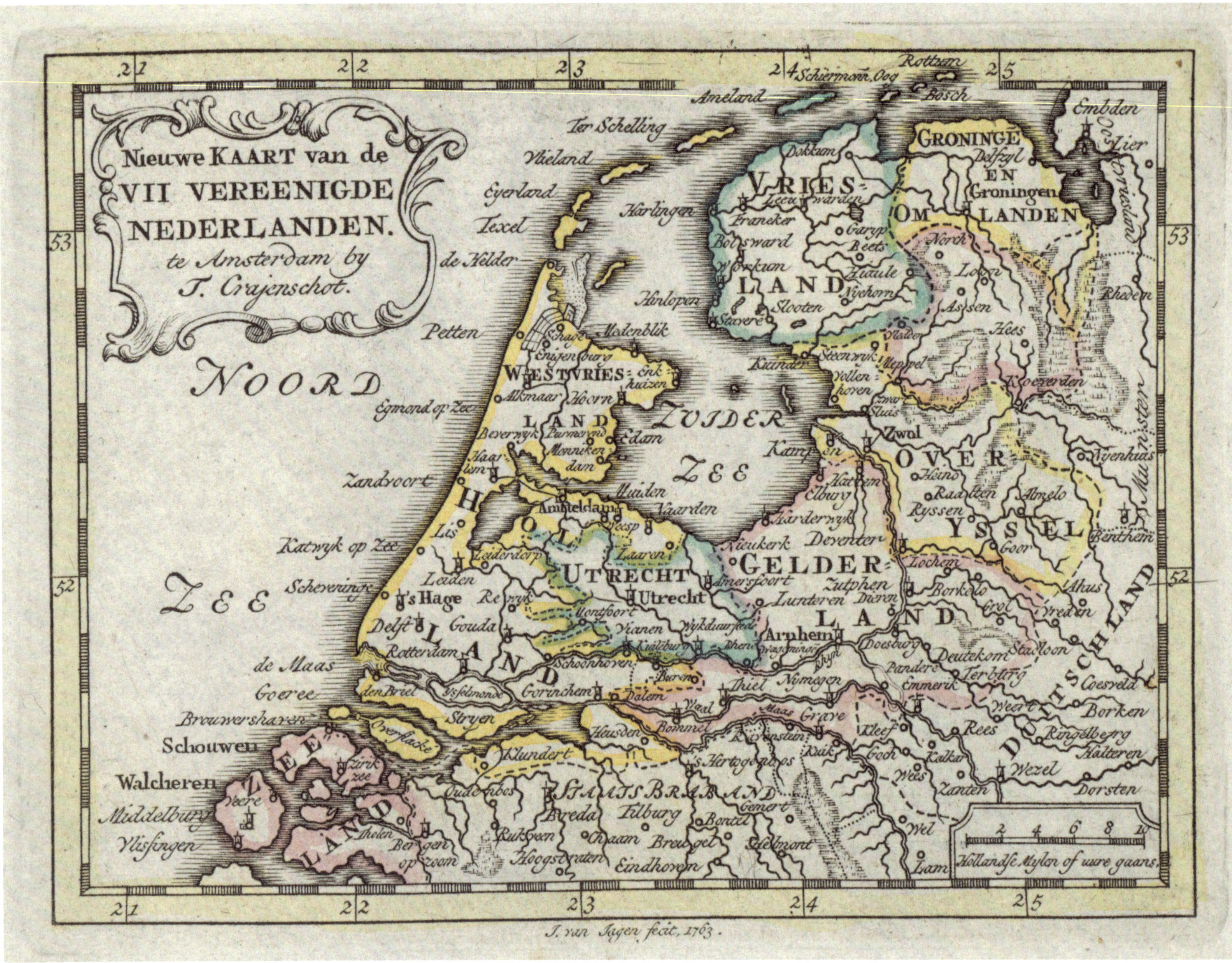

Nieuwe Kaart van de VII Vereenigde Nederlanden
Map of the Northern Netherlands
From: Jan van Jagen, *Zak-atlas, of Leidsman des reizigers*, after 1793
ANTWERP, THE PHOEBUS FOUNDATION

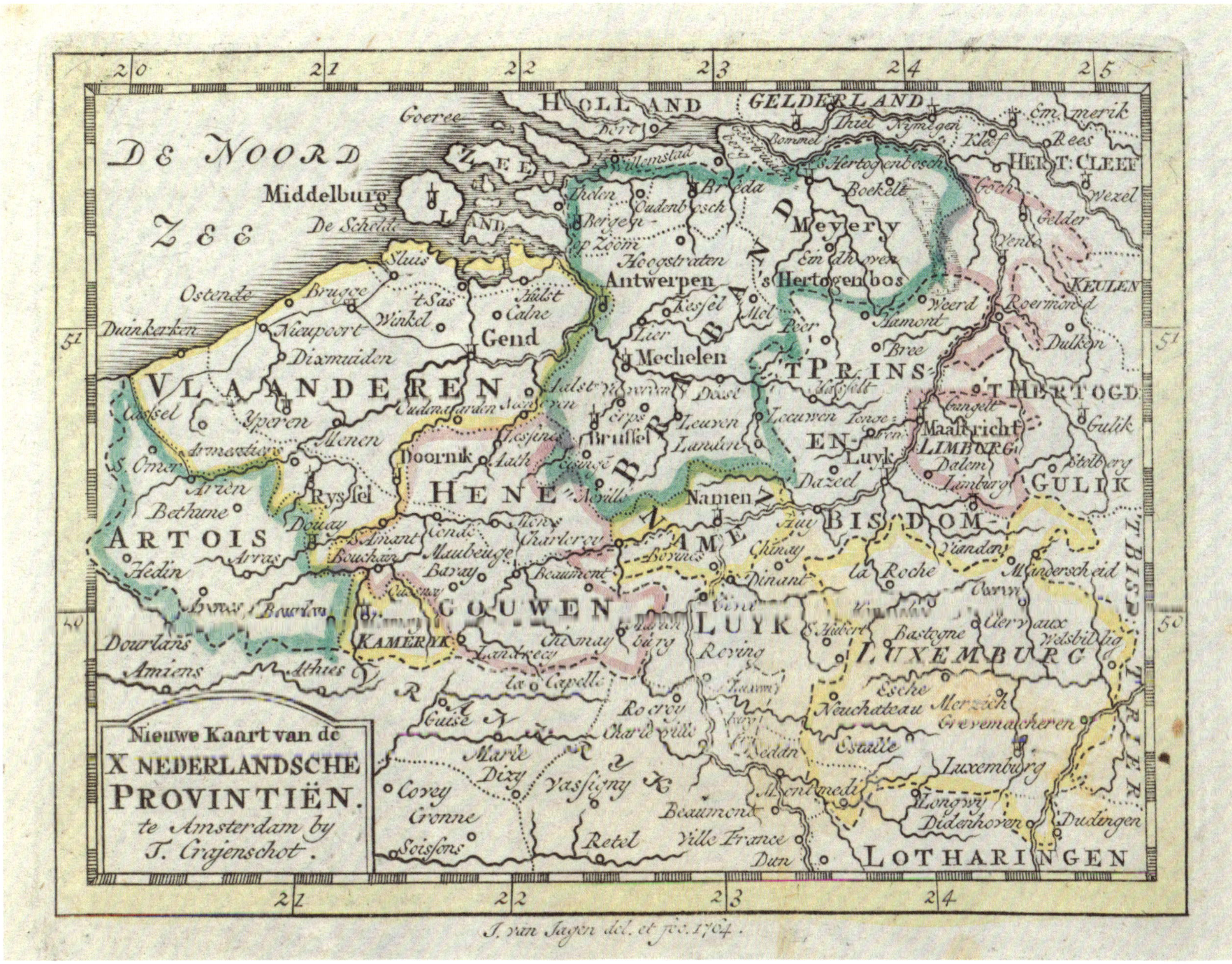

Nieuwe Kaart van de X Nederlandsche Provintiën
Map of the Southern Netherlands
From: Jan van Jagen, *Zak-atlas, of Leidsman des reizigers*, after 1793
ANTWERP, THE PHOEBUS FOUNDATION

LITERATURE

W. Blaeu & R. Skelton, *The Light of Navigation. Amsterdam 1612*, Amsterdam, 1964.

M. Donkersloot-de Vrij, *Drie generaties Blaeu: Amsterdamse cartografie en boekdrukkunst in de zeventiende eeuw*, Amsterdam, 1992.

C. Koeman, *The History of Abraham Ortelius and His Theatrum Orbis Terrarum*, Lausanne, 1964.

R. Skelton, *County Atlases of the British Isles, 1579-1703*, London, 1970.

M. van den Broecke, P. van der Krogt & L. Voet, *Abraham Ortelius and the First Atlas: Essays Commemorating the Quadricentennial of His Death, 1598-1998*, 't Goy, 1998.

P. van der Krogt, *Globi Neerlandici: The production of globes in the Low Countries*, Utrecht, 1993.

P. van der Krogt, *Koeman's Atlantes Neerlandici; volume II: The Folio Atlases Published by Willem Jansz. Blaeu and Joan Blaeu*, 't Goy, 2000.

P. van der Krogt, *Atlas Maior van 1665*, Cologne, 2010.

D. Wildeman, *De Wereld in het klein. Globes in Nederland*, Amsterdam / Zutphen, 2006.

K. Zandvliet, *De wereld van de familie Blaeu*, Zutphen, 2022.

The prevailing belief in classical antiquity was that the world and everything in it were made up of four elements: fire, water, air and earth. Humans were similarly thought to be composed of four 'temperaments', which balanced one another not only physically but also spiritually. Fire represented desire, water the emotions, earth stability and air thought. Following in his father's footsteps and in collaboration with Frans Francken II (1581-1642), Jan Brueghel II produced a series of allegorical paintings on the theme. We are struck in his painting *Air* by the armillary sphere held by the central female figure. This instrument, which contemporaries of Brueghel and Francken used to calculate their position on the earth with reference to the stars, has become an iconographic symbol of the universe and knowledge of it.

It appears again, accompanied by other instruments, in another painting by Francken. The man holding the globe is none other than Abraham Ortelius, while the other, raising a meaningful finger, is Justus Lipsius (1547-1606), a Southern Netherlandish humanist and Latin scholar at the university in Leuven. They are surrounded by contemporary paintings, sculptures, flowers and other 'naturalia' or specimens from the natural world, such as shells, fish and shellfish. The coloured atlas lying open by the window on the right might well refer to Ortelius's *Theatrum Orbis Terrarum*. Set in a remarkable interior, this is a scene that never really took place but is an imaginary representation of the typical Renaissance *Kunstkammer*: a collection of natural and man-made wonders of the world, a cosmos in miniature assembled by wealthy art lovers.

CHAPTER XI

MOVING HEAVEN AND EARTH

SCIENCE, INSTRUMENTS AND THE COSMOS

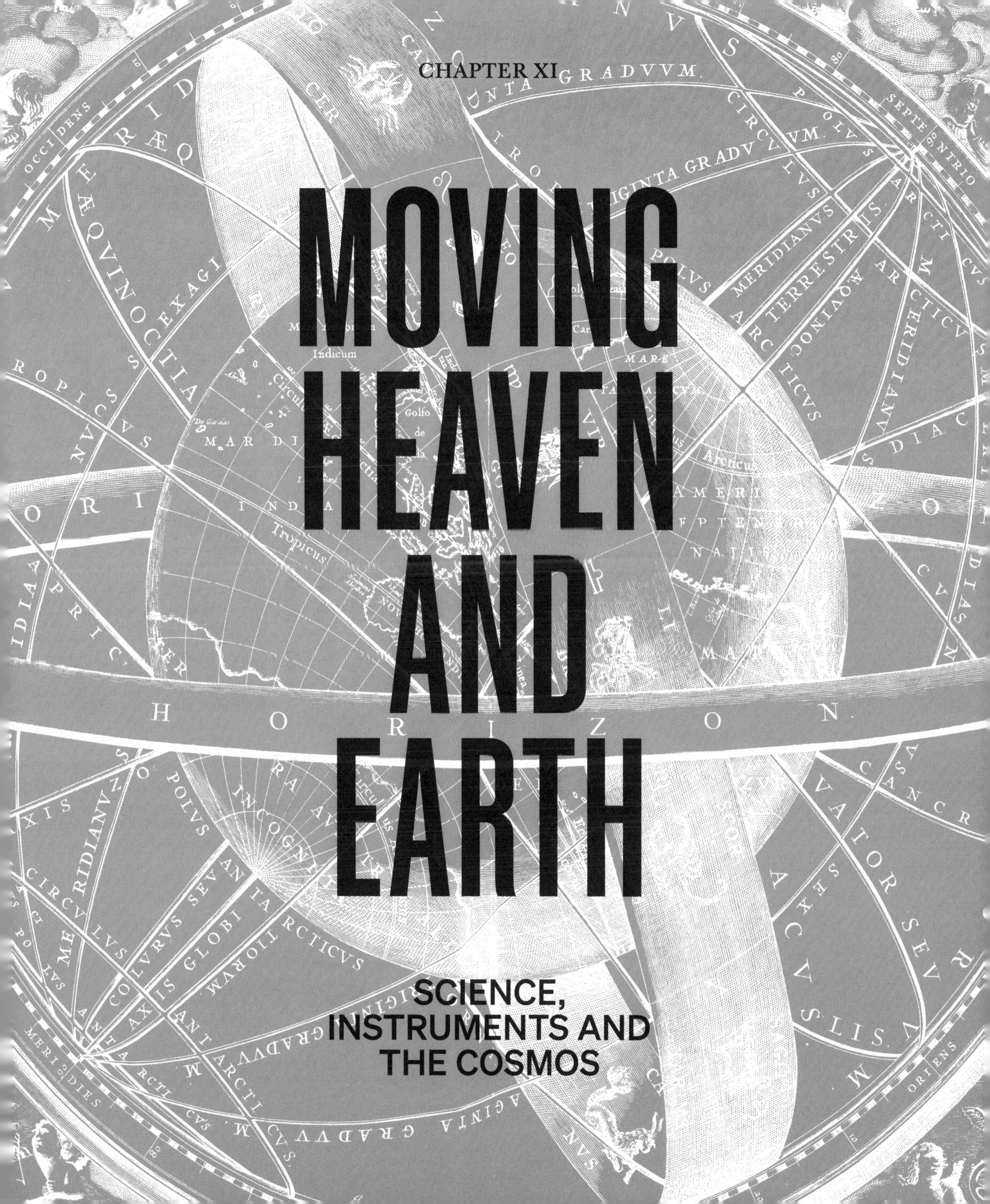

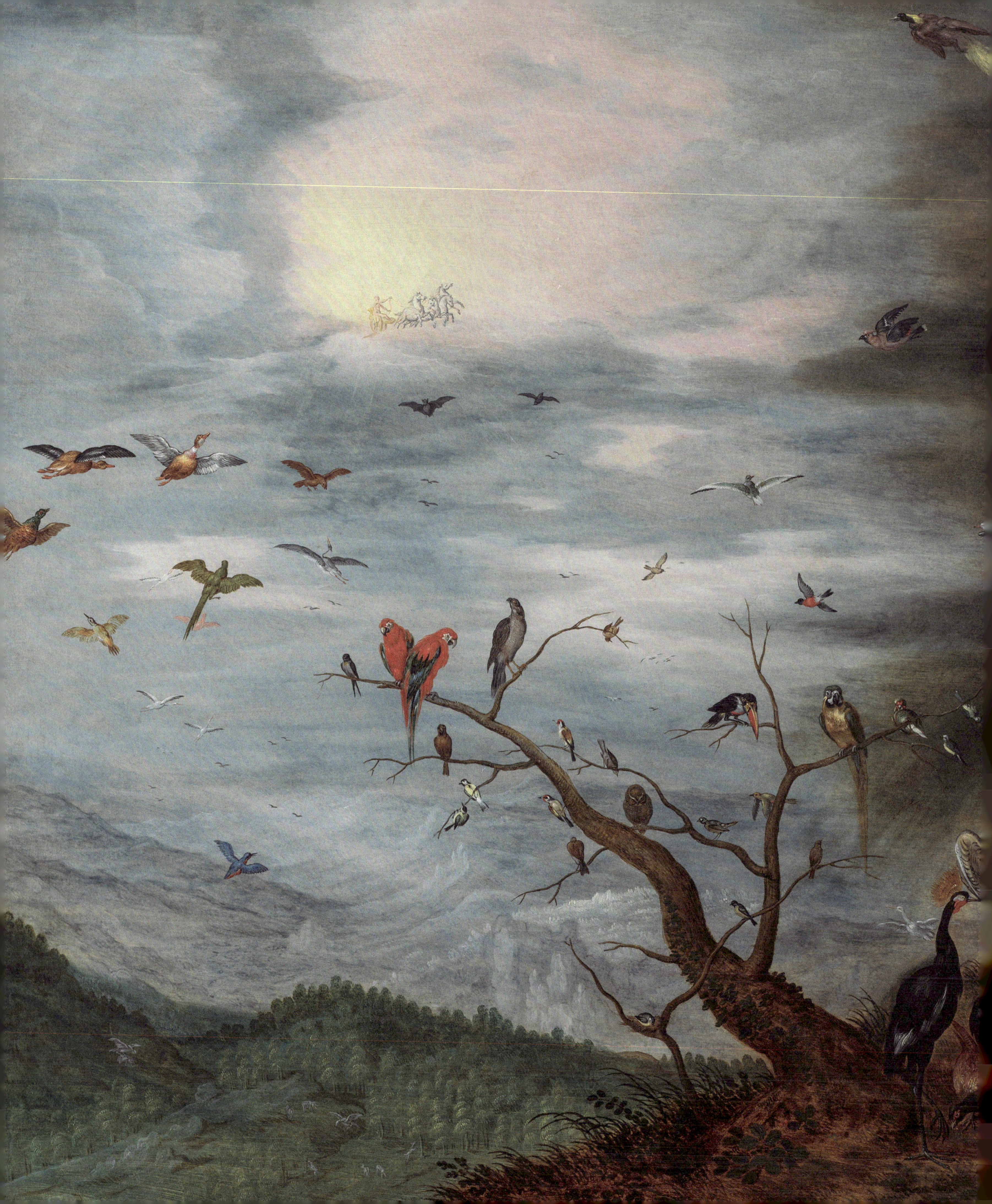

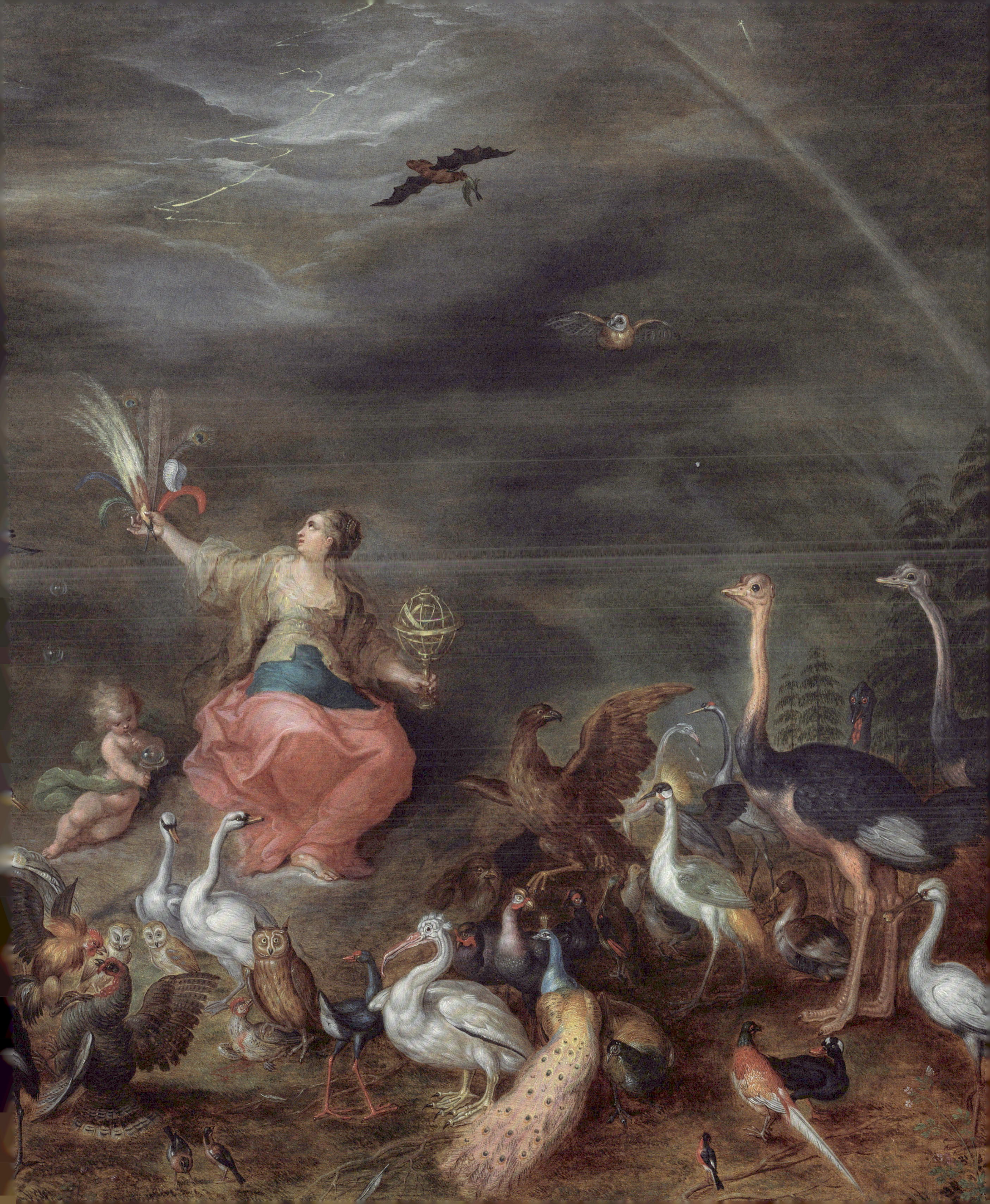

THE SILICON VALLEY OF THE LOW COUNTRIES

Leuven was the scientific centre in the sixteenth-century Low Countries. Where Antwerp was the capital of commercial cartography and the leading city for printers, publishers and merchants, Leuven's interests were more intellectual. Scholars from all over met there to share knowledge and technical innovations, making the city a veritable 'Silicon Valley of the Low Countries', frequented by the likes of Gemma Frisius, Gerard Mercator and Jacob van Deventer. New discoveries and insights were promptly transformed into improved maps and more effective instruments. There was little specialism at the time: surveying required a knowledge of astronomy, while cosmography, with all the calculations it entailed, was more a branch of mathematics. A material culture emerged in Leuven too in which research and mathematical knowledge were applied in useful maps and practical instruments.

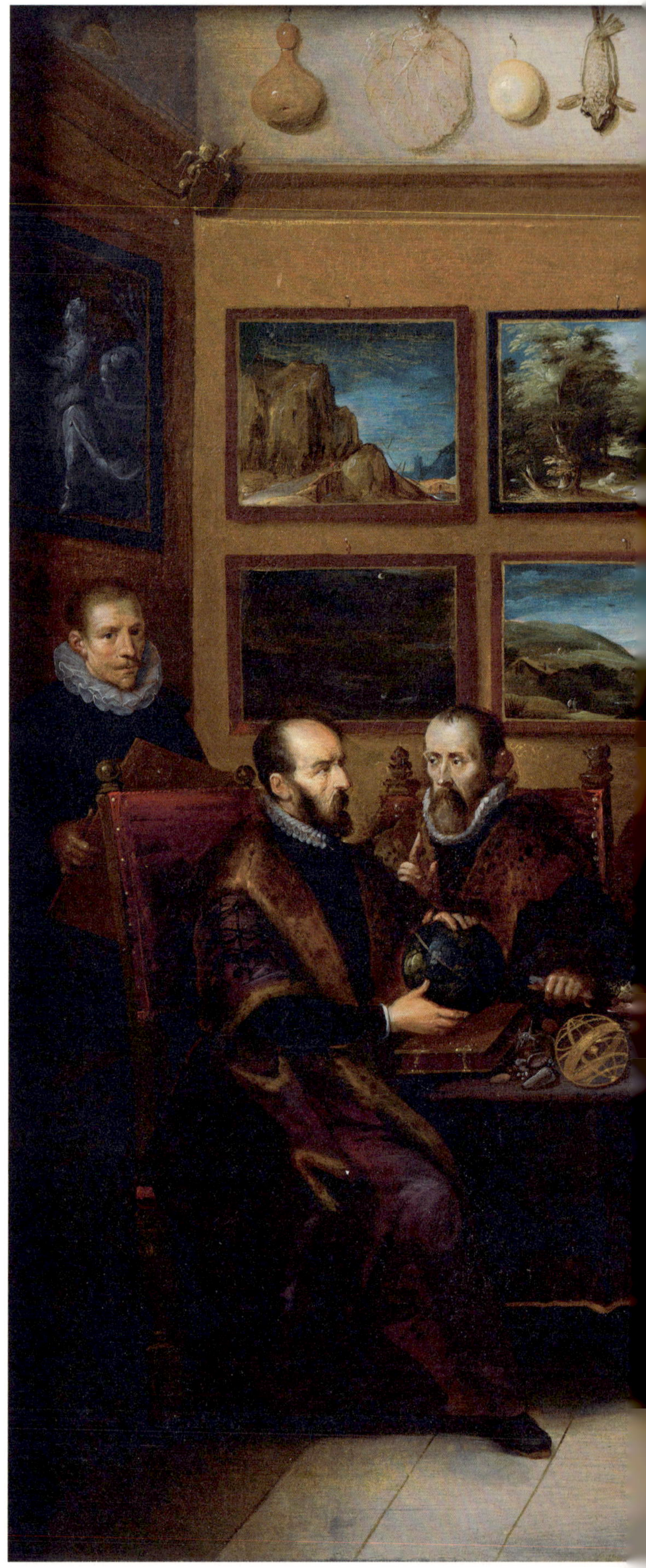

pp. 286-287
Jan Brueghel II and Frans Francken II
Four Elements: Air, c.1632-1642
Oil on canvas, 57.5 × 94 cm
ANTWERP, THE PHOEBUS FOUNDATION

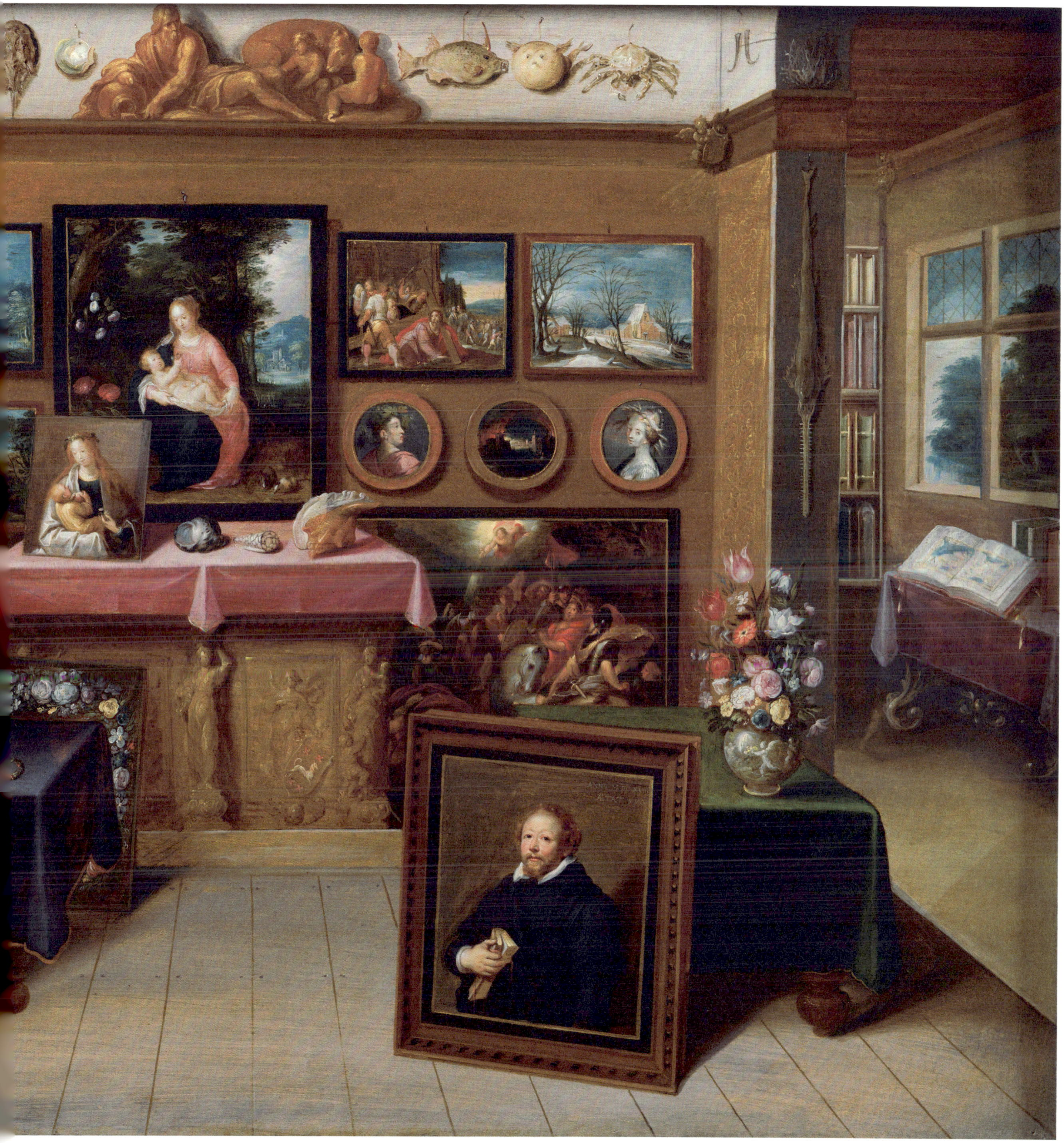

Frans Francken II
Collector's cabinet with Abraham Ortelius and Justus Lipsius, 1617
Oil on canvas (transferred from panel), 53 × 74 cm
ANTWERP, THE PHOEBUS FOUNDATION

Cosmography

Although cosmography refers to the description of the heavens and earth, at the time of the Renaissance the discipline existed more at the intersection of cosmology (the study of the structure and evolution of the universe), astronomy, geography and mathematics. The terms 'cosmography' and 'geography' also tended to be used interchangeably, although strictly speaking a geographer is concerned solely with the description of the earth. Sebastian Münster was one such Renaissance cosmographer, and his *Cosmographia* contains images and descriptions of places around the world but not of the stars and the universe (⟶ Chapter IX).

The science of cosmography consisted mainly of mathematical measurements and astronomical principles: cosmographers did not set out to explore and record the world themselves. Science continued to develop in the early modern period through the 'discovery' of previously unknown countries and territories, travel reports, improved measuring instruments and the production of accurate maps. In this way, the broad, overarching field of 'cosmography' split into the separate branches of geography, surveying, cartography and astronomy.

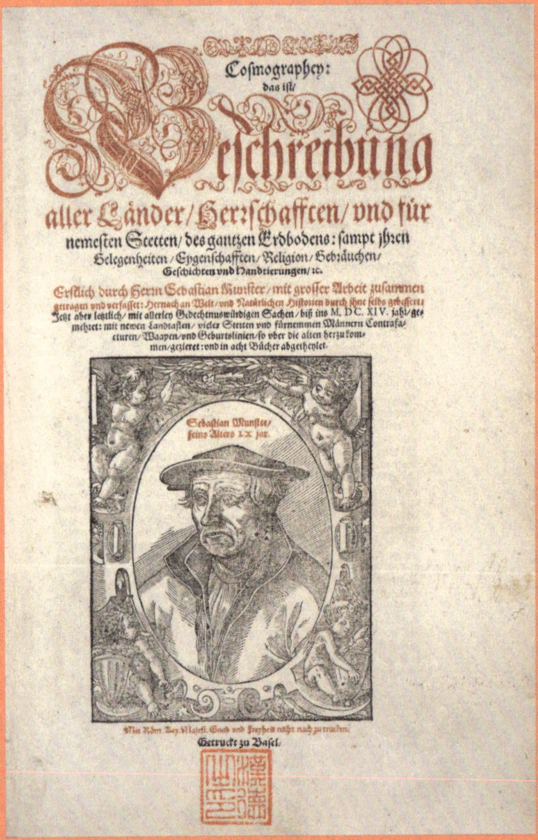

Title page
From: Sebastian Münster, *Cosmographey: das ist Beschreibung aller Länder, herzschafften, und fürnemesten Stetten, des gantzen Erdbodens*, 1614
ANTWERP, THE PHOEBUS FOUNDATION

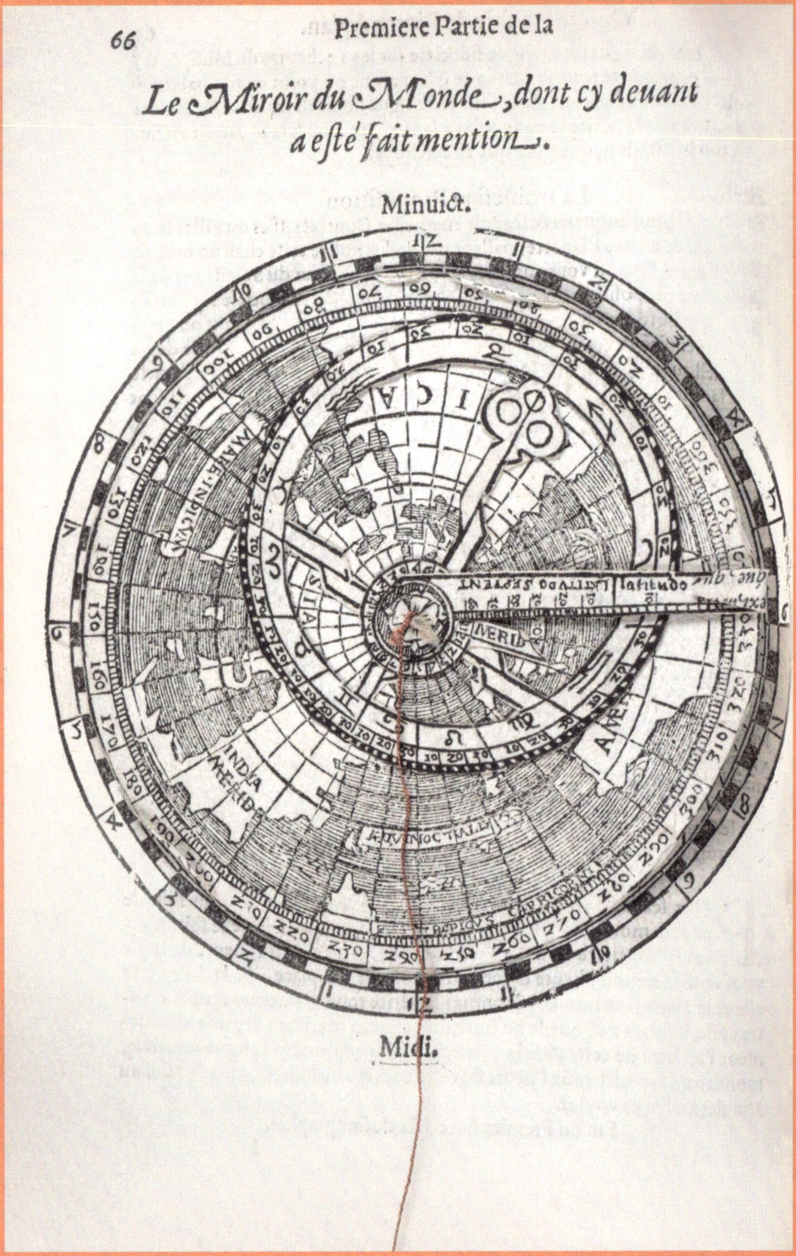

Volvelle
From: Peter Apian and Gemma Frisius, *Cosmographie, ou description des quatre parties du Monde*, 1581
ANTWERP, THE PHOEBUS FOUNDATION

PAPER INSTRUMENTS

One of the books circulating in Leuven was the German mathematician and cosmographer Peter Apian's influential *Cosmographicus Liber* (first edition: Landshut, 1524), which built on Ptolemy's *Geographia* or *Cosmographia* (⟼ Chapter IX) to create a milestone in the study of geography, mathematics and astronomy. The main body of the text consists chiefly of instructions on measuring latitude and longitude (⟼ p. 182) and other methods for determining human beings' place in the cosmos. Although Apian himself studied in Vienna and Leipzig, his work was seized on in Leuven. The fact that it was a purely scholarly publication is evident from its many amended and revised editions. The *Cosmographicus Liber* only achieved true fame, however, when Gemma Frisius, a professor in Leuven, published a revised edition in 1529-1530 with an appendix on triangulation (⟼ p. 28). The Phoebus Foundation's copy is the 1581 French translation, published in Antwerp, which also contains this important appendix together with new material on America that had previously been incorporated in the Spanish translation (printed in Antwerp in 1575).

A noteworthy feature of Apian and Frisius's treatise is the use of volvelles – paper instruments that readers had to put together themselves with string and glue and stick to the book. They could then rotate and move them to gain a better understanding of the astronomical principles in an interactive and playful manner. The volvelle's system of movement can already be found in medieval manuscripts. Apian added four more, mimicking the division of the earth into climate zones, the use of parallels and meridians, the calculation of latitude and various methods for determining longitude. An example is the volvelle of the polar projection of the earth's northern hemisphere ‘the mirror of the world' (‘*le miroir du monde*') as it is called – which indicates the latitude and longitude of locations on the globe but also simulates the earth's rotation around the polar axis. The paper tool is related to the astrolabe and allowed the relative time in different parts of the earth to be read off.

It took no little imagination to achieve anything like an understanding of cosmography, so embedding a rotating paper instrument like this in the book as a visual aid meant that even the less privileged student could get to work on the material. The real thing – often made of brass – could cost as much as a house. Paper instruments might not have produced the same results, but they offered a way to practise and master astronomical methods. The individual parts of the volvelle were usually printed as woodcuts in the text for the reader to cut out, assemble and use.

Little volvelles were also placed inside other instruments, such as the ivory pocket sundial. The rotating disc and pointer are silver, while the ivory part features an engraved table of latitudes for twelve cities (including Antwerp, Ghent, Enkhuizen and London).
An anonymous Flemish instrument maker created the little piece in 1590. Its small lightweight format suggests that it was intended for travel use.

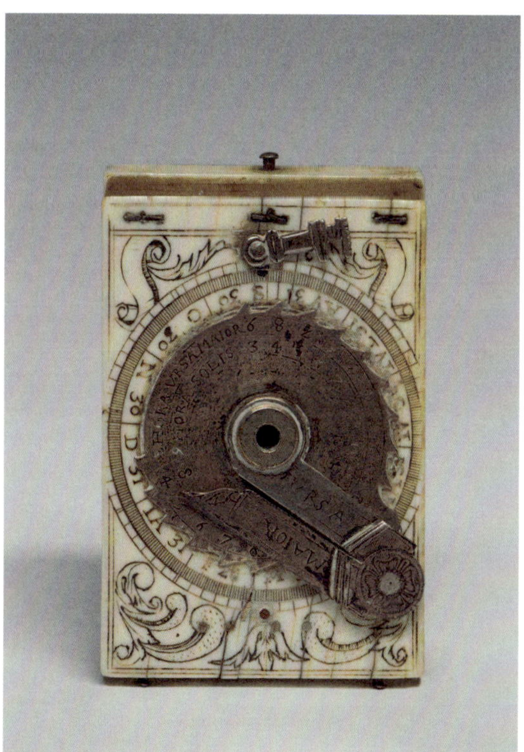

Sundial, 1590
Ivory and silver, 7.5 × 5 × 2 cm
ANTWERP, THE PHOEBUS FOUNDATION

Astrolabe

Arab astronomers were already making and using astrolabes – small, simplified versions of the night sky – in the Middle Ages. They are sophisticated, multifunctional instruments, the main purposes of which are telling time, determining location and calculating latitude. Astrolabes mostly comprise a number of rotating discs fixed to a central peg representing the north celestial pole. The thick disc at the bottom is called the *mater*, while the one in the middle is called the *tympanum* (some have several *tympana*) and indicates the horizon. The topmost, cutout disc is called the *rete* and represents the constellations. Rotating the discs allows the daily movement of the stars to be simulated. The time could be worked out using the astrolabe to measure the position of the stars at two intervals. Based on the difference, it was possible to determine the hour.

The back of the instrument (and hence the *mater* disc) was marked with degrees and was used to measure the height of buildings, trees or the sun, for example. Hence the rotating sight or *alidade*, which could be pointed at the sun in the daytime or the pole star at night, allowing the angle to the horizon to be read off against a plumb line. This angle then made it possible to read the elevation to that point and hence also the latitude. The ring attached to the instrument enabled the user to hold it up in the air to measure vertical angles.

The user of the astrolabe was able, as it were, to hold the very earth and the cosmos in their hand. It was, however, also an extremely complex and expensive instrument, engraved in Latin letters to boot, which meant it was reserved for a wealthy, educated elite. Although astrolabes remained in use until the eighteenth century, other, simpler instruments that focused on a single function became more desirable.

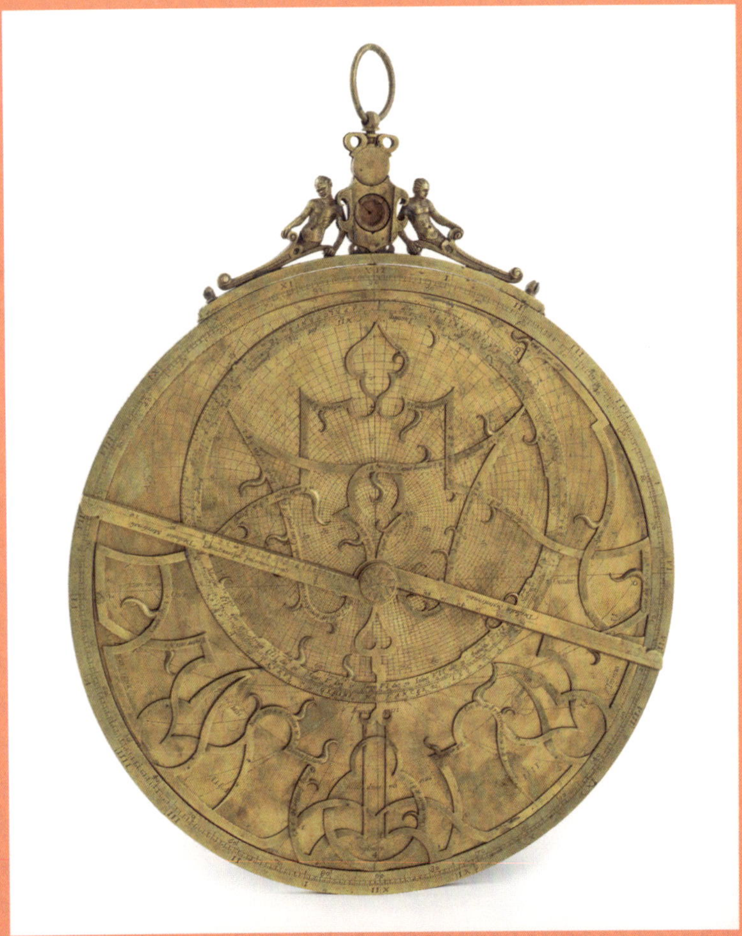

Gualterus Arsenius
Astrolabe, 1579
Engraved brass, 29.8 cm (diameter)
ANTWERP, THE PHOEBUS FOUNDATION

ARSENIUS'S ASTROLABE

The volvelle in Apian's treatise mimics the principles of an astrolabe. In its three-dimensional form, ingeniously expanded to include a variety of complex functions, the astrolabe was a measuring and calculating instrument that could be used to calculate longitude, latitude, time and much more besides. One of the prints in the *Nova Reperta* series shows the invention of the astrolabe. Another print in the same series depicts Amerigo Vespucci setting foot in Amerika holding one. Instruments of this kind were made of metal or wood and were expensive to produce. Some astrolabes were engraved with a map of the hemisphere or a star chart.

The Phoebus Foundation has one such sophisticated and ingenious instrument, made in Leuven by Gualterus Arsenius (1530-1580), born in Flanders and educated in the university city, where he studied under his uncle Gemma Frisius, among others. In the section on Leuven in his *Descrittione*, Guicciardini (→ Chapter VI) calls Arsenius 'a highly skilled artist in mathematical instruments' and numbers him among the university's great scholars. Arsenius is thought to have collaborated with Gerard Mercator, who made globes and instruments in addition to his celebrated maps, and he performed commissions for the king of Spain. This rare astrolabe is one of the 21 surviving pieces made by him.

A square diagram known as the '*Quadratum nauticum*' is engraved on the inside of the solid baseplate (*mater*). It was probably devised by Frisius, who included it in his reprints of Apian's treatise, and was a modified, extended version of Apian's original diagram. It was intended for use at sea to determine without further calculation the course between two points of known latitude and longitude. The difference in degrees was indicated using wires, which the mariner had to stretch across the longitudes and latitudes: the line between the two intersections of the wires then indicated the course.

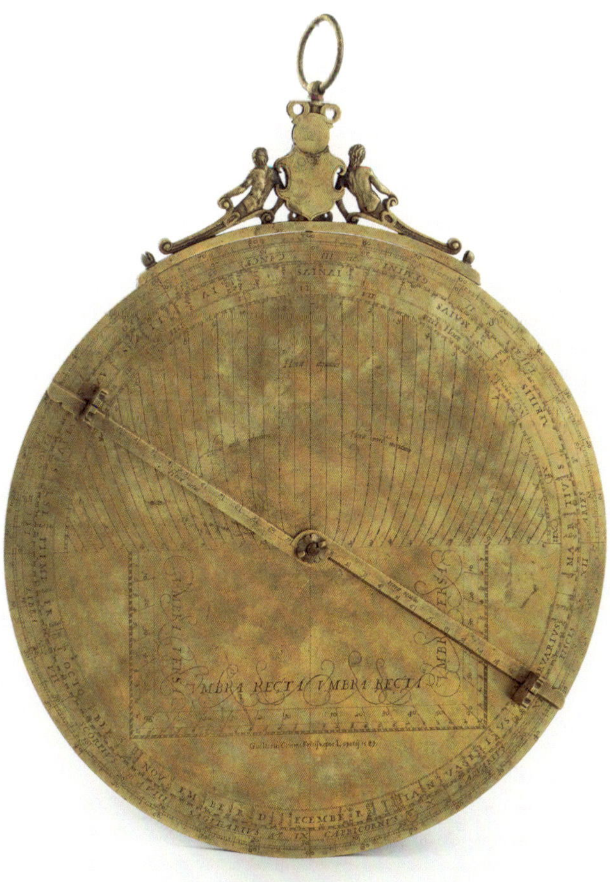

Gualterus Arsenius
Astrolabe, 1579
The inside of the astrolabe is engraved with a square diagram, the *Quadratum nauticum*

ASTRONOMICAL RING

Frisius and Apian's treatise is packed with descriptions of and instructions for other instruments. The 1539 edition added a supplement titled *Usus annuli astronomici*, a guide to using the astronomical ring, which comprised three rotating rings attached to each other to form a sphere. It was used, among other things, to determine the time at nightfall and to calculate the length of the day. In a similar manner to a sundial, the time can be read off using the sun's rays and the shadow they cast on a certain place in the ring.

The tool could also be used, however, for orientation and for measuring the height of buildings by shadow. The appendix that Frisius added to Apian's treatise includes a picture of a ring and of a person using it to measure the height of a building. The brass astronomical ring in The Phoebus Foundation's collection was attributed to the school of Leuven (the maker stays anonymous) and was probably made in the late sixteenth century. A 24-hour scale ([0]-3-6-9-12-9-3-[0]) is engraved on it, together with a calendar with the months indicated by zodiac signs.

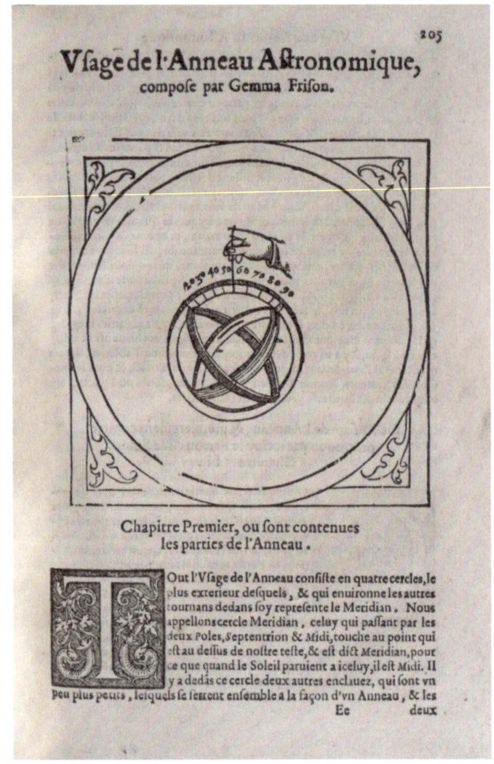

205

Vſage de l'Anneau Aſtronomique,
composé par Gemma Friſon.

Chapitre Premier, ou ſont contenues les parties de l'Anneau.

TOut l'Vſage de l'Anneau conſiſte en quatre cercles, le plus exterieur deſquels, & qui enuironne les autres tournans dedans ſoy repreſente le Meridian. Nous appellons cercle Meridian, celuy qui paſſant par les deux Poles, Septentrion & Midi, touche au point qui eſt au deſſus de noſtre teſte, & eſt dict Meridian, pour ce que quand le Soleil paruient a iceluy, il eſt Midi. Il y a dedās ce cercle deux autres enclauez, qui ſont vn peu plus petits, leſquels ſe ſerrent enſemble a la façon d'vn Anneau, & les

Ee deux

Illustration of an astronomical ring
From: Peter Apian and Gemma Frisius, *Cosmographie, ou description des quatre parties du Monde*, 1581
ANTWERP, THE PHOEBUS FOUNDATION

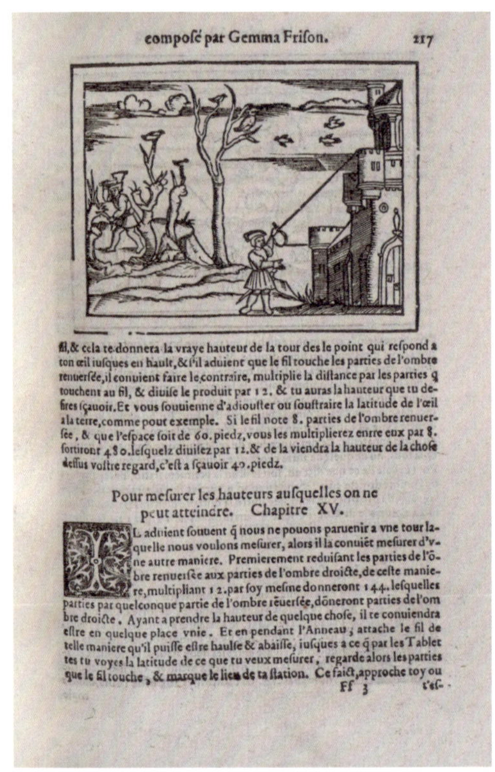

composé par Gemma Friſon. 217

fil, & cela te donnera la vraye hauteur de la tour des le point qui reſpond a ton œil iuſques en hault, & ſ'il aduient que le fil touche les parties de l'ombre renuerſée, il conuient faire le contraire, multiplie la diſtance par les parties q̃ touchent au fil, & diuiſe le produit par 12. & tu auras la hauteur que tu deſires ſçauoir. Et vous ſouuienne d'adiouſter ou ſouſtraire la latitude de l'œil a la terre, comme pour exemple. Si le fil note 8. parties de l'ombre renuerſée, & que l'eſpace ſoit de 60. piedz, vous les multiplierez entre eux par 8. ſortiront 480. leſquelz diuiſez par 12. & de la viendra la hauteur de la choſe deſſus voſtre regard, c'eſt a ſçauoir 40. piedz.

Pour meſurer les hauteurs auſquelles on ne peut atteindre. Chapitre XV.

IL aduient ſouuent q̃ nous ne pouons paruenir a vne tour laquelle nous voulons meſurer, alors il la conuiēt meſurer d'vne autre maniere. Premierement reduiſant les parties de l'ōbre renuerſée aux parties de l'ombre droicte, de ceſte maniere, multipliant 12. par ſoy meſme donneront 144. leſquelles parties par quelconque partie de l'ombre rēuerſée, dōneront parties de l'ombre droicte. Ayant a prendre la hauteur de quelque choſe, il te conuiendra eſtre en quelque place vnie. Et en pendant l'Anneau, attache le fil de telle maniere qu'il puiſſe eſtre hauſſé & abaiſſé, iuſques a ce q̃ par les Tablettes tu voyes la latitude de ce que tu veux meſurer, regarde alors les parties que le fil touche, & marque le lieu de ta ſtation. Ce faict, approche toy ou

Ff 3 t'eſ-

Illustration of a person using an astronomical ring to measure a building
From: Peter Apian and Gemma Frisius, *Cosmographie, ou description des quatre parties du Monde*, 1581
ANTWERP, THE PHOEBUS FOUNDATION

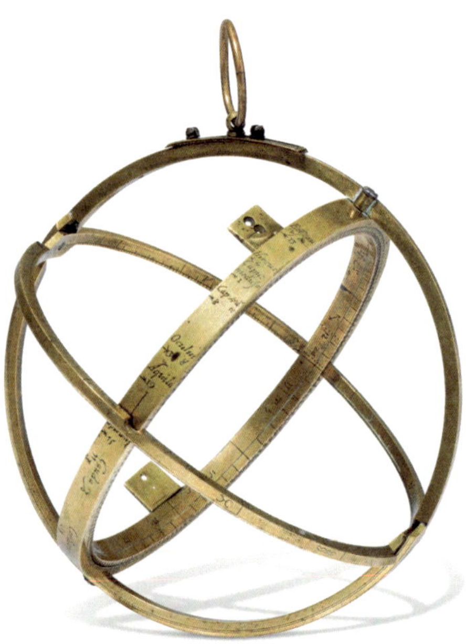

Astronomical ring, late sixteenth century
Brass, 15.5 × 12.4 × 0.8 cm
ANTWERP, THE PHOEBUS FOUNDATION

CELESTIAL REALM

Cosmography also caught the imagination of mapmakers and artists. The description of the earth and the universe was expressed in imposing atlases. Around the same time that Joan Blaeu launched his *Atlas Maior* (Amsterdam, 1662), the world's 'most beautiful celestial atlas' also made its first appearance. The *Harmonia Macrocosmica* by the German cosmographer Andreas Cellarius (*c.*1596-1665) was published in 1660 by Johannes Janssonius. It has come to be seen as the artistic and scientific pinnacle of celestial cartography, with 29 large-format maps depicting the night sky in high artistic quality.

Yet this was not the most progressive or innovative view of the universe: Cellarius's atlas was a combination of traditional constellations and new views of the galaxy. The publication *De revolutionibus orbium coelestium* by the scientist Nicolaus Copernicus (1473-1543) demonstrated that the sun rather than the earth lay at the middle of our solar system. Ptolemy's geocentric model, in which the earth was the centre of the universe, had been rejected by scholars but as yet without opprobrium. It co-exists in Cellarius's atlas with the new heliocentric system of Copernicus and the 'transitional system' proposed by Tycho Brahe. This might have been motivated by practical considerations too, in that it meant the book was not banned by the Catholic Church, which still adhered to the Ptolemaic view of the world.

The Phoebus Foundation's copy of the celestial atlas is a complete republication by Gerard Valk (1652-1726) and Pieter Schenk I, the publishers who had bought Johannes Janssonius's plates in 1694. They again did not merely incorporate the latest developments in astronomy. The book was evidently considered more of a decorative, eye-catching coffee-table production than a scientifically useful reference. All the same, the tension between the stubbornly conservative Ptolemaic view of the universe on the one hand and innovative astronomical research on the other persisted into the eighteenth century.

New constellations were added, for instance, such as the 'Toucan' and the 'Chameleon', which Petrus Plancius (⟶ Chapter VIII) devised based on observations by seafarers and inspired by the flora and fauna of the southern hemisphere. The plates are adorned with hovering figures, festive putti, astronomers in discussion, artful depictions of constellations, and other celestial and terrestrial scenes. The sumptuously decorated borders, meanwhile, once again incorporate the instruments that gave rise to this knowledge of the universe. The book came to be admired as a tribute to and celebration of the development of cosmography in Europe, from ancient times to the present.

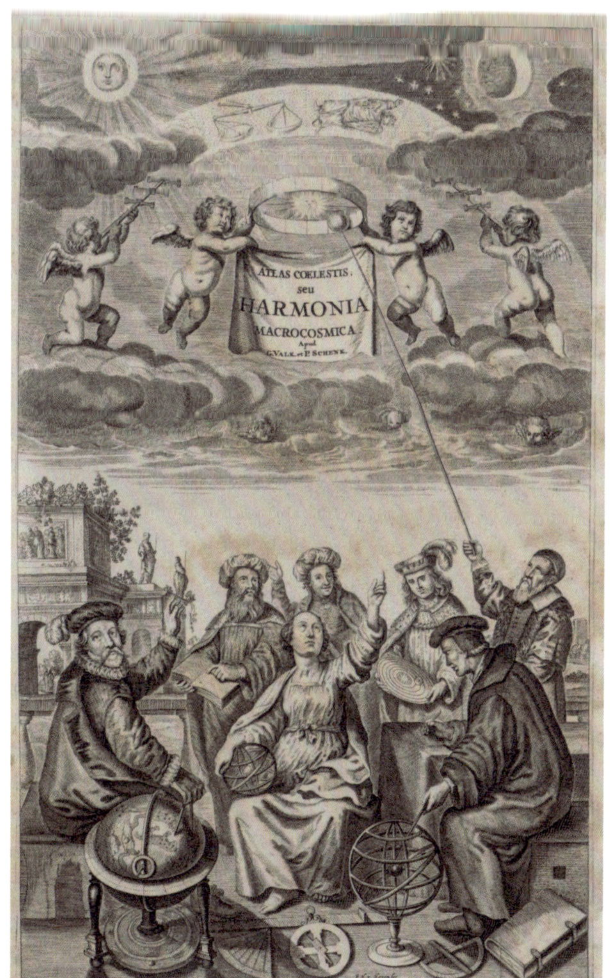

Title page
From: Andreas Cellarius, *Harmonia Macrocosmica*, 1708
ANTWERP, THE PHOEBUS FOUNDATION

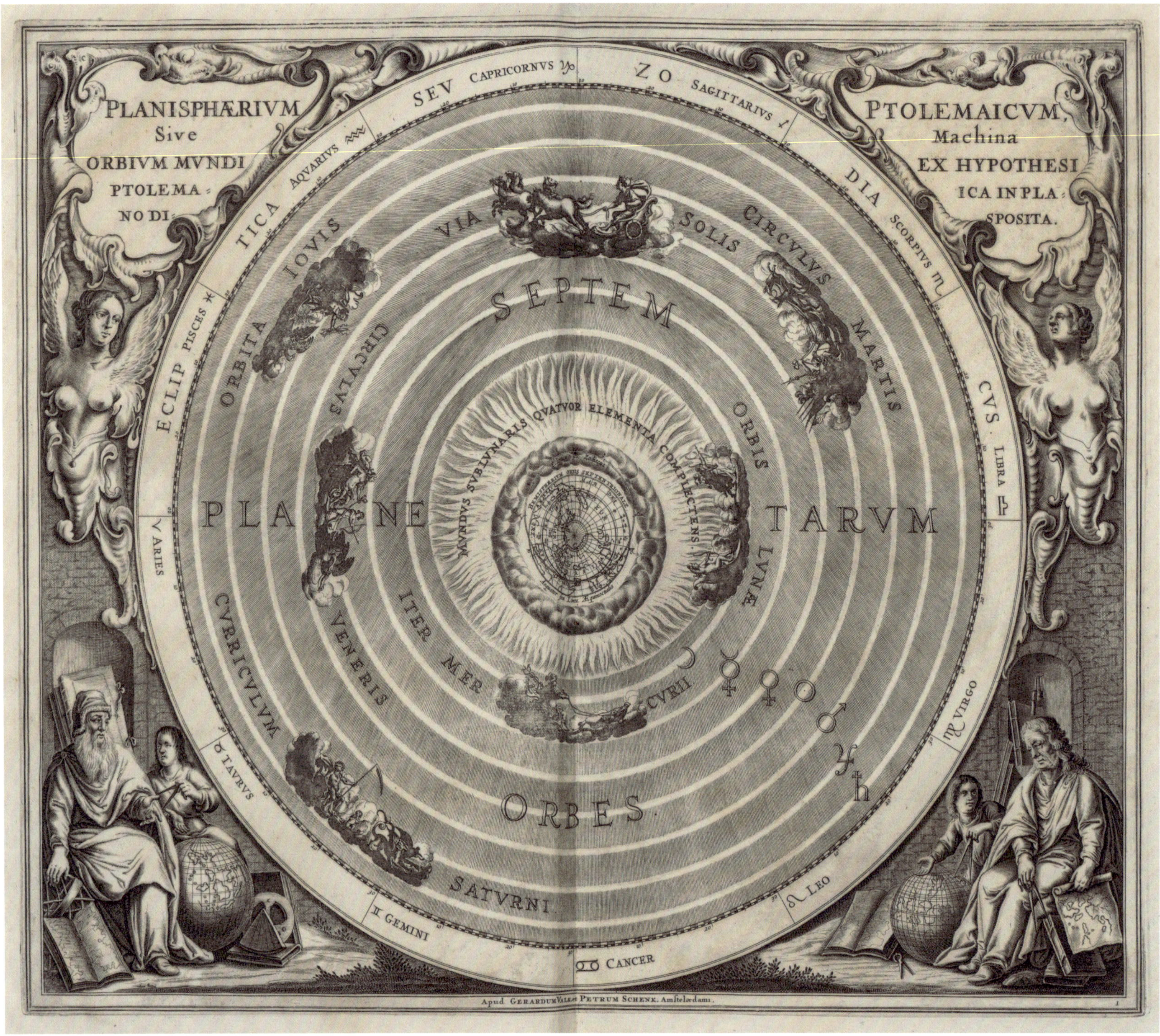

Planisphaerium Ptolemaicum, Sive Machina Orbium Mundi Ex Hypothesi Ptolemaica In Planno Disposita
Celestial map of the Ptolemaic system of the universe
From: Andreas Cellarius, *Harmonia Macrocosmica*, 1708
ANTWERP, THE PHOEBUS FOUNDATION

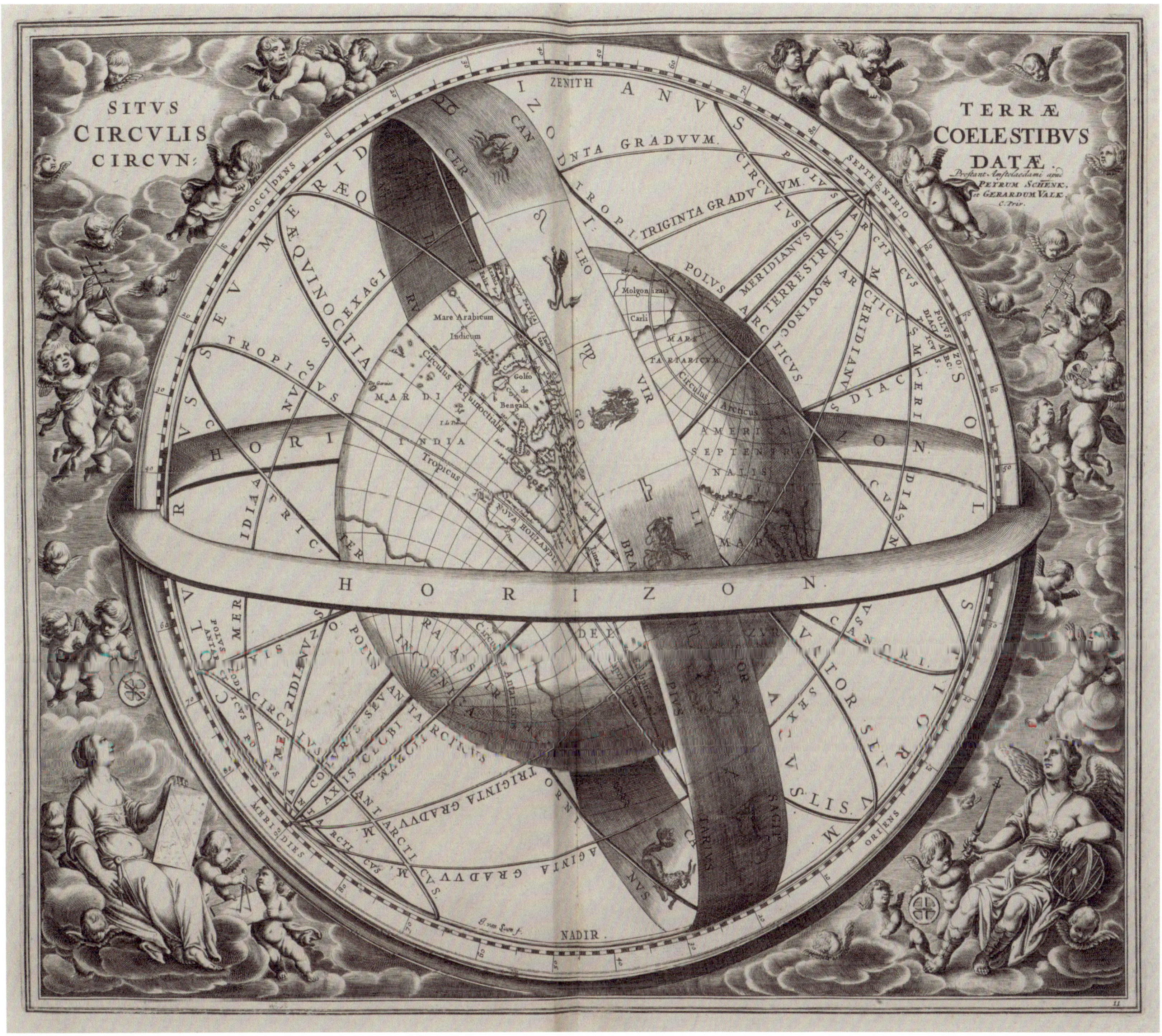

Situs Terrae Circulis Coelestibus Circundatae
Celestial map of the Ptolemaic system of the universe on which the earth is visible
From: Andreas Cellarius, *Harmonia Macrocosmica*, 1708
ANTWERP, THE PHOEBUS FOUNDATION

LITERATURE

G. Crupi, 'Volvelles of Knowledge. Origin and Development of an Instrument of Scientific Imagination (13th-17th Centuries)', *JLIS.it: Italian journal of library and information science*, 10, 2, 2019: 1-27.

S. Schmidt & K. Nichols, *Altered and Adorned: Using Renaissance Prints in Daily Life*, Chicago, 2011.

K. Van Cleempoel, 'The Production of Scientific Instruments in Louvain in the 16th Century'. In: *Scientific Instruments in the Sixteenth Century: The Spanish Court and the Louvain School*, Madrid, 1997-1998: 59-74.

K. Van Cleempoel, *A Catalogue Raisonné of Scientific Instruments from the Louvain School: 1530 to 1600*, Turnhout, 2002.

R.H. van Gent, 'De hemelatlas van Andreas Cellarius: Het meesterwerk van een vergeten Hollandse kosmograaf', *Caert-Thresoor*, 19, 1, 2000: 9-25.

R.H. van Gent, *Andreas Cellarius: Harmonia Macrocosmica, The Finest Atlas of the Heavens / Der prächtigste Himmelsatlas / L'atlas céleste le plus admirable*, Cologne, 2006.

S. Vanden Broecke, 'The Use of Visual Media in Renaissance Cosmography: the Cosmography of Peter Apian and Gemma Frisius', *Paedagogica Historica*, 36, 1 (2000): 130-150.

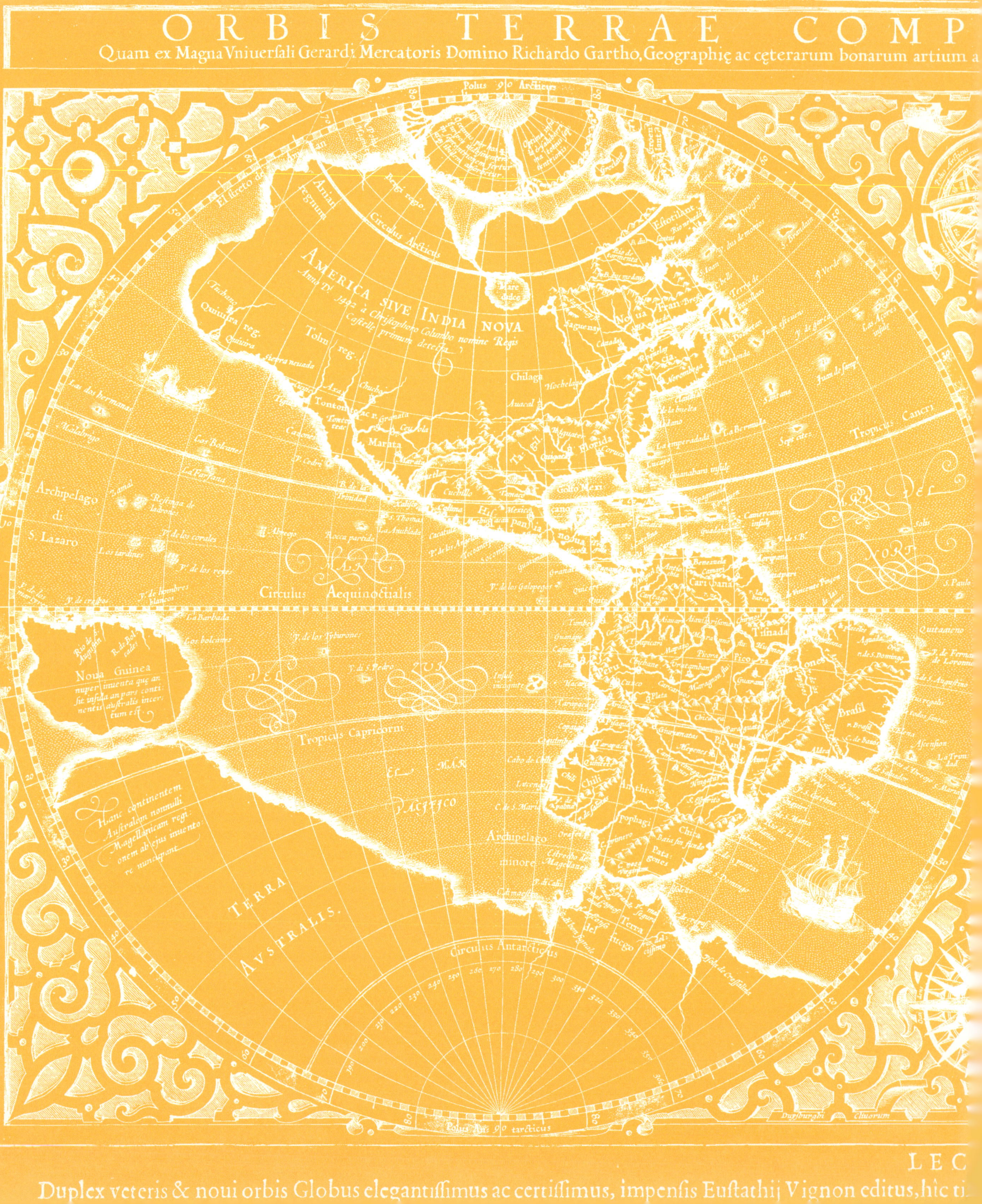
ORBIS TERRAE COMP
Quam ex Magna Vniuersali Gerardi Mercatoris Domino Richardo Gartho, Geographię ac cęterarum bonarum artium a
Polus Arcticus
Circulus Arcticus
AMERICA SIVE INDIA NOVA
Anno D. 1492 à Christophoro Colombo nomine Regis Castellę primum detecta
Tropicus Cancri
Archipelago di S. Lazaro
MAR DEL NORT
Circulus Aequinoctialis
Noua Guinea
MAR DEL SUR
Tropicus Capricorni
MAR PACIFICO
Archipelago minore
TERRA AUSTRALIS
Circulus Antarcticus
Polus Antarcticus
LEC
Duplex veteris & noui orbis Globus elegantissimus ac certissimus, impensis Eustathij Vignon editus, hic ti

NDIOSA DESCRIPTIO

ri ac fautori summo, in veteris amicitię ac familiaritatis memoriã Rumoldus Mercator fieri curabat A°. M.D.Lxxxvii.

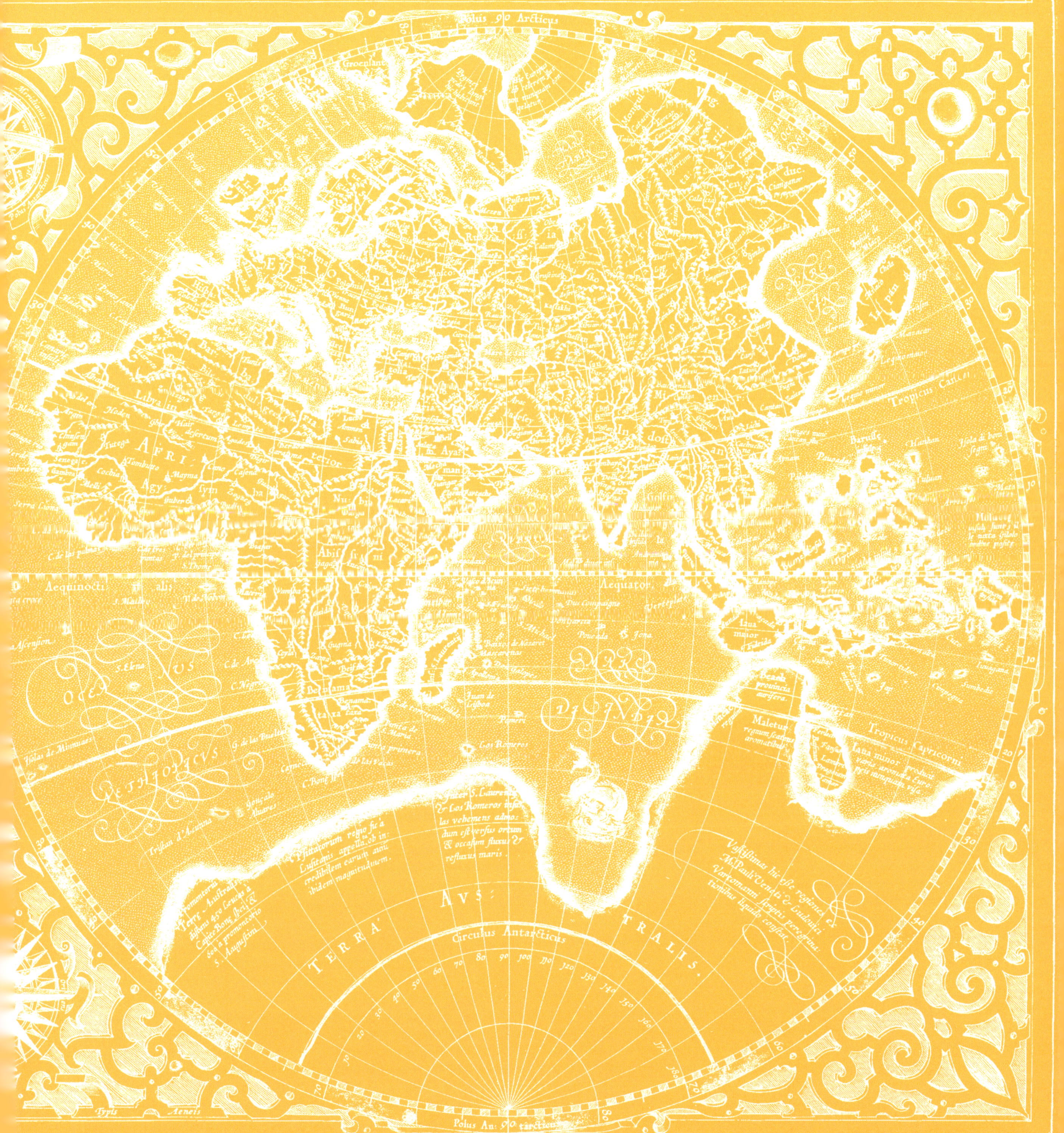

RI S. P.

rtur, vt nuper ab eo absolutam maximis impensis & laboribus Geographiam Strabonis, Manu-

ACKNOWLEDGEMENTS

It seems like an eternity since Bram Vannieuwenhuyze (Professor by special appointment of Historical Cartography at the University of Amsterdam) and I arrived at the offices of The Phoebus Foundation in Antwerp. Two months later, on 15 November 2021, I spent my first day at the Foundation's impressive storage facility. Who would have thought that two years later the end result would be a magnificent book? Or that I would go on to make old maps my profession? Prof. Bram Vannieuwenhuyze is currently also PhD supervisor for my doctoral research into narrative cartography in the early modern Low Countries.

So it goes without saying that my first thanks are due to Bram who, having paid a visit to The Phoebus Foundation's exceptionally varied and high-quality map collection, initiated conversations about a research fellowship. I am grateful to its chief of staff, Dr Katharina Van Cauteren, for the opportunity of completing this grand and exciting adventure as the Foundation's first research fellow from November 2021 until February 2022. A fine collaboration followed, during which I spent three months alternating between a week in the depot and a week in Amsterdam. At the storage facility, I was able to explore over five hundred maps, prints, atlases, books, instruments and globes, while at the map room at the Allard Pierson in Amsterdam (UvA special collections), I followed up my observations with the relevant secondary literature. I was supported and kept company throughout by the other researchers at the Explokart group. Peter van der Krogt in particular helped me on several occasions to spot differences between states and to correctly use cartobibliographies.

I received a fantastic reception in Antwerp from Niels Schalley, who disclosed with great clarity the tasks entailed by the research fellowship and what The Phoebus Foundation wished to achieve, which enabled me to get off to a flying start. I was also able to approach him with questions and for advice for the duration of the fellowship. Laura Bijnens helped with the technical challenges of doing my work remotely and with setting up an appropriate cataloguing system. Thanks also to Luk Van Hove for making it practicable to commute between Amsterdam and Antwerp. I immensely enjoyed the friendly lunch every day with the other conservators and staff at the storage facility and the cheerful presence of the art handlers.

Once the fellowship had been completed, the concept and structure of this book were born during a brainstorming session with Dr Katharina Van Cauteren, Gautier Platteau, Dr Leen Kelchtermans and Niels Schalley. Later on, during the writing stage, Leen proved very helpful. Her immense publishing experience and knowledge of early modern Netherlandish (art) history, combined with her helicopter view, really helped forge this book into a whole. Throughout the different drafts, she provided the thorough and useful feedback I needed. Anke Van Achter and Katrijn Van Bragt helped collate the photographs (by Adri Verburg, Dominique Provost and Michel Wuyts, among others) and object specifications as I was writing the book in Amsterdam. They made it extremely easy for me to work remotely.

I am grateful to Gautier Platteau, Sofie Meert and Stephanie Van den bosch of Hannibal Books for their pleasant cooperation, especially Stephanie who made a great job of managing the project. My thanks go to Jan Vangansbeke and Xavier De Jonge for copy editing the texts, Ted Alkins for the English translation and Paul Boudens for his exquisite design. I am also indebted to Joop Koopmans and Djoeke van Netten for reading some of the chapters. And lastly, all my thanks and love of course to Minke, not only for eagerly reading and correcting some of the chapters but above all for her sober gaze and unfailingly solid confidence.

Anne-Rieke van Schaik

ABOUT THE AUTHOR

Anne-Rieke van Schaik was The Phoebus Foundation's Topography and Cartography Research Fellow from November 2021 to February 2022. At present she is engaged in PhD research on narrative cartography in the early modern Low Countries at the University of Amsterdam (UvA). She is co-author of the book *Oude kaarten lezen. Handboek voor historische cartografie* (2023) and a member of the Explokart research group, affiliated to the Allard Pierson in Amsterdam and the UvA.

SCIENTIFIC EDITOR
Katharina Van Cauteren

AUTHOR
Anne-Rieke van Schaik

CONTENT EDITING & COORDINATION
Leen Kelchtermans

EDITING
Wannes Wets

IMAGE RESEARCH
Séverine Lacante
Katrijn Van Bragt

TRANSLATION
Ted Alkins

COPY EDITING
Xavier De Jonge

PROJECT MANAGEMENT
Stephanie Van den bosch

GRAPHIC DESIGN
Paul Boudens

PRINTING & PHOTOENGRAVING
die Keure, Bruges, Belgium

BINDING
Van Mierlo, Nijmegen, The Netherlands

PUBLISHER
Gautier Platteau

ISBN 978 94 6494 104 3
D/2024/11922/16
NUR 646

© Hannibal Books, 2024
www.hannibalbooks.be

© The Phoebus Foundation Public Benefit Foundation |
Stichting van Openbaar Nut, 2024
www.phoebusfoundation.org

All rights reserved for all countries. No part of this publication may be reproduced, stored in a database and/or made public in any form or by any means, electronic, mechanical or otherwise, without the prior written permission of the publisher.

The Phoebus Foundation Public Benefit Foundation and Hannibal Books have made every effort to apply the laws on copyright to all the images that appear in this publication and exhibition. Whosoever should still wish to assert copyright is requested to contact the publisher.

COVER
Pieter Van den Keere
Hydrographica descriptio, in qua Hispaniae orae maritimae à Capite S. Martini usq'ad Caput Dragonis / Beschryvinge vande zeecusten van Spaengen van Cabo S. Martin tot Cabo Dragonis
From: Willem Barentsz., *Caertboeck Vande Midlandtsche Zee*, 1609
&
Rumold Mercator after Gerard Mercator
Orbis terrae compendiosa descriptio, 1587
ANTWERP, THE PHOEBUS FOUNDATION

PICTURE CREDITS
p. 81 © Museum de Lakenhal, Leiden
pp. 98-99 © Rijksmuseum Amsterdam
p. 279 © Allard Pierson UvA, Amsterdam

All other images © The Phoebus Foundation

The Foundation would like to thank all the institutions and photographers who have provided visual material, especially Adri Verburg and Michel Wuyts.

WITH THE SUPPORT OF COMPANIES IN THE KATOEN NATIE AND INDAVER GROUP

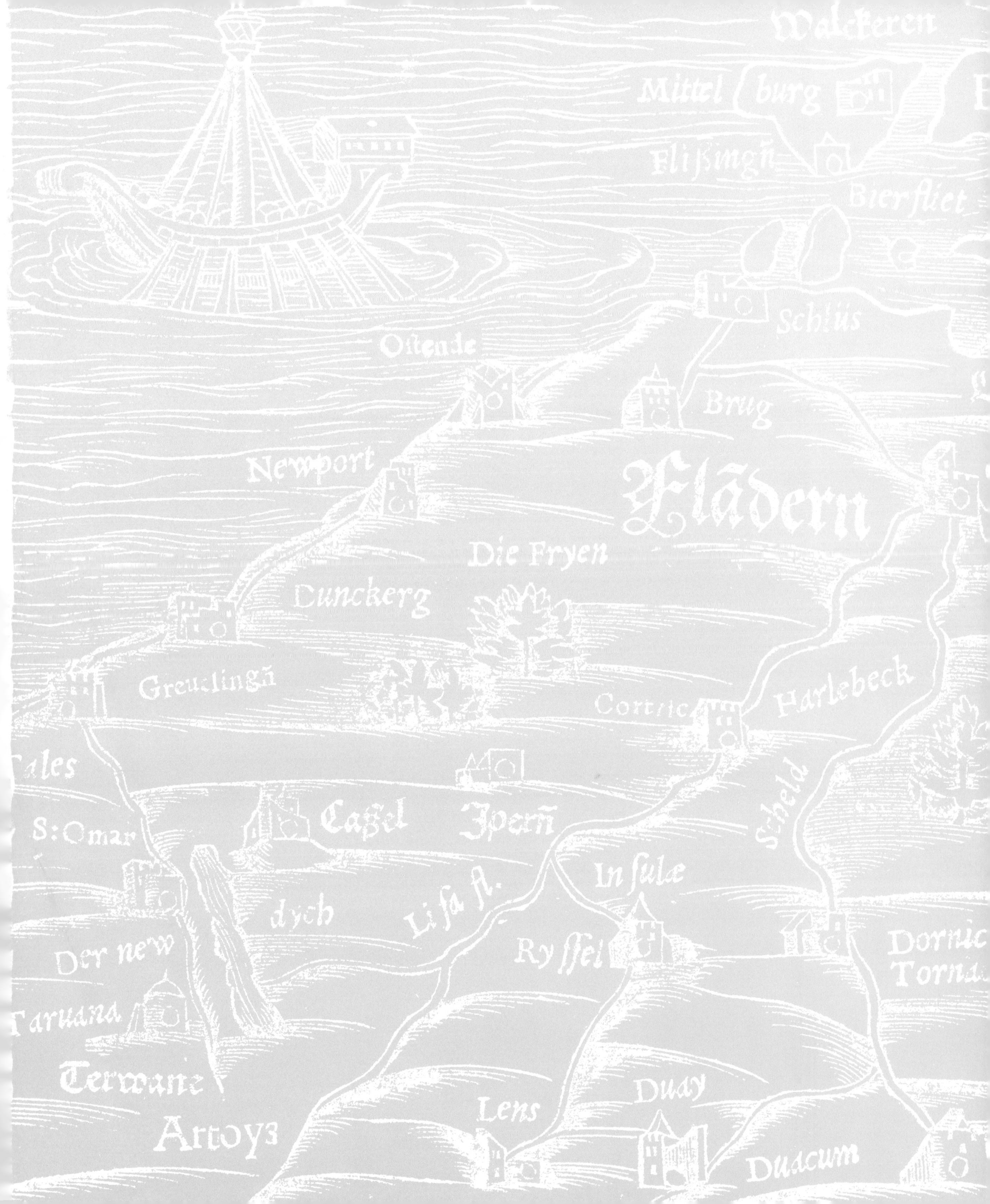
Walckeren
Mittel burg
Flißingē
Bierfliet
Schlüs
Ostende
Brug
Newport
Flādern
Die Fryen
Dunckerg
Greualingā
Cortric
Harlebeck
S:Omar
Caßel
Ipern
Scheld
In sulæ
Lisa fl.
dych
Der new
Ryssel
Taruana
Terwane
Artoys
Lens
Duay
Duacum

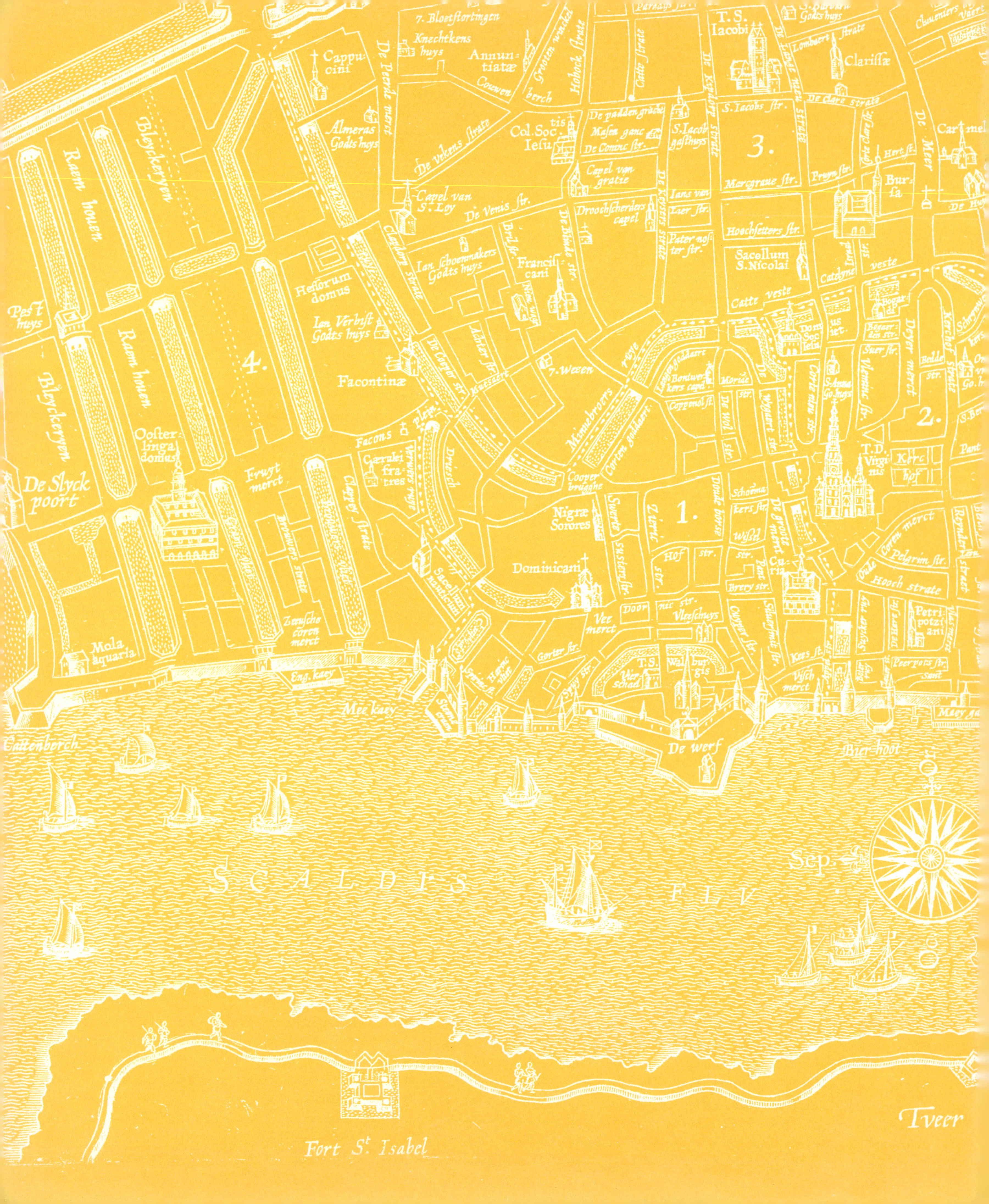

7. Bloetfortingen
Knechtkens huys
Cappucini
Annuntiatæ
T.S. Iacobi
Clarissæ
Carmel
Almeras Godts huys
De vekens strate
Col. Soc. Iesu
S. Iacobs str.
3.
Capel van S. Loy
De Venis str.
Capel van gratie
Mergreue str.
Bursa
Hochstetters str.
Sacellum S. Nicolai
Catte veste
Raam houen
Bleyckeryen
Pest huys
Hessorum domus
Ian Verbist Godts huys
Franciscani
4.
Facontinæ
7. Weeen
2.
Oosterlinga domus
De Slyck poort
Fruyt merct
Cærulei fratres
Coopersbrugghe
Nigræ Sorores
1.
T.D. Virginis
Kerc hof
Dominicani
Pelgrim str.
Hooch strate
Vee merct
Vleeschuys
Mola aquaria
Zeeusche coren merct
Eng. kaey
T.S. Walburgis
Vischmerct
Mee kaey
De werf
Bier hoot
Cattenberch
SCALDIS FLV.
Sep.
Tveer
Fort St. Isabel